PRINCIPLES OF DAIRY SCIENCE

PRINCIPLES OF DAIRY SCIENCE

SECOND EDITION

G. H. Schmidt
The Ohio State University

L. D. Van Vleck
Cornell University

M. F. Hutjens
University of Illinois

PRENTICE HALL, Englewood Cliffs, New Jersey 07632

Library of Congress Cataloging-in-Publication Data

Schmidt, G. H. (Glen Henry) (date)

Principles of dairy science / G. H. Schmidt, L. D. Van Vleck, M. F. Hutjens.
p. cm.
"All chapters have been updated"—P.
Bibliography: p.
Includes index.
ISBN 0-13-709818-9
1. Dairying. I. Van Vleck, L. Dale (Lloyd Dale), 1933– . II. Hutjens, M. F., 1945– III. Title.
SF239.S37 1988 87-18621
637—dc19 CIP

Editorial/production supervision and interior design: TKM Productions
Cover design: Suzanne Bennett & Associates
Manufacturing buyer: Peter Havens

Printed in the United States of America

10 9 8 7 6 5 4 3 2 1

ISBN 0-13-709818-9

Prentice-Hall International (UK) Limited, *London*
Prentice-Hall of Australia Pty. Limited, *Sydney*
Prentice-Hall Canada Inc., *Toronto*
Prentice-Hall Hispanoamericana, S.A., *Mexico*
Prentice-Hall of India Private Limited, *New Delhi*
Prentice-Hall of Japan, Inc., *Tokyo*
Simon & Schuster Asia Pte. Ltd., *Singapore*
Editora Prentice-Hall do Brasil, Ltda., *Rio de Janeiro*

Contents

PREFACE xiii

1 THE ROLE OF DAIRY CATTLE IN WORLD FOOD PRODUCTION 1

1.1 Value of Milk in Human Nutrition 1

1.2 Milk and Health Issues 2

1.3 The Role of Milk and Milk Products in the Human Diet 4

1.4 The Role of Dairy Cattle in the Production of Animal Protein 6

1.5 History of the Dairy Industry 7

2 DAIRY INDUSTRY IN THE UNITED STATES 10

2.1 History 10

2.2 Development of the Dairy Industry 11

2.3 Production of Milk 13
2.4 Consumption of Milk 16
2.5 Use of the Milk Supply 18
2.6 Breeds of Dairy Cattle 18

3 THE MAMMARY GLAND **32**

3.1 Anatomy of the Mammary Gland 32
3.2 Histology of the Mammary Gland 41
3.3 Growth and Development of the Mammary Gland 43

4 PHYSIOLOGY OF MILK SECRETION **46**

4.1 Cell Cytology and Milk Secretion 46
4.2 Comparative Composition of Blood Plasma and Milk 53
4.3 Udder Pressure and Secretion Rate 54
4.4 Milking Intervals and Frequencies 55
4.5 Drying Off Cows 57
4.6 Incomplete Milking 58
4.7 Milk Ejection 58

5 FACTORS AFFECTING YIELD AND COMPOSITION OF MILK **63**

5.1 Changes Occurring During a Normal Lactation 63
5.2 Day-to-Day Variations 65
5.3 Dry Period and Body Condition 66
5.4 Age of Cow at Calving 67
5.5 Body Weight 68
5.6 Gestation 68
5.7 Environment 69
5.8 Season of Year 70
5.9 Feeds and Feeding 71
5.10 Hormones 73

6 HARVESTING THE MILK CROP: MACHINE MILKING 78

6.1 The Milking Machine 78

6.2 Machine Milking 84

6.3 Rate of Milking Dairy Cows 90

6.4 Production of High-Quality Milk 91

7 MILKING SYSTEMS 95

7.1 Comparison of Milking Systems 96

7.2 Factors to Consider in Constructing a Milking Parlor 97

7.3 Evaluation of Milking Systems 100

7.4 Milking Parlor Types 103

7.5 Automation of Milking Parlors 107

7.6 Choosing a Milking Parlor 112

8 UDDER DISEASES 115

8.1 Mastitis 115

8.2 Udder Edema 129

9 REVIEW AND APPLICATION OF MENDELIAN GENETICS 133

9.1 Transmission of Genetic Material 134

9.2 Combinations of Traits 137

9.3 Genetic Tests for Parentage 139

9.4 Detecting Carriers of Genes for Abnormal Traits 142

10 KEYS TO GENETIC PROGRESS 149

10.1 The Keys 149

10.2 Variation: The Key to the Unexpected 154

10.3 Heritability 159

10.4 Appendix 161

11 RELATIONSHIPS AND INBREEDING **168**

11.1 Importance of Knowing Relationships 169
11.2 Computing Relationships 170
11.3 Inbreeding Depression 174
11.4 Levels of Inbreeding 175
11.5 Appendix 176

12 CORRECTION OF RECORDS FOR NONGENETIC FACTORS **179**

12.1 Adjustment for Age of Cow 179
12.2 Adjustment for Body Weight 189
12.3 Adjustment for Days Dry 189
12.4 Adjustment for Calving Interval 189
12.5 Adjustment for Herd Management Level 190
12.6 Adjustment for Times Milked per Day 191
12.7 Stage of Lactation 191
12.8 Example of Use of Adjustment Factors 191
12.9 Summary of Adjustments 193

13 COW EVALUATION AND SELECTION **194**

13.1 Evaluation and Selection of Cows for Real Producing Ability 195
13.2 Evaluation and Selection for Transmitting Ability 199

14 BULL EVALUATION AND SELECTION **203**

14.1 Modified Contemporary Comparison 204
14.2 The Sire Comparison Method 208
14.3 Use of Dam–Herdmate Difference in Sire Evaluation 211
14.4 Natural Service Sire Evaluation 214
14.5 Young Bull Testing 215
14.6 Success of Selection 220

14.7 Accuracy and Confidence Ranges for a Sire's Transmitting Ability or for a Daughter's Genetic Value or Record 221

14.8 Risk 225

15 SELECTION FOR ECONOMIC VALUE AND ECONOMICS OF SELECTION **229**

15.1 Total Economic Value 229

15.2 Selection for Production and Type 233

15.3 Index for Product or Dollar Value 236

15.4 A Net Product Value and Type Index 239

15.5 Selection for Fertility and Production 239

15.6 Selection for Milk Yield and Herd Life 241

15.7 Selection for Milk Yield and Components of Milk 243

15.8 Economics of Selection 245

16 CROSSBREEDING **258**

16.1 Reasons for Crossbreeding 258

16.2 Theory of Heterosis 261

16.3 Experimental Results 263

16.4 Systems of Crossbreeding 266

16.5 Which Breeds to Use in Crosses 268

16.6 Results Obtained in Tropical Countries 268

17 BASIC PRINCIPLES OF DAIRY CATTLE NUTRITION **271**

17.1 Digestion in the Ruminant Stomach 271

17.2 Digestion in the Intestines 274

17.3 Nutrient Utilization 275

17.4 Voluntary Feed Intake 278

18 NUTRITIVE REQUIREMENTS OF DAIRY CATTLE **281**

18.1 Energy 281

18.2 Protein 286

18.3 Fats and Oils 290
18.4 Minerals 291
18.5 Vitamins 296
18.6 Water 297

19 NUTRITIONAL VALUE OF FEEDS 299

19.1 Methods of Analysis 299
19.2 Forages 304
19.3 Concentrates 305

20 PRODUCTION OF HIGH-QUALITY FORAGE 313

20.1 Legume–Grass Forages 313
20.2 Pastures and Green Chop Systems 321
20.3 Corn Silage 324
20.4 Development of a Forage Program 327
20.5 Forage Testing Programs 328
20.6 Forage Inventory 329

21 FEEDING CONCENTRATE TO DAIRY COWS 331

21.1 Functions of Concentrate 331
21.2 Substitution of Concentrate for Forage 332
21.3 Balancing Dairy Rations 333
21.4 Rules of Thumb 337
21.5 Concentrate Feeding Guidelines 339
21.6 Phase Feeding 341
21.7 Concentrate Feeding Systems 343
21.8 Economic Considerations 347

22 NONINFECTIOUS DISEASES 350

22.1 Ketosis 350
22.2 Bloat 354
22.3 Milk Fever 356

22.4 Displaced Abomasum 360
22.5 Fatty Liver Syndrome 361
22.6 Acidosis 362

23 REPRODUCTION: PROCESSES AND PROBLEMS **364**

23.1 Female Reproductive Tract 365
23.2 Male Reproductive Tract 372
23.3 Attaining High Reproductive Efficiency 374
23.4 Reproductive Diseases 377
23.5 Other Reproductive Problems 381

24 REPRODUCTION MANAGEMENT TOOLS **382**

24.1 Artificial Insemination 382
24.2 Embryo Transfer 390
24.3 Estrous Synchronization 392
24.4 Other Technologies 394

25 HERD RECORDS **396**

25.1 Dairy Records 398
25.2 Dairy Herd Improvement Program 402
25.3 Use of Computers in Dairy Records 406

26 RAISING CALVES AND HEIFERS **408**

26.1 Digestive System of the Calf 409
26.2 Management at Calving 410
26.3 Colostrum 411
26.4 Raising Calves from Birth to Weaning 413
26.5 Raising Calves from Weaning to One Year of Age 420
26.6 Raising Heifers from One to Two Years of Age 422
26.7 Veal Calf Production 424

26.8 Dairy Beef 425
26.9 Contract Raising of Heifers 426

27 DAIRY CATTLE HOUSING **428**

27.1 Major Components of a Housing System 428
27.2 Types of Housing Arrangements 436
27.3 Planning New or Remodeled Dairy Barns 442

28 MILK MARKETING **445**

28.1 Methods of Selling Milk 446
28.2 Development of Milk Cooperatives 446
28.3 Milk Pricing Systems 447
28.4 Government Regulation of Milk Marketing 449

INDEX **453**

Preface

This book is intended as a textbook for students taking courses in dairy cattle production and management. Commercial dairy farmers will also find the book useful for the care and management of their herds. The book will provide information to extension agents, veterinarians, feed company representatives, Dairy Herd Improvement supervisors, and artificial insemination technicians.

The authors have taught courses in dairy production, management, and genetics for a number of years and have conducted extension education programs. The authors' approach to teaching has been to emphasize the basic principles of dairy science and their application to dairy farm situations; this book reflects that emphasis. No textbook is ''exactly right'' for any instructor; however, this book provides the basic principles so that an instructor can spend most of the class time covering the application of principles to a particular region or set of conditions. The basic principles apply to dairying all around the world, even though feeds and climatic conditions in different parts of the United States and regions of the world are not the same.

Dr. M. F. Hutjens has been added as an author in this edition. He has an extensive background in dairying and is regarded as one of the outstanding extension specialists in the United States. He has rewritten the chapters on nutrition and feeding as well as the one on raising calves and heifers. These chapters contain examples for a number of practical situations that should be of value to dairy farmers and that should help students understand the basic principles of nutrition and calf raising.

All chapters have been updated, with those of a more practical nature receiving the greatest revision. The most recent National Research Council recommendations on the nutrient requirements of cattle have been used. Whenever applicable, the economic implications of management recommendations are included.

Certain chapters—those on the mammary gland, physiology of milk secretion, genetics, basic principles of nutrition, and fundamentals of reproduction—are similar to those found in more basic textbooks. These chapters are included to serve as foundations for further discussion of principles more closely applied to the care and management of cows.

Discussion of some concepts may be oversimplified; however, most students realize that textbooks have to deal with general principles and cannot describe all variations that might be found on dairy farms. These variations are characteristic of most biological systems. Our goal is for the student not only to grasp the basic concepts but also to realize that variations frequently exist.

This book would not be possible without the help of a number of people. We would especially like to thank the following for their help, support, and encouragement: Debbie Gallagher, Carol Ann Hutjens, Rosemary Judy, Joy Pate, Peter Spike, Susan Herbert, and certainly not least, the many students who have tested the material and who have provided valuable feedback to the authors. We also would like to thank the many persons and organizations that provided the pictures and illustrations for the book. Without them, the book would be much less readable.

1

The Role of Dairy Cattle in World Food Production

The dairy industry is based on the ability of mammals to produce milk in excess of that required to nourish their young. The excess milk, which has a high nutritional value, is used for human consumption. Cattle are the primary producers of milk in the world, producing about 91% of the world's supply. In some countries sheep, goats, and water buffalos are the principal producers of milk. The camel and reindeer are also used for milk production. Over the centuries, cows and some other mammals have been bred and selected for their ability to produce large quantities of milk.

1.1. VALUE OF MILK IN HUMAN NUTRITION

Milk, described as nature's most perfect food, is the sole source of food for newborn mammals. For the human infant, milk is the only source of nutrients for the first several months of life, and in many countries milk plays a major role in the diet of the growing child. Milk or milk substitutes are important during early growth of most domesticated mammals as well. In addition, milk can be a valuable source of nutrients for mature humans, especially the elderly.

Milk is very important in the human diet because of two principal ingredients: protein and calcium. Protein provides many of the essential amino acids often deficient in the cereal grains used for food. In addition, the proteins are easily digested and are relatively inexpensive. The common recommended quantity of 1 quart of milk per day supplies all of the protein requirements for children up to 6 years of age and more than 60% for growing children from 6 to 14 years of age. For humans from 14 to 20 years of age, 1 quart of milk supplies approximately one-half of the daily protein requirements—except during lactation, when it supplies approximately 44% of the requirements. In the United States in 1980, dairy foods, excluding butter, provided 20% of the human protein requirements.

Calcium is the nutrient most likely to be lacking in diets of persons who do not consume milk or milk products. Adequate calcium intake is difficult to attain in the human diet if milk or milk products are excluded. The daily calcium requirement of a nursing mother is 2.0 g, whereas that of pregnant women, infants, children, adolescents, young adults, and the elderly is 1.5 g. The requirement for other adults is 1.0 g per day. One quart of milk supplies approximately 1.15 g of calcium; therefore, 1 quart of milk per day supplies a large percentage of the daily requirements for most people except for lactating women.

Calcium and vitamin D are extremely important in the diet of the elderly, especially women after the age of menopause. Approximately 25% of postmenopausal women have senile osteoporosis severe enough to cause vertebral, hip, or long bone fractures. Osteoporosis is the severe loss of calcium from the bones resulting in porous, weak, and fragile bones. About 20% of the calcium in the adult bone is replaced yearly. If calcium is not consumed daily, it will diffuse out of the bones and lead to osteoporosis. Consumption of adequate calcium and continued physical activity are important factors in preventing this disease. Milk supplies approximately 72% of the human calcium requirements in the United States. Vitamin D, which is important in the absorption of calcium from the intestinal tract, is routinely added at 400 IU per quart to 98% of the fluid milk marketed in the United States.

Riboflavin and vitamin A are the vitamins most likely to be deficient in the human diet. One quart of milk per day supplies all of the riboflavin requirements of growing children and adults, except pregnant and lactating women. One quart of milk supplies almost all of the vitamin A requirements of infants less than 1 year of age, roughly 72% of the requirements for children between 1 and 2 years of age, more than 40% of those for children between 8 and 10 years of age, and 29% of those of adults. Dairy products, excluding butter, now supply 36% of the riboflavin and 12% of the vitamin A in the human diet in the United States. In addition, milk and milk products, excluding butter, supply 10% of the energy, 33% of the phosphorus, 20% of magnesium, 18% of vitamin B_{12}, and 11% of vitamin B_6.

1.2 MILK AND HEALTH ISSUES

Atherosclerosis and Coronary Heart Disease

It has been postulated that increased intakes of cholesterol and saturated fatty acids, which are in higher concentration in animal products than in plant products, elevate blood cholesterol and in turn increase the incidence of atherosclerosis (formation of fatty plaques in the lining of blood vessels) and coronary heart disease. Major dietary changes have been recommended by the American Heart Association, the Intersociety Commission for Heart Disease Resources, and the Senate Select Committee on Nutrition and Human Needs. These recommendations center around a reduction of total fat, saturated fat, and cholesterol intake. Although a number of studies have been conducted with both humans and laboratory animals, research has been unable to confirm a significant relationship between dietary saturated fat and cholesterol intake and elevated blood cholesterol. In fact, there is some evidence that whole milk, skim milk, and yogurt exhibit a cholesterol-lowering effect. Atherosclerosis and coronary heart disease appear to result from a complex mechanism that involves a multitude of risk factors. Most of the evidence suggests that for the general healthy population, dairy products consumed as part of a nutritionally adequate diet with energy to maintain body weight do not contribute to heart disease or mortality. Dairy products generally are not particularly high in saturated fatty acids or cholesterol.

Milk Intolerance

Even though milk has an even greater potential for supplying nutrients in food-deficient countries, there may be limitations of milk usage in populations in which a large proportion of adults have lactose intolerance. This results from a deficiency in the intestinal enzyme lactase, which converts lactose to glucose and galactose. The enzyme is lost as children approach puberty and the consumption of milk results in abdominal bloating, rumbling, cramping, and diarrhea, which in turn may result in a decreased absorption of other nutrients from the intestinal tract.

Low amounts of lactase are found in the digestive tract of 70% of the black persons and 6 to 12% of the white persons in the United States. Throughout the world, lactose intolerance is quite high among nonwhite populations. Lactose intolerance is usually measured by giving test subjects relatively large clinical test doses (50 to 100 g, the amount in 1 or 2 quarts of milk) and observing either a small rise in blood glucose or the symptoms of lactose intolerance. Even though persons may be diagnosed as lactose intolerant by the foregoing measure, most persons can tolerate the amount of lactose found in a typical serving of 8 ounces of milk. Although the amount of lactose broken down in the intestinal tract may be small, the calcium, protein, and vitamins supplied by the milk are important in the diet. Alternatives are available for individuals who are truly intolerant. Among these are: consumption of small amounts of milk but frequently during the day, consumption of most cheeses and many fermented products such as yogurt, or consumption of lactose-hydrolyzed milk.

Obesity and Other Problems

Although milk and other dairy products have been incriminated as a contributing factor to obesity, there is no evidence that any particular food or group of foods leads to obesity per se. The main factor in obesity is the amount of energy consumed in relation to the energy expenditure. In 1980, dairy foods, not including butter, supplied only 11% of the total fat in the U.S. civilian diet. If total calorie consumption is a problem, low-fat milk and milk products can be consumed to reduce energy consumption, yet retain the desirable protein and calcium intakes.

Milk protein allergy presents a problem for some infants; however, the incidence is probably less than 1% in infants and children in the industrialized countries. The condition is difficult to diagnose because of inadequate diagnostic tools, its wide spectrum of clinical symptoms, and its transient nature. Children usually outgrow the condition by 2 years of age.

The association between diet and cancer is highly controversial but is receiving a lot of research attention. Total fat intake has been implicated in contributing to some types of cancer (i.e., breast, colon–rectum, prostate); however, the evidence is far from conclusive from either laboratory animal or human epidemiological studies. It appears that total energy intake may be more important as a contributing factor for some types of cancer than dietary fat. There is some evidence to suggest that dietary calcium may protect against colon cancer. In addition to a possible beneficial effect of milk through its calcium content protecting against colon cancer, milk also appears to be related to reducing the number of dental caries (cavities).

1.3 THE ROLE OF MILK AND MILK PRODUCTS IN THE HUMAN DIET

Foods of animal origin, including milk, contribute greatly to human nutrition by effectively balancing and supplementing the nutrients from plant foods. The population, milk production, and per capita milk consumption in the regions of the world are shown in Table 1.1. Milk supplies a significant portion of the animal protein consumed in the United States, Canada, most of the European countries, New Zealand, and Australia. Current world supplies are adequate to furnish every person in the world with 20 g of animal protein per day, but the problem is distribution. Generally, there is no shortage of animal protein in the world; yet the amount available in developing countries is only 9 g per person per day, compared to more than 44 g per person in the developed countries.

There are more than 3.8 billion head of livestock in the world and over 6.7 billion domesticated fowl. Although approximately 65% of the livestock and 52% of domestic fowl are in developing countries, these countries produce only 37% of the world's meat, 17% of the world's milk, and 38% of the world's chickens and eggs.

Protein deficiencies are generally found in countries with caloric deficiencies, especially some countries in Africa and Asia. Usually, countries deficient in total protein are also deficient in animal protein. In India milk protein makes up nearly all of the animal protein consumed, because of the objection to the slaughter of animals.

TABLE 1.1 POPULATION, TOTAL MILK PRODUCED, AND PER CAPITA MILK CONSUMPTION BY REGIONS IN THE WORLD

Region	Population 1980 (millions)	Projected annual growth rate, 1980–1993 (%)	Milk produced, 1979–1981 (MMT)[a]	Per capita consumption, 1978–1980 (kg)
North Africa, Middle East	246	2.5	16.3	92.8
Sub-Saharan Africa	369	3.2	9.7	27.8
European Economic Community	254	0.3	113.7	386.6
Other western European countries	77	0.6	23.3	278.7
USSR	262	0.9	92.9	363.7
Eastern Europe	131	0.6	44.8	337.6
South Asia	869	2.2	42.5	48.7
East Asia	448	1.7	7.3	19.9
Asia, centrally planned	1069	1.4	7.9	7.2
Oceania	31	1.2	12.2	290.4
Latin America	354	2.2	34.1	102.0
North America	246	0.7	66.1	263.0
World, total	4356	—	470.8	107.1

[a] Million metric tons.

Source: Data from Winrock International, 1983, *World Agriculture: Review and Prospects into the 1990s*, Winrock International, Morrilton, AR.

Problem of Supplying Sufficient Animal Protein

A number of problems hinder providing sufficient animal protein in the developing countries. Malnutrition is most prevalent in the low-calorie, developing countries in which the population is increasing most rapidly and in which most of the arable land is devoted to production of cereal grains. Animal production must depend on land that cannot be used for cultivation, land that is used for forage in a crop-rotation system, or grains in excess of human requirements. The projected growth rate in human population is highest in many areas of the world where the per capita availability and consumption of milk are already low. Another problem is that most of the developing agricultural countries are located in the tropical or subtropical regions of the world, in which climatic and feed conditions are not as favorable for animal production as those in temperate regions.

A third problem is low per capita income in many developing countries. There is a direct correlation between consumption of animal protein and per capita income. In general, doubling the per capita income in many of the developing nations results in an increase of 12.5 g of animal protein consumption per day.

1.4 THE ROLE OF DAIRY CATTLE IN THE PRODUCTION OF ANIMAL PROTEIN

The major function of animal production, including dairy animals, is to provide protein, energy, minerals, and vitamins to supplement the cereal grains and pulses in the human diet. Animal products also contribute to gustatory pleasure and add variety to the human diet. Ruminant animals can obtain much of their energy and protein from feeds not directly convertible into food suitable for human consumption, such as food and liquor by-products, grains not meeting standards for human consumption, forages, pasture, urea, and other nonprotein nitrogen sources. In the United States, dairy cattle receive approximately 67% of their feed nutrients from forages. Between 60 and 90% of the protein consumed by cattle in the United States is also obtained from forages.

In most countries, especially the food-deficient countries, large areas of land not suitable for food-crop production can be used for growing forages that may be harvested directly by the animals. For example, about 25% of the total land area in developing countries is used for permanent pastures; the figure for Africa is 28%, compared with 28% for Europe, 14% for North America, 54% for Oceania, and 17% for the USSR.

Dairy cattle are rather efficient converters of feed protein into food protein. The overall conversion of feed protein into milk protein can be more than 50% for high-producing dairy cows, but it is approximately 30% for cows of slightly-above-average productivity (Table 1.2). Of concern in feeding the world population, however, is the efficiency of converting the protein and energy of grains that can be used for human consumption into milk protein and energy. An example is shown in Table 1.2, which compares the protein and energy inputs from grain and forages with protein and energy outputs in milk. The gross efficiencies for energy and protein are about 20% and 30%, respectively. When the efficiency is expressed in terms of energy and protein in grain that can be used for human consumption, the efficiencies increase to approximately 60% and 90%. If one considers the biological value (nutritional value) of grain and milk proteins, the efficiency becomes even greater. The average biological value of grain protein is about 60%, giving about 142 kg of grain protein (236 kg × 60%). The biological value of milk protein is about 100%; thus the example in Table 1.2 gives 213 kg of milk protein. The efficiency of converting grain protein corrected for biological value into milk protein is 150% (213 ÷ 142). Further efficiencies could be attained by substituting nonprotein nitrogen sources, such as urea, for some of the grain protein fed to the cows. The efficiency of conversion of high-producing cows, which require considerably more grain than cows of average productivity, is somewhat lower, because more of the protein input comes from grain.

Dairy animals are also important sources of meat. In Israel, Turkey, and many European countries, much of the meat produced comes from the same breeds of cattle that are used for milk production. In the United States, 11% of the beef consumed comes from cows and bulls discarded from domestic dairy herds. In addition, dairy steers make up 15 to 20% of the total steer and heifer slaughter.

TABLE 1.2 EFFICIENCY OF CONVERTING FEED ENERGY AND PROTEIN INTO MILK ENERGY AND PROTEIN BY DAIRY COWS [636 kg (1400 lb) DAIRY COW]

	Amount fed [kg (lb)]	Energy (Mcal)	Protein [kg (lb)]
Grain	1,818 (4,000)	8,000	236 (520)
Forage	3,636 (8,000)	16,000	473 (1,040)
Total	5,454 (12,000)	24,000	709 (1,560)
Milk produced	6,364 (14,000)	4,760	213 (468)
Gross efficiency		19.8%[a]	30.0%[b]
Efficiency of grain energy and protein		59.5%[c]	90.3%[d]

[a] 4760/24,000.

[b] 213/709.

[c] 4760/8000.

[d] 213/236.

Source: Adapted from D. L. Bath, 1977, *Hoard's Dairyman*, 122:1349.

Dairy cattle are most prevalent in the cooler and relatively humid regions of the temperate zones. About 80% of the world's milk is produced in North America, Europe, the USSR, Oceania, and South Africa. The milk available per capita by regions of the world is shown in Table 1.1. Many Asian and African countries have a very small amount of milk available per capita, which contributes to a protein deficiency in many of these areas.

Some countries, notably Ireland, Denmark, and New Zealand, have high per capita milk production; however, per capita consumption of milk is considerably below production. These and other countries in the same category are responsible for the small amount of milk that is exported. Slightly over 3% of the world's total milk production crosses national boundaries. The European Economic Community and New Zealand are responsible for about 80% of the world's exports. The United States produces about 4.6% of the world's trade in cheese and 11.0% of skim milk powder but negligible quantities of exported butter and casein.

1.5 HISTORY OF THE DAIRY INDUSTRY

The domestication of cattle and the use of their milk for human food began somewhere in Asia or northeast Africa between 6000 and 8000 B.C. Before the cow was domesticated, it was probably hunted by primitive man. Over the years, the cow has been used as a beast of burden and has been a source of food, an object of worship, a source of sacrificial offering, and a subject of mythology. Cow's milk and its products have been used for food, sacrificial offerings, cosmetics, and medicants.

The oldest written records are believed to go back to the Sumerians of Mesopotamia in approximately 6000 B.C. Dairying was highly developed at that time.

A mosaic frieze on one of the oldest buildings unearthed in the Euphrates Valley shows the milking of the cows from behind, the pouring of milk through a strainer into a vessel, and the presence of two attendants wearing fleece petticoats that are believed to have been the dress of priests and priest-kings. The frieze dates back to at least 3100 B.C.

The people of India were raisers of cattle as early as 2000 B.C. Butter was used as a food and as a holy offering to the gods. The butter was changed to ghee (butter oil or clarified butter). The cow at that time was considered holy.

Egyptian records dating back to approximately 3000 B.C. indicate that milk, butter, and cheese were used extensively. Cows were milked from the side, instead of from behind as was done by the Sumerians; in both Egypt and Sumeria, however, the calves were placed in front of the cows during milking.

Greek and Roman records go back to approximately 1550 B.C. and 750 B.C., respectively, and in both areas, milk and cheese were important parts of the human diet. In ancient Greece, most of the milk came from goats; in Rome, it came from sheep. In both countries butter was used in ointments and medicants but not as food. Butter was used as a food by some people in Europe, but it was not used extensively until the eighth century, when it became an important food in Norway.

References to cattle and milk are frequently made in the Old Testament and the New Testament alike. The Old Testament has 40 references to cattle. Milk and milk products were very desirable foods and are mentioned at least 50 times. Palestine is referred to approximately 20 times as a "land flowing with milk and honey."

The major developments in the dairy industry from the beginning of the Christian era to the middle 1850s occurred in Europe. Most of the currently important dairy cattle breeds in the United States and Europe originated in Europe.

REFERENCES AND FURTHER READINGS

BATH, D. L. 1977. Dairy cattle are great protein and energy converters. *Hoard's Dairyman*, 122:1349.

BYERLY, T. C. 1966. The role of livestock in food production. *J. Anim. Sci.*, 25:552.

DUNKLEY, W. L., and C. L. PELISSIER. 1981. Relationship of United States dairy industry to dairying internationally. *J. Dairy Sci.*, 64:975.

HODGSON, R. E. 1969. Animal agriculture can contribute to needed world food supply. *J. Dairy Sci.*, 52:1881.

ISAAC, E. 1962. On the domestication of cattle. *Science*, 137:195.

MARQUART, W. R. 1964. Dairy beef in the packing industry. *J. Dairy Sci.*, 47:1145.

NATIONAL DAIRY COUNCIL. 1976. Role of the dairy cow in providing food. *Dairy Counc. Dig.*, 47(1).

NATIONAL DAIRY COUNCIL. 1986. Diet, nutrition, and cancer: New findings. *Dairy Counc. Dig.*, 57(2).

NATIONAL MILK PRODUCERS FEDERATION. 1985. *Dairy Producers Highlights*, National Milk Producers Federation, Arlington, VA.

PETERSEN, W. E. 1950. *Dairy Science*, 2nd ed. J. B. Lippincott Company, Philadelphia.

PHILLIPS, R. W. 1969. Factors favoring animal production. *Proc. 2nd World. Conf. Anim. Prod.*, College Park, MD. Bruce, St. Paul, MN, p. 15.

ROSENWEIG, N. S. 1969. Adult human milk intolerance and intestinal lactase deficiency: A review. *J. Dairy Sci.*, 52:585.

RUSSOFF, L. L. 1955. The miracle of milk. *J. Dairy Sci.*, 38:1057.

RUSSOFF, L. L. 1970. Milk: Its nutritional value at a low cost for people of all ages. *J. Dairy Sci.*, 52:1296.

SPECKMAN, E. W., M. F. BRINK, and L. D. MCBEAN. 1981. Dairy foods in nutrition and health. *J. Dairy Sci.*, 64:1008.

WINCROCK INTERNATIONAL. 1983. *World Agriculture: Review and Prospects into the 1990s.* Winrock International, Morrilton, AR.

2

Dairy Industry in the United States

2.1 HISTORY

Cattle were first imported into the western hemisphere with Columbus's second voyage. The four calves and two heifers were taken either to Cuba or to South America. There were no cattle in North America prior to that time. The Spanish expedition headed by Coronado brought approximately 150 cattle to North America. The longhorn cattle of the Southwest probably originated from this importation.

The first importation of cattle by settlers into the United States occurred in the Jamestown colony in 1611. Cattle were not brought by the Pilgrims on the *Mayflower*; the first cattle arrived in the Plymouth colony in 1624. During the first winter more than half of the first settlers that arrived on the *Mayflower* died of malnutrition; some historians contend that some of these deaths could have been avoided if dairy cattle had been present. These first two cattle importations were followed by many more. After the cattle population became established, cattle moved westward with the pioneers to all parts of the United States.

The dairy industry from colonial times to the mid-1850s was based on the family cow. Most families had one or two cows that supplied milk and milk products to the immediate family. Only small amounts of milk or milk products were sold. Milk that was available to consumers was of two types. The major portion came from cows fed "swill," a cheap by-product from liquor distilleries. The milk was usually of low quality because it was produced under very unsanitary conditions and contained various amounts of added water. Other milk came from dairy farmers who kept cows in rural areas. The milk was brought into the cities daily by wagons or boats. This milk was also diluted, in most cases, before it reached the consumer. Milk from both sources was distributed by persons on foot or from wagons. Milk was dipped from one large container into receptacles supplied by the consumer. Milk produced and sold did not have to meet any health regulations. No permits or licenses were required of the distributors of this milk.

The first rail shipment of milk occurred in 1841 between Orange County, New York, and New York City, allowing farmers to ship fluid milk and cream as far as 80 miles to market. Refrigeration was not used, so the time that milk could be kept in shipment was limited. At first farmers shipped their own milk to the city, where it was received and distributed by dealers. Later, milk receiving stations were established in rural areas, either by dealers, independent shippers, or cooperative groups of farmers.

2.2 DEVELOPMENT OF THE DAIRY INDUSTRY

The establishment of a dairy industry in any country depends on three main factors; (1) a suitable market for milk and milk products; (2) facilities and trained personnel to collect, process, and distribute the milk; and (3) a continuing supply of milk of satisfactory quality. These factors are present in the United States as well as in other parts of the world, notably western Europe and Oceania.

The dairy industry has changed dramatically since 1900. Many developments have helped to make the industry what it is today; some of the major ones will be discussed briefly.

The development of bacteriology established the procedures for controlling diseases and harmful bacteria. These procedures are also important in the production of milk and the processing of milk products, including such cultured products as yogurt, sour cream, and cheese. Experiments by Louis Pasteur, the French scientist who is considered the father of modern bacteriology and immunology, led to the process of pasteurization, or the partial sterilization of foods. Pasteur used this process to prevent the acid, bitter, and "ropy" tastes in wines and the sour taste in beer. He is credited with saving the French wine and beer industries. The general principles of pasteurization were later used to destroy the harmful bacteria in milk. Robert Koch, a German scientist who was a contemporary of Pasteur, is credited with proving the germ theory of disease. His first experiments were done with anthrax, but he subsequently worked with many other diseases.

The first cheese factory in the United States was established in 1851 in Oneida, New York. This provided dairy farmers in the area with an additional outlet for their milk. The first creamery provided an even larger market for dairy farmers' milk since the creamery produced dairy products in addition to cheese. In 1856, Gail Borden received the first patent on condensed milk, and the first successful condensery was built in 1857. This provided yet another way of processing milk and widened the market for it.

Breeders' associations for owners of purebred cattle were started in approximately 1860. They played significant roles in the improvement of cattle in addition to establishing herd books for registration of animals. Among the improvement programs were establishment of official production-testing programs and type-classification programs, encouragement of public exhibition of cattle, and recognition of outstanding cattle and breeders.

The Land Grant Act of 1862 established colleges of agriculture in every state. The first college of agriculture had been founded in Michigan in 1855. Land-grant colleges have been important in educating dairy farmers to apply scientific principles of feeding, breeding, and management to their dairy cattle. In 1887 the colleges were strengthened by the Hatch Act, which established experiment stations in each state. This act led to much-needed research on dairy cattle production and milk processing, which are of importance in the production and distribution of milk. The colleges of agriculture and their sister colleges, the colleges of home economics, also gathered and disseminated information on the value of milk and milk products in the human diet.

The establishment of the extension service by the Smith-Lever Act in 1914 resulted in a more direct route of passing the information gathered in the scientific laboratories to the dairy farmer, the milk processor, and the consumer. The extension service also serves as a means of relating the problems on the farm and in the milk-processing plant to research workers. Not only have the colleges of agriculture and home economics been involved in research and education, but credit must also be given to many commercial companies that gathered and disseminated information of value to the producer, the processor, and the consumer.

The centrifugal separator was invented almost simultaneously in 1871 by Gustav DeLaval and Winstrup and Nielsen. The separator provided a mechanical means of separating the cream from the milk. Previously, this had to be done by allowing the cream to rise to the top of milk held in shallow pans and then skimming off the cream. Separators were soon in use on dairy farms.

The fat content of milk has been an important determinant of the price of milk. A chemical test for the determination of fat content was perfected by S. M. Babcock in 1890. The Gerber test for fat content, which is used extensively in Europe, was developed at approximately the same time. The fat tests allow payment for milk in relation to its energy content. Some present pricing systems also use protein content, solids–not fat, or total solids content in addition to fat content. Automated machines are now used routinely in Dairy Herd Improvement and milk plant laboratories to determine the composition of milk. Other chemical methods have been developed for use to determine the purity, adulteration, and deterioration of milk.

Control of diseases in animals, especially those diseases communicable to humans, was a necessary development for the dairy industry. The tuberculin testing of dairy herds was introduced in 1890. The first commercial pasteurizing machines for the destruction of pathogenic bacteria in milk appeared in 1895. Even though some work on the control of brucellosis in dairy herds had been done earlier, the systematic testing of dairy herds for the disease and the practical application of methods for its control did not take place in the United States until the 1940s.

In addition to these specific developments, many other inventions and discoveries served to promote growth of the industry. These include refrigeration, milking machines, and related equipment, such as homogenizers, churns, clarifiers, ice cream freezers, glass and paper milk containers, automatic bottling equipment, and bulk tanks and bulk tank milk trucks.

2.3 PRODUCTION OF MILK

The numbers of dairy cows and the annual milk production by states is shown in Figure 2.1. The dairy industry is concentrated in the Great Lakes area and the northeastern part of the United States. Slightly over one-half of the milk is produced by the eight states touching the Great Lakes. The other major milk-producing state is California.

The establishment of the dairy industry in these areas was based on several factors. First, these areas contain the greatest human population. Because milk is bulky (i.e., low market value in relation to its volume), most of the milk used for fluid milk and cream must be produced near its market. The central and northeastern areas include the hay and pasture belt. Much of this land is well adapted to growing forages and has limited alternative use for other crops. Hay and silages are bulky and generally cannot be shipped great distances economically. The major exception is alfalfa hay, which is shipped relatively long distances to the

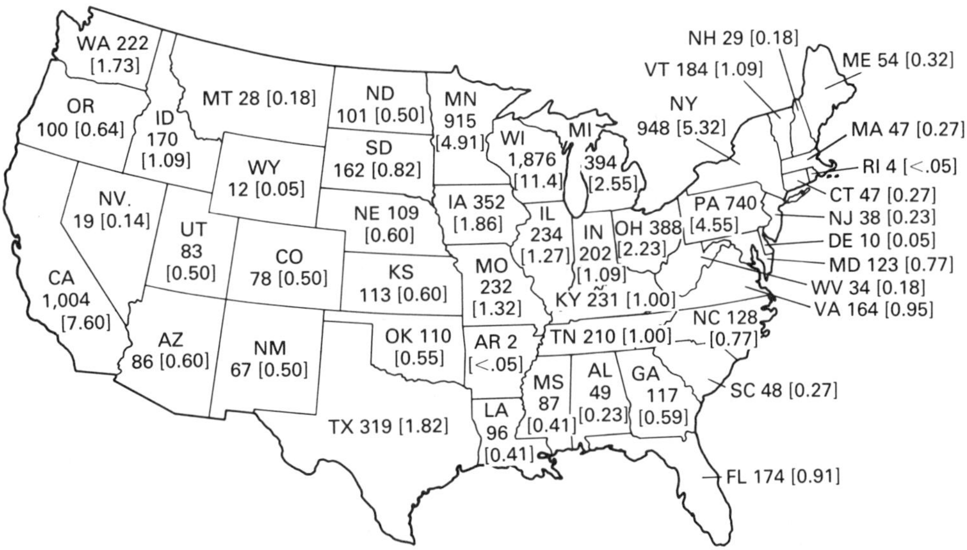

Figure 2.1 Numbers of dairy cattle (thousands) and total milk production (billion kilograms in brackets) by states in 1985.

milk-producing areas concentrated in southern California. The climate of the Great Lakes and northeastern states is similiar to that of northern Europe, the native home of the dairy breeds in the United States.

Three of the four top dairy states in terms of cow numbers and total milk production border the Great Lakes. The leading dairy state is Wisconsin, which has almost twice as many cattle and produces one and one-half times as much milk as California, the next-leading state. A large amount of the milk produced in the western part of the corn belt is used for manufactured milk products because of limited demand for fluid milk and cream in that area.

Significant changes have occurred in the dairy industry during the last half-century (Figure 2.2). The number of milk cows reached a peak of approximately 25 million cows in 1945, after which there was a substantial yearly reduction until

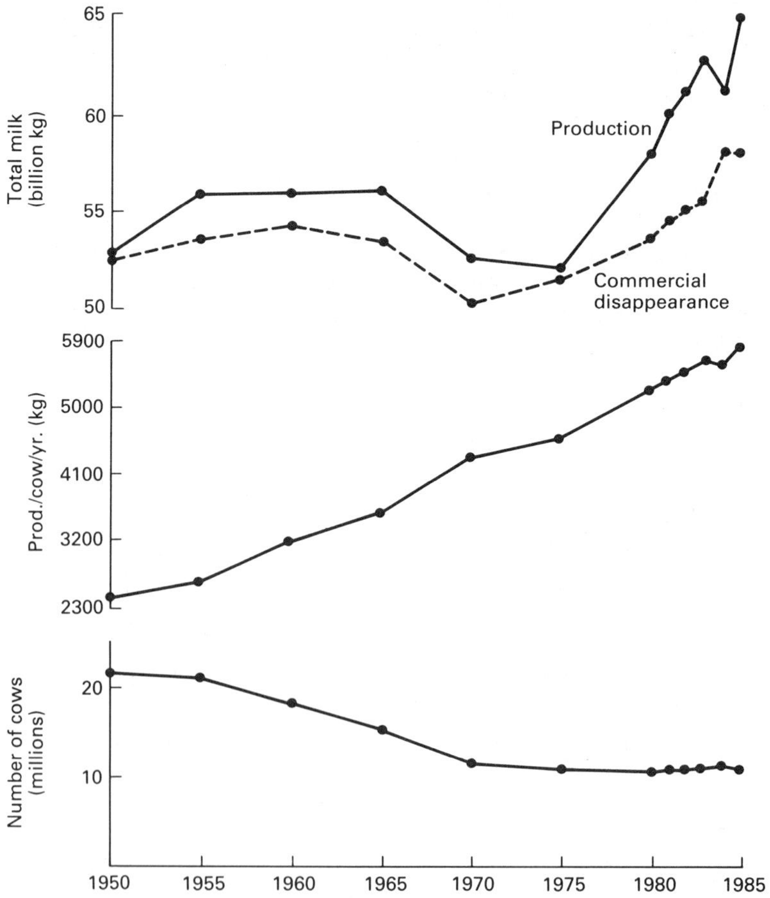

P-H Schmidt, Fig. 2.2, ⅓ off

Figure 2.2 Number of cows, milk production per cow, total milk production, and commercial demand for milk from 1950 to 1985.

1978. At that time the price of milk was very favorable (caused in part by high federal government support price for milk) and overproduction of milk in relation to demand occurred. The surplus milk is purchased by the Commodity Credit Corporation of the U.S. Department of Agriculture (USDA) to support the price

TABLE 2.1 SHIFTS IN MILK PRODUCTION BY REGIONS AND THE TOP DAIRY STATES FROM 1975 TO 1985 (BILLIONS)

Area and state[a]	1975 lb	1975 kg	1985 lb	1985 kg	Percent change
Northeast					
New York	9,904	4,502	11,746	5,339	+18.5
Pennsylvania	7,140	3,245	9,983	4,538	+39.8
CT, MA, ME, NH, NJ, RI, VT	4,775	2,170	5,169	2,350	+8.3
Total	21,819	9,917	26,898	12,227	+23.3
Appalachian					
DE, KY, MD, NC, TN, VA, WV	9,583	4,356	10,518	4,781	+9.8
Delta					
AR, LA, MS	2,637	1,199	2,630	1,195	−0.3
Southeast					
AL, FL, GA, SC	4,348	1,976	4,471	2,032	+2.8
Lake					
Michigan	4,434	2,015	5,568	2,531	+28.6
Minnesota	8,946	4,066	10,840	4,927	+21.2
Wisconsin	18,900	8,591	25,107	11,412	+32.8
Total	32,280	14,672	41,515	18,870	+28.6
Corn Belt					
Iowa	3,916	1,780	4,058	1,845	+3.6
Ohio	4,254	1,934	4,870	2,214	+14.5
IL, IN, MO	7,791	3,541	8,099	3,681	+4.0
Total	15,961	7,255	17,027	7,740	+6.7
Northern plain					
KS, NE, ND, SD	5,128	2,331	5,534	2,515	+7.9
Southern plain					
Texas	3,221	1,464	3,968	1,804	+23.2
OK	1,060	482	1,183	538	+11.6
Total	4,281	1,946	5,151	2,342	+20.3
Mountain					
AZ, CO, ID, MT, NM, NV, UT, WY	5,093	2,315	7,837	3,562	+53.9
Pacific					
California	10,853	4,933	16,734	7,606	+54.2
Washington	2,322	1,055	3,750	1,705	+61.5
OR	990	451	1,438	654	+45.3
Total	14,165	6,439	21,922	9,965	+54.8
Noncontiguous					
AK, HI	163	74	164	75	+0.6
Total	115,458	52,481	143,667	65,303	+24.4

[a] The top 10 states are spelled out.

of milk as determined by Congress. In 1984–1985 a milk diversion program was used to reduce the volume of milk produced and in 1986, a herd-buyout program was initiated to reduce the size of the national dairy herd.

Major shifts have also occurred in the location of cows in the United States during the last several decades (Table 2.1). Wisconsin continues to be the leading dairy state; however, California has moved into the number 2 position in terms of number of milk cows and total milk produced. In general, many of the Corn Belt states, the delta states, the southeastern states, and the Appalachian states have lost large numbers of cows and total milk production has decreased or increased only slightly since 1975 (Table 2.1). Dairying has become more concentrated in the Pacific and Great Lakes states. Even though the percentage increase in the mountain states is very large, it represents a small portion of the total milk supply. States with large percentage increases in milk production from 1975 to 1985 include New Mexico (192%), Washington (61%), Arizona (61%), and California (54%). Most of this increase has been associated with the movement of the human population.

The milk production per cow increased only slightly from 1900 to 1950, but between 1950 and 1985, the production has increased almost 2.5 times. During the last 10 years, the average production per cow has increased approximately 123 kg (270 lb) per year. Many factors are responsible for this increase, but most are related to the higher genetic potential of present-day animals combined with marked improvements in their feeding and management. The reduced number of dairy cows combined with a marked increase in the number of cows per farm has resulted in a tremendous reduction in the number of dairy farms. In 1950 there were 3,648,000 farms in the United States that reported having dairy cows; in 1982 this figure had dropped to 278,000, only 7.6% of the 1950 figure. The number of commercial dairy farms (more than 20 cows) was 180,895 in 1983. Over the years, most of the dairy farmers leaving the industry were the smaller farmers. In general, dairy farmers who have remained in the business are the better farmers and their operations are becoming larger in scale and more efficient.

Cash receipts from farm marketings of dairy products in 1984 were more than $17.9 billion, which represented approximately 12.7% of the total farm income for that year. In addition to this income, dairy farming produces income from the sale of cull and surplus animals.

2.4 CONSUMPTION OF MILK

The increase in per capita consumption of milk from 1975 is shown in Table 2.2. These values include total domestic disappearance and include milk products distributed from USDA supplies. During this period, the per capita consumption of total milk equivalent increased from 245.5 kg (540 lb) to 264.5 kg (582 lb). Prior to 1975 the consumption of milk had decreased steadily from a peak of 372.3 kg in 1945. A continuing drop in fluid milk and cream consumption is taking place; however, part of this decrease is offset by an increase in 2% milk and skim milk consumption.

TABLE 2.2 CHANGE IN PER CAPITA CONSUMPTION OF MILK AND MILK PRODUCTS (KILOGRAM OF PRODUCT)

Component	1975	1984	Percent change
Total milk equivalent	245.5	264.5	+7.7
Fluid milk and cream	83.9	60.4	−28.0
Low-fat and skim milks	35.3	48.9	+38.5
Butter	2.0	1.7	−15.0
Total cheese	6.4	8.7	+35.9
Cottage cheese	2.1	1.9	−9.5
Ice cream	8.5	8.2	−3.5
Evaporated and condensed milks	3.9	3.3	−15.4
Nonfat dry milk	1.4	1.2	−14.3

Source: Data from National Milk Producers Federation, 1985, *Dairy Producers Highlights*, National Milk Producers Federation, Arlington, VA.

Much of the drop in the consumption of milk equivalent other than fluid milk and cream was caused by a decrease in butter consumption, which dropped from 4.2 kg per person in 1950 to 1.7 kg per person in 1984. It takes approximately 23 kg of milk with a 3.5% milk-fat content to produce 1 kg of butter; thus a drop of 2.4 kg in per capita butter consumption is equal to a drop of 55.2 kg in per capita milk consumption.

Decreased butter consumption is attributed to several factors, but the most important one is the competition from margarine. Sales of margarine have increased markedly; margarine is in direct competition to butter because of its lower price and its somewhat comparable nutritional and culinary properties. Consumers have become increasingly calorie conscious, a situation that has led to a reduction in consumption of products with high fat contents. In addition, the intake of relatively large amounts of saturated fatty acids, as found in animal fats, has been implicated as one of the factors contributing to higher human blood cholestrol. High blood cholesterol is associated with atherosclerosis, the accumulation of fatty plaques in blood vessels and a contributing factor to heart disease. A clear-cut relationship between the consumption of saturated fatty acids and cholesterol to high blood cholesterol has not been established for most persons, but the publicity surrounding the issue has probably led to decreased consumption of butter and whole milk. The decrease in butter consumption has been offset in part by a marked increase in cheese consumption (Table 2.2).

The decrease in the consumption of fluid milk and cream is probably caused by the same factors that are responsible for decreased butter consumption. In addition, competition from other beverages has had an effect on consumption. Sales of soft drinks and beer have increased dramatically during the last 20 years (Table 2.3).

Sex, age, and economic status influence the consumption of milk. In 1965 a one-day survey was taken of the food intake of a sample of 14,500 men and women in the United States (Miller, 1970). Excluding infants, daily intake of milk and milk drinks was highest among males between 9 and 19 years of age. Con-

TABLE 2.3 PROPORTION (PERCENT) OF TOTAL BEVERAGE CONSUMPTION PER CAPITA IN THE UNITED STATES, 1962 AND 1982

Beverage	1962	1982
Milk	29.0	20.3
Coffee	33.4	18.3
Soft drinks	14.1	29.7
Beer	13.2	18.3
Fruit juices/tea	8.2	10.3
Wine/liquor	2.0	3.1
Total	100.0	100.0
Per capita beverage consumption (liters)	432.4	503.3

Source: R. E. Jacobson, 1986, *J. Dairy Sci.*, 69:1447.

sumption fell off for females after their early teens and for males after the age of 19. Adult women drank the least amount of milk. Per capita consumption of ice cream and cream was quite uniform for all age–sex groups. Per capita consumption of milk was generally lower among the lower-income groups, especially for children from 3 to 19 years of age.

Although race was not considered in the study above, it is well known that a high percentage of descendants from most central African and most Asian countries have some lactose intolerance (Chapter 1). These people cannot tolerate large amounts of milk, especially when consumed in considerable quantities at one time. This has a marked effect on the consumption of milk by these persons.

2.5 USE OF THE MILK SUPPLY

Over 80% of the milk in the United States is used for fluid milk and cream, butter, and cheese. In 1984, fluid milk and cream accounted for 37.4% of the total use, butter for 15.9%, and cheese for 28.6%. The other products, such as ice cream, evaporated and condensed milk, yogurt, nonfat dry milk, and sherbert, accounted for the remaining 18.1%. The percentage utilization for fluid milk and cream has decreased somewhat from 1950 (45.1 to 37.4%); however, the percentage use for butter has decreased markedly (28.1 to 15.9%), whereas that for cheese has increased from 10.1% to 28.6% since 1950.

2.6 BREEDS OF DAIRY CATTLE

A breed of dairy cows is a group of animals that have a common ancestry and were developed for a special function or functions. The three most distinguishing characteristics of the breeds of dairy cattle currently important in the United States are size, color of hair coat, and differences in yield and composition of milk.

Origin of the Breeds

The six major dairy breeds in the United States—Ayrshire, Brown Swiss, Guernsey, Holstein–Friesian, Jersey, and Milking Shorthorn—were developed in western Europe. There was little communication among the people living in these areas because they were usually banded together in relatively small groups for protection from outside invaders. Consequently, there was little crossbreeding with cattle from other areas. After cattle were domesticated, they were bred with an eye to the selection of specific characteristics, usually related to the market or use of the milk. Attention was also directed to the color of the animals as well as to their physical conformation. After years of breeding and selection and adaptation to the environment, cattle in one locality had fairly uniform characteristics and thus became a breed.

Zoological Classification of Cattle

Cattle belong to the family Bovidae, a subdivision of the suborder Ruminantia (ruminants, or cud-chewing animals) of the order Artiodactyla (even-toed ungulates, or animals with cloven hooves). The members of the family Bovidae, which includes sheep, goats, and true antelopes, are distinguished from other ruminants, such as deer and camels, by the possession of permanent horns with a bony core. The family comprises several genera, one of which, the genus *Bos*, includes all cattle, both wild and domestic. The domestic cattle belong to the subgenus *Taurus* of the genus *Bos*. Members of this subgenus roamed wild over Europe, Asia, and northern Africa until relatively recently. These animals have cylindrical horns set wide apart at the poll, a flat forehead, long nasal bones, and eyes that are far below the horn. The back is straight and the hair is short. The subgenus *Taurus* consists of two species, *Bos (Taurus) typicus* and *B. (T.) indicus*. The former includes all animals without humps and the latter, all animals with humps. *B. (T.) indicus* includes the zebu cattle, which are distinguished by a hump over the withers, large drooping ears, and a grunting cry. *B. (T.) typicus* includes most of the modern breeds of cattle in Europe and the United States. The four subspecies of *B. (T.) typicus* were the forerunners of the modern dairy cattle breeds. Two of these, *B. (T.) primigenius* and *B. (T.) longifrons*, were involved in the development of the dairy breeds. The members of the first subspecies were large, standing 6 to 7 ft in height at the withers. Their horns were long and strong and made a semicircular curve forward. The forehead and face were long and narrow. In contrast, the members of the second subspecies were much smaller than the other subspecies, the horns and face were shorter, and the forehead was longer and broader.

Classification by Use and Purity

Cattle can be classified according to use as dairy, beef, and dual-purpose. Milking Shorthorn used to be classified as dual-purpose, but within recent years the breeders have decided to emphasize the dairy characteristics of the breed. In addition to the Milking Shorthorn, the other five major dairy breeds in the United States are Ayrshire, Brown Swiss, Guernsey, Holstein–Friesian, and Jersey. A few Dexter and Kerry cattle, which are also considered dairy breeds, are found in North America. The Jamaican Hope is a dairy breed that was developed on the island of Jamaica. In Europe, the Red Dane and several Red and White breeds are classified as dairy breeds. In India, the Red Sindhi is primarily a dairy breed. Devon and Red Polled cattle are considered dual-purpose breeds.

Classification can also be made according to the "purity" of the animal. A purebred is defined as an animal whose ancestry can be traced back through all lines to the foundation animals of the breed. After the breed was developed, a herd registry book was usually established and animals possessing certain breed characteristics were allowed to be registered. Minor variations now exist, because all but the Holstein–Friesian breed association allow the registration of an animal whose parents are not registered through their open herdbook programs. These stipulate identification of animals for specified generations if an animal has the characterisitcs of the breed as well as other desirable qualities. Such animals are technically not purebreds by the previous definition. In 1960 it was estimated that 1.44 million (8.2%) of the dairy cattle in the United States were registered with the breed associations (Rock, 1986). In 1985, the estimate was 1.58 million (14.6%). Although the number of registered cows increased, the marked increase in the percentage was due to the large reduction in the number of dairy cows from 17.65 million in 1960 to 10.8 million in 1985.

A grade animal is defined as a nonpurebred animal that possesses the major characteristics of a breed. In many cases, a grade is a descendant of purebred animals that has not been registered with the breed association. A crossbred is an animal resulting from the mating of two distinct breeds. Crosses frequently involve more than two breeds, but the animals are still considered to be crossbred. The term "scrub" has several different meanings. It is usually defined as a noncrossbred animal that does not possess the major characteristics of a breed; however, the term is also used to designate an animal with undesirable characteristics, even though it may be a purebred.

Distribution of the Breeds

No accurate measure of the breed distribution on farms in the United States has been made in recent years, because counting of dairy cattle breeds in the census has been discontinued. Therefore, only indirect estimates are available. The estimated distribution of the dairy breeds from 1960 through 1985 is shown in Table 2.4. The striking trend in the table is the increase in the percentage of Holstein cattle and the decrease in the Guernsey and Jersey breeds. Two major reasons are present for the shift. The most important is the market for milk. The pricing system in the United States favors large volumes of milk and consequently the Holstein breed has prevailed. Also, many Jerseys were kept in the southern states because they are more heat-tolerant than Holsteins. Under good management, however, Holsteins were found to adapt quite well to the hotter climate; consequently, Holsteins have replaced a large number of Jerseys in the south. The

TABLE 2.4 PERCENTAGE OF COWS OF THE SIX DAIRY BREEDS

Breed	Status	Year 1960	1970	1980	1985
Ayrshire	Registered	5	4	3	2
	Grade	3	2	0.5	0.5
	Total	3	2	1	0.5
Brown Swiss	Registered	6	5	4	3
	Grade	3	2	0.5	0.5
	Total	4	2	1	1
Guernsey	Registered	20	12	6	5.5
	Grade	10	5	1.5	1.5
	Total	11	6	2.5	2
Holstein	Registered	51	64	76	78.5
	Grade	71	84	95	95
	Total	69	82	92	93
Jersey	Registered	15	13	10	10
	Grade	11	6.5	2.5	2.5
	Total	11	7.5	3.5	3.5
Milking Shorthorn	Registered	3	2	1	1
	Grade	2	0.5	<0.25	<0.25
	Total	2	0.5	<0.25	<0.25

Source: J. M. Rock, 1986, *Hoard's Dairyman*, 131:305.

Holsteins and Jersey breeds are distributed quite evenly throughout the United States; the Guernsey and Ayrshire breeds are found mostly in the northeastern states. Milking Shorthorn and Brown Swiss are most prominent in the midwest, although New York is one of the leading states in Brown Swiss numbers.

Characteristics of Individual Breeds

Some of the characteristics of the six dairy breeds found in the United States are given in Table 2.5. The ideal size of the cow and bull and some of the color markings are taken from the Unified Dairy Score Card. The weight is considered the ideal for the breed and does not necessarily represent the average weight of the breed, especially for the Holsteins, which tend to be somewhat less.

Holstein–Friesian. Holstein–Friesian is the official name of the breed, although they are commonly referred to as Holsteins in the United States. In Europe "Friesian" is usually used to designate those of European descent and "Holstein" for those with North American ancestors. It is the largest breed and the most prevalent one in the United States. An outstanding cow is shown in Figure 2.3. The ideal size of the mature cow is 1500 lb and that of the bull is 2200 lb. The average for the cows of the breed is somewhat less. Holsteins are noted for producing the largest volume of milk with the lowest milk-fat percentage. The lowest milk-fat percentage is also accompanied by the lowest nonfat solids per-

TABLE 2.5 CHARACTERISTICS OF THE SIX MAJOR DAIRY BREEDS IN UNITED STATES

Characteristic	Ayrshire	Brown Swiss	Guernsey	Holstein	Jersey	Milking Shorthorn
Ideal size						
Cow						
(lb)	1,200	1,400	1,100	1,500	1,000	1,400
(kg)	545	636	500	682	454	636
Bull						
(lb)	1,850	2,000	1,700	2,200	1,500	2,100
(kg)	841	909	773	1,000	682	955
Color	Red or mahogany and white	Brown	Fawn and white	Black and white or red and white	Fawn with or without white	Red, white, or roan
Grazing ability	Excellent	Excellent	Good	Fair	Good	Good
Maturity	Moderate	Late	Early	Late	Early	Late
Weight of calves						
(lb)	75	90	75	95	60	75
(kg)	34	41	34	43	27	34
Value for beef and veal	Good	Excellent	Poor	Excellent	Poor	Good
Milk fat[a] (%)	3.96	4.10	4.67	3.65	4.85	3.65
Milk yield[a]						
(lb)	12,507	13,065	11,452	15,939	10,906	11,277
(kg)	5,685	5,939	5,205	7,245	4,957	5,126
Native home	Scotland	Switzerland	Guernsey	Holland	Jersey	England

[a] Average for cows on official DHI tests in 1984.

centage; in general, if the nonfat solids percentage of a breed is low (or high), the milk-fat percentage will be low (or high) also.

Registered Holsteins must have clearly defined black-and-white or red-and-white markings. Those colored red and white must carry the term "Red" as a suffix in their registration names.

The Holstein is well adapted for beef purposes because it makes relatively fast growth and produces a very good carcass. The calves at birth are the largest of the dairy breeds, and this, in addition to the white color of the body fat, makes them well suited for veal purposes. The Holstein does not mature as early as the Jersey, Guernsey, or Ayrshire.

The Holstein originated in the provinces of northern Holland and western Friesland in Holland. It was developed from the wild ancestor *Bos (Taurus) typicus primigenius* and has been bred in Holland for at least 2000 years. Holland is noted for its fertile soil and abundant rainfall, which result in excellent pastures. Thus the poor grazers could compete as well there as the good grazers, so the Holstein was not forced, by circumstance, to develop the grazing ability of some of the other breeds, especially the Ayrshire and Brown Swiss breeds. The winters in Holland are rather cold, so that the cows need to be stabled. The barns and homes frequently were under the same roof, which necessitated the development of docile animals. In addition, cows were often milked on pasture, which again dictated the necessity of having docile animals. When the breed was being developed, the

Figure 2.3 Northcroft Ella Elevation is an outstanding Holstein–Friesian cow. She is the all-time All-American aged cow, was All-American four times, is classified excellent at 97 points, and produced over 21,800 kg of milk in one lactation. Photograph courtesy of the George Knight family, Airville, PA.

milk was used primarily for cheese making. Cheese requires relatively large quantities of milk; consequently, the development of the breed emphasized cows with high yields and low milk-fat percentages.

Jersey. The smallest of the dairy breeds is the Jersey and because of this and its yellow body fat, it does not have a very high value for meat and veal. The milk of Jerseys has the highest milk-fat percentage and nonfat solids percentage of all dairy breeds. Jerseys are noted for having a highly developed nervous system; they are sensitive and react quickly to stimuli. Jerseys are the earliest-maturing dairy breed and reach peak milk production earlier than do cattle of the other breeds.

Jersey are descendants of the wild ancestor *Bos (Taurus) typicus longifrons*; they are believed to have descended from cattle of that subspecies from Brittany and Normandy. The breed was develped on the island of Jersey, which is one of the Channel Islands located between France and England. The area of the island is approximately 62 square miles. Much of the agriculture is devoted to the production of potatoes, tomatoes, and greenhouse products. The chief feed for cows is pasture, and the grasses grow very luxuriantly. Cows are usually tethered on

Fgiure 2.4 This outstanding Jersey cow, Rocky Hill Debbie Rockal, is appraised excellent at 92% and has a lifetime production of over 121,300 kg of milk. Her highest record was over 14,860 kg of milk. Photograph courtesy of Foundation Associates, Newberry, SC.

pasture and receive a limited quantity of concentrates. During the development of the breed, butter was the major product produced from milk; consequently, emphasis was placed on selecting cattle that produced a high percentage of fat in the milk. The milk of the Jersey cow has a characteristic yellow color because of a high carotene content. Breeds other than Jersey and Guernsey have the ability to convert more carotene to vitamin A and thus their milk does not have the yellow color characteristic of the two smaller breeds.

Jerseys are noted for being uniform in type, because registration of animals that did not have desirable type characteristics was not allowed on the island of Jersey. The excellent type conformation of this breed is shown in Figure 2.4.

Guernsey. The Guernsey is slightly larger than the Jersey, but its type and milk characteristics are similiar to those of the Jersey, as illustrated in Figure 2.5. The milk-fat percentage and the total nonfat solids percentage are slightly lower than those of the Jersey, but the color of the milk is more yellow because of a higher carotene content. The Guernsey calf is slightly larger at birth than the Jersey calf (Table 2.5). Their small size and the yellow color of their body fat cause Guernseys to be relatively undesirable for beef and veal. Guernseys mature slightly later than Jerseys, and their grazing ability is not quite as good.

Guernseys were developed from wild cattle of the subspecies *Bos (Taurus) typicus longifrons* on the island of Guernsey, another Channel Island between France and England. The foundation stock of Guernseys, like that of the Jerseys, is believed to have come from Brittany and Normandy. The island of Guernsey, which is only 24 square miles in area, is located approximately 22 miles northwest of Jersey. The island slopes to the north; consequently, the climate is slightly colder than that on the island of Jersey. The main agricultural enterprises on the

Figure 2.5 Cleverlands Nances Judy Jean, an outstanding Guernsey cow, is a three-time winner of the Liebers Trophy, which recognizes the living Guernsey cow with the highest lifetime milk production credits. She has five lactations with over 13,600 kg of milk and six sons in artificial insemination organizations. Photograph courtesy of Ray Orisio, Woodburn, OR.

island are flowers and greenhouse crops. The pasture conditions and the management of the cows are comparable to those on the island of Jersey. The herds are small and the cows are tethered. Guernseys are characteristically rather docile. Guernseys are slightly less refined than Jerseys, but the breed is also noted for its uniformity of type. In 1830 a scale of points was adopted on the island of Guernsey for the evaluation of animals prior to their registration. Milk was used primarily for butter; therefore, high milk-fat percentage was important in the development of the breed.

Brown Swiss. The Brown Swiss, oldest of the dairy breeds, descended from the wild subspecies *Bos (Taurus) typicus longifrons*. Bones that closely resemble the skeleton of the modern Brown Swiss cow were found in ruins in Switzerland dating back to 4000 B.C. The Brown Swiss is the second-heaviest dairy breed, not quite reaching the size of the Holstein. The color is brown and varies from very light to very dark. White and off-color spots above the underside of the belly or a white core in the switch are not desired by purebred breeders. The Brown Swiss is a docile animal with a notable tendency to stubbornness. Because of its relatively large size, the high birth weight of its calves, and its

Figure 2.6 This outstanding Brown Swiss cow, Idyl Wild Improver Jinx, was the All-American 2-year-old, 3-year-old, and aged cow. She has two records of over 11,300 kg of milk. Photograph courtesy of Forest Lawn Farm, Wausaw, WI.

white body fat, the Brown Swiss is similar to the Holstein for beef and veal purposes. Brown Swiss tend to mature and reach peak production at a later age than do cattle of the other dairy breeds. An outstanding Brown Swiss cow is shown in Figure 2.6.

Brown Swiss were developed on the mountain slopes of Switzerland, where grazing begins at the foot of the mountain in the spring and proceeds up the rugged mountain slopes during the summer. In the fall, the cows are returned to the valleys and are housed in barns throughout the winter. During the spring, summer, and fall, their feed is limited to the grass on the mountain slopes. Grazing on mountain slopes has produced excellent grazing ability in this breed. Because the breed was developed primarily for cheese production, large volumes of milk with low milk-fat percentages have been emphasized.

Ayrshire. The wild ancestors of the Ayrshire breed were *Bos (Taurus) typicus primigenius* and *B. (T.) typicus longifrons*. The Ayrshire breed was the latest breed to develop; its chief development took place after 1750, long after the other breeds were well established. The breed was developed in the county of Ayr in southwestern Scotland. Cattle from Holland, the Channel Islands, and the Teeswater cattle from northern England were the foundation animals of the breed.

Figure 2.7 Matoon Pond Bell's Pansy Girl is an outstanding Ayrshire cow that is classified excellent at 91. She has a lifetime production of over 47,200 kg of milk. Photograph courtesy of Sycamore Meade Farm, Millersburg, PA.

The climate in the county of Ayr is cold and damp during a large part of the year. There was relatively little forage available during the development of the breed, because of the infertile soils in the area. This circumstance selected animals of extreme hardiness and good foraging ability.

The Ayrshire is larger than the Guernsey and Jersey and its color varies from red and white to mahogany and white. The latter color combination is more prevalent among bulls. The birth weight of the calves is relatively low, so they have only moderate value for veal purposes; cull cows and surplus animals do have moderate beef value, because the animals fatten quite well and have white body fat. The milk-fat percentage is approximately 4%, and nonfat solids percentage is lower than that of the Guernsey and Jersey.

The Ayrshire breed has been noted for its style and symmetry, its characteristic upturned horns, and the symmetry and attachments of the udder (Figure 2.7). Most breeders now remove the horns; therefore, the upturned horns are rarely seen. In the development of the breed, emphasis was placed on the shape of the udder, and the Ayrshire excels in this characteristic in comparison to most other breeds.

Milking Shorthorn. The Milking Shorthorn was considered a dual-purpose breed for many years, but in 1969, the breeders declared it to be a dairy breed. In 1972 the Milking Shorthorn Association became part of the Purebred Dairy Cattle Association.

Milking Shorthorns are not a distinct breed but are part of the Shorthorn breed, which is one of the oldest recognized breeds. Shorthorn cattle were developed in northeastern England in the valley of the Tees River. Their wild ancestors were *Bos (Taurus) typicus primigenius*. Many of the early imported cattle into the United States were referred to as Durham cattle, named after one of the counties in England where early improvement of the breed took place. Up to 1935, no distinction was made between dairy and beef types in the registration of cattle in the United States, but at that time a breeder could request that "Milking" be added to "Shorthorn" on the registration certificate. In 1948, the American Milking Shorthorn Society was incorporated and started to register cattle and promote the breed. In 1967, Australian Illawarra Shorthorn semen was introduced into the United States and many Milking Shorthorns today have part of their ancestry from the Illawarra animals.

The breed is known as a versatile breed for milk and beef production. Their relatively large size gives them an advantage for beef production and high salvage value for cull cows. The relatively low birth weight of calves [34 to 36 kg (75 to

Figure 2.8 Woodview Flora WMP, an outstanding Milking Shorthorn cow, has two records that average 9538 kg of milk and 358 kg of fat on a mature equivalent basis. She is the first recipient of the High Cow Performance Index Award. She is classified very good at 87. Photograph courtesy of Bierhup Brothers, Wellston, OH.

80 lb)] makes them less desirable for veal than are Holstein and Brown Swiss calves. Their milk is well suited for fluid use and cheese production. The color of the animals ranges from nearly white to all red, and many have a mixture of the two colors, referred to as roan color.

Since the breed was selected for both beef and dairy purposes for a number of years, the uniformity of dairy type is less than that for the other breeds. However, marked improvement in dairy type has occurred since the breeders decided to make it a dairy breed. The animals are most numerous in the midwestern states. An outstanding cow is shown in Figure 2.8.

Breed Registry Associations

The maintenance of purebred cattle required the development of breed registry associations. Most of the associations in the United States were formed during the last quarter of the nineteenth century. The primary purpose of the associations was to register animals that were distinctive in their breeding, to protect importers of cattle from unethical practices by other importers and dairy farmers. The original herd books were developed primarily for the recording of pedigrees of animals. Most of the animals that were originally recorded were imported from Europe.

Breed associations are made up of owners of purebred dairy cattle. The associations are financed by membership fees and the collection of fees for participation in various programs, such as registration of animals and for production testing, type classification, and promotion programs conducted by the associations.

A major function of the breed associations is the preservation of the purity of the breed by requiring the registration of animals. Generally, only animals whose parents are registered with an association can be registered; however, all associations but the Holstein–Friesian association allow for the registration of a few outstanding animals whose parents are not registered. The breed associations also record the transfer of ownership of purebred cattle. In addition, the associations also keep records on the type classification and the production of cows in the breed.

A major objective of a breed association is to improve the breed. Two basic types of programs are carried out. One is the keeping of production records, which are now basically the use of Dairy Herd Improvement (DHI) records that meet certain standards. Breed associations played a significant role in the early development of the production-testing programs. The second is the improvement of type by sponsoring a type classification or appraisal program. In these programs, cows are given a rating in comparison to the ideal type conformation of the breed. Within recent years a linear-type program had been added in which the various parts of the animal are evaluated. This allows the dairy farmers to see the strengths and weaknesses in individual cows and yields more useful information for genetic evaluation, particularly of bulls. Breed associations have also encouraged the showing of animals in local, state, and national fairs as a means of improving type.

Another major objective of a breed association is promotion of the breed. In addition to the type classification and production testing programs, most associations have special awards and programs to improve and promote the breed. Among these are animal awards and breeder awards, the publication of a breed magazine, and employment of one or more field persons to help farmers improve and promote their cattle.

The Purebred Dairy Cattle Association

The Purebred Dairy Cattle Association (PDCA) was formed in 1940 by members of the dairy breed associations. The purpose of PDCA is to make a united effort to promote purebred dairy cattle. In addition, they have developed uniform programs of concern and interest to the purebred dairy cattle breeders. Among the programs instituted are (1) unified dairy cow and bull score cards; (2) a set of unified rules for official testing; (3) a set of unified rules and regulations for the registration of artificially bred animals; (4) a list of classes for local, state, and national dairy shows; and (5) a code of ethics for private and public sales.

Choosing a Breed

Several factors should be considered in selecting a breed. The most important consideration is to choose a breed that meets present and future market conditions. In the past, breeds with high yields and low milk-fat tests have been favored. Personal preference is very important to some dairy farmers. If persons can be happy with only one breed, that breed should be chosen. Personal satisfaction is very important in working with dairy cattle.

Choosing the predominant breed in the area is advantageous when buying and selling surplus animals; however, with modern-day transportation and communication, this is of only limited importance. Climate may be important in some regions, particularly the south; however, the least-tolerant breeds to heat are now performing satisfactorily with good management practices. Grazing ability should be considered if cows will be on pasture for part of the year. Vigor of the calves is important and appears to be related to the birth weight of the calves—the larger the calves, the better their chances for survival. Again with good management, the calves from the smaller breeds survive well.

REFERENCES AND FURTHER READINGS

Crop Reporting Board. 1986. *Milk Production* (Feb. 14). U.S. Department of Agriculture Statistical Reporting Service.

Dunkley, W. L., and C. L. Pelissier. 1981. Relationship of United States dairy industry to dairying internationally. *J. Dairy Sci.*, 64:975.

Jacobson, R. E. 1986. Review of current and future consumption trends for milk and dairy products. *J. Dairy Sci.*, 69:1447.

Johnson, J. E. 1965. Problems of producing milk around the world. *J. Dairy Sci.*, 48:1550.

Miller, R. R. 1970. Milk consumption by age, sex, and income, including away-from-home use. *Dairy Situation*, DS-331:25, U.S. Department of Agriculture.

Morley, L. W. 1956. Dairy cattle breed associations. *J. Dairy Sci.*, 39:712.

National Milk Producers Federation. 1985. *Dairy Producers Highlights*, National Milk Producers Federation, Arlington, VA.

Petersen, W. E. 1950. *Dairy Science*, 2nd ed. J. B. Lippincott Company, Philadelphia.

Rock, J. M. 1986. There have been many changes in dairy breeds. *Hoard's Dairyman*, 131:305.

Russoff, L. L. 1955. The miracle of milk. *J. Dairy Sci.*, 38:1057.

3

The Mammary Gland

The dairy industry is based on the ability of the mammary glands of some mammals to produce milk in excess of that required to nourish their young. The anatomy, histology, growth, and development of the mammary gland are discussed in this chapter.

3.1 ANATOMY OF THE MAMMARY GLAND

General Appearance of the Udder

The udder of the dairy cow is composed of four mammary glands located in the inquinal region. The udder is connected to the body wall by a series of attachments and is joined to the body cavity by the inguinal canal, through which pass arteries, veins, lymph vessels, and nerves. The shape of the udder has been described as that of a teacup handle with the fullest part of the handle directed toward the rear quarters, where the udder is partially hidden by the rear legs. The udder is divided into right and left halves by the medial suspensory ligament. The fore and rear quarters of each half have separate duct systems, but no membrane separates them. The rear quarters produce approximately 60% of the milk and the forequarters produce the remaining 40%.

The size and shape of udders vary considerably with the producing ability, age, and inheritance of the cow. In addition to having the necessary tissue for secretion of milk, the udder must be large enough to hold the secreted milk between milkings. A high-producing cow secreting 23 kg (50.6 lb) of milk in a 12-hour period is not uncommon; together, the milk and secretory tissue of such an animal before milking would weigh well over 50 kg. Thus a strong suspensory apparatus is needed to support the udder.

Each of the four mammary glands is drained by a teat, which is located on the ventral surface of the udder. Some cows have extra teats (supernumerary teats), which are usually located posterior to the two rear teats. Extra teats are usually removed surgically while the animal is young because they tend to become infected with bacteria and thus to develop mastitis.

Suspensory Apparatus of the Udder

Seven different tissues suspend and support the udder. The skin is the most superficial of these and serves in a minor capacity to suspend and stabilize the udder. The skin is attached to the underlying udder by a second tissue—a fine, areolar, subcutaneous tissue that also serves in a minor capacity to suspend the udder. A cordlike tissue forms a loose bond between the upper surface of the forequarters and the lower abdominal wall.

The two halves of the mammary gland are divided by two heavy sheets of elastic tissue that arise from the abdominal wall and attach to the flat surfaces of the glands (Figure 3.1). These two sheets, which make up the medial suspensory ligament, send out numerous lamellae that pass into the gland to become continuous with the connective tissue framework of the udder. The lamellae are more prevalent in the bottom part of the udder.

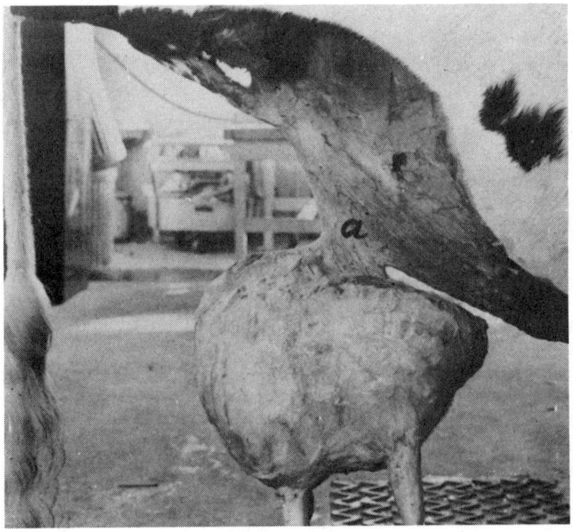

Figure 3.1 Medial suspensory ligament (a) of the udder gives a nearly perfectly balanced suspension of the udder. From W. W. Swett et al., 1942, *J. Agric. Res.*, 65:19.

The subpelvic tendon shown in Figure 3.2 arises from the ventral surface of the pelvic bone. It does not support the udder directly, but it gives rise to the lateral suspensory ligaments. One lateral ligament is located on each side of the udder (Figure 3.3). Each lateral suspensory ligament is divided into two parts, a superficial layer and a somewhat thicker, deeper layer. The lateral ligaments are more fibrous than the medial suspensory ligaments. The superficial layers of each lateral suspensory ligament extend downward and forward over the udder and reflect off the inner surface of the thigh. The deep layers of the lateral suspensory ligaments extend over the udder and attach to the lateral surface of the udder. They also send out numerous lamellae that pass into the udder and become continuous with the interstitial framework of the udder.

If the medial and lateral suspensory ligaments become elongated and the cordlike tissue that attaches the forepart of the udder to the body wall lengthens, the udder "breaks away" from the body. If the condition becomes too severe, a pendulous udder results, as shown in Figure 3.4.

Duct System

The teat for each quarter drains only the milk secreted in that quarter. The small canal located in the end of each teat (teat meatus, streak canal) is approximately

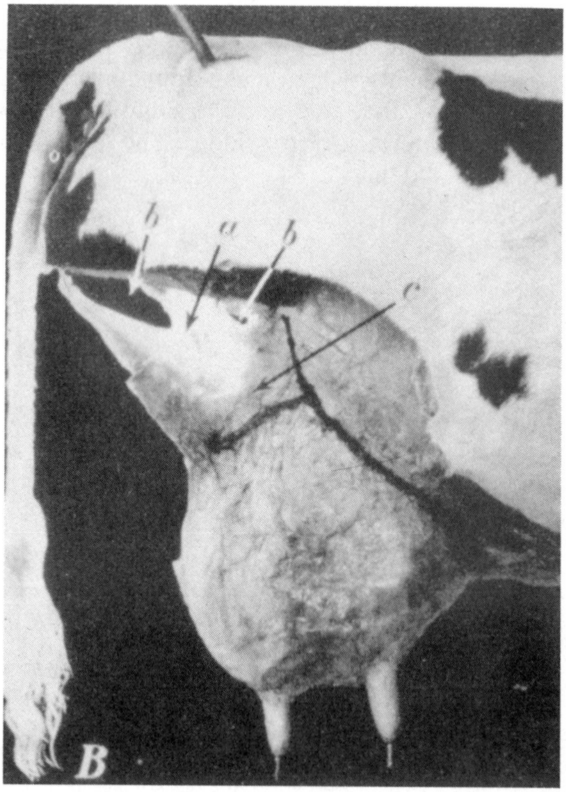

Figure 3.2 Subpelvic tendon (a), which gives rise to the two sheets of the lateral suspensory ligament on each side of the udder. From W. W. Swett et al., 1942, *J. Agric. Res.*, 65:19.

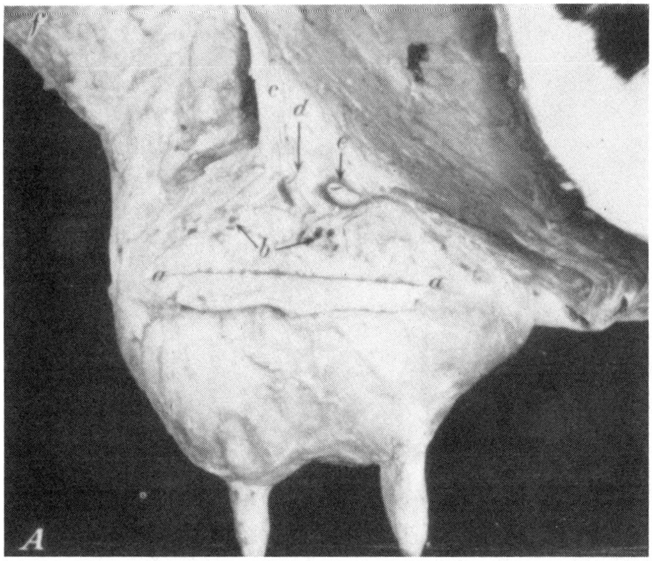

Figure 3.3 Suspensory apparatus of the udder. Deep layer of the lateral suspensory ligament (a). A portion of the medial suspensory ligament (b). From W. W. Swett et al., 1942, *J. Agric. Res.*, 65:19.

$\frac{1}{4}$ to $\frac{1}{2}$ in. long and is the only sphincter in each gland (Figure 3.5). The teat meatus is made up of three to five convex epithelial projections that lie close together to make a star-shaped slit. The projections are held closed by involuntary sphincter muscles. The teat meatus prevents the escape of milk between milkings and serves as a barrier to bacteria and foreign material. The rate of milking of a cow is partially dependent on the size of the teat meatus; faster-milking cows usually have a meatus of larger diameter. Located just above the teat meatus are seven or eight loose folds of membrane known as Fürstenburg's rosette. Pressure caused by accumulation of milk from above causes the folds to smooth out, which aids in

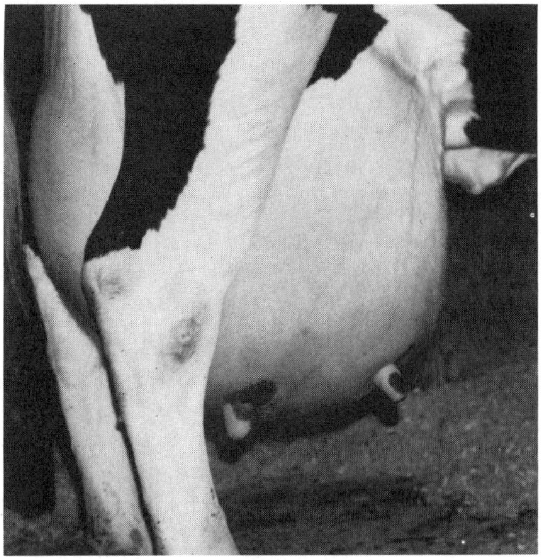

Figure 3.4 A pendulous udder resulting from the lengthening of the medial and lateral suspensory ligaments. This condition makes the udder more difficult to milk and causes it to be more susceptible to injury. From *Principles of Dairy Science* by G. H. Schmidt and L. D. Van Vleck. Copyright © 1974, W. H. Freeman and Company. Used by permission.

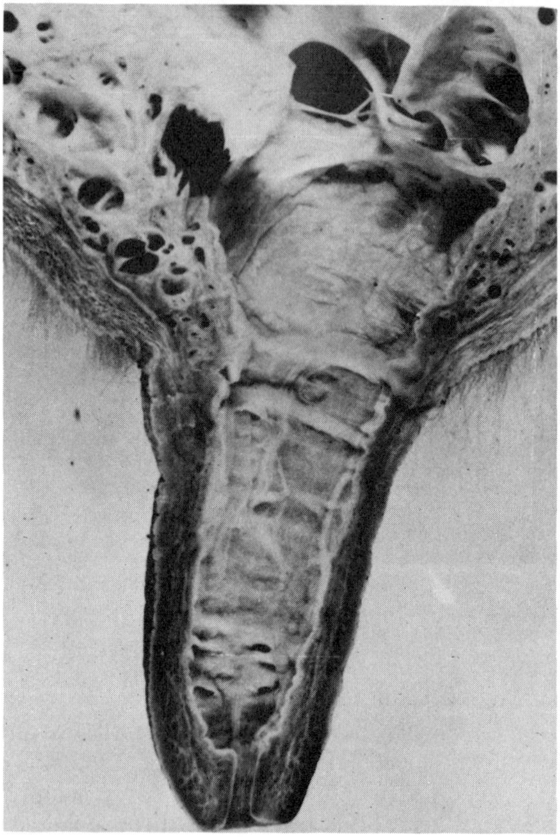

Figure 3.5 Longitudinal section of the teat, showing the teat meatus and the many circular and longitudinal folds. Photograph courtesy of Babson Bros. Co., Oak Brook, IL.

blocking the escape of milk. The teat cistern, the cavity within the teat, holds from $\frac{1}{2}$ to $1\frac{1}{2}$ ounces of milk, depending on the size of the teat. The walls of the teat cistern have many circular and longitudinal folds (Figure 3.5), which have a tendency to smooth out under pressure. These folds make a number of pockets and pouches in the interior wall of the teat. In some animals, the folds may form a complete membrane across the teat, thus blocking the escape of milk and producing permanent "blindness" of the quarter. A small constriction at the top of the teat cistern, the annular fold, contains dense connective tissue that can also restrict milk flow.

The gland cistern is located within the mammary gland and is continuous with the teat cistern (Figure 3.6). It varies greatly in size, shape, and capacity and functions as a collecting vessel for the 15 to 20 major ducts that lead to it.

The major ducts branch and rebranch many times and finally give rise to small ducts that lead to the secretory tissue, composed of alveoli. Milk secreted within the alveoli is carried via the duct system to the teat cistern, where it can be obtained either by hand or machine milking. At the time of milking, approximately 40% of the milk is in the teat and gland cisterns and the major ducts and the remaining 60% is in the alveoli.

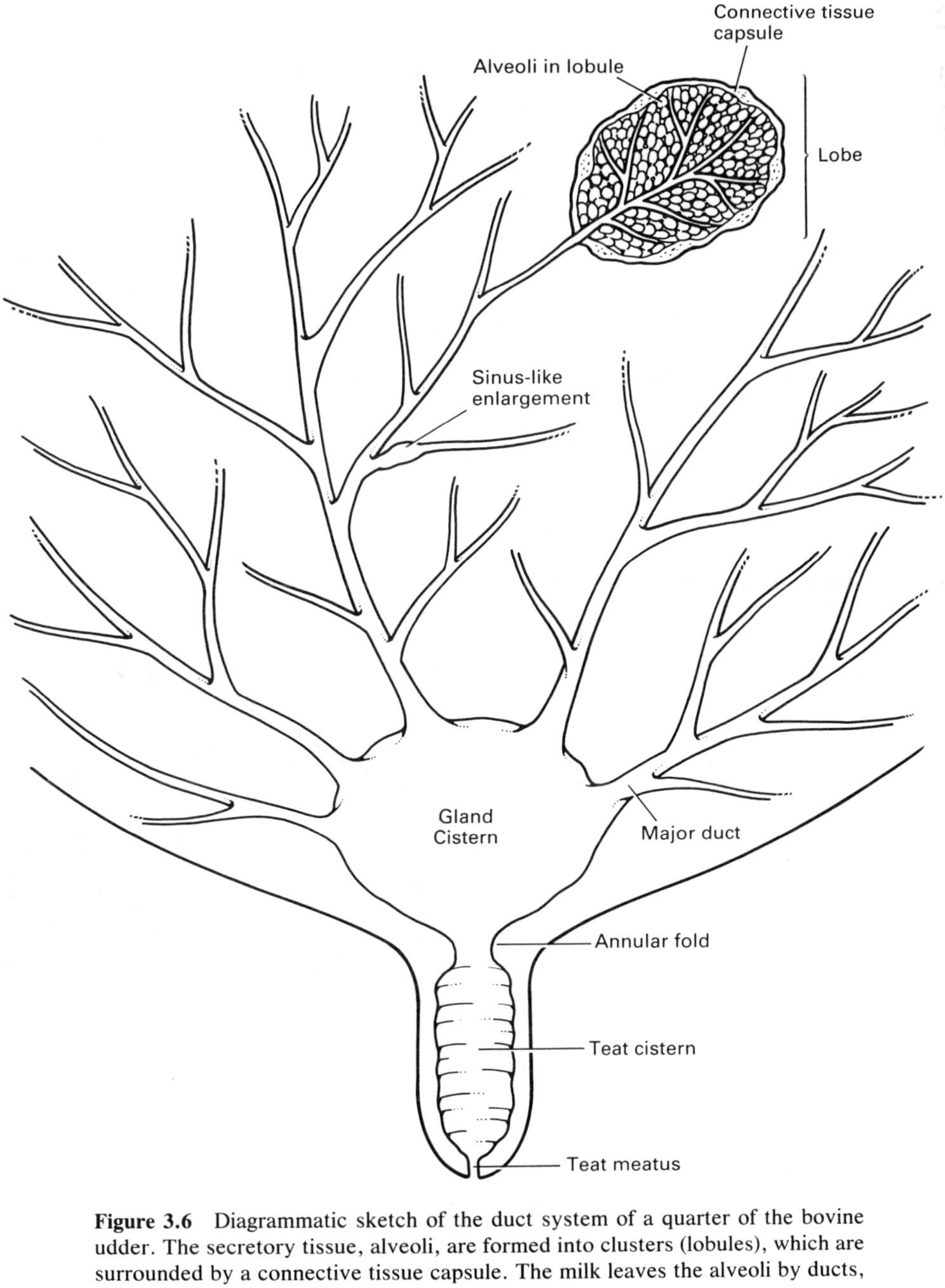

Figure 3.6 Diagrammatic sketch of the duct system of a quarter of the bovine udder. The secretory tissue, alveoli, are formed into clusters (lobules), which are surrounded by a connective tissue capsule. The milk leaves the alveoli by ducts, which increase in size and drain into the gland cistern.

The teat cistern, gland cistern, and the large ducts are lined with a two-layer epithelium that does not secrete milk. The alveoli and small ducts are lined with a single layer of milk-secreting epithelial cells. At the point at which a duct sends off smaller branches, there is a constriction in the diameter of the small duct that swells into a sinuslike enlargement before the duct branches again (Figure 3.6). This constriction prevents milk from flowing freely in the duct system; therefore, a positive force must be applied to the alveoli to move the milk from the secretory tissue through the duct system into the teat and gland cisterns. The positive force is brought about by the milk ejection process.

Blood Supply

Almost all of the blood supplied to the udder is carried by two major arteries, one for each half of the udder. These arteries, the external pudendal arteries, enter the udder through the inguinal canal. They are branches of the external iliac arteries, which are branches of the abdominal aorta (Figure 3.7). The external pudendal arteries become the mammary arteries as they pass through the inguinal canal. The mammary arteries divide into caudal and cranial branches, which

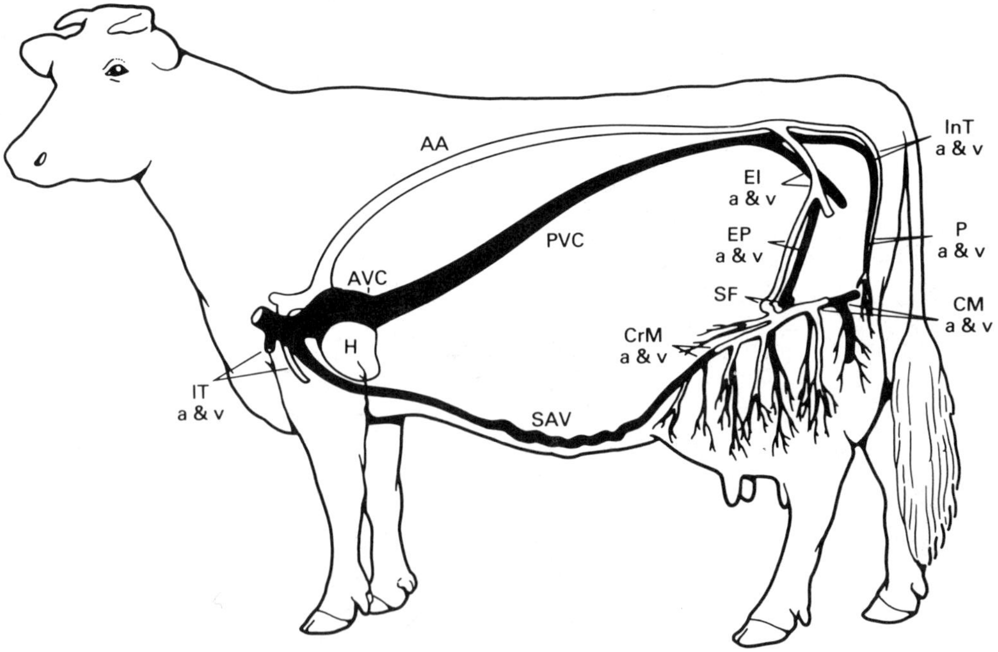

Figure 3.7 Blood circulation to and from the udder. H, heart; IT a&v, internal thoracic artery and vein; AVC, anterior vena cava; AA, abdominal aorta; PVC, posterior vena cava; SAV, subcutaneous abdominal vein; EI a&v, external iliac artery and vein; InI a&v, internal iliac artery and vein; EP a&v, external pudendal artery and vein; P a&v, perineal artery and vein; SF, sigmoid flexure of EP a&v; CM a&v, caudal mammary artery and vein; CrM a&V, cranial mammary artery and vein.

branch and rebranch many times and end in small capillaries surrounding each alveolus. The pudendal arteries make S-shaped curves (sigmoid flexures) as they emerge from the inguinal canal. This probably allows for lengthening of the arteries as the mammary gland fills with milk and extends downward. Perineal arteries supply blood to a small portion of the posterior dorsal part of the udder.

Each half of the udder is supplied with two major veins. The external pudendal vein follows the same course as the external pudendal artery. After entering the udder, the vein branches into cranial and caudal portions, and these branch and rebranch many times to provide a network of venuoles that drain blood from each alveolus. The external pudendal veins are branches of the external iliac veins, which are branches of the posterior vena cava. The second major vein in each half of the udder is a continuation of the cranial branch of the mammary vein that leaves the udder at its anterior border and continues along the ventral surface of the abdomen as the subcutaneous abdominal vein. This vein enters the body cavity posterior to the front legs and unites with the internal thoracic vein. The milk well is the opening in the abdominal cavity through which the subcutaneous abdominal vein passes.

Lymphatic System

The secretory and connective tissues within the udder are surrounded by a colorless tissue fluid, lymph, that arises from the arterial blood. Part of the fluid is reabsorbed by the venous system; a portion, however, leaves the lymphatic system. The lymphatic vessels carry lymph from the tissues toward the heart. The lymphatics are prevalent in the teats and under the skin of the udder. They are also found within the glandular tissue of the udder. Because of the colorlessness of the lymph and because of the thin walls of lymphatic vessels, identification of the vessels is difficult.

The lymphatics drain toward the upper posterior part of the udder, where they converge on the convex surface of the supramammary lymph node (Figure 3.8). Most cows have two supramammary lymph nodes, one on each side of the midline. In some cows, as many as seven small lymph nodes are found on each side of the midline. Lymph enters the lymph nodes, passes through a network of sinuses, and emerges from the concave portion of the nodes. The lymph nodes filter the lymph and remove or destroy any bacteria and foreign material present, thus helping to localize infections in the body. Lymph leaving the supramammary lymph nodes travels via the larger lymph vessels through the inguinal canal and joins lymph from many other lymph vessels from other parts of the body. Lymph is returned to the venous blood through the thoracic duct, which enters the anterior vena cava.

The lymph vessels also contain valves that point in the direction of the heart, causing the lymph to flow in that direction. The main factor affecting lymph flow is the difference in pressure between the lymph vessels and the anterior vena cava. In addition, the blood pressure in the capillaries aids in the movement of lymph in the vessels, as do contraction of muscles and the movements of the cow, particularly udder movements.

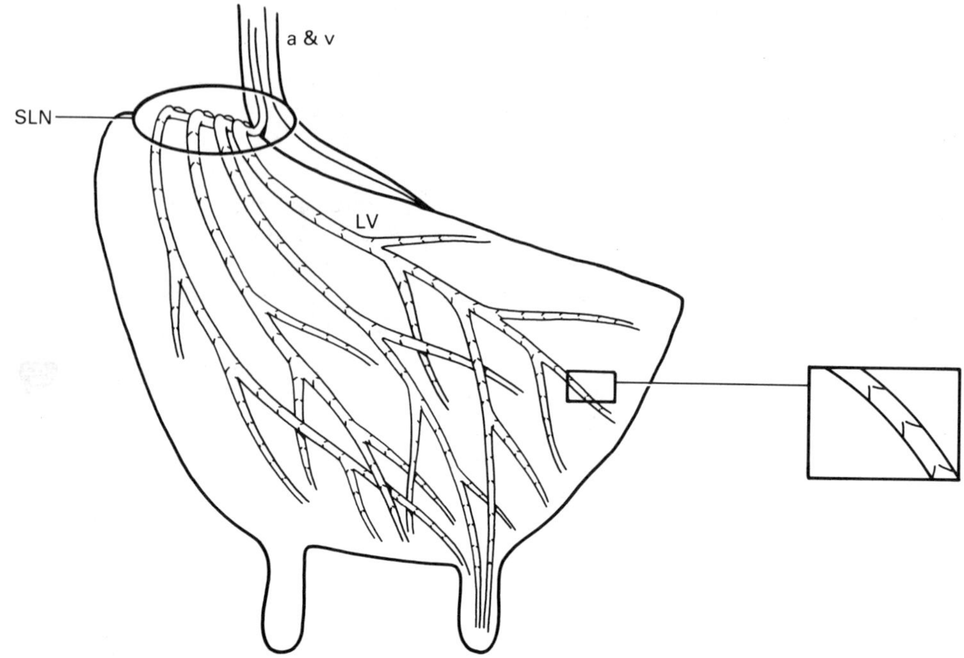

Figure 3.8 Diagrammatic sketch of the subcutaneous lymph vessels (LV). The lymph in the vessels travels in a dorsal-caudal direction and is filtered by the supramammary lymph node (SLN) before traveling to the thoracic duct and the anterior vena cava. The insert shows the small valves in the vessels allowing lymph to travel in one direction only. a&v, external pudendal artery and vein.

Interference with the lymphatic drainage of the udder causes the accumulation of lymph, a common condition known as edema. Edema is most prevalent in first-lactation cows and occurs in older cows with pendulous udders at the time of calving. The lymph accumulates between the skin and the secretory tissue of the udder, but it may also accumulate around the secretory tissue.

Nervous System

The udder contains two types of nerves, the afferent (or sensory) nerves and the efferent (or motor) nerves. The afferent nerve fibers originate from the dorsal root of the spinal cord and the efferent fibers arise from its ventral root. The major nerve extending to each half of the udder, the inguinal nerve, is composed of fibers that originate from the second, third, and fourth lumbar vertebrae. The rear portion of each udder half is supplied with a perineal nerve that arises from branches of the second, third, and fourth sacral vertebrae. Both of these nerves are part of the sympathetic nervous system. The sensory nerves play an important role in milk ejection in that they carry the sensory stimuli from the udder to the hypothalamus of the brain. The efferent nerve fibers apparently play no direct role in milk secretion, but they can have an inhibitory effect by causing constriction of the blood vessels that carry blood to the udder.

3.2 HISTOLOGY OF THE MAMMARY GLAND

Most of the mammary gland is made up of secretory and connective tissues. The secretory units are small grapelike objects called alveoli (Figure 3.9). A group of

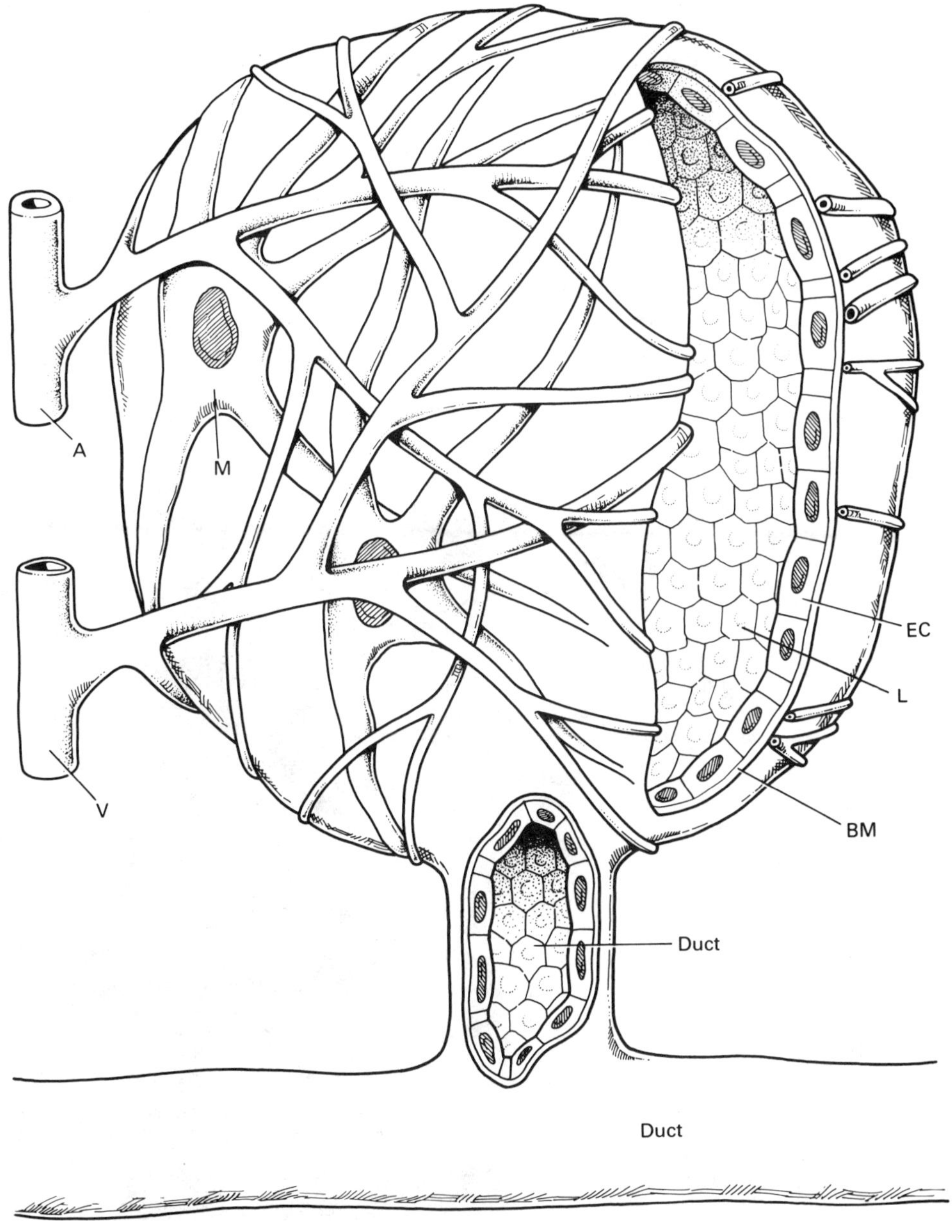

Figure 3.9 Diagrammatic sketch of an alveolus and surrounding blood vessels and myoepithelial cells. A, arteriole; V, venuole; M, myoepithelial cell; L, lumen; EC, epithelial cell on basement membrane (BM).

them are joined together in a cluster with a common duct, so that they resemble grapes attached to a stem. Each alveolus has a small duct that leads into larger ducts, and all of the alveoli in a cluster (lobule) are drained by a common duct. Each lobule has between 150 and 220 alveoli and measures roughly 0.75 mm^3. Each lobule is surrounded by a connective tissue capsule (Figure 3.6). A number of lobules come together to form a lobe, which is drained by a larger duct and surrounded by a connective tissue capsule. The udder is made up of lobes that are attached to the duct system.

The basic component of an alveolus is a single layer of epithelial cells surrounding a central cavity, the lumen (Figure 3.9). The epithelial cells have a single nuclei and rest on a basement membrane. Each alveolus is supplied with small capillaries and venuoles, which supply blood to the alveolus and drain away unused blood. The capillaries and venuoles connect with the larger arteries and veins in the udder. Also surrounding each alveolus are a series of specialized cells, the myoepithelial cells, that are responsible for milk ejection. The epithelial cells absorb nutrients from the capillaries, convert the precursors into milk components, and release them into the lumen of the alveolus. Most of the milk remains in the lumen until the time of milking. At milking a stimulus causes the release of oxytocin from the posterior pituitary, which causes the myoepithelial cells to

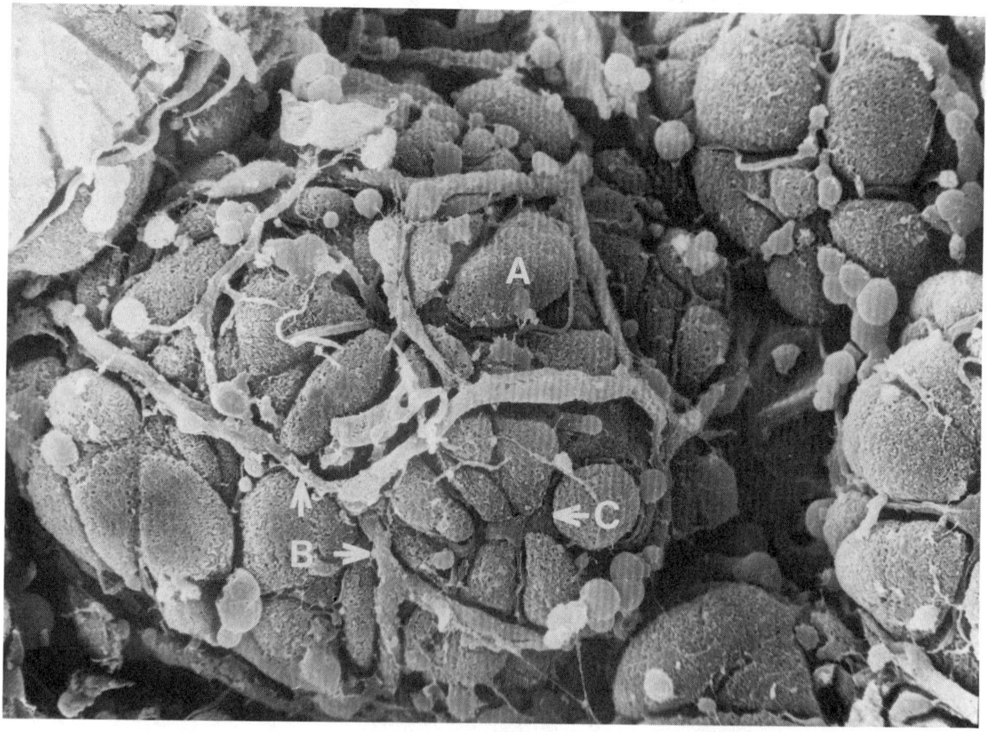

Figure 3.10 Scanning electron microphotograph of the surface of an alveolus (A) with a blood vessel (B) and a myoepithelial cell (C). From E. V. Carulo, 1980, *J. Dairy Sci.*, 63:1887 and courtesy of E. V. Carulo, North Carolina State University.

constrict and force most of the milk out of the lumen of the alveolus, through the small duct at the end, into the larger ducts. A scanning electron microphotograph of an alveolus, a myoepithelial cell, and blood vessels is shown in Figure 3.10.

3.3 GROWTH AND DEVELOPMENT OF THE MAMMARY GLAND

The development of the mammary gland in the bovine fetus begins approximately 4 to 6 weeks after conception. The first development occurs as a single layer of ectodermal cells behind the umbilicus. This single layer of cells converges into two mammary lines, one along each side of the midline in the inguinal region. Shortly thereafter, each mammary line develops into two distinct buds, the mammary buds. Each of these buds then forms a solid core of cells, the primary sprout, that penetrates the underlying cells to give rise to the teat and gland cisterns (Figure 3.11). A rapid growth of cells in the area of the primary sprout causes the growth to protrude from the surface and form the teat. The cells of the primary sprout separate to form the canals for the teat meatus and the teat and gland cisterns. After the rudimentary teat and gland cisterns have developed, each of the primary sprouts gives off secondary and tertiary sprouts that become the rudimentary duct system of the udder (Figure 3.12). At the same time, the underlying mesenchyme tissue gives rise to fat tissue and the basic vascular, lymphatic, and nervous systems of the udder. Except for further deposition of fat, most of the fetal development is completed by 6 months after conception.

At birth, the calf has well-defined teats, teat cisterns, and gland cisterns. Several secondary sprouts have developed from the gland cistern, some of which may have canals but are still rather rudimentary in form. Development of the mammary gland between birth and puberty results primarily from a deposition of fat in the gland; there is slight further development of glandular tissue.

The recurring estrous cycles after puberty cause a further development of the mammary gland. Release of estrogen and progesterone from the ovaries during the estrous cycles is responsible for further duct growth and some growth of

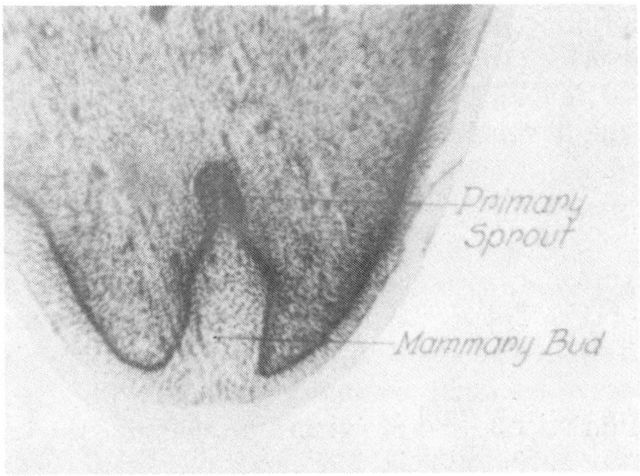

Figure 3.11 Development of the primary sprout in the bovine fetus. The primary sprout develops into the teat and gland cisterns and gives rise to the secondary sprouts. From C. W. Turner, 1930, *Mo. Agric. Exp. Stn. Res. Bull.* 140.

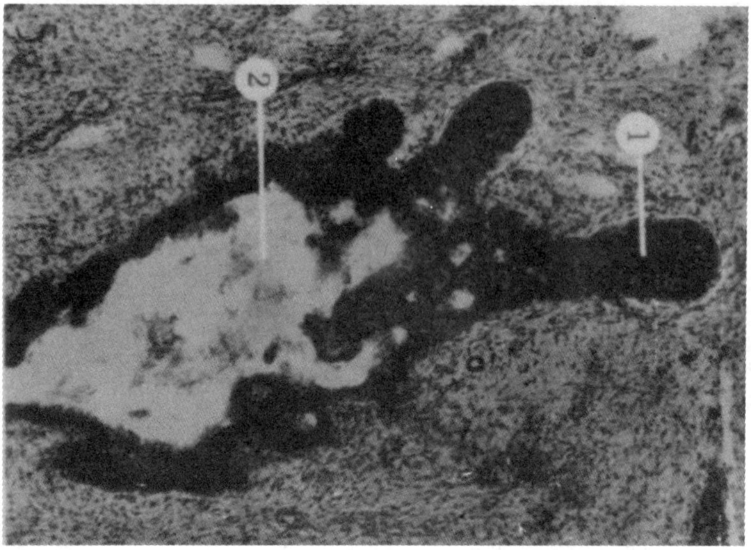

Figure 3.12 Development of several secondary sprouts (1) from the primary sprout. The secondary sprouts will become the major ducts in the udder. The gland cistern (2) results from the canalization of the primary sprout. From C. W. Turner, 1930, *Mo. Agric. Exp. Stn. Res. Bull.* 140.

secretory tissue. The mammary gland grows most during pregnancy. A wave of growth occurs near the time of parturition, followed by a period of slight growth shortly after parturition. This growth appears to be completed 3 to 6 weeks after parturition. Further growth of the mammary gland takes place during the dry period, particularly after the first lactation. The tremendous increase in size of the mammary gland shortly before parturition is caused by accumulation of milk in the alveoli.

Hormonal Control of Mammary Gland Development

The exact combination of hormones responsible for the growth of the mammary gland is not known, but estrogen, progesterone, prolactin, and growth hormone play important roles in normal development of the gland. Much of the information on the role of hormones in the development of the mammary gland comes from laboratory animals. Estrogen and growth hormone are primarily responsible for the development of the duct system. The development of the secretory tissue, the lobuloalveolar growth, requires a combination of progesterone, prolactin, growth hormone, and estrogen. Estrogen and progesterone are produced by the ovary; growth hormone and prolactin are produced by the anterior pituitary. The placenta secretes some of these hormones, or hormones with similar properties, and may play an important role in normal mammary gland growth.

The initiation of lactation at the time of parturition is under the control of prolactin and the adrenal corticoid hormones. The onset of lactation at the time of parturition is believed to result from a marked increase in prolactin in the blood, or a marked decrease in blood progesterone, or a combination of the two. In most species, lactation does not occur during pregnancy; the cow is an exception. In animals in which lactation and pregnancy do not occur simultaneously, it is believed that the high amounts of estrogen and progesterone present during pregnancy inhibit the effect of prolactin or reduce the amount of circulating prolactin in the blood. Apparently, the prolactin in the dairy cow is high enough to overcome the inhibitory effects of estrogen and progesterone.

FURTHER READINGS

ANDERSON, R. R. 1985. Mammary gland. In *Lactation*, B. L. Larson, ed. Iowa State University Press, Ames, IA.

COWIE, A. T., ed. 1961. *Milk: The Mammary Gland and Its Secretion*, Vol. 1. Academic Press, New York, Chap. 1.

LARSON, B. L. 1985. *Lactation*. Iowa State University Press, Ames, IA.

SCHMIDT, G. H. 1971. *Biology of Lactation*. W. H. Freeman and Company Publishers, San Francisco.

TURNER, C. W. 1952. *The Mammary Gland*. Lucas Brothers, Columbia, MO.

4

Physiology of Milk Secretion

4.1 CELL CYTOLOGY AND MILK SECRETION

The major function of the mammary gland is the secretion of milk, which takes place in the epithelial cells of the alveolus. A drawing of an epithelial cell is shown in Figure 4.1. The cells have relatively large nuclei with one or more nucleoli. The nucleus has two membranes, an outer one that is continuous with the endoplasmic reticulum and the Golgi apparatus and an inner one that has pores for the movement of material from the nucleus to the cytoplasm. Chromatin material in the nucleus contains information for cell duplication and the synthesis of milk protein, including the production of enzymes required for synthesis of other milk constituents.

Each of the secretory cells of a lactating animal has an abundant network of endoplasmic reticulum and an enlarged Golgi apparatus. There are numerous protein droplets in the Golgi apparatus and the lumina of the alveoli, and a large number of fat droplets exist in the cell apices and the lumina. A marked increase in the number of mitochondria occurs in preparation for secretion. An electron microphotograph of an epithelial cell is shown in Figure 4.2.

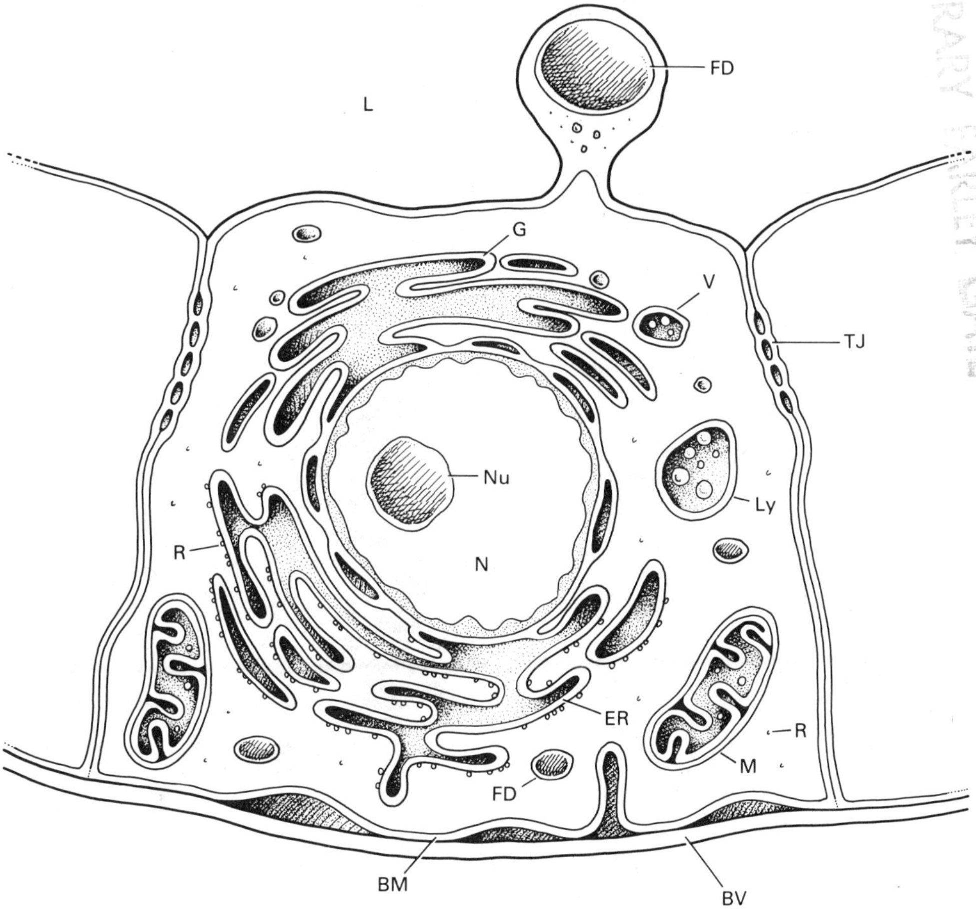

Figure 4.1 Diagrammatic cross section of an epithelial cell. L, lumen; FD, fat droplet; Ly, lysosome; G, Golgi apparatus; V, vacuole; TJ, tight junction; R, ribosome; N, nucleus; Nu, nucleolus; ER, endoplasmic reticulum with ribosomes; M, mitochondria; BM, basement membrane; BV, blood vessels.

Mitochondria

Mitochondria located throughout the cell are responsible for breaking down various molecules, especially glucose and acetate, and capturing the released energy in the form of adenosine triphosphate (ATP). The energy in ATP is then utilized in the synthetic reactions that are necessary for the secretion of milk. The mitochondria are known as the "powerhouses" of the cell.

Endoplasmic Reticulum and Protein Synthesis

The endoplasmic reticulum (ER) and the Golgi apparatus are continuous with the cell membrane and the lumen of the alveolus. Part of the ER is lined with dense

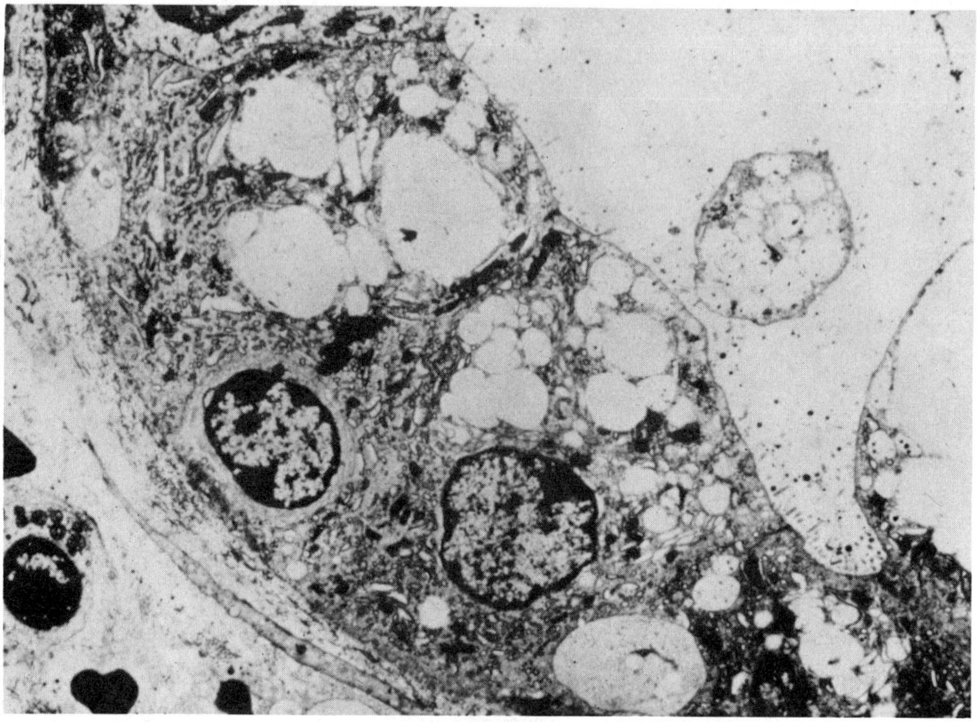

Figure 4.2 Electron microphotograph of an epithelial cell. L, lumen; N, nucleus; FD(c), fat droplet in cell; FD(l), fat droplet in lumen. Photo courtesy of F. L. Schanbacher, Ohio Agricultural Research and Development Center, Wooster, Ohio.

particles known as ribosomes, which are attached to the outer surface of the membrane and are responsible for most of the protein synthesis of the cell. Some ribosomes are not attached to the ER but are scattered throughout the cytoplasm. That part of the ER not lined with ribosomes is primarily responsible for the transfer of materials within the cytoplasm.

Three types of ribonucleic acid (RNA) molecules are involved in protein synthesis. All of these are formed in the nucleus and are transported through the cytoplasm to the sites where protein synthesis occurs. Messenger RNA molecules from the nucleus are transported to the ribosomes, which contain a second type of RNA, ribosomal RNA. Messenger RNA contains the code for the sequence of amino acids that will make up a particular protein molecule. The code in the messenger RNA was transcribed from the code contained in the deoxyribonucleic acid (DNA) molecules of the nucleus. Individual amino acids in the cytoplasm are activated and picked up by a third RNA molecule, transfer RNA, and carried to the ribosomes to which the messenger RNA has been attached. The transfer RNA recognizes the code on the messenger RNA and places its amino acid in its proper location. A second transfer RNA places another amino acid in position according to the code on the messenger RNA, and a peptide bond is formed between the two amino acids. This process is repeated until the protein molecule is formed. The protein molecule then moves through the lumen of the ER to the smooth ER and thence to the Golgi apparatus, in which the milk proteins are condensed into vacuoles to be exported to the alveolar lumina.

A vacuole is a membrane-lined cavity containing protein and lactose droplets that moves to the apex of the cell and there deposits its materials into the lumen of the alveolus. Several vacuoles are shown in Figure 4.1.

Lysosomes

The lysosomes in the cell contain enzymes for the breakdown of larger molecules into smaller molecules. The lysosomes also contain degradative enzymes, which when released from their membranes, cause breakdown and death of the cell. The maintenance of lactation is partially controlled by the lysosomal membrane, which prevents degradative enzymes from being released into the cell. The lysosomes probably play a major role in the involution of the mammary gland.

Milk-Fat Secretion

When sections of the mammary gland are stained with fat-absorbing dyes, small fat droplets are noted at the base of the epithelial cells (Figure 4.1). As these fat droplets move toward the apex of the cell, they become progressively larger. The droplet then presses against the wall of the cell, and the cell membrane enlarges and continues to surround it. When the fat globule is released, the cell membranes join together beneath the fat globule, thus preventing the cytoplasm from being

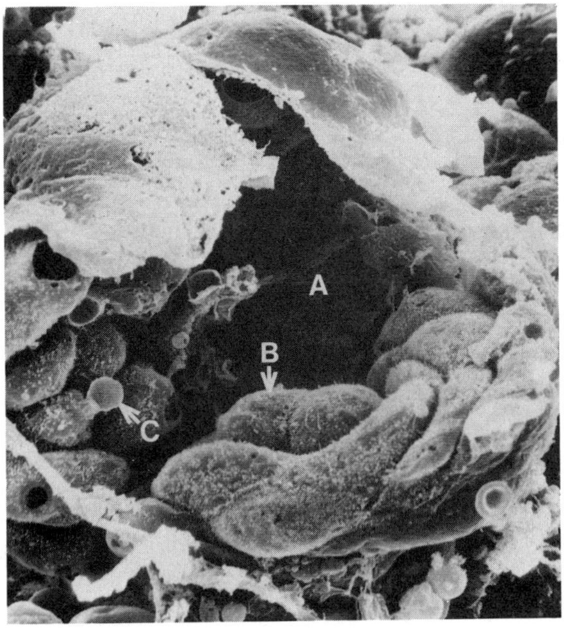

Figure 4.3 Scanning electron microphotograph of the interior of a lumen (A) of an alveolus with a fat droplet (C) about to be secreted. B, microvilli on the apical surface of an epithelial cell. From E. V. Carulo, 1980, *J. Dairy Sci.*, 63:1887, and courtesy of E. V. Carulo, North Carolina State University, Raleigh, NC.

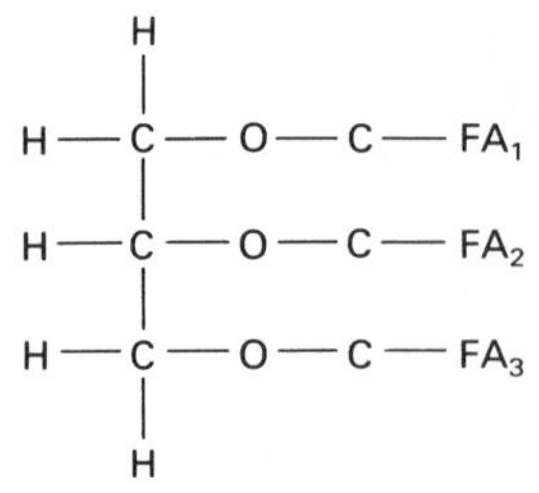

Figure 4.4 Formula for a triglyceride. The glycerol molecule on the left has three fatty acids (FA_1, FA_2, FA_3) of varying carbon lengths and degrees of unsaturation attached to it.

exposed to the lumen of the alveolus; and the fat droplet enters the lumen completely surrounded by membrane from the cell that secreted it. The small amount of protein that is found in milk fat is probably the result of some protein droplets being entrapped within the membranes as the fat droplets are secreted. Figure 4.3 shows a scanning electron microphotograph of a fat droplet.

Almost all milk fat is made up of triglycerides. A triglyceride is composed of three fatty acids attached to a glycerol molecule (Figure 4.4). The fatty acids have different carbon chain lengths; the carbon chain lengths of the milk fat of four species are shown in Table 4.1. Ruminant milk fat contains a higher proportion of short-chain fatty acids than does milk fat from nonruminants. Ruminant milk fat also contains a smaller amount of unsaturated fatty acids (oleic and linoleic acids).

The fatty acids that are esterified to glycerol arise from two major sources. Most of the short-chain fatty acids up through the C_{14} acids, and some of the C_{16} acids, are synthesized within the mammary gland. They are built up from acetate units, which contain two carbon atoms, and β-hydroxybutyric acid molecules, which contain four carbon atoms. These short-chain carbon units come from the rumen fermentation process. Much more acetate than β-hydroxybutyric acid is used for milk-fat synthesis.

TABLE 4.1 PERCENTAGE OF FATTY ACIDS IN TRIGLYCERIDES OF MILK FAT

		Percent moles in triglycerides			
Fatty acid	Carbon length	Human	Pig	Goat	Cow
Saturated					
Butyric	4	—	2	7	10
Caproic	6	—	2	5	3
Caprylic	8	—	2	4	1
Capric	10	2	2	13	2
Lauric	12	8	2	7	3
Myristic	14	9	2	12	9
Palmitic	16	23	29	24	21
Stearic	18	9	6	5	11
Unsaturated					
Oleic	18:1	34	35	17	31
Linoleic	18:2	7	14	3	5
Other	—	8	12	3	4

Almost all of the C_{18} fatty acids, and many of the C_{16} fatty acids, come from the fatty acids that are absorbed as such from the blood. Some of the C_{14} and C_{16} fatty acids also arise from C_{16} and C_{18} through the removal of two-carbon or four-carbon units. The mammary gland can remove hydrogen ions from the C_{18} fatty acids to unsaturate them. The mammary gland absorbs smaller amounts of oleic and linoleic acids from the blood than are found in the milk; therefore, some of the fats in the mammary gland are unsaturated. Most of the glycerol molecules are synthesized from glucose, but some of them come from the glycerol portion of the triglycerides that are absorbed from the blood.

Lactose Secretion

Lactose is made up of two simple sugars, glucose and galactose (Figure 4.5). In the process of lactose formation, a glucose molecule is converted to galactose; the galactose molecule is then united with a second glucose molecule to form lactose. Approximately 50% of the glucose entering the udder is used in lactose formation and the remainder is used for energy purposes. The enzyme catalyzing the last step in lactose formation is lactose synthetase, which is composed of two subunits. One subunit is α-lactalbumin, one of the common milk proteins. In the rat, the formation of α-lactalbumin is under hormonal control. Prior to parturition, the α-lactalbumin content of the rat mammary gland is very low and it increases only after the progesterone of the blood decreases prior to, or at, parturition. It is believed that the initiation of milk secretion in the rat is partially caused by the formation of α-lactalbumin, which then becomes part of the lactose synthetase molecule. Whether the same mechanism operates in the dairy cow is not known.

Lactose formation apparently occurs in the Golgi apparatus because α-lactalbumin is synthesized at the ribosomes and moves through the lumen of the ER to the Golgi apparatus, where the second subunit (galactosyltransferase) of lactose synthetase is present. Because lactose synthetase is formed in the Golgi apparatus, it is believed that lactose formation takes place there and that the lactose is exported with the protein particles through the vacuoles.

The lactose content of milk is relatively constant. Lactose is the major component responsible for the osmotic pressure of milk, but chloride, potassium, and sodium ions also play a role in this respect. Water is transferred to the lumen of the alveolus until the osmotic pressure of the milk is similar to that of the blood. Most of the water passes through the epithelial cells and into the milk by filtration.

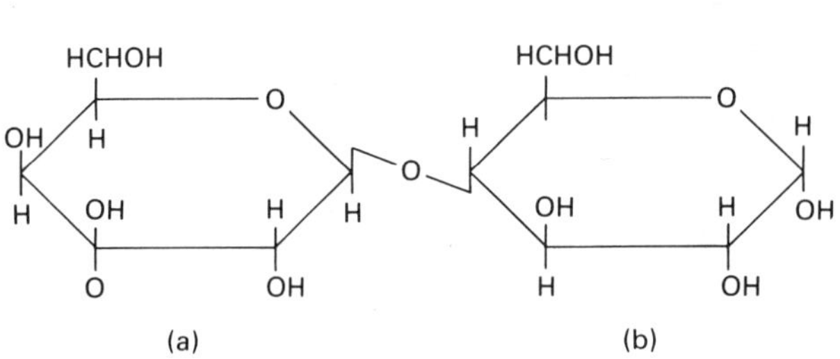

Figure 4.5 Formula of lactose. A molecule of galactose (a) is joined to a molecule of glucose (b).

Other Secretion Processes

The mineral and vitamin contents of milk are governed by the filtration process; however, the epithelial cells act as a membrane barrier or carrier of the particles from the blood to the lumen of the alveolus. In the healthy mammary gland, little, if any, filtration occurs between the epithelial cells due to the presence of tight junctions between the cells (Figure 4.1). The epithelial cells combine some minerals with organic compounds; for example, 75% of the calcium in milk is in chemical or physical combination with casein, phosphate, and citrate, and more than half of the phosphorus in milk is combined with casein. Vitamin molecules in milk are transferred unchanged from the blood. Concentrations of some vitamins, especially the fat-soluble vitamins, can be increased by increasing the vitamin content of the blood plasma.

Thus two types of secretory processes operate in the mammary gland; the filtration process accounts for the secretion of water, vitamins, and minerals. True cell metabolism accounts for the production of fat, lactose, and most of the protein.

Cellular components are also found in milk. Nineteenth-century investigators believed that the secretion process was one of cell synthesis and degeneration. Since then, the secretion of fat, lactose, and protein have all been shown to be true secretion processes. The presence of cytoplasmic and nuclear components in milk is due to the breakdown and degeneration of epithelial cells, which are discarded in the milk. Few of these cells are replaced during lactation, and the

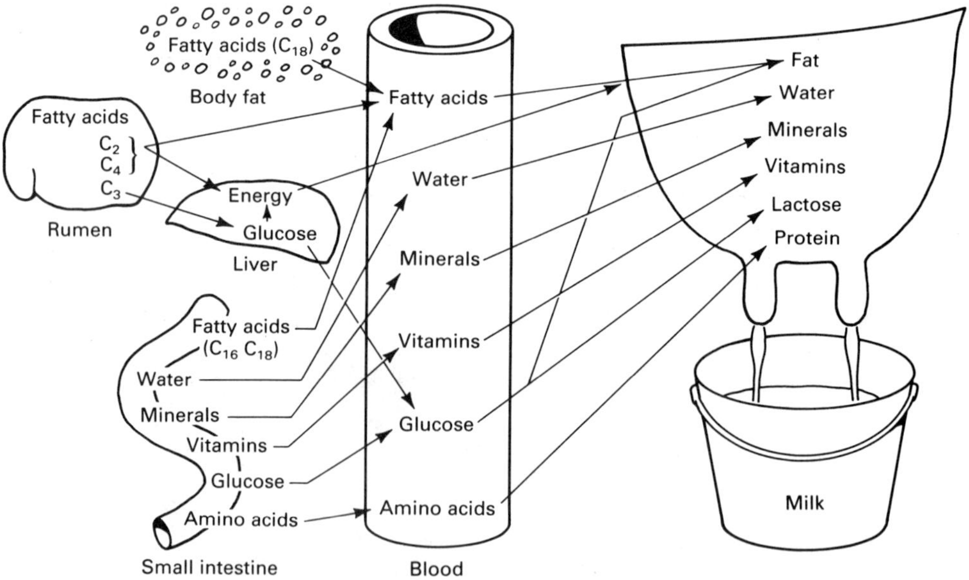

Figure 4.6 Diagrammatic sketch of the precursors and their origins for the synthesis of milk by the ruminant animal.

decrease in production occurring after peak lactation is caused, in large part, by a decrease in cell numbers.

The precursors of the various constituents of ruminant milk are shown in Figure 4.6. Similar precursors would be used by most nonruminant animals for the production of milk, except that the fatty acids would come from the intestines and body fat and the short-chain fatty acids would not be available.

4.2 COMPARATIVE COMPOSITION OF BLOOD PLASMA AND MILK

The comparative composition of milk and blood plasma is shown in Table 4.2. The water content of milk is slightly lower than that of blood plasma; therefore, a certain concentration of water must take place during milk secretion. The glucose concentration in blood plasma is very low in comparison with the lactose concentration in milk. The concentration of fat in the milk is much higher than that in the blood.

The proteins of milk are made up of four casein components, plus lactoglobulin, lactalbumin, blood serum albumin, and immune globulins. The latter two protein components amount to only 1 to 2% of total milk protein. Blood serum albumin and the immunoglobulins are filtration products from the blood and are not synthesized in the epithelial cells. The other proteins are synthesized from amino acids within the epithelial cells.

Milk contains much more calcium, phosphorus, and potassium than does blood plasma; blood plasma, in turn, has much more sodium and chloride. It is for this reason that milk has a tendency to become salty during mastitis and other mammary gland disturbances.

TABLE 4.2 COMPARATIVE COMPOSITION OF BLOOD PLASMA AND COW'S MILK

Blood plasma		Milk	
Constituent	Percent	Constituent	Percent
Water	91.0	Water	87.0
Glucose	0.05	Lactose	4.9
Serum albumin	3.2	Lactalbumin	0.52
Serum globulin	4.4	Lactoglobulin	0.20
		Casein	2.9
Neutral fat	0.06	Neutral fat	3.7
Phospholipids	0.24	Phospholipids	0.10
Calcium	0.009	Calcium	0.12
Phosphorus	0.011	Phosphorus	0.10
Sodium	0.34	Sodium	0.05
Potassium	0.03	Potassium	0.15
Chlorine	0.35	Chlorine	0.11
Citric acid	Trace	Citric acid	0.20

Source: Maynard and Loosli, 1969, *Animal Nutrition*, McGraw-Hill Book Company, New York.

4.3 UDDER PRESSURE AND SECRETION RATE

Approximately 15% of the milk present in the udder at the beginning of milking remains in the udder after normal milking as residual milk. The only way it can be obtained is to inject oxytocin into the blood after a normal milking and then to remilk the cow, which is not a practical procedure. Inheritance appears to play a role in the amount of residual milk left in the udder. Cows with a high percentage of residual milk have been found to have a lower persistency than do cows with a smaller percentage of residual milk.

As milk accumulates in the udder, particularly within the lumen of the alveolus, the epithelial cells must secrete against increasing pressure. When the pressure within the lumen of the alveolus becomes too great, a reduction in the secretion rate occurs. Attempts have been made to measure the relationship between udder pressure and secretion rate, but there is no scientific method now available to measure the pressure within the lumen of the alveolus. Consequently, measurements of udder pressure are made by cannulating the teat and measuring the pressure within the teat and gland cisterns. The pressures in the teat and gland cisterns are assumed to reflect those in the alveolar lumina.

The general relationships between pressure in the teat and gland cisterns, milk secretion rate, and milking intervals are shown in Figure 4.7. The marked increase in udder pressure within the first hour after milking results from the hydrostatic pressure of milk that drains from the secretory tissue to the gland and teat cisterns. After this initial marked rise, the pressure rises gradually until approximately 9 to 10 hours after the last milking, at which time there is another rise. Thereafter, the pressure again increases at a gradual rate until 35 hours after

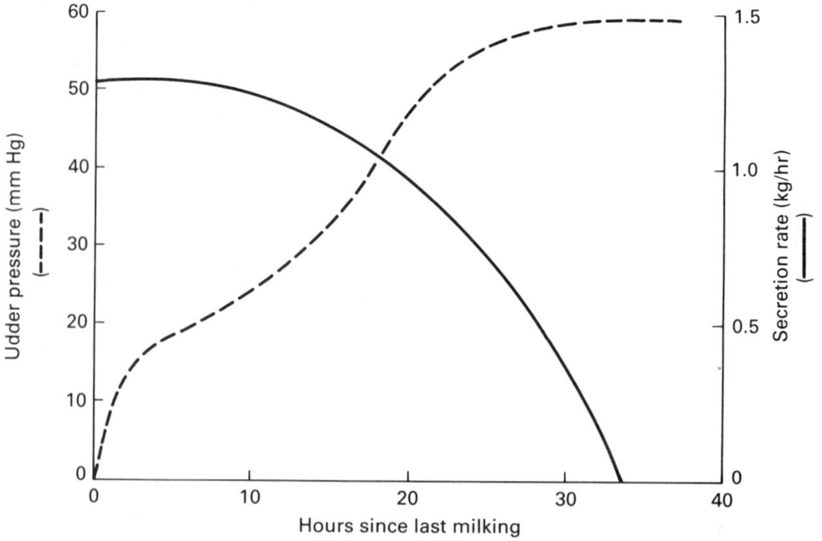

Figure 4.7 Relationship of pressure in the udder to the milk secretion rate with hours after the last milking.

the last milking, at which time the secretion stops. During the period of highest pressure, some of the milk is absorbed into the blood, which accounts for the presence of lactose and casein in the blood of lactating cows. The rate of milk secretion remains the same for approximately 10 hours after the last milking. Then a gradual decrease in the rate of secretion occurs. After 14 hours, there is a marked reduction in the secretion rate that continues until secretion stops altogether at approximately 35 hours after the last milking. At a given time after milking, most cows have the same udder pressure. High-yielding cows thus produce a much larger volume of milk per unit of pressure than do lower-producing cows.

The reduction in the milk secretion rate during long intervals between milkings is the reason that equal milking intervals have been advocated. One of the most important points to be borne in mind in the management of dairy cattle is the extent of the decrease in the secretion rate when cows are milked at longer intervals.

4.4 MILKING INTERVALS AND FREQUENCIES

Two studies have been conducted with average- to high-producing cows to determine the effect of unequal milking intervals for complete lactations on milk and fat production. These results are shown in Table 4.3. The cows milked after alternate 10- and 14-hour intervals in the first experiment produced approximately the same amount of milk as those milked after equal intervals. Cows milked after alternate 16- and 8-hour intervals produced slightly less milk (1.2%) than did those milked after equal intervals. In the second experiment, the reduction resulting from the wider milking intervals was 3.5%. Half of the lactations in the second experiment were from first-lactation cows. Apparently, cows in first lactation are more adversely affected by longer milking intervals than are older, higher-producing cows because the udders of first-lactation cows are smaller and the increase in udder pressure per unit of milk is greater than in older cows.

Many studies indicate that milking cows three or four times per day causes an increase in milk production in comparison with twice-daily milking. Compar-

TABLE 4.3 EFFECT OF UNEQUAL MILKING INTERVALS ON MILK AND MILK-FAT PRODUCTION

Interval (hr)	Number of cows	Length of record (days)	Milk kg	Milk lb	Source
12–12	35	305	6,255	13,760	Schmidt and Trimberger (1963)
14–10	35	305	6,234	13,716	Schmidt and Trimberger (1963)
16–8	35	305	6,174	13,582	Schmidt and Trimberger (1963)
12.5–11.5	82	266	4,921	10,824	Ormiston et al. (1967)
14.5–9.5	82	266	4,811	10,584	Ormiston et al. (1967)

TABLE 4.4 LACTATION MILK YIELDS OF COWS MILKED TWO AND THREE TIMES A DAY

Times milked per day	Age of cows	Number or cows or (herds)	Milk kg	Milk lb	Percent increase over twice a day	Source
2	First lactation	7	5,578	12,272	—	Amos et al. (1985)
3	First lactation	7	6,987	15,371	25.2	Amos et al. (1985)
2	Mature	17	6,903	15,187	—	Amos et al. (1985)
3	Mature	16	8,179	17,994	18.5	Amos et al. (1985)
2	First lactation	7	7,859	17,290	—	DePeters et al. (1985)
3	First lactation	8	8,315	18,293	5.8	DePeters et al. (1985)
2[a]	Mature	12	7,744	17,037	—	DePeters et al. (1985)
3[a]	Mature	13	9,033	19,873	16.6	DePeters et al. (1985)
3[b]	Mature	13	8,738	19,224	12.8	DePeters et al. (1985)
2	First lactation	(7)	6,439	14,166	—	Allen et al. (1986)
3	First lactation	(7)	7,691	16,920	19.4	Allen et al. (1986)
2	Second lactation	(7)	7,474	16,443	—	Allen et al. (1986)
3	Second lactation	(7)	8,482	18,660	13.5	Allen et al. (1986)
2	Third lactation	(7)	7,800	17,160	—	Allen et al. (1986)
3	Third lactation	(7)	8,710	19,162	11.7	Allen et al. (1986)
2	Four or more lactations	(7)	7,791	17,140	—	Allen et al. (1986)
3	Four or more lactations	(7)	8,838	19,444	13.4	Allen et al. (1986)

[a] Cows switched from high- to medium-energy diet at 28 kg of milk per day and from medium- to low-energy diet at 23 kg/day.

[b] Cows switched from high- to medium-energy diet at 31 kg of milk per day and from medium- to low-energy diet at 25 kg/day.

ison of records of cows on official testing programs shows that cows milked three times a day produce 15 to 25% more milk than cows milked twice a day. Recent controlled experiments indicate that milking cows three times a day results in a 6 to 25% increase in milk yield. Results of these studies are shown in Table 4.4. Results with second or later lactation cows show an 11 to 19% increase in production with three-times-a-day milking; however, the results with first-lactation cows were markedly different. Part of the difference might be related to the amount of milk production, since first-lactation cows in the experiment by Amos et al. (1985) and Allen et al. (1986) produced less milk and had much higher percentage increases in milk yield than those in the experiment by DePeters et al. (1985). In the experiment by Amos et al., both the first-lactation and mature cows consumed more feed, whereas in the experiment by DePeters, three-times-a-day milking did not increase feed intake. In the experiment by DePeters et al., first-lactation cows milked three times daily had reduced weight gains over the lactation. In all experiments there were small differences in milk composition between milking frequencies. Reproductive performance and udder health of cows milked twice and three times daily varied among experiments, but where these were adversely affected by three-times-a-day milking, the differences were not large. If three-times-a-day milking is to be done, it should be carried out only in well-managed herds where the extra nutrients and care required by higher-producing animals are provided.

How much of the increased milk production resulting from three-times-a-day milking that is due to decreased udder pressure is difficult to determine, but results from controlled experiments would indicate that 5 to 10% is due to the decreased udder pressure and the remainder may be due to other factors, such as preferential utilization of feed nutrients. Under field conditions the increase in production may also be a result of better feeding and management of cows milked three times a day. All of the increase in production resulting from milking four times a day in comparison to three times a day is probably due to the better feeding and management of these cows, with little increase resulting from reduced udder pressure. Whether a dairy farmer practices three-times-a-day milking will depend primarily on the available labor and cost of it in comparison to the expected increase in milk yield resulting from three-times-a-day milking.

Once-daily milking results in marked decreases in milk yield. Milking older cows once daily results in approximately a 40% decrease in milk yield, whereas the decrease for first-lactation cows is approximately 50%. Milking first- and second-lactation cows 13 times a week was shown to result in 89.2% and 95.4% of that produced when they were milked 14 times a week. This procedure appears to have little practical importance for most dairy farmers.

4.5 DRYING OFF COWS

The purpose of drying off cows is to give them a rest before the next lactation so that they can attain maximum milk production during the next lactation. The rest period gives the mammary gland a rest between lactations and allows the cow to replenish some of her body supplies. The drying-off process should be done as rapidly as possible without injuring the udder. Drying off is accomplished by allowing the udder pressure to reach the point at which milk secretion is stopped; eventually, the milk remaining in the udder is absorbed by the blood. The best method for most cows is simply to stop milking them and to allow the absorption to take place as rapidly as possible. Periodic milking of the cows, because it relieves the udder pressure, prolongs the drying-off process. Complete cessation of milking is the method of choice for the cow that is free from mastitis or for the cow with a mastitis infection that is being treated at drying off. For very high producing cows or those with previous severe clinical mastitis, it might be desirable to proceed more slowly by gradually omitting more and more milkings until the cow is completely dried off. Incomplete milking is not recommended because the decrease in udder pressure at each milking allows more secretion to take place, which greatly prolongs the process. A major aid in drying off cows is to reduce their feed intake. Water restriction also helps. After the last milking, the teats should be thoroughly cleaned and dipped in an effective teat end dip and treated with an effective antibiotic, which reduces the rate of new mastitis infections during the dry period.

4.6 INCOMPLETE MILKING

Incomplete milking must be distinguished from leaving the normal amount of residual milk in the udder. Incomplete milking refers to leaving milk in the udder that could have been obtained during a normal adequate milking procedure. Incomplete milking has a detrimental effect on milk production—the severity is dependent on the degree of incomplete milking and the length of it. Incomplete milking may also play a role in mastitis. Incomplete milking increases the incidence of clinical mastitis of quarters infected with *Streptococcus agalactiae*, but it does not influence the quarters infected with other organisms.

4.7 MILK EJECTION

After the cow has been prepared for milking, the teat and gland cisterns become distended with milk, which results in a hardening of the udder. This condition is usually referred to as "let-down" of milk. At times, such a condition fails to occur, and this is known as "hold-up" of milk. The preferred term is "milk ejection," which refers to the movement of milk from the lumina of the alveoli into the major ducts and thence to the bottom part of the udder. A failure of milk ejection results in the holding up of milk; the cow, however, cannot willfully do this. Because of the spongy condition of the secretory tissue in which most of the milk is held prior to milking and because of the constrictions that mark the points where smaller ducts drain into larger ones, some positive ejection force is required to move the milk from the secretory tissue. This process involves both the nervous system and a hormone. The concept of milk ejection was formulated by Petersen and his co-workers at the University of Minnesota in 1941. Petersen was also instrumental in arousing the interest of dairy farmers in the milk ejection concept and in getting them to apply the principles of it.

The neurohormonal process of milk ejection begins with a nervous impulse arising from a stimulus that the cow associates with suckling or milking (Figure 4.8). In most milking routines, washing of the cow's udder prior to the attachment of the milking machine serves as the stimulus. If washing is not done, the appli-

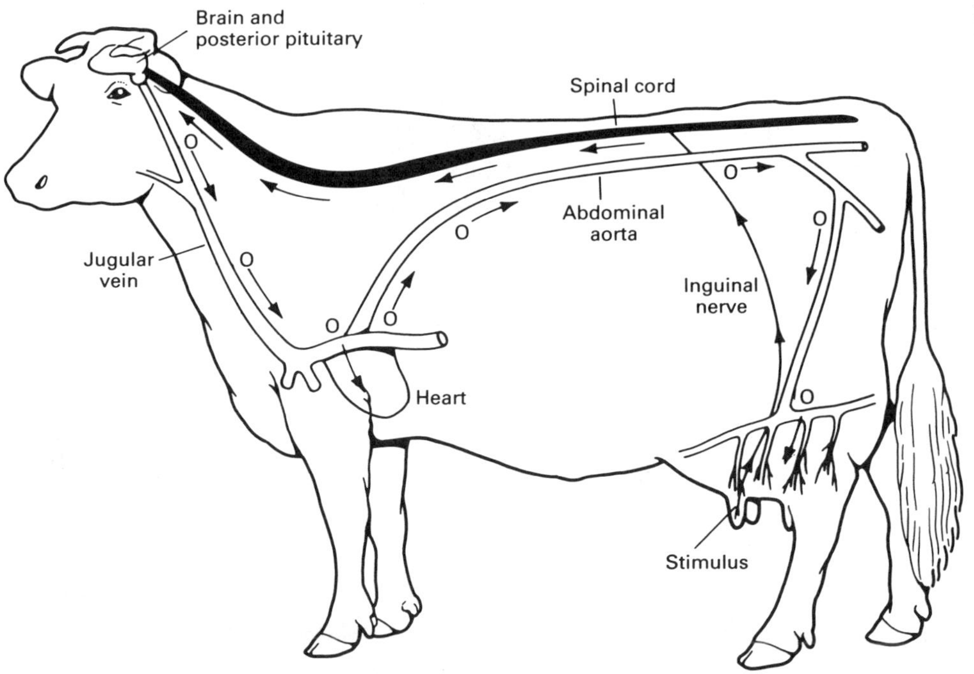

Figure 4.8 Neurohormonal reflex of milk ejection. A stimulus applied to the udder sends a nerve reflex (arrows) to the brain, which causes the posterior pituitary gland to release oxytocin (o) into the blood. The oxytocin is carried to the heart and then to all parts of the body. At the udder it causes contraction of the myoepithelial cells.

cation of the teat cups can act as the stimulus. Any phenomenon that a cow associates with milking can also serve as the stimulus; among these are feeding, the noise of milker pails, and the sound of the vacuum pump being turned on. The nerve impulse travels to the hypothalamus of the brain, which causes the posterior lobe of the pituitary gland to release oxytocin into the blood. Oxytocin is produced in the hypothalamus but is stored in the posterior pituitary. Oxytocin is carried to all parts of the body, and when it reaches the mammary gland, it causes the myoepithelial cells surrounding each alveolus to contract and to decrease the size of the lumen of the alveolus, forcing the milk out into the ducts (Figure 4.9). The small ducts also shorten and widen to facilitate the movement of milk from the alveoli into the larger ducts. A period of approximately 45 to 60 seconds elapses between the time oxytocin is released into the blood and the time it arrives at the udder.

The oxytocin in the blood of cows prior to, during, and after milking is shown in Figure 4.10. The amount of oxytocin in the blood prior to stimulation for milk ejection is very small and then increases very markedly to reach a peak in about 2 minutes after stimulation. After the peak, the oxytocin decreases and reaches the prestimulation level by 10 minutes after stimulation. When no udder washing

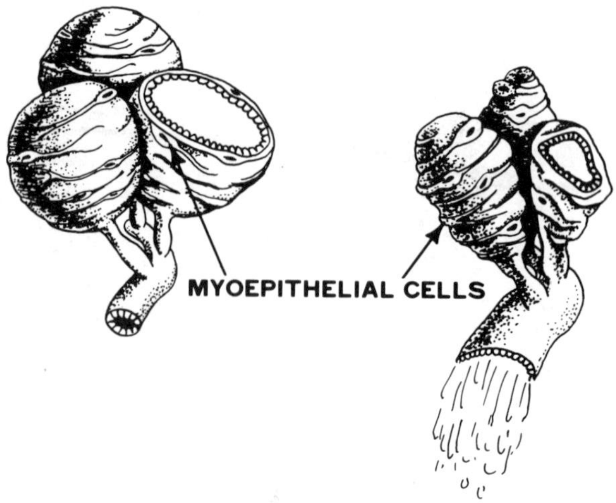

Figure 4.9 Contraction of the myoepithelial cells by oxytocin forces milk from the lumen of the alveolus into the ducts.

is done, the time of peak oxytocin concentration is delayed, resulting in a longer milking time.

The rapid decrease in oxytocin concentration in the blood after stimulation is important in the milking routine. The milking machine should be applied and

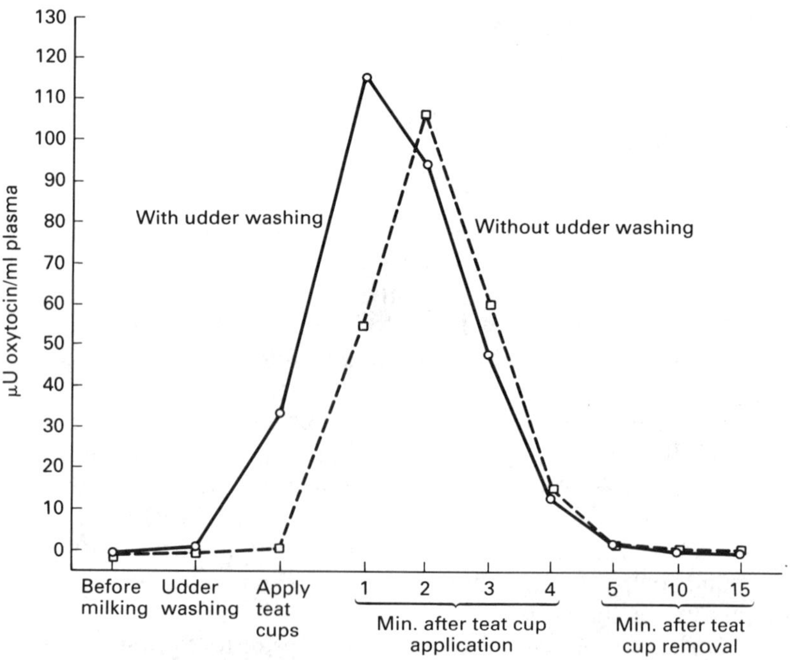

Figure 4.10 Oxytocin concentration in the blood of cows before, during, and after milking. The dashed lines indicate the oxytocin concentration when no manual stimulation for milk ejection was given. From V. G. Momongan and G. H. Schmidt, 1970, *J. Dairy Sci.*, 53:747.

milk removed when the oxytocin is most active. If the milking machine is applied several minutes after stimulation, a decrease in milk yield may occur because the effect of the oxytocin stimulation of the myoepithelial cells may have worn off and the milk moved back into the lumina of the alveoli. If this happens, a second milk ejection would be required for complete milk removal. It appears that most cows are refractory to a second milk ejection for some period of time after the first one.

Because they are associated with the hormone adrenaline (epinephrine), fear and excitement interfere with milk ejection. Adrenaline that is released into the blood before the stimulus for milk ejection completely inhibits milk ejection; adrenaline release after milk ejection causes partial inhibition. The inhibition occurring before the release of oxytocin is caused partially by the constriction of blood vessels, which prevents oxytocin from reaching the udder, and partially by the interference of oxytocin release from the posterior pituitary. Inhibition after oxytocin release is probably related to the interference with the action of oxytocin on the myoepithelial cells. For maximum milk production, fright and excitement should be avoided during milking.

REFERENCES AND FURTHER READINGS

Allen, D. B., et al. 1986. Three times a day milking: Effects on milk production, reproductive efficiency, and udder health. *J. Dairy Sci.*, 69:1441.

Amos, H. E., et al. 1985. Influence of milking frequency on productive and reproductive efficiencies of dairy cows. *J. Dairy Sci.*, 68:732.

Brachet, J. 1961. The living cell. *Sci. Am.*, 205(3):1.

Claesson, O. 1962. A 6½-day week for cows. *Dairy Farmer Ipswich*, 9:36.

Claesson, O., et al. 1959. Studies on monozygous twins: 17. Once-a-day milking compared with twice-a-day milking. *Acta Agric. Scand.*, 9:38.

DePeters, E. J., et al. 1985. Three or two times daily milking of older cows and first lactation cows for entire lactations. *J. Dairy Sci.*, 68:123.

Gorewit, R. C. 1979. Method for determining oxytocin concentrations in unextracted sera; characterization in lactating cattle. *Proc. Soc. Exp. Biol. Med.*, 160:80.

Korkman, N. 1959. A study of the milk pressure in the teat cisterns of the cow. *K. Lantbrukshogsk. Ann.*, 20:303.

Larson, B. L. 1985. Biosynthesis and cellular secretion of milk. In *Lactation*. B. L. Larson, ed., Iowa State University Press, Ames, IA.

Momongan, V. G., and G. H. Schmidt. 1970. Oxytocin levels in the plasma of Holstein–Friesian cows during milking with and without a premilking stimulus. *J. Dairy Sci.*, 53:747.

Ormiston, E. E., et al. 1967. Effects of milking at unequal intervals for a complete lactation on milk yield and composition. *J. Dairy Sci.*, 50:1597.

Schmidt, G. H. 1960. Effect of milking intervals on the rate of milk and fat secretion. *J. Dairy Sci.*, 43:213.

SCHMIDT, G. H. 1971. *Biology of Lactation*. W. H. Freeman and Company, Publishers, San Francisco.

SCHMIDT, G. H., and G. W. TRIMBERGER. 1963. Effect of unequal milking intervals on lactation milk, milk fat, and total solids production of cows. *J. Dairy Sci.*, 46:19.

TUCKER, H. A. 1985. Endocrine and neural control of the mammary gland. In *Lactation*, B. L. Larson, ed. Iowa State University Press, Ames, IA.

TURNER, C. W. 1952. *The Mammary Gland*. Lucas Brothers, Columbia, MO.

5

Factors Affecting Yield and Composition of Milk

The daily yield and composition of milk are affected by many factors, which can be divided into two broad areas, physiological and environmental. The physiological factors are governed in part by inheritance of the animal and in part by such nonhereditary factors as age, number of previous lactations, and pregnancy. The heritable factors are discussed in Chapter 10. In general, the dairy farmer has little control over the physiological factors but has some control over the environmental factors. A thorough understanding of the factors that affect the yield and composition of milk enables the dairy farmer to evaluate changes that occur and to manage the environment in order to produce desired changes in milk yield and composition.

5.1 CHANGES OCCURRING DURING A NORMAL LACTATION

Normal Lactation Curve

The relationship of milk yield to the milk fat and protein percentages of a cow during a normal lactation period is shown in Figure 5.1. The cow reaches her peak production approximately 3 to 6 weeks after parturition, and then a gradual decline in the yield takes place. A lactation period of 305 days is recommended

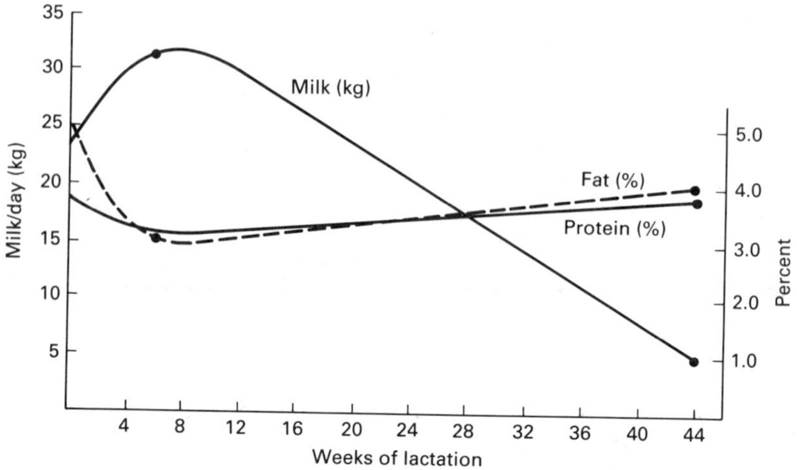

Figure 5.1 Milk yield and fat and protein percentages of Holstein cows with advancing lactation.

to take advantage of a 60-day dry period and a yearly calving interval. The peak yield of the cow is dependent on her body condition at calving, her inherited potential, her freedom from metabolic and infectious diseases, and the feeding regime after calving. Good body condition at calving and an adequate feeding program after calving tend to increase peak milk production. Peak milk production plays an important role in determining lactation milk production, since there is a high correlation between these two factors. Freedom from metabolic diseases, particularly milk fever and ketosis, allows the cow to attain her inherited level of milk production.

The rate of decline in yield after calving is called persistency. Cows that have a sharp decline in milk yield after the peak have a low persistency. Cows must have high persistency as well as high peak milk production for high-lactation milk yields. The rate of decline is accentuated by pregnancy, particularly starting at the twenty-second week of pregnancy, which occurs during the seventh to eighth months of the normal lactation period if the cow is bred back 60 days after calving.

There is a general inverse relationship between milk yield and protein and milk-fat percentages (Figure 5.1). As yield increases, the percentage composition of these two elements decreases. The protein and milk-fat percentages are at a low point during the peak of lactation and then gradually increase toward the end of lactation. Lactose content shows a very slight decline toward the end of lactation and ash content shows a very slight increase with advancing lactation.

Colostrum

The first milk obtained after parturition is colostrum, which has a markedly different composition from that of normal milk (Table 5.1). The most striking dif-

TABLE 5.1 COMPARISON OF THE COMPOSITION OF COLOSTRUM AND NORMAL MILK OBTAINED TWO WEEKS AFTER CALVING

	Colostrum (%)	Normal milk (%)
Total solids	23.9	12.9
Minerals	1.1	0.7
Protein	14.0	3.1
Fat	6.7	4.0
Lactose	2.7	5.0

Source: D. B. Parrish et al., 1950, *J. Dairy Sci.*, 33:457.

ference is the high percentage of protein in colostrum. Most of the increase in protein is attributable to the globulins, particulary immune globulins, which carry the immune bodies, or antibodies. Antibodies can be absorbed by the intestinal tract during the first 24 hours of the calf's life, which gives the newborn calf passive immunity from its dam. The fat and ash percentages are also somewhat higher in colostrum than in normal milk. These differences cause the total solids content to be approximately double that of normal milk. The increase in fat and total solids contents may be related to the fact that the newborn calf needs a rich energy source, and this is one way in which nature has provided for it. Not shown in Table 5.1, but very significant, is a marked increase in the vitamin A content of colostrum. There is approximately 10 times more vitamin A in colostrum than in normal milk. Again the vitamin A stores of the young calf are very low at birth and colostrum consumption is a way of rapidly increasing the vitamin A content. Vitamin A is important because it aids in the control of infectious diseases.

5.2 DAY-TO-DAY VARIATIONS

Both milk yield and percentage composition of milk vary considerably from day to day. In general, daily variation in milk yield is related to the completeness of evacuation of milk from the udder. Variations caused by disease, undernutrition, cows being off-feed, and related factors are usually of longer duration than those caused by udder evacuation, incomplete milking, estrus, or excitement. Depressions in milk yield that last for several days are usually accompanied by higher fat tests because there is a general inverse relationship between milk yield and fat test. High yields as a result of daily variations are accompanied by increases in the milk-fat percentages and low yields are accompanied by low milk-fat content. An incomplete evacuation of milk fails to obtain the last-drawn milk, which is extremely high in milk fat. First-drawn milk, or foremilk, may be as low as 1% milk fat, whereas the last-drawn milk may be as high as 8 to 15% (Figure 5.2). The milk yield at the next milking is higher in milk-fat content since it contains the leftover milk from the last milking with a high fat content and the normal yield and fat content from the present milking.

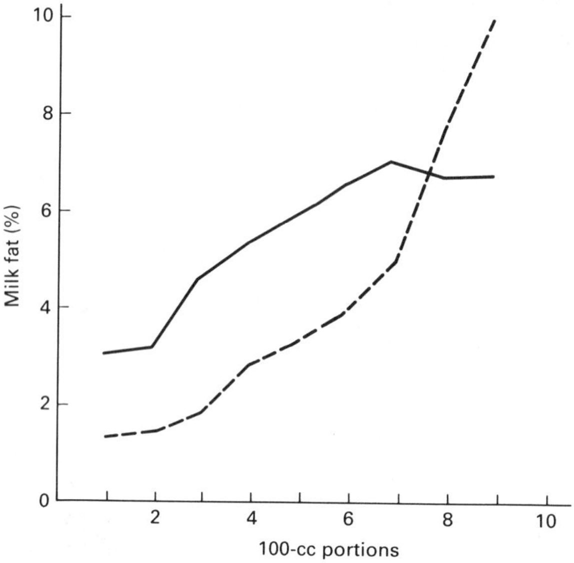

Figure 5.2 Percentage of milk fat present in successive 100-cc portions of milk obtained from a Jersey cow on entering the barn (solid line) and after standing in the barn for 2 hours (dashed line). Data from A. C. Ragsdale et al., 1921, *J. Dairy Sci.*, 4:448.

The milk-fat test may vary as much as 1% between the night and morning milkings, when cows are milked at 14- and 10-hour intervals. Part of this variation is caused by the relationship of pressure to milk and fat secretion, in which milk produced during a long interval and against a higher udder pressure has a lower fat test. Part of the variation results from exercise, which has a tendency to increase the milk-fat percentage. Consequently, the milk from the evening milking of cows milked at equal intervals has a slightly higher fat test than does that obtained from the morning milking.

5.3 DRY PERIOD AND BODY CONDITION

The length of the dry period and body condition at calving are related. Cows must be in good body condition at calving and must have had a dry period to attain maximum production. Cows having a thin body condition at the end of lactation require a dry period to replenish their body supplies. Body fat can be utilized for milk production: 1 kg of body fat supplies energy to produce approximately 7 kg of milk. Cows in good condition at calving also have a slightly higher milk-fat percentage than do those in fair or poor body condition. Overconditioning may be detrimental, especially for heifers entering the dairy. A cow that is overconditioned at calving is also more susceptible to metabolic diseases, particularly ketosis.

The dry period is important for replenishing body supplies if the cow is in poor body condition at calving. Cows also need a dry period to regenerate secretory tissue. In an experiment by Smith and Dodd (1966), two quarters of two cows were dried off 10 weeks before the expected date of calving and the other

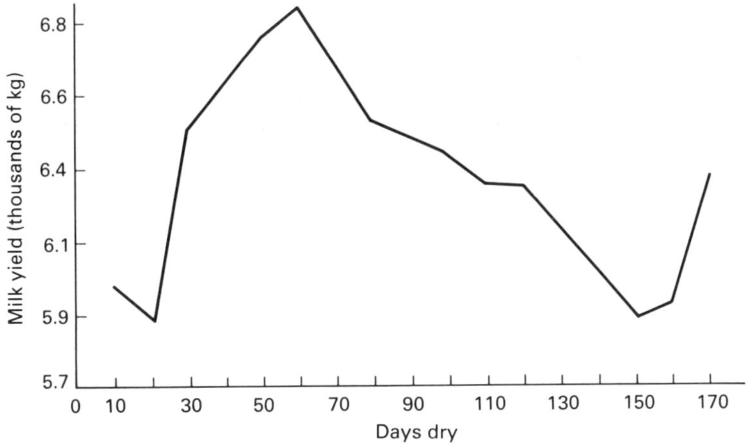

Figure 5.3 Average 305-day milk production following dry periods of various lengths. From L. R. Schaeffer and C. R. Henderson, 1972, *J. Dairy Sci.*, 55:107.

two quarters were milked twice daily throughout pregnancy. In the first 3 months of the second lactation, the continuously milked quarters of the cows gave 56% and 62% of the milk yields of quarters that were dried off.

An increase in milk yield results with dry periods up to approximately 60 days. Results of one trial are shown in Figure 5.3. Cows with dry periods of 50 to 59 days had the highest subsequent lactation production, whereas cows with dry periods of 40 to 49 or 60 to 69 days had only slightly less production. The reason for the reduction in production caused by long dry periods is not known, but it could be due to an exceptional population of animals in the study instead of a random one.

5.4 AGE OF COW AT CALVING

It is well known that cows produce more milk as they become older. First-lactation cows freshening at 24 months of age produce approximately 75% of the milk produced by mature cows. Average figures for 3-year-old cows (second lactation) indicate that they produce approximately 85% of the milk produced by a mature cow; the figures for 4- and 5-year-old cows are 92% and 98%, respectively. Cows of most breeds are considered mature as 6-year-olds. There is some variation among breeds; the Brown Swiss matures somewhat slower than the other breeds. The use of age correction factors to convert records made by cows at different ages is discussed in Chapter 12. When cows are 8 to 9 years of age, a slight reduction in the amount of milk produced occurs and the reduction continues until their death. In addition to the increase in milk production with age, there is a slight decrease in nonfat solids and milk-fat percentages through the fifth lactation, beyond which there is little change.

5.5 BODY WEIGHT

A general relationship exists between the body weight of cows and amount of milk produced. Larger cows have more udder secretory tissue and larger digestive systems. The energy required for the maintenance of the cow is proportional to the metabolic size or surface area, which is body weight to the 0.73 power (body weight$^{0.73}$). Two cows, each weighing 360 kg, will generally produce more milk than one weighing 720 kg; however, the maintenance requirements of the two cows would be somewhat higher than that of the 720-kg cow.

5.6 GESTATION

A marked drop in milk production occurs toward the end of pregnancy (Figure 5.4). The exact reason for this decrease in not known. One hypothesis states that there is an increase in the nutrients required for fetal development; however, this appears to be only 1 to 2% of the daily requirements of the cow. A more plausible explanation is that a change in hormone production (in which large amounts of estrogen and progesterone flow into the blood) occurs at this time and may be detrimental to milk secretion. An increase in the nonfat solids content of milk occurs during the fourth to fifth months of gestation.

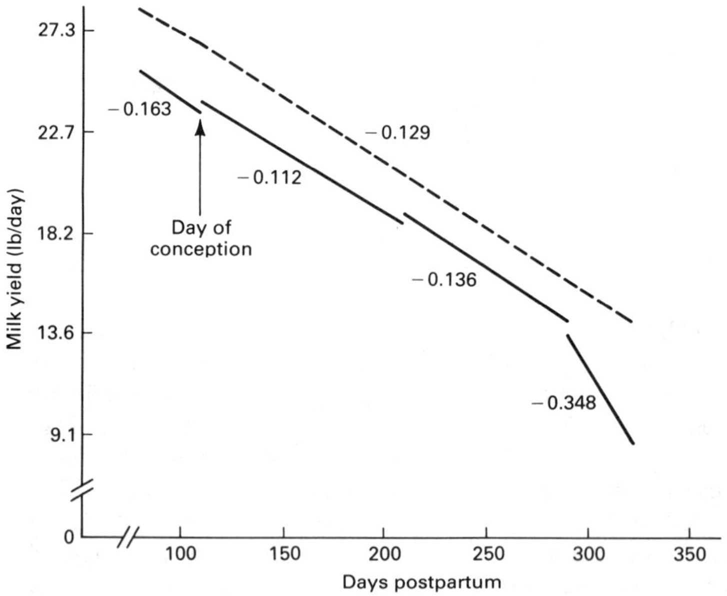

Figure 5.4 Average daily milk yield of pregnant cows (solid line) and nonpregnant cows (dashed line) for the same period. Numbers indicate the average daily decrease in milk yield for the various parts of the lactation. From R. E. Erb et al., 1952, *J. Dairy Sci.*, 35:224.

5.7 ENVIRONMENT

Temperature and Humidity

Temperatures between 40 and 75°F have no effect on the milk production of most dairy animals. In this range (known as the comfort zone), no body processes are directly involved in maintaining body temperature. When the temperature drops below 40°F, there is no adverse effect on milk production if extra feed is supplied and protection is provided from the elements. Cold temperatures (5°F or lower) may have a detrimental effect on milk yield. Larger breeds are more tolerant of low temperatures than are smaller breeds. Milk-fat content, as well as nonfat solids and total solids percentages, increases with decreasing temperatures.

Conversely, with increasing temperatures, there is a slight decrease in milk yield and slight decrease in milk-fat, nonfat solids, and total solids percentages. At temperatures above 75°F, a decrease in milk yield occurs (Figure 5.5). This

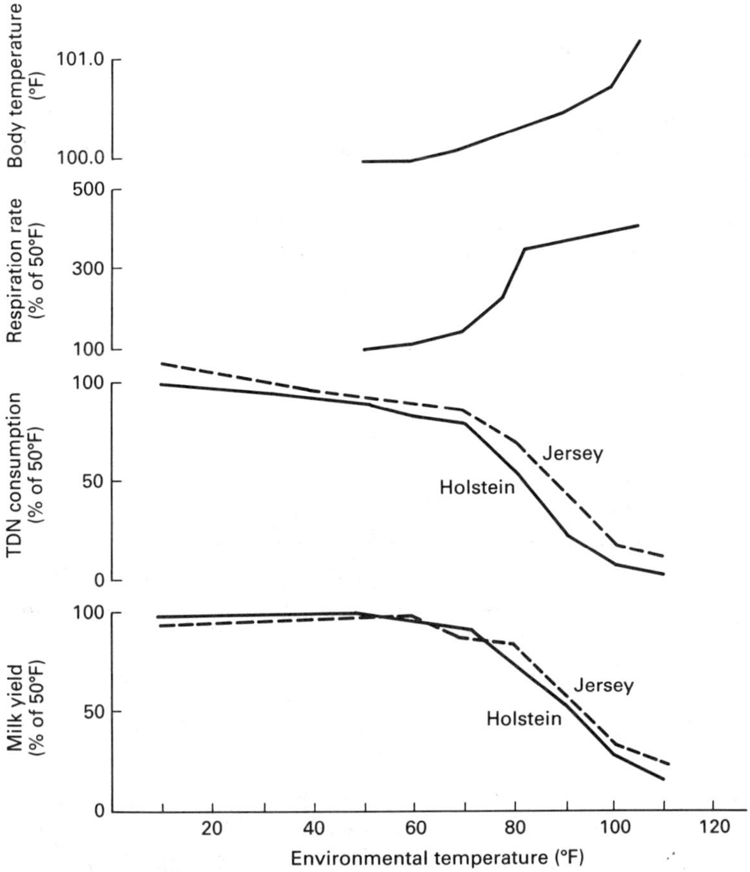

Figure 5.5 Effect of environmental temperature on milk yield, TDN consumption, respiration rate, and body temperature of Holstein and Jersey cows. Adapted from *Mo. Agric. Exp. Stn. Bull.* 425, 435, 460, 471.

decrease in yield may cause an increase in the milk-fat composition. A decrease in feed consumption, an increase in water intake, an increase in body temperature, and an increase in respiration rate also take place as a result of higher temperatures. The smaller breeds, particularly the Jersey, are more tolerant of high temperatures than are the larger breeds, particularly the Holstein. The smaller breeds have a larger surface area per unit of body weight and apparently can dissipate heat somewhat more rapidly than do larger cows. The Brown Swiss is an exception; it is much more tolerant of heat than is the Holstein.

Other factors may modify the effect of temperature. High relative humidity accentuates the problem of high temperatures. Air velocity at higher temperatures helps to cool the animals. Radiation from the sun increases the stress of the animal caused by higher temperatures.

Photoperiod

Tucker and his co-workers at Michigan (Peters et al., 1981; Stanisiewski et al., 1985) have studied the effect of various lighting regimes on milk production. They found that cows subjected to 16 hours of light and 8 hours of darkness produced 7 to 10% more milk than did those exposed to normal winter hours of 9 to 12 hours of light per day. In controlled studies they found that feed intake also increased with longer periods of exposure to light.

In a field trial involving 13 dairy herds, cows were exposed to 16 to 16.25 hours of light per day in comparison to 9 to 12 hours for the control animals. Cows exposed to the longer periods of light produced an average of 2.2 kg more milk per day, but had a drop of 0.16% milk-fat percentage.

5.8 SEASON OF YEAR

In the northern part of the United States, cows freshening in the fall and early winter produce considerably more milk and higher milk-fat yields than do those freshening in late winter, spring, and summer. In past years, differences were as high as 680 kg; however, the differences have been reduced over the years as a result of better feeding and management of cows throughout the year. There are several reasons for the higher production of cows freshening in fall and early winter. Milk production is usually lower during the summer because of the higher environmental temperatures and the undernutrition that exists on some farms, especially where poor pastures are the major source of feed. Cows freshening in the fall have their peak production during winter, when feeding and management are usually better. Most dairy farmers have more time to spend with the cows and more emphasis is placed on the feeding program at this time. In addition, a boost in milk production occurs when the cows are turned out on pasture during May and June. This boost occurs during the declining phase of lactation and has a significant effect on lactation milk yield. The cows are also dry during part of the summer, when high environmental temperatures and poor feeding cause a decrease in milk production.

The increase in fat yield results from the higher milk-fat percentage that occurs during the winter months when the milk yield is higher. The lower milk-fat percentages occur during the summer months when the yield is lower.

5.9 FEEDS AND FEEDING

Milk-Fat Content

The fat content of milk can be increased somewhat by feeding. The increases are usually temporary in nature and rather impractical for dairy farmers. Feeds high in fat, such as linseed oil, cottonseed oil, and tallow, increase the fat content of milk; however, feeding cod liver oil causes a marked depression in the fat percentage.

Some rations cause a decrease in the milk-fat percentage. Among these are high-concentrate, low-roughage rations; grass from lush spring pastures; finely ground hay; heat-treated feeds; and feeds in pelleted form, especially forages. These rations depress acetic acid production and increase propionic acid production in the rumen. Generally, the molar percentages of the volatile fatty acids in the rumen are: acetic acid, 65%, propionic acid, 20%; butyric acid, 12%; and others, 3%. Fat-depressing rations cause the acetic acid to drop by 10% or more and the propionic acid to increase by a similiar amount.

The exact reason for the milk-fat depression is not known; however, changes in the acetic/propionic ratio affect the fat tissue in the body by causing it (1) to compete directly with the mammary gland for milk-fat precursors, especially acetic acid; (2) to take up and esterify more long-chain fatty acids; and (3) to decrease the mobilization of long-chain fatty acids from the fat tissue for milk-fat synthesis.

The fat depression caused by feeding can be partially overcome in several ways. Among these are the daily feeding of sodium or potassium bicarbonate, magnesium oxide, magnesium carbonate, partially delactosed whey, sodium bentonite, and calcium carbonate. These compounds decrease rumen propionate production either by increasing the rumen pH or by increasing the dilution rate of the rumen fluid or both. Even though these compounds partially restore the milk-fat percentage, most of them are unpalatable and have to be masked by other feeds in order for cows to consume them.

Following are some dietary guidelines that help to prevent a milk-fat depression: (1) feed unground forage at a minimum rate of 1.5% of body weight per day; (2) feed a ration containing at least 19% acid detergent fiber; (3) use a screen that is more than $\frac{1}{8}$ in. in diameter if ground forage is used; (4) feed forages and concentrates together; and (5) limit corn to one-third of the total ration. When concentrate mixtures are fed in the form of meal, the fat test is usually 0.1% higher than when the same concentrate ration is fed in pelleted form.

The fat content of rations normally fed on dairy farms has no influence on either milk yield or milk composition. The variation in fat content of milk from cows fed mostly home-grown grains and protein supplements is very small. The effect of fat content in the feed on milk production is primarily a result of the energy value of the feed. High-fat feeds result in slightly higher milk production than low-fat feeds. Recent research by Palmquist and his associates (Palmquist and Jenkins, 1980) shows that fat can be added at 3% of the total ration and that 5% added fat may be beneficial to high-producing dairy cows. The 5% added fat then substitutes for starch in the ration, which helps to prevent the fat-depressing effect of a high starch–low fiber diet. High-fat diets tend to cause a decrease in the casein content and thus the protein content of milk.

Protein and Nonfat Solids Contents

The protein and nonfat solids composition of milk can be changed only slightly, and the variation is considerably less than that of fat. Most of the changes in the nonfat solids composition of milk result from changes in protein content. Mineral and lactose contents are quite unchangeable. Underfeeding of dairy cows results in a 0.2% reduction in protein and nonfat solids percentages and a depression in milk yield. Increasing the plane of nutrition to 25% above normal standards results in an increase in nonfat solids and protein percentages of the same magnitude. An increased plane of nutrition usually results in an increase in milk production; however, a cow cannot produce above her inherited potential. The increase in milk protein percentage with higher planes of nutrition results from an increase in propionic acid production in the rumen.

Mineral and Vitamin Contents

The mineral content of milk is relatively stable, and the content of the major elements (calcium, phosphorus, potassium, chloride, and sodium) cannot be changed by altering these elements in the diet of the cow. The trace minerals of milk, with the exception of iron and copper, can be increased by increasing the minerals in the ration. For most of the elements, the amount transferred from the ration to the milk is very low; the highest degree of transfer is for iodine, and only 3 to 5% of the iodine in the ration appears in the milk.

The content of some vitamins in milk can be changed by increasing them in the feed. This is true for the fat-soluble vitamins, A, D, and E. Vitamins A and E must be supplied by the feed because they are not synthesized in the body. Rumen microorganisms synthesize the B-complex vitamins, and consequently they are available to the animal independent of the feed. Marked seasonal variations occur in the vitamin A and D contents of milk. The vitamin A content of milk can be raised by increasing the carotene content of the ration, either by increasing the proportion of green leafy roughages or by pasture grazing. The vitamin A content of cow's milk is 1.5 to 15 times higher during the grazing season than during the winter months. The vitamin D content of milk can be increased by providing sun-cured roughages or by exposing the cows to direct sunlight. Milk produced during the summer when cows are on pasture contains approximately 35 units of vitamin D per quart, whereas that produced during the winter months contains only 14 units. This is of little consequence in providing vitamin D for human consumption since dairy plants routinely add 400 units per quart of milk.

Off-Feed

When an animal goes off-feed there is a decrease in the volume of milk produced, accompanied by increases in the fat, mineral, protein, and total solids contents and marked reductions in lactose content and specific gravity of milk. The decrease in lactose content and specific gravity produces the same characteristics as if water were added to the milk. If a relatively large number of cows are off-feed at the same time, milk plants might wrongly accuse the dairy farmer of adding water to the milk.

5.10 HORMONES

Several hormones have been used commercially and experimentally to increase milk production and change the composition of milk. The two most common ones are thyroprotein (iodinated casein) and growth hormone (somatotropin, STH).

Thyroprotein

Thyroprotein is made by placing casein in the presence of iodine under controlled conditions. Some of the tyrosine amino acids of casein combine and take up iodine to form thyroxine. Iodinated casein contains approximately 1% thyroxine. Thyroxine is secreted by the thyroid gland and one of its major functions is to control the metabolic rate. Cows receiving iodinated casein have elevated body temperatures, increased heart rates, and increased respiration rates.

The short-term feeding of thyroxine at 10 to 15 g per animal per day results in an increase in milk production, which may be as high as 15 to 20%. A transitory increase in the milk-fat test results, but the test soon returns to normal. To make thyroprotein feeding effective, additional feed must be fed for the synthesis of milk and maintenance of the animal.

Several problems result from the feeding of thyroprotein to dairy cattle. One is that thyroprotein appears to be effective for periods of 60 to 120 days. When the thyroprotein is withdrawn after feeding or it become unprofitable, a marked drop in milk production results. During feeding of thyroxine the animal's thyroid gland becomes inactive. When the thyroxine is withdrawn, the thyroid gland requires time to return to an active state. During this time a drop in milk production occurs. In many cases, the decrease in milk production after thyroxine withdrawal is equal to the additional amount produced during thyroxine feeding. Several research workers have withdrawn thyroxine at a gradual rate instead of abruptly, but no consistent beneficial effects have been obtained. Thyroxine does not increase the efficiency of milk production when the loss in body weight is calculated. In most cases, a loss in body weight occurs when cows are fed thyroxine, and

TABLE 5.2 EFFECT OF FEEDING IODINATED CASEIN ON LACTATION MILK YIELD, MILK COMPOSITION, AND REPRODUCTIVE PERFORMANCE

Measure[a]	Animals per treatment	Control		Iodinated casein	
		kg	lb	kg	lb
Milk (ME)					
All cows	63	7,265	15,983	7,237	15,921
Older cows	47	7,155	15,741	7,007	15,415
First lactation	16	7,586	16,689	7,971	17,536
Fat (%)	63	3.38		3.36	
Protein (%)	63	3.24		3.24	
FCM	63	6,623	14,571	6,587	14,491
Services per conception	61	1.79		2.38	
Calving interval (months)					
Cows	36	12.6		13.1	
First lactation	13	12.6		15.0	

[a] ME, mature equivalent; FCM, 4% fat-corrected milk.

Source: G. H. Schmidt et al., 1971, *J. Dairy Sci.*, 54:481.

additional feed must be fed either to maintain body weight or to restore body weight after thyroxine has been withdrawn.

Feeding thyroprotein during most of the lactation period does not appear to be beneficial. The results of one trial are shown in Table 5.2. The increase in milk-fat percentage lasted only 2 to 4 weeks. Marked increases in body temperature, respiration rate, and heart rate occurred in addition to body weight losses (Figure 5.6). Feeding of iodinated casein caused an increase in services per conception for all cows and significantly increased the calving interval between the first and second lactations.

Feeding thyroprotein might be effective in herds where cows are sold after they have been milked for one lactation. Feeding the hormone may also be beneficial during the last 3 to 4 months of lactation to increase milk production. In addition, feeding thyroprotein may be profitable if a short-term stimulation in milk yield is required: for example, to establish or maintain a milk production base in the market. Thyroprotein might be fed for a 3- to 4-month period before cows are turned onto pasture, since pasture feeding prevents the precipitous drop in production that occurs when thyroprotein is withdrawn. The same prevention of drop in production should be attainable by feeding extra energy after thyroprotein withdrawal.

Growth Hormone (Somatotropin)

The growth hormone stimulation of milk production has been known for about 50 years, but its use was impractical because of the high cost of the naturally produced compound that was extracted from the pituitary glands of animals after slaughter. Much interest has developed in its use after it was possible to produce rather large amounts with bacteria through synthetic production by using bioengineering methods of recombinant DNA. A number of short-term studies indicate that increases of 10 to 40% could be attained by growth hormone injections and that the milk production returned to preinjection yields after the hormone was withdrawn, instead of the precipitous drop seen with feeding thyroprotein.

Bauman and his colleagues at Cornell University (Bauman et al., 1985) conducted a longer-term study of high-producing cows to determine the effect on milk yield and composition. Cows were started on the growth hormone injections at approximately 84 days after calving and continued receiving the injections for 188 days. Three amounts of recombinant-derived growth hormone were compared to control animals and animals receiving pituitary-derived growth hormone. The results are shown in Table 5.3. Recombinant-derived growth hormone increased fat-corrected milk in a dose-dependent manner from 23 to 41% over controls. The pituitary-derived growth hormone resulted in a 16% increase over controls. There were no differences in fat, protein, or lactose contents of the milk.

The control cows were in positive energy balance throughout the trial. The marked increase in milk production of the growth hormone–treated cows caused

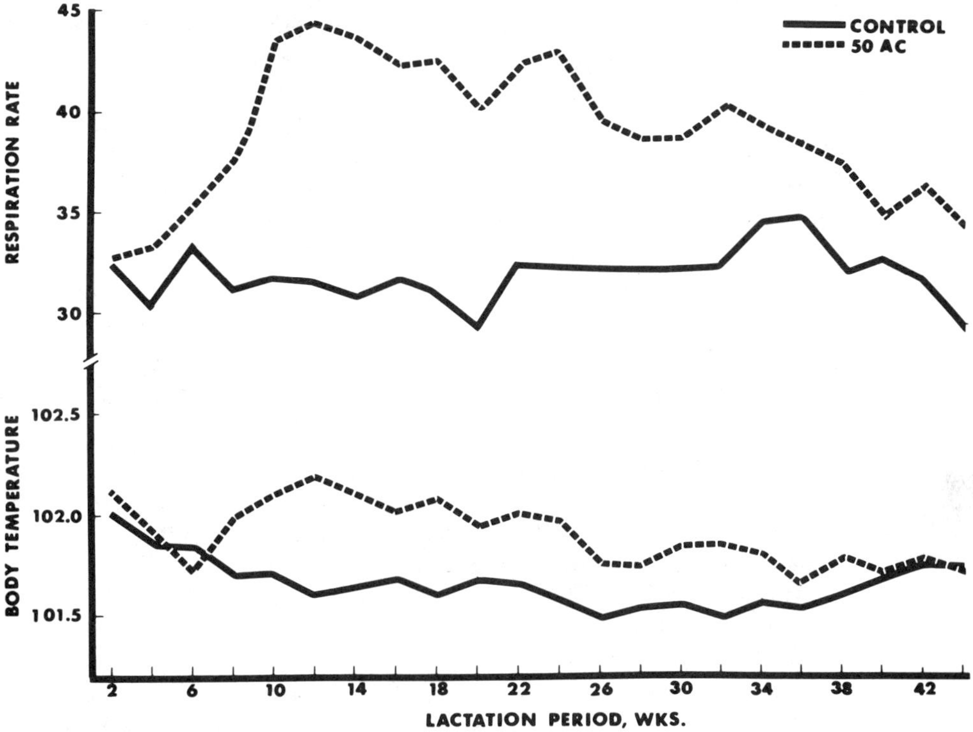

Figure 5.6 Body temperature and respiration rate of control cows (solid lines) and cows fed thyroprotein starting 50 days after calving (dashed lines). From G. H. Schmidt et al., 1971, *J. Dairy Sci.*, 54:481.

TABLE 5.3 EFFECT OF SOMATOTROPIN (STH) ON MILK YIELD AND COMPOSITION

Variable	Control	Pituitary STH: 27.0 mg/day	Synthetic STH 13.5 mg/day	Synthetic STH 27.0 mg/day	Synthetic STH 40.5 mg/day
Number of cows	6	6	6	6	6
kg FCM[a]/day	27.9	32.5	34.4	38.0	39.4
Percent increase over control	—	16.5	23.3	36.2	41.2
Fat (%)	3.6	3.3	3.8	3.6	3.6
Protein (%)	3.4	3.4	3.4	3.4	3.4
Lactose (%)	4.8	4.8	4.9	4.8	4.9

[a] FCM, 4% fat-corrected milk.

Source: D. E. Bauman et al., 1985, *J. Dairy Sci.*, 68:1352.

them to be in negative energy balance, but they were all restored to a positive energy balance by the tenth week after receiving growth hormone. Gross lactation efficiency (milk produced per unit of net energy intake) was improved with the use of growth hormone. It is believed that growth hormone produces its beneficial action by serving as a homeorhetic control that coordinates the physiological processes by partitioning more nutrients to the udder for milk secretion and enhancing the mammary gland to secrete milk.

Growth hormone has not yet been approved by the Food and Drug Administration for use in dairy cattle. In addition, the current method of administration requires the daily injection of the hormone. This might present some problems with nervous cows. Work is under way to find methods of administration other than daily injections. Before it is used commercially, it must be cleared for use and the economic benefits of its use determined.

REFERENCES AND FURTHER READINGS

Bauman, D. E., et al. 1985. Responses of high-producing dairy cows to long-term treatment with pituitary somatotropin and recombinant somatotropin. *J. Dairy Sci.*, 68:1352.

Dodd, F. H. 1957. Factors affecting the rate of secretion of milk and lactation yields. In *Progress in the Physiology of Farm Animals*, Vol. 3, J. Hammond, ed. Butterworth & Company (Publishers) Ltd., London, p. 962.

Huber, J. T., and R. L. Boman. 1966. Nutritional factors affecting the solids-not-fat content of milk. *J. Dairy Sci.*, 49:816.

Johansson, I., and O. Claesson. 1957. Factors affecting the composition of milk. In *Progress in the Physiology of Farm Animals*. J. Hammond, ed. Butterworth & Company (Publishers) Ltd., London, Chap. 21.

Laben, R. C. 1963. Factors responsible for variation in milk composition. *J. Dairy Sci.*, 41:452.

Palmquist, D. L., and T. C. Jenkins. 1980. Fat in lactation rations: Review. *J. Dairy Sci.*, 63:1.

Peters, R. R., et al. 1981. Milk yield, feed intake, prolactin, growth hormone, and glucocorticoid response of cows to supplemental light. *J. Dairy Sci.*, 64:1671.

Schaeffer, L. R., and C. R. Henderson. 1972. Effects of days dry and days open on Holstein milk production. *J. Dairy Sci.*, 55:107.

Smith, A., and F. H. Dodd. 1966. Effect of milking throughout pregnancy on milk yield in the succeeding lactation. *J. Dairy Sci.*, 49:895.

Stanisiewski, E. P., et al. 1985. Effect of photoperiod on milk yield and milk fat in commercial dairy herds. *J. Dairy Sci.*, 68:1134.

6

Harvesting the Milk Crop: Machine Milking

On the average dairy farm more time is spent milking cows than with any other farm operation. Between 50 and 80 hours are spent with each cow each year, and the majority of this time is spent in the milking operation. Thus, the milking machine is probably used more hours per year than any other equipment. The milking process is the harvesting operation performed after most of the other requisites for production of milk (breeding, feeding, housing, calf raising, sanitation, and disease control) have been provided. A flaw in any one of these operations can limit the amount of milk produced by cows as well as the efficiency of production. A number of studies have shown that the milking process can be a major limitation to high and efficient production.

6.1 THE MILKING MACHINE

Before the details of a recommended milking routine are discussed, an understanding of the operation of the milking machine is desirable. The modern milking machine was first assembled in 1903 by Alexander Gillies of Australia. He added a teat cup and liner to a machine that had been manufactured in Great Britain. His machine was the first to employ all of the principles of the modern milking machine. The essential components of the machine are: a source of vacuum, a receptacle for milk, a pulsator, and a teat cup and liner for each teat. In addition, hoses are necessary to attach the teat cups to the pulsator and milk receptacle and to attach the machine to a vacuum source.

Vacuum

All milking machines operate by vacuum. A partial vacuum is created when air is removed from a confined space. The atmosphere constantly exerts a pressure on all surfaces, which at sea level is equal to 14.7 pounds per square inch. This decreases slightly with increasing elevation above sea level. This pressure also causes mercury (Hg) to rise in a tube to a height of 30 in., which is called the barometric pressure. A perfect vacuum is created in the complete absence of air; however, present-day equipment cannot produce a perfect vacuum. Vacuum pumps remove air from a closed space and thus reduce the pressure. This reduction in pressure causes the column of mercury to rise to a shorter height; thus the difference between the atmospheric pressure and the reduced pressure is expressed in inches or millimeters (mm) of mercury. Most milking machines operate between 13 and 15 in. Hg. Vacuum in now expressed in kilopascals (kPa). Atmospheric pressure is equal to 30 in. Hg, 760 mm Hg, or 100 kPa. Fifteen inches Hg is equal to $\frac{1}{2}$ atmosphere, 380 mm Hg, or 50 kPa.

Milk is removed from the udder by application of a vacuum to the end of the teat. A positive pressure of approximately 5.5 kPa exists in the interior of the udder due to the ejection of milk from the alveoli into the teat and gland cisterns. When the milking machine is applied, the pressure below the teat is reduced to the vacuum level of the milking machine. This reduction in pressure and the positive pressure within the udder create a pressure differential across the teat meatus, which causes it to open and milk to flow (Figure 6.1, A). The rate of milk flow is dependent on the pressure differential; thus machines with a higher vacuum level cause a faster rate of milk flow than do lower-vacuum machines.

A constant milk flow occurs when the vacuum is continually applied to the teat; however, this causes irritation and damage to the teat and teat ends. To overcome this problem, a liner is inserted into the teat cup to form a double-action milking machine. The purposes of the liner are to massage the teat, keep the blood flowing in the teat, and prevent irritation to it.

In designing a milking installation, the vacuum pump should be large enough so that the vacuum level will not drop more than 1 in. Hg during normal operation and the vacuum level should return to normal within 10 seconds if a drop does occur. Serious vacuum fluctuations can increase the incidence of mastitis infections and should be avoided.

The pulsator introduces atmospheric air and vacuum intermittently between the teat cup liner and shell. A constant vacuum is found on the inside of the liner and on the end of the teat. When the same vacuum is introduced to the space between the liner and shell, the liner takes its normal open position. It is during this open position that milk flows (Figure 6.1, A). Milk begins to flow when the liner is half open and continues to flow until the liner is half closed. When the pulsator introduces air into the space between the liner and shell, the higher

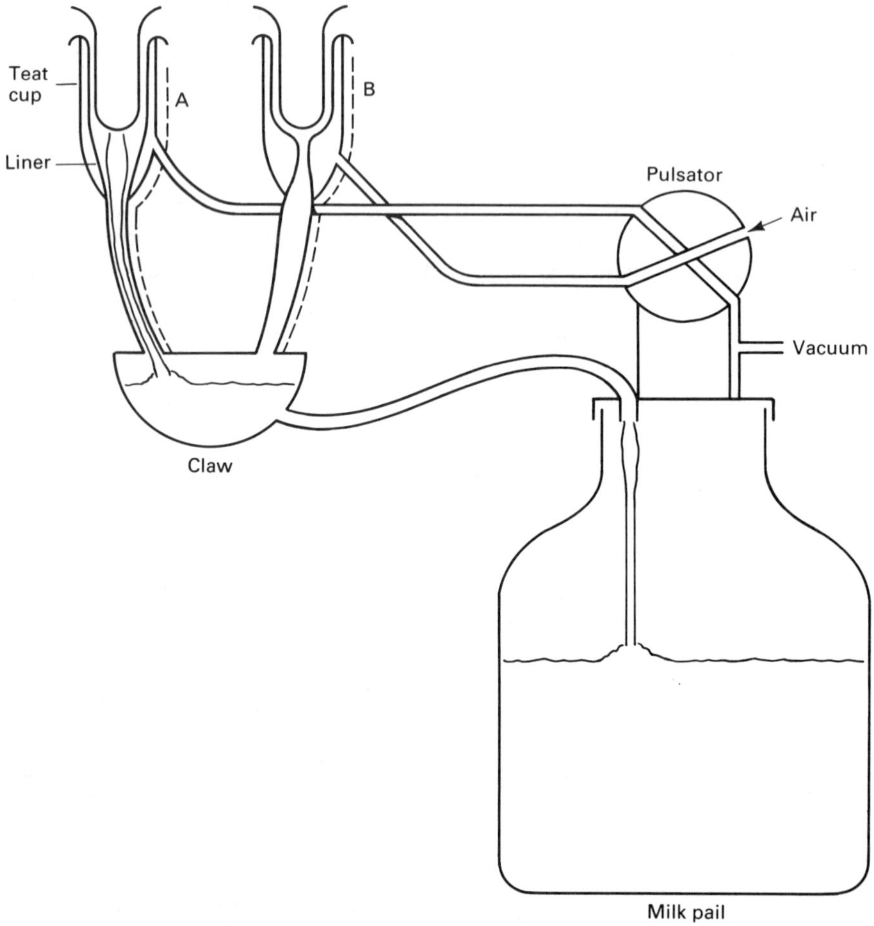

Figure 6.1 Components of a milking machine. Milk is removed from the teat by vacuum when the liner is open (A). Milk moves from the liner into the claw and into the milk pail or receiver. When the pulastor introduces air into the space between the liner and teat cup shell (B), the liner collapses, which massages the teat and causes milk flow to stop.

pressure on the outside of the liner causes it to collapse, which results in the squeezing and massaging of the teat (Figure 6.1, B). The closing of the liner causes both the teat meatus and liner below the teat to close, resulting in the stoppage of milk flow. The opening and closing of the teat cup liner constitute one pulsation cycle. Most milking machines operate between 50 and 80 pulsation cycles per minute.

Milking Machine Components

The vacuum to the milking machine is supplied by a vacuum pump, which is rated in cubic feet of air per minute (cfm) at a specified vacuum level. The size of the vacuum pump to be used for a milking installation depends on the number of milking units, the size and length of the pulsator lines, the type of pulsators, the type of system (bucket or pipeline), and the number and types of other vacuum-operated equipment in the milking system. The cfm ratings of pumps are expressed either as the American Society of Mechanical Engineers (ASME) standard or the New Zealand standard. The ASME standard is based on cfm at atmospheric pressure, whereas the New Zealand standard is based on cfm at 50 kPa (Figure 6.2). Two volumes of air at 50 kPa are equal to one volume of air at atmospheric pressure; therefore, the New Zealand standard requirements are twice as high as the ASME standard (Table 6.1).

Some milking machines expose all four teats to the milking phase at the same time, whereas other machines simultaneously expose two teats to the collapsed phase and two teats to the milking phase of the liner. In milking machines that have a pulsation ratio wider than 1:1 (milking/collapse phases), all four teats are exposed to the milking and collapse phases at the same time.

The pulsation ratio is the length of time that the teat cup liner is in the milking phase in comparison to the time that it is in the collapse phase. Until the 1960s most milking machines operated at a 1:1 pulsation ratio; however, in recent years machines operating at 2:1 and $2\frac{1}{2}$:1 milking/collapse ratios have been used. Wider ratio milking machines result in faster rates of milk flow because the teat is exposed to the milking phase for a longer period during each cycle.

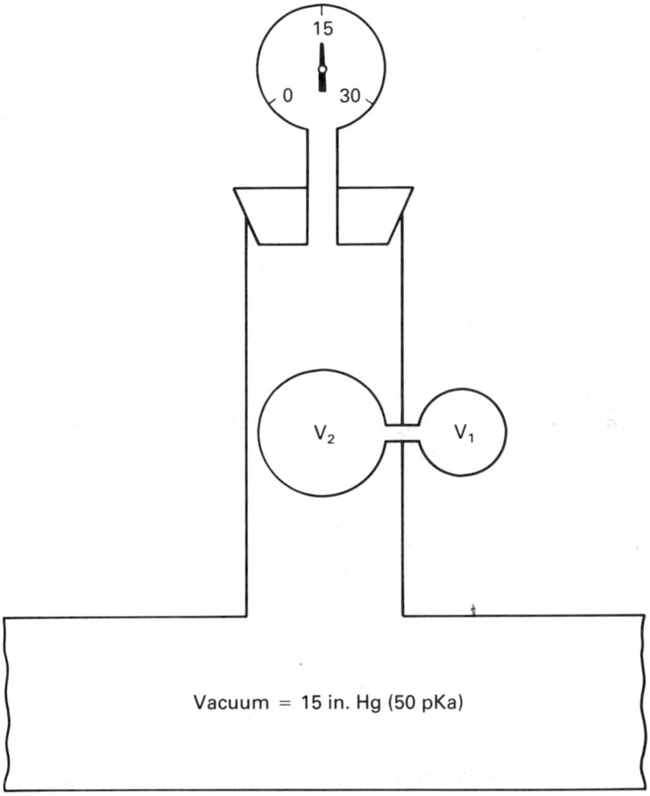

Figure 6.2 Relationship between the New Zealand and the American Society of Mechanical Engineers (ASME) standards for measuring airflow. The New Zealand standard measures airflow at 50 pKa and the ASME at atmospheric pressure; thus the volume of air of the New Zealand standard (V_2) is twice that of the ASME standard (V_1).

TABLE 6.1 RECOMMENDED VACUUM PUMP CAPACITIES IN CUBIC FEET PER MINUTE (CFM) FOR PIPELINE SYSTEMS IN CONVENTIONAL PARLORS USING AMERICAN SOCIETY OF MECHANICAL ENGINEERS (ASME) AND NEW ZEALAND (NZ) STANDARDS

Number of units	CFM (ASME)		CFM per system (NZ)
	Per unit	Per system	
4	9.0	36	72
6	8.7	52	104
8	8.3	66	132
10	8.0	80	160
12	7.5	90	180
16	7.0	112	224
20	6.3	126	252
24	5.5	132	264
30	5.1	152	304

Source: S. B. Spencer, 1980, *Natl. Milking Cent. Design Cong.*, Rochester, NY.

Two basic types of pulsators are available: unit pulsators and master pulsators with slave pulsators. A unit pulsator is located at each milking unit and works independently of other pulsators, whereas the master pulsator serves more than one unit. In some cases, a master pulsator is used to drive two or more slave pulsators, one located at each unit. Both types of pulsators can be operated by vacuum or by electricity.

Two basic types of milking machine receptacles are used with bucket milking machines. One type rests on the floor and the second type is suspended from either the cow by a strap (sursingle) or a mechanical arm. The floor type (Figure 6.1) has a claw that collects the milk from the four teat cups and in turn transfers it to the bucket by a hose. In the suspended type, the teat cups empty directly into the lid of the suspended pail. In the bucket type, the claw should be large enough to handle the milk from high-producing cows without flooding or blocking milk flow. Flooding and blocking milk flow can cause serious vacuum fluctuations on the end of the teat. Modifications of these two types of receptacles are found in milking parlors and in around-the-barn pipeline milking installations. Milk can be transferred through a claw, regularly found on a floor-type machine, to a weigh jar or a pipeline. Machines that use the suspended-type unit usually have the four teat cups empty into a small breaker cup, from which the milk is transferred to a weigh jar or pipeline. In both cases, the milk is transferred to the bulk tank by pipelines instead of being hand carried.

Many types of teat cup liners are available with various claims made for their ability to produce fast rate of milk flow, reduce machine stripping, and prevent trauma to the teats. The research on the desirable qualities of liners is limited. Most liners fit into two basic types, molded liners and extruded liners. Molded liners have a mouthpiece, a barrel, and a milk tube molded into one piece. The extruded liner is made from a straight tube and the mouthpiece is formed by inserting a metal ring that is somewhat larger than the liner in the upper end. A milk tube, usually made from the bottom part of the molded liner, is attached to the bottom of the liner. The advantages of extruded liners are that the size of the mouthpiece can be varied and the tension on the liner can be increased by stretching the tube. Molded liners are easier to assemble than extruded liners; however, extruded liners appear to produce faster milking rates than do molded liners.

Although the evidence is limited, it is generally accepted that the geometry of the barrel influences the milking rate and the mouthpiece affects the completeness of milk out. The depth of teat penetration into the liner at the start of milking is dependent on the bore of the liner, but by the end of milking, friction of the teat and liner is most important. As milking progresses the teat gradually moves farther into the liner, due to decreased pressure within the teat. If the teat penetrates too far into the liner, the teat and gland cisterns becomes occluded and milk flow stops. For this reason machine stripping has been practiced.

Liners should be selected that minimize stripping time, fall-offs, air slips, and machine on-time. In addition, they should not cause any discomfort or trauma to the cow.

Factors That Influence Rate of Milk Flow

Measurements of milk flow rate. Rate of flow is dependent on the pressure differential across the teat meatus, which is primarily dependent on the vacuum level of the milking machine and partially dependent on the internal udder pressure. Internal udder pressure is related to the amount of milk in the udder at the beginning of milking and decreases as milk is removed from the udder during milking. Other factors that influence rate of milk flow are size of the teat meatus and the tautness of the sphincter muscle surrounding the teat canal.

Several measurements are used to describe rate of milk flow. The rate is usually expressed as the average rate of flow or the maximum rate of flow. The average rate is obtained by dividing the total yield per milking by the total milking time. Another way to calculate average rate is to divide the yield before machine stripping by the time before machine stripping. The latter measure of average rate eliminates the variation in machine stripping time and yield. The maximum rate of flow is the highest yield obtained during a time interval of the milking operation, such as 1 minute. The maximum rate gives the best measure of the inherited potential milking rates of cows.

Alternate measures of milk flow rate include milk yield during the first minute or the first 2 minutes of milking. All measures are quite highly related to each other, with milk yield during the first 2 minutes being highly correlated with maximum rate of flow. The yield during the first 2 minutes probably has the most practical application for field use to determine the inheritance of milking rate.

Machine stripping time or yield is defined as the time or yield after milk has reached an arbitrary low level, such as 0.1 kg per 15 seconds. This measurement will always assure some machine stripping time and yield because the measurement begins before milk flow stops. Work is needed to measure the amount of machine stripping required, if any, after milk flow has stopped.

TABLE 6.2 EFFECT OF VACUUM AND PULSATION RATE AND RATIO ON MILK FLOW RATES AND MACHINE STRIPPING

Measure	Time before stripping (min)	Stripping time (min)	Maximum rate of flow	
			kg	lb
Vacuum				
In. Hg kPa				
12 40	4.0	0.6	3.7	8.1
18 60	3.0	0.8	4.9	10.7
24 80	2.8	1.6	4.8	10.5
Pulsation rate				
40	3.4	1.1	4.4	9.7
80	3.3	1.0	4.5	9.9
120	3.0	1.1	4.4	9.7
Pulsation ratio				
1:1	3.5	0.9	4.3	9.4
3:1	3.0	1.2	4.9	10.7

Source: G. H. Schmidt et al., 1963, *Cornell Agric. Exp. Stn. Bull.* 983.

Mechanical factors that influence milking rate. A number of research studies have been conducted to determine the effect of vacuum level, pulsation rate, and pulsation ratio on rate of milk flow and machine stripping time and yield. Typical results are shown in Table 6.2. In general, an increase in vacuum causes a marked increase in milk flow rate, measured either as maximum rate or average rate of flow. In most cases, an increase in machine stripping time and yield also occurs, especially if the vacuum is increased above 50 kPa. Widening the milking/collapse ratio of the teat cup liners also increases the rate of milk flow and in some cases, but not all, increases machine stripping yield. Increasing the pulsation rate has a slight beneficial effect in increasing the rate of milk flow, but the amount does not warrant the extra air movement required to operate a machine with high pulsation rates. In most cases a minimum of 40 pulsations per minute is suggested to prevent damage to the teats. When a new machine is used, cows take some time to become accustomed to it. At least one week is required before maximum rates of milk flow are attained.

A number of studies have been conducted to relate the changes in vacuum level, pulsation rate, pulsation ratio, and many other machine factors to somatic cell count, udder infection, and clinical mastitis. In most cases, no relationship between these milking machine factors and mastitis measures has been found in controlled laboratory conditions. Whether this is the case on dairy farms is not known.

6.2 MACHINE MILKING

A common misconception is that efficient milking practices preclude individual attention per cow. Since the objective of most dairy farmers is high and efficient milk production to obtain high net or labor income, sufficient attention must be given to each cow to ensure high production under all circumstances. Poor or sloppy milking management and lack of individual attention to dairy cows may go hand in hand, yet it is possible to give a minimum amount of individual attention to cows and still have adequate milking procedures for high production.

An adequate milking procedure with a minimum amount of individual attention to cows requires (1) removal of all of the milk obtainable after a normal milk ejection, (2) production of milk free from dirt and other debris, (3) prevention of injury to the udder or teats, (4) prevention of the transfer of mastitis-causing bacteria from an infected quarter to a noninfected one, and (5) efficient use of labor and equipment.

Most recommendations for conducting a successful milking operation are based on one or more of these requirements and on knowledge of the basic physiology of the dairy cow and how she reacts to different conditions. They merit consideration, even though some of them are supported by numerous field observations, but lack experimental proof.

Basic Milking Routine

Udder washing and drying. The first step in a good milking routine is washing and massaging the udder for 15 to 30 seconds with a hot-water sanitizing solution (Figure 6.3a). The water should be as hot as the milker's hand can stand for the first cow and stay at approximately the same temperature throughout the entire milking operation. A separate towel or towels should be used for each cow to prevent spreading dirt and bacteria from one cow to the next. Cloth towels can be used if they are thoroughly washed and sterilized between milkings; however, this is not practical on most dairy farms. Paper towels can be used and

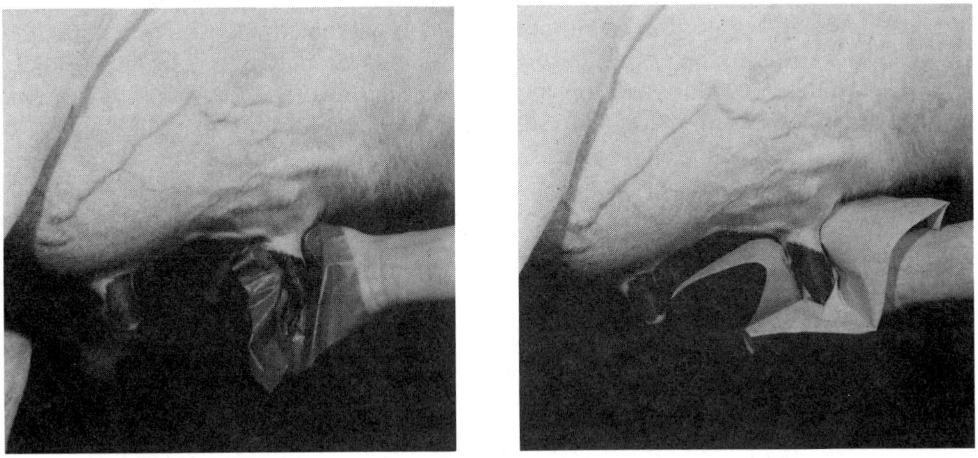

Figure 6.3 The initial steps in a good milking routine are to (a) wash the udder with individual paper towels and a sanitizing solution and (b) thoroughly drying them with individual paper towels after washing. Photographs courtesy of the Ohio Cooperative Extension Service.

thrown away after each cow at a relatively low cost, but a large number of them are needed to clean dirty cows adequately. A common sponge should not be used because sanitizing a sponge cannot be done and it spreads mastitis-causing bacteria from cow to cow. After washing the udder, the teats should be dried with individual paper towels (Figure 6.3b).

There are two basic reasons for the first step in the milking routine. First, most milking machines cause milk to splash around the teat during milking, which in turn removes the dirt from the teat and passes it into the milk. A clean teat will prevent this. Second, the washing of the udder produces a massage that is usually associated with milking and the cow uses it as a signal for milk ejection. Although a great deal of research has been conducted on the length of udder washing required for an adequate milk ejection, it appears that the consistency of time from one milking to the next is most important. This is especially true of high-producing Holstein cows and probably applies to all breeds in the United States. If 10 seconds are adequate to clean the teats and bottom of the udder and this time is consistent from one milking to the next, it will produce an adequate milk ejection. It is desirable to have the cow respond to the same stimulus at each milking. In addition, it is important to attach the milking machine to the cow within a short time after milk ejection. Since the washing operation of the cow can be performed at a specified time before machine application, the cow can be trained to respond to the washing stimulus for milk ejection. If all cows have a milk ejection in response to the feeding of the first cow or the starting of the vacuum pump, the milking machine can be applied to only two or three cows within a couple of minutes after milk ejection. The remaining cows would need a second ejection during milking for normal milk removal. Whether a cow can have a second milk ejection within minutes after the first one is not known.

Drying of the udder removes the remaining wash water from the teats and bottom of the udder and thus prevents it from contaminating the milk. Drying the teats and udder is very important when mechanical washing is done and water is still dripping from the teats when the milking machine is to be applied. The contaminated water from the udder can seep into the teat cups during milking and may lead to an increase in udder infections caused by mastitis-causing organisms that live in the environment.

Use of a strip plate. The use of a strip cup or plate (Figure 6.4) should follow the washing of the cow's udder and teats. Several streams of milk from each teat are squirted onto a fine mesh screen or black plate and examined for abnormal color, flakes, and clots. Abnormal color is easier to detect on the plate than on a screen. Detection of abnormal milk calls attention to quarters showing either mastitis or injury and can be treated. Abnormal milk can be kept out of the milk to be sold. No great loss occurs if the first two or three streams of milk from each quarter are discarded because the milk has a relatively low fat content. Also, the bacteria count of the first milk is higher than that obtained later; therefore, the bacteria count of the bulk milk can be reduced. This strip-plate milk can be fed to calves. The removal of the milk also produces additional massage of the teats and reinforces the milk ejection reflex. This is especially important if

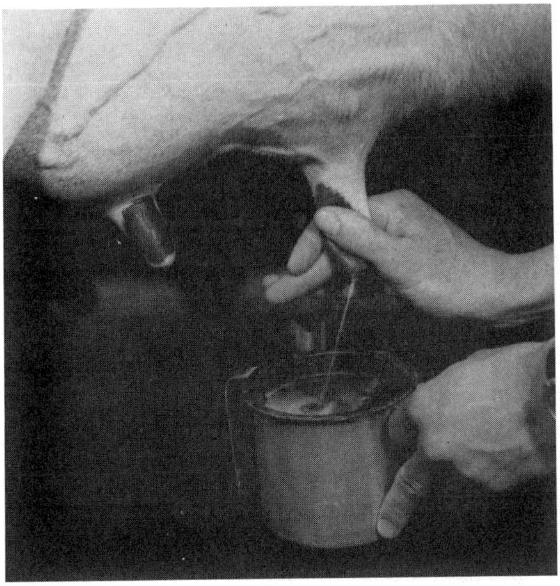

Figure 6.4 After udder washing and drying, two or three streams of milk should be removed from each teat into a strip cup or plate to detect abnormal milk. Photograph courtesy of the Ohio Cooperative Extension Service.

the udder washing time is minimal or if the cow is washed by mechanical means. The first three steps should not require more than 25 to 30 seconds per cow. Studies with high-producing Holstein cows have shown that they will have a normal milk ejection and milk yield even if they are not stimulated by udder washing; however, the milk flow rate is lower and the milking machine must be on the cow for a longer period. In this case, the application of the teat cups and their massage of the teats apparently serve as the stimuli for milk ejection.

Application of the teat cups. The fourth step is the application of the teat cups (Figure 6.5). The time interval between milk ejection and machine application should be minimal and not more than 3 to 4 minutes. An interval of less than 1 minute is desirable. Oxytocin in the blood starts to decrease after several minutes and the myoepithelial cells start to relax so that if the milk is not removed in 7 or 8 minutes, the milk will reenter the lumen of the alveoli. If 5 to 6 minutes are required to remove the milk by machine and the machine is not applied until 3 or 4 minutes after milk ejection, a decrease in milk production occurs. Also, the milking machine should be warm when it is applied to the cow's teats. Applying cold teat cups, especially in winter, could produce shock to the cow and interfere with the milk ejection reflex.

Machine stripping. Machine stripping is pulling down on the teat cup claw assembly with one hand and massaging the udder in a downward fashion with the other hand to aid the removal of milk during the end of milking (Figure 6.6). It helps to open the space between the teat and gland cisterns if teat cup crawl has closed them. Most measurements of machine stripping begin before milk flow stops completely. In controlled experiments, machine stripping after

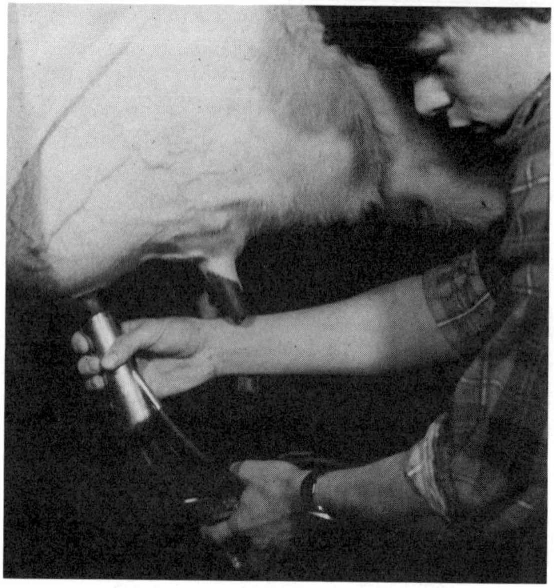

Figure 6.5 The milking machine should be applied to the cow within several minutes after the udder has been washed and stimulated for milk ejection. Photograph courtesy of the Ohio Cooperative Extension Service.

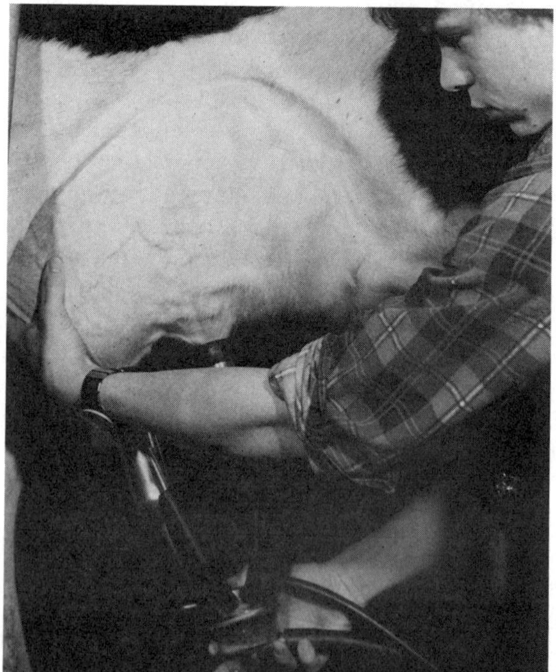

Figure 6.6 The udder should be checked for completeness of milking. If machine stripping is required, a minimum amount should be done. This is accomplished by pulling down on the claw and massaging the udder in a downward direction. Photograph courtesy of the Ohio Cooperative Extension Service.

milk flow stops completely has resulted in very little additional milk production. Also, in milking parlors with automatic takeoff units, machine stripping is no longer done. It appears that machine stripping can be eliminated without any loss of milk production or harm to the cows. In milking with bucket units or in milking parlors without automatic takeoff units, udder checking should be substituted for routine machine stripping. Machine stripping should be done only for the few cows that do not milk out completely without machine stripping or those with unequal milk out of quarters where one or two may need a little extra attention.

Elimination of machine stripping is of primary importance in attaining high labor efficiency since stripping requires from 20 to 60 seconds per cow and is usually the most time-consuming step in the milking process. Reducing this time to an average of 5 or 10 seconds per cow greatly increases the number of cows milked per hour in larger milking parlors.

Machine removal. The sixth step in the milking routine is the removal of the milking machine (Figure 6.7). The machine should be removed as soon as milk flow stops. Overmilking may play a contributing role in udder infection and clinical mastitis; however, the experimental evidence is far from convincing. Overmilking results in inefficient use of the milking machine and there is no reason to leave the machine attached to a cow when it is not removing milk and could be used to milk another cow. Overmilking is usually the result of a dairy farmer using too many units. The farmer has too many jobs to perform between the time

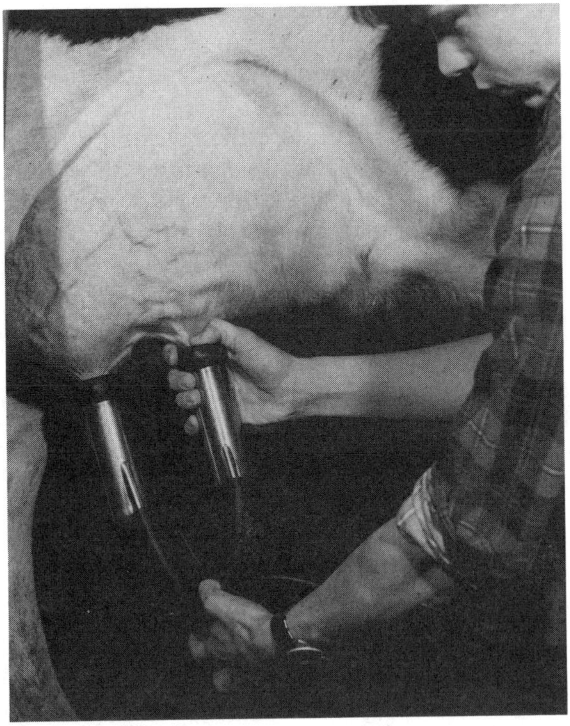

Figure 6.7 The milking machine should be removed from the cow as soon as milk flow stops. The vacuum to the teats should be turned off and air admitted to one of the teat cups before gently removing the teat cup cluster from the cow. Photograph courtesy of the Ohio Cooperative Extension Service.

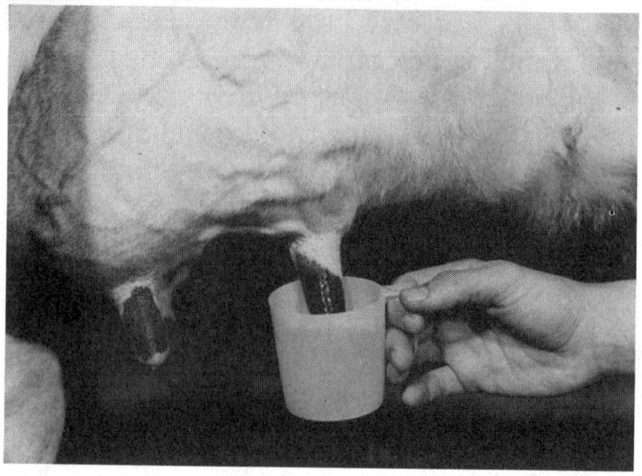

Figure 6.8 The teats of each cow should be dipped in an effective disinfectant solution after milking to remove bacteria from the ends of the teats and to disinfect them.

the machine is applied to a cow and the time it should be removed. In many cases, using one less unit may result in milking the same number of cows per hour and reducing the amount of overmilking. The number of units that can be handled depends on the type of operation, the vacuum of the milking machine, and the ability of the dairy farmer.

Teat end dipping. The final step in the milking routine is dipping the teats in a sanitizing solution after milking to reduce the number of new mastitis infections (Figure 6.8). Teat end dipping should be done as soon as possible after the machine is removed. This procedure eliminates the milk left on the teat after milking, destroys the bacteria on the teat end, and leaves a chemical film on the teat that destroys pathogenic bacteria between milkings.

The steps outlined here should not take more than 2 minutes per cow. To accomplish this, the milking routine must be well organized and the machines must be readily available when needed. If machine stripping is minimal, the routine can be reduced to $1\frac{1}{4}$ to $1\frac{1}{2}$ minutes per cow, and more cows milked per hour. These are the minimal steps required for maximum and clean milk production, and no additional steps or time is needed.

Some dairy farmers still practice hand stripping after machine stripping. Hand stripping is not recommended because cows can be adequately and completely milked by machine. Hand stripping is also very time-consuming and greatly reduces labor efficiency. In addition, the milker's hands can transfer mastitis-causing bacteria from an infected cow to a noninfected cow.

6.3 RATE OF MILKING DAIRY COWS

The factors that determine the rate of milking can be divided into two categories: the physiological factors of the cow and the mechanical factors of the milking machine. The most important physiological factor is the size of the teat opening, which is controlled by the tautness of the sphincter muscles surrounding the teat meatus, as demonstrated by Baxter et al. (1950). The milking rate of eight quarters ranged from 0.5 kg (1.1 lb) to 1.3 kg (2.9 lb) of milk per minute when milked at a vacuum of 47 kPa (15.9 in. Hg). A cannula of uniform size was then inserted into each teat and the quarters were milked again by the same machine with the same vacuum. The cannula made the diameter of the teat opening the same for all cows. The milking rates then varied between 1.2 kg (2.7 lb) and 1.4 kg (3.0 lb) per minute, clearly demonstrating that the size of the teat opening is a major determinant of the rate of milk flow.

The second physiological factor affecting milk flow rate is the amount of the milk present in the udder which increases the internal udder pressure at the beginning of milking and increases the pressure differential across the teat opening. High-producing cows have a faster rate of milk flow because of increased udder pressure, but their larger volume of milk increases the total milking time above that of lower-producing cows. The mechanical factors—vacuum level and pulsation rate and ratio—that influence rate of milk flow were discussed previously (see Section 6.1).

6.4 PRODUCTION OF HIGH-QUALITY MILK

For a successful dairy operation, not only must the dairy farmer have relatively high production per cow, but high-quality milk must be produced to ensure an immediate market for the milk as well as a long-term demand by the consuming public. Regulatory agencies use bacteria counts, sediment tests, and somatic cell counts as measures of quality; however, milk with acceptable measures of these can still contain rancid, oxidized, weedy, and other off-flavors. The consumer uses flavor as the most important indicator of desirable milk. Good-quality milk must have a low bacteria count, a low content of foreign material, a low somatic cell count, and a desirable flavor. The flavor of desirable milk is described as bland, slightly sweet, and odorless.

The production of high-quality milk requires clean, healthy, properly fed cows; clean, dry, and well-ventilated barns; and clean, sanitized, smooth, copper-free utensils. Milk should be quickly cooled and constantly protected after it comes from the cow. Healthy cows are free from tuberculosis, brucellosis, mastitis, and any other disease or condition that prevents normal milk production. Cows having tuberculosis or brucellosis should be sold immediately and steps taken to control the spread of these diseases in the herd. Abnormal milk resulting from mastitis should be discarded and not added to the bulk milk supply.

The cows, especially the udders, should be clean. Maintenance of clean cows requires a clean barn and barn lots. The barn should also be well ventilated, for some flavor-producing odors can be inhaled through the lungs, diffuse into the blood, and pass into the milk. Flavor-producing feeds should be fed after milking so that the flavors do not appear in the milk.

The production of high-quality milk also requires the use of smooth utensils that are clean and sanitized to minimize the presence of bacteria in the milk. Milk should not come in contact with copper or iron because these metals can induce oxidized flavors in milk. Dairy equipment should be installed to minimize air admission and agitation of milk, both of which can induce a rancid flavor in milk.

Milk should be cooled to 40°F within 2 hours after the completion of milking to maintain the flavor and extend the keeping quality. This temperature keeps bacterial growth at a minimum. The milk should be constantly protected against bacterial contamination, dust, dirt, extreme hot or cold temperatures, daylight, or a strong artificial light, all of which tend to decrease milk quality.

Undesirable Flavors in Milk

Among the flavors that impair the quality of milk from a taste standpoint are oxidized, rancid, feed and weed, unclean, malty, and salty flavors. The oxidized flavor is also described as cardboardy, tallowy, or metallic. The usual causes of this flavor are exposure of milk to copper or iron, poor nutrition of the cow, or excessive exposure of milk to daylight or artificial light. For this reason, all milk in transparent pipelines, weigh jars, and receiver jars should be protected from strong light. Because copper and iron increase the incidence of oxidized flavor, stainless steel, glass, plastic, or rubber should be used for all milk-handling equipment. Water used for cleaning equipment should contain minimum amounts of iron and copper. The oxidized flavor is more prevalent during the winter because the green feeds consumed in summer supply substances (mainly, vitamin E) that delay the oxidation of milk.

Rancid flavor is another common off-flavor and is probably the most objectionable one found in milk. It is also described as bitter or soapy. The flavor is more prevalent in milk obtained from cows in late lactation and from low-producing cows. It is also found to a large extent in milk that is handled in pipelines that contain risers and in milk that passes through pipeline milkers that contain excessive air leaks. Some cows are more likely than others to produce milk with a rancid flavor. Milk pumps that continue to run without milk contribute to the off-flavor.

Rancidity is induced by the enzyme, lipase, which breaks down milk fats. This enzyme is present in milk, but fats of normal milk are protected by a membane surrounding the fat globules. Excessive agitation breaks the membrane and exposes the fats to lipase. Pasteurization destroys the enzyme; therefore, no problem occurs after pasteurization unless raw milk is added to the pasteurized milk. To avoid this off-flavor, excessive agitation of milk should be avoided. Milk lines should not contain risers and all leaky fittings and running pumps in a "starved" condition should be avoided. Cows near the end of lactation should be watched carefully. Milk from those producing less than 4.5 kg (10 lb) per day or from cows less than 6 weeks before parturition should not be used. Adding warm milk to cold milk also contributes to the problem if the milk rises to 80°F before cooling. The blend temperature should be kept below 50°F if at all possible.

Feed flavors are caused by sudden changes in feed, grass or corn silage, and such vegetables as cabbage and turnips. Common weed flavors are caused by the consumption of garlic, wild onion, and some other weeds. These flavors can arise if the cows eat these feeds or weeds or breathe the odor from them.

The primary causes of feed and weed flavors are the feeding of silage before milking and the inhalation of strong barny or feed odors by the cow. The best methods of preventing these flavors are to withhold the causative feeds from the ration of the cow 2 to 4 hours prior to milking and to provide adequate barn ventilation. If the flavor-producing weeds and vegetables cannot be controlled, the cow should be kept off the pasture entirely. Sudden changes in the ration should be avoided and cows should be gradually shifted from one feed to another.

The unclean flavor is also described as a "cowy" or a "barny" flavor. These flavors are highly undesirable to the consumer. The primary cause of the unclean flavor is inadequate barn ventilation. Cows inhale the odors of damp, dirty barns and impart this flavor to the milk. The best ways of preventing this flavor are to avoid these barn conditions and to wash the udder adequately prior to milking. The milking equipment should be clean and dry and sanitized before use. Ketosis also produces an unclean flavor as a result of the large amount of ketones in the milk.

Malty or acid flavor in milk is also described as a Grapenuts-like flavor or a sour flavor. This flavor is caused either by unclean milk-handling equipment or by poor cooling of the milk and occurs when a large number of bacteria are present in milk. To avoid this flavor, milking equipment should be clean and sanitized before use and milk should be cooled to 40°F within 2 hours after milking.

Another flavor is the salty flavor that occurs in milk from cows in late lactation or cows with mastitis infections. Flat milk is tasteless as a result of low levels of fat or nonfat solids. Other flavors may be caused by the accidental addition of medication, insecticides, or disinfectants to the milk. Extreme care should be taken to avoid such additions.

REFERENCES AND FURTHER READINGS

Baxter, E., et al. 1950. Factors affecting the rate of machine milking. *J. Dairy Res.*, 17:117.

Goff, K. R., and G. H. Schmidt. 1967. Effect of eliminating machine stripping of dairy cows on milk production, residual milk, and mastitis. *J. Dairy Sci.*, 50:1787.

Gorewit, R. C., and K. B. Gassman. 1985. Effects of duration of udder stimulation on milking dynamics and oxytocin release. *J. Dairy Sci.*, 68:1818.

Heald, C. W. 1985. Milk collection. In *Lactation*, B. L. Larson, ed. Iowa State University Press, Ames, IA.

Mahle, D. E., et al. 1982. Effects of vacuum and pulsation ratio on udder health. *J. Dairy Sci.*, 65:1252.

Milking Machine Manufacturers Council of the Farm and Industrial Equipment Institute. *The Modern Way to Efficient Milking* (Bulletin).

Natzke, R. P., et al. 1982. Effect of overmilking on udder health. *J. Dairy Sci.*, 65:117.

Schmidt, G. H. 1971. *Biology of Lactation*. W. H. Freeman and Company, Publishers, San Francisco.

Schmidt, G. H., et al. 1963. Effect of teat cup liner diameter and mouthpiece on the milking rate, machine stripping, and mastitis of dairy cows. *J. Dairy Sci.*, 46:1064.

Schmidt, G. H., et al. 1964. Relationship of milking times, procedures, and installations to level of milk production and incidence of mastitis. *Cornell Agric. Exp. Stn. Bull.* 996.

Thiel, C. C., and F. H. Dodd, eds. 1977. *Machine Milking*. National Institute for Research in Dairying, Shinfield, Reading, England.

Thompson, P. D. 1981. Milking Machine—The past twenty-five years. *J. Dairy Sci.*, 64:1344.

Thompson, P. D., and R. E. Pearson. 1983. Milk droplet impacts during induced vacuum fluctuations: Influence of claw and other characteristics. *J. Dairy Sci.*, 66:562.

Thompson, P. D., et al. 1983. Milking cows with positive pressure stimulation in late lactation. *J. Dairy Sci.*, 66:1167.

Zinn, S. A., et al. 1982. Milking responses of cows given premilking stimulation at four lactational stages. *J. Dairy Sci.*, 65:668.

7

Milking Systems

Although some changes have occurred in the ways that cows are housed and milked, the majority of cows in the United States are still milked in stanchion barns, but an increasing number are being milked in milking parlors. Most cows milked in milking parlors are housed in free-stall barns. The use of milking parlors has produced advantages to the dairy farmer in terms of greater labor efficiency and less effort and drudgery in milking cows. On the other hand, the milking parlor has created some problems, the major one being the inability of high-producing cows to consume sufficient grain in the parlor. In addition, some dairy farmers feel that they cannot give as much individual attention to cows in a milking parlor as they can to cows milked in a stanchion barn.

Regardless of the type of milking system used, it should meet the following minimum requirements: (1) allow for the complete milking of cows without teat or udder damage, (2) provide for the production of clean milk, (3) provide for comfort of the person in the parlor, and (4) have equipment that can be cleaned thoroughly and quickly with a minimum of effort. If cows are fed in the parlor, sufficient time should be available for high-producing cows to consume adequate grain.

7.1 COMPARISON OF MILKING SYSTEMS

Cows are milked either in their regular stalls or in a milking parlor. Milking cows in a stall barn (stanchion barn) can be accomplished in one of three ways: (1) milking in regular buckets and hand carrying the milk to the bulk tank, (2) milking in regular buckets and dumping the milk into a dumping or transfer station, and (3) milking directly into a pipeline.

Bucket System

The bucket system has several advantages: (1) it is an inexpensive system that requires only milking units and a vacuum line or portable vacuum source, (2) it does not require movement of cows from their stalls, (3) milk can be strained into cans in the milk room without the necessity of a bulk tank (this is not very common today), (4) cows can eat all the grain that they require during milking, and (5) milk production from each cow can be checked at each milking.

The disadvantages of the bucket milking system are: (1) it requires a large amount of physical effort, since cows are on the same level as the milker, which requires a lot of stooping; (2) the milk has to be transported to the milk room or dumped into a transfer system; and (3) it requires full use of the barn for milking during the pasture season.

Dumping or Transfer System

The dumping or transfer system has the advantages of the bucket system and in addition milk does not have to be hand carried to the milk house. Also, it is cheaper than a pipeline system and the production of each cow can be checked at each milking.

Its disadvantages are: (1) milking still requires a large amount of physical effort, (2) the herd size cannot be increased without additional stalls, (3) the barn has to be used during the pasture season, and (4) equipment washing time is increased, although circulation washing is possible.

Around-the-Barn Pipeline

Advantages of this system are: (1) lifting of milk by the dairy farmer is eliminated, (2) movement of cows is not required, (3) adequate grain intake is possible, and (4) an additional building for a milking parlor is not required.

Disadvantages of the system are: (1) one-level milking and consequently high physical effort is still required; (2) investment and annual costs are higher than with the bucket system; (3) water and sanitizer requirements are high; (4) some barns have ceilings that are too low for proper installation, which can result in rancid milk if pipeline risers are used; (5) daily measurements of milk weights are more difficult; (6) a high lift from the cow's udder to the pipeline can result in vacuum drops at the teats; (7) it requires a larger vacuum pump and larger-diameter vacuum lines than other systems; and (8) the barn must still be used for milking during the pasture season.

Milking Parlor

The advantages of the milking parlor are: (1) it requires the least amount of physical effort to milk cows; (2) more flexibility in herd size since more cows can be milked without increasing the building and equipment costs; (3) cleaner milk can be produced since no bedding is under the cows during milking; (4) shorter vacuum and milk lines impose fewer vacuum and flavor problems; (5) it is the most labor-efficient system since more cows can be milked per person per hour than any other system; and (6) the barn is not needed to milk cows during the pasture season.

Disadvantages of the milking parlors include: (1) initial cost is high because of the extra building and equipment needed, (2) the additional space of the milking parlor and holding area must be cleaned, (3) slow-milking cows and dry cows are difficult to handle in some types of parlors, and (4) inadequate grain intake of cows in most milking parlors.

7.2 FACTORS TO CONSIDER IN CONSTRUCTING A MILKING PARLOR

Height of Cows above Milker

One major advantage of the milking parlor is that it eliminates stooping during milking because the cow's udder is at the level of the milker's hands. The optimal height depends on the height of the milker and the type of parlor. In a parlor where the cow has more freedom to move around, the cows need to be a little lower than in a herringbone parlor, where the udder is close to the milker's hands. Tall milkers need parlors where the cows are slightly higher than the level required for shorter milkers. For most persons, the cows should be 32 to 36 in. above the floor of the milker.

One-Sided versus Two-Sided Parlors

The distance that the milker walks to milk each cow is less in a two-sided parlor than in a one-sided parlor. It is also easier to use one milking unit for two milker stalls in a two-sided parlor than in a one-sided parlor, especially where pipelines are located above the height of the udders.

Number of Stalls per Unit

Milking parlors usually have either one or two stalls per milking unit; occasionally, parlors have one more stall than units, such as four stalls for three units. This affects labor efficiency and the initial cost of the parlor. With one stall per milking unit, the unit is idle while being removed from the first cow, while that cow leaves the parlor, and while a second cow enters and is prepared for milking. When two stalls are present per milking unit, the unit can be switched directly from the first to the second cow; consequently, there is a marked reduction in time that the milking machine is idle. Because of this, more cows can be milked per unit per hour with two stalls per unit than with one stall per unit. Another advantage of two stalls per milking unit is that the cow remains in the parlor while she is being milked plus the time required to milk another cow. This is an important consideration in allowing additional time for cows to consume grain in parlors. Most parlors being constructed today, however, have a milking unit at each stall. With low-level milk lines, it is less cumbersome to have the hoses located at each stall instead of being moved across the parlor. Also, with automatic takeoff units, it is much easier to use one milking unit at each stall than trying to switch the claw and takeoff from one stall to the next.

Feeding Grain in the Parlor

The amount of grain that a cow can consume in the parlor depends on her appetite, milk yield, amount of time in the parlor, and her previous training. In general, cows consume between 0.3 and 0.34 kg ($\frac{2}{3}$ and $\frac{3}{4}$ lb) of air-dry grain per minute. Adding water to the grain can increase the consumption rate to 0.45 to 0.68 kg (1 to $1\frac{1}{2}$ lb) per minute. Pelleting increases rate of grain consumption by 20 to 30%. Coarse grains are also consumed at a slightly faster rate than are finer grains.

Under average conditions, a cow milked in a properly designed parlor can consume between 9.1 and 10.9 kg (20 to 24 lb) of grain per day in the parlor, assuming that she can eat 0.45 kg per minute and stays in the parlor 10 to 12 minutes per milking twice daily. This requires a milking parlor with one milking unit for two stalls and a milking time between 5 and 6 minutes. A higher intake would require either three-times-a-day milking or feeding grain outside the parlor.

Because of inadequate grain intake by high-producing cows in most parlors, dairy farmers are moving away from grain feeding in the parlor. Since additional grain must be fed outside the parlor anyway, these farmers plan to feed all grain outside the parlor, either in a total mixed ration or with a computer feeder. Elimination of grain feeding in the parlor also reduces the grain dust and the amount of urination and defecation of cows in the parlor. In addition, less cost is involved in the initial construction as well as maintenance of the parlor. Without grain feeding in the parlor, cows are more hestitant to enter the parlor.

Production Testing

Measurement of milk yield and determination of milk composition from individual cows are requirements for cows on production testing programs, such as the Dairy Herd Improvement (DHI) program. It is also desirable for herds not on a testing program to have production records from a management standpoint. Parlors should accommodate the measurement of these without much additional effort on test days. Several alternative methods are available to accomplish this.

A weigh jar (Figure 7.1) is used in a number of parlors. The determination of milk yield is on a volumetric basis and a sample of milk is obtained from the bottom of the jar at the end of milking for milk composition determination. Weigh jars can also be used for making daily observations of milk yield. They have the added advantage of producing a more stable vacuum on the teat end, especially if the milk is elevated to a high milk line. A further advantage of weigh jars is that abnormal milk can be diverted from the bulk milk supply after a cow is milked. Before weigh jars are purchased, contact should be made with the local DHI supervisor to make sure that the weigh jar has been approved for official use.

A number of milk metering devices have been approved for use with cows on the DHI testing program. Some of these are based on a volumetric determination and others on a weight basis. All have a device for taking a representative sample of milk for milk fat and protein determinations as well as somatic cell counts. Again, the DHI supervisor should be contacted before a metering device is purchased to make sure that it has been approved for official use.

A third alternative is the use of regular milking buckets on test day or when a milk weight is desired. This is not a practical method on most dairy farms.

Figure 7.1 A weigh jar to measure the milk yield from individual cows is one of several methods approved for the DHI testing program. Photograph courtesy of Donald Pritchard. The Ohio State University.

Entrances and Exits

Cows should have a straight-line entrance to the parlor from the holding area to minimize the time required for cows to enter the parlor. This is also desirable when cows leave the parlor, but it is rather difficult to accomplish with most milking parlors. If turns are required when cows enter or leave the parlor, it is better to have them in the exits. Entrance and exit gates should be power operated instead of manually operated to minimize the time and effort of the milker. For larger parlors, it is also desirable to have two return lanes instead of one. Also with one return lane, one group of cows has to cross over in front of the parlor in order to leave.

Holding Area

To provide a straight entrance into the parlor, the holding area must be next to the cows' entrance to the parlor. Cows move into the parlor more readily if the holding area is part of the milking parlor. An overhead garage door can be used to separate the parlor from the holding area between milkings. The holding area should be large enough to hold the number of cows from a housing unit or corral. Approximately 15 square feet are required per cow. The number of cows in a group should be related to the throughput of the parlor; for example, a double-8 herringbone should hold a number that is a multiple of 8. Cows should not remain in the holding area more than 2 hours at any milking and no more than 1 hour during hot weather. If the holding area is used to sort cows for breeding, treatment, or changing of production groups, an area adjacent to the holding pen must be provided.

Stray Voltage

Some dairy farmers experience health problems and reduced milk production resulting from "stray" voltage. This results when a voltage of undesirable intensity occurs between two animal points, usually in the milking parlor. The causes and sources of stray voltage are varied and complex, but increasing problems are expected with increased size of dairy operations, deteriorating farm electrical wiring systems, and increased electrical loads on rural wiring systems. When a problem occurs, the farmer should contact the local electrical power company for help. The company should be consulted for electrical recommendations before a new parlor is constructed.

7.3 EVALUATION OF MILKING SYSTEMS

Performance in milking parlors can be evaluated in a number of ways. Pounds of milk harvested per hour is probably the best measure, but since dairy farmers have a certain number of cows to milk at each milking, they are more interested in the number of cows that can be milked per person per hour. This is the measure that will be used.

The three major factors that influence labor efficiency are (1) routine time per cow, (2) unit time per cow, and (3) number of milking units. These factors are used primarily to evaluate milking parlors; however, the principles can be used for stall barns as well.

Routine Time

Routine time is the amount of time spent with each individual cow regardless of the number of cows or milking units. Components of the routine time include: time for the cow to enter and leave the parlor, time required to feed the cow, udder preparation time, time for applying and removing the milking machine, machine stripping time, time to dip teat ends, and time for other jobs performed on each cow, such as recording milk yield and dipping teat cup clusters between cows. An additional 15 to 20% of time should be added to the routine time to allow for emergency jobs, such as replacing kicked-off units, treating cows, and encouraging hesitant cows to enter the parlor. This is planned idle time. The routine time plus the planned idle time is defined as the modified routine time (MRT). The minimum time to carry out a good milking routine without automation is $1\frac{1}{4}$ minutes per cow. This time does not include machine stripping. With it, the MRT would be about 2 minutes per cow. If the average MRT is 1.25 minutes, only 48 cows can be milked per hour, regardless of the number of units (60 minutes divided by 1.25 = 48). If MRT averages 2 minutes, only 30 cows can be milked per hour (60/2 = 30).

Unit Time

The unit time per cow (UT) is the sum of the time that the machine is on each cow (machine on-time) and the time that the machine is idle between cows (machine idle time). The machine on-time is dependent on the milk yield of the cow, the inherited milking rate of the cow, and machine factors such as vacuum level and pulsation ratio. In most herds this should be between 5 and 6 minutes per cow, but with high-producing cows, it could be higher.

The machine idle time is largely dependent on the number of stalls per milking unit. It also depends on the operations performed on the unit between cows. If two stalls are present for each milking unit, the unit can be switched directly from one cow to the next, whereas with one milker stall per unit, the unit has to remain idle during the time that the first cow leaves the parlor and the second one enters the stall and is prepared for milking. With two stalls per milking unit, the machine idle time can be as low as 30 seconds, and with one stall per unit, the machine idle time can be 2 to 3 minutes. Emptying the weigh jar and dipping and sterilizing the teat cup clusters after each cow also increase the machine idle time.

The number of cows that can be milked by a milking unit per hour is equal to 60 (minutes per hour) divided by UT. If UT = 6, 10 cows can be milked per unit per hour (60/6 = 10). If UT = 8, only an average of 7.5 cows can be milked per unit per hour (60/8 = 7.5).

Units per Person

The number of units per person (N) is determined primarily by the size and type of milking parlor. If the parlor was designed as a double-4 herringbone parlor, the dairy farmer will use four or eight units. In designing a new parlor or evaluating the efficiency of a current parlor, the best estimate of the number of units that a person can handle is the machine on-time divided by the routine time. This is based on the assumption that the person should be available to remove the unit when the cow is milked out. This becomes less critical when automatic takeoff units are used.

Combination of Factors

The performance of a parlor (P) in terms of cows milked per hour can be estimated from the MRT, the UT, and the number of units (N). Two formulas are involved: (1) $P = 60/\text{MRT}$ and (2) $P = 60/\text{UT} \times N$. If one factor in one formula changes to reduce or increase P, a compensatory change in another component of the second formula also takes place. For example, assume that UT = 6 and $N = 3$. The performance in formula (2) becomes $60/6 \times 3 = 30$. If MRT = 2, then P from formula (1) stays the same ($60/2 = 30$); therefore, 30 cows can be milked per hour. If the MRT were 3, only 20 cows could be milked per hour ($60/3 = 20$). In this case P in formula (2) would also be 20, which would mean that the UT was increased to about 9 ($20 = 60/9 \times 3$). In this case, either the milking machines were left on the cows or the machines were idle for longer periods of time.

Another type of imbalance occurs if the farmer uses a milking parlor with three units, has a UT of 6 minutes, and has an MRT of 1.5 minutes. Only 30 cows can be milked per hour because of the limitation of formula (2) ($60/6 \times 3 = 30$); whereas formula (1) would allow the milking of 40 cows per hour ($40 = 60/1.5$). In this case the milker would be idle waiting for the machines to complete milking, thus increasing the MRT to 2 minutes per cow. If four units are used instead of three, formula (2) would allow for 40 cows per hour ($40 = 60/6 \times 4$), which would balance with the MRT of 1.5 minutes.

The number of cows milked per milking unit is fixed since the average machine on-time cannot be changed very much. To construct parlors for high performance, attention must be given to the number of units and the MRT. The number of units per person can be increased with little difficulty when the parlor is constructed. The units must be located so that the travel distance between cows is minimized; otherwise, the MRT is markedly increased. The MRT can be decreased by automation and elimination of some steps, such as machine stripping. Automatic udder washing eliminates from 15 to 30 seconds per cow. To milk 60 cows per person per hour would require a MRT of 1 minute or less per cow, which requires a high degree of automation. To milk 90 cows per person per hour would require an MRT of $\frac{2}{3}$ minute or less per cow.

7.4 MILKING PARLOR TYPES

Milking parlors vary in the number of stalls per unit, the number of milking units, the arrangement of cows in the parlor, and the way cows enter and leave the parlor. Even though there are many types of parlors, only the more common ones will be discussed. They are the side-opening parlor, the herringbone parlor, the polygon parlor, and the trigon parlor. Others that are still being used are the abreast type, the chute type, and circular parlors. Circular parlors came onto the scene in the 1960s; however, they did not live up to early expectations and few, if any, new ones are being constructed. Only the first four will be discussed.

Side-Opening Parlors

The side-opening parlor is so named because each stall has two side opening gates—one for the cow to enter and one to exit. Usually two, three, or four stalls are lined up on one side of the milker alley (Figure 7.2). The parlors may have one side or two sides. The two-sided parlor is more labor efficient since the milker has to walk a shorter distance between cows. The side-opening parlor has the major advantage of treating cows as individuals. A slow-milking cow does not hold up the rest of the group and at the same time she can be left in the parlor for more grain consumption. Because cows are handled as individuals, more time is spent in letting cows in and turning them out in comparison to parlors where cows are handled as a group. The side-opening parlor has the lowest labor efficiency in comparison to the other parlors to be discussed. Expected average performance values are shown in Table 7.1.

Herringbone Parlors

Herringbone parlors were introduced into the United States in the 1950s and the double-4 herringbone has become the most popular parlor for medium-sized dairy farms. Cows are located in the parlor at a 30° angle to the milker's pit (Figure 7.3). Usually only the part of the cow behind the hipbones is accessible to the milker. Cows are closer together, the udders are closer, and considerably less travel distance is required per cow in comparison to the side-opening parlor. Additional advantages are: cows are let in and out as a group, thus saving MRT time; construction equipment is cheaper; first lactation cows are easier to train; and labor efficiency is highest (Table 7.1). Some disadvantages include: different-sized cows present a problem, identification of cows is more difficult, slow-milking cows hold up the group, and dry cows must be handled separately if labor efficiency is to be maintained. The time required for cows to enter and leave the parlor can be excessive in the large herringbone parlors, where 8 to 12 cows are handled at a time. Double-4 and double-6 herringbone parlors are most common for medium-sized farms, and double-8, double-10, and double-12 parlors are used for the larger herds (Table 7.2).

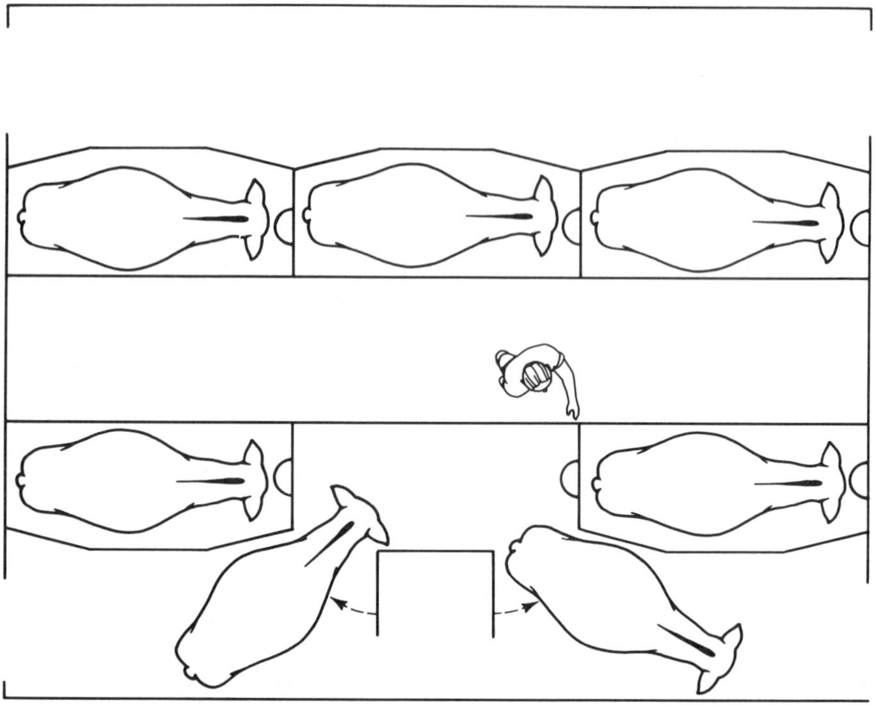

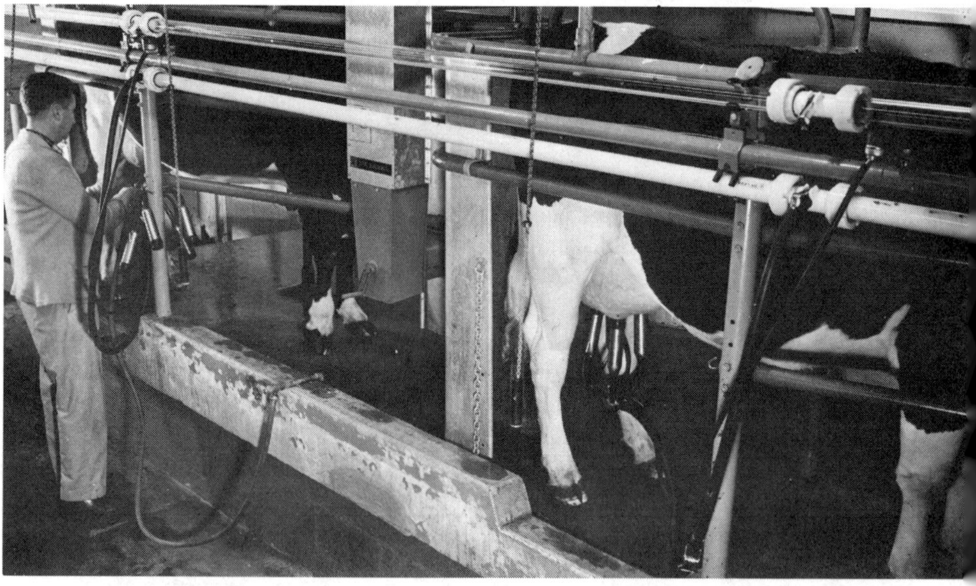

Figure 7.2 Side-opening milking parlor: (a) arrangement of cows in the parlor; (b) photograph of cows in the parlor.

TABLE 7.1 STEADY-STATE THROUGHPUTS (COWS/HOUR) FOR MILKING PARLORS

Parlor type	Size	With crowd gate	With crowd gate, feed bowl covers, detachers
Herringbone	Double-4	34–47 (1)[a]	39–52 (1)[a]
	Double-6	55–71 (2)	57–73 (1)
	Double-8	69–87 (2)	70–88 (1)
	Double-10	88–97 (2)	83–92 (1)
Trigon	12	59–79 (2)	59–78 (1)
	16	74–96 (2)	73–95 (1)
	18	83–103 (2)	78–98 (1)
Polygon	16	78–104 (2)	81–107 (1)
	24	110–136 (2)	96–122 (1)
	32	131–167 (3)	137–173 (2)
Side-opening	Double-2	28–38 (1)	40–50 (1)
	Double-3	52–65 (2)	50–63 (1)

[a] Number of persons in parlor.

Source: W. G. Bickert, 1983, *Am. Soc. Agric. Eng. Conf.*

TABLE 7.2 COWS MILKED WITH VARIOUS MILKING PARLORS AND MILKING TIMES

Parlor type	Cows/hour	Milking time[a] 3 hours	4 hours	6 hours
Herringbone				
Double-4	40	80	110	190
Double-6	65	114	179	293
Double-10	87	152	218	370
Trigon				
12 stalls	68	119	187	289
16 stalls	84	147	210	357
18 stalls	88	154	220	374
Polygon				
16 stalls	94	165	235	400
24 stalls	109	191	273	436
32 stalls[b]	155	233	349	581
Side-opening				
Double-2	45	90	124	203
Double-3	56	112	154	252

[a] Time available at each milking including 1 hour of setup and cleanup time and 15 minutes of change time for each 100 cows

[b] Two persons in parlor in 32-stall polygon parlor.

Source: Adapted from W. G. Bickert, 1983, *Am. Soc. Agric. Eng. Conf.*

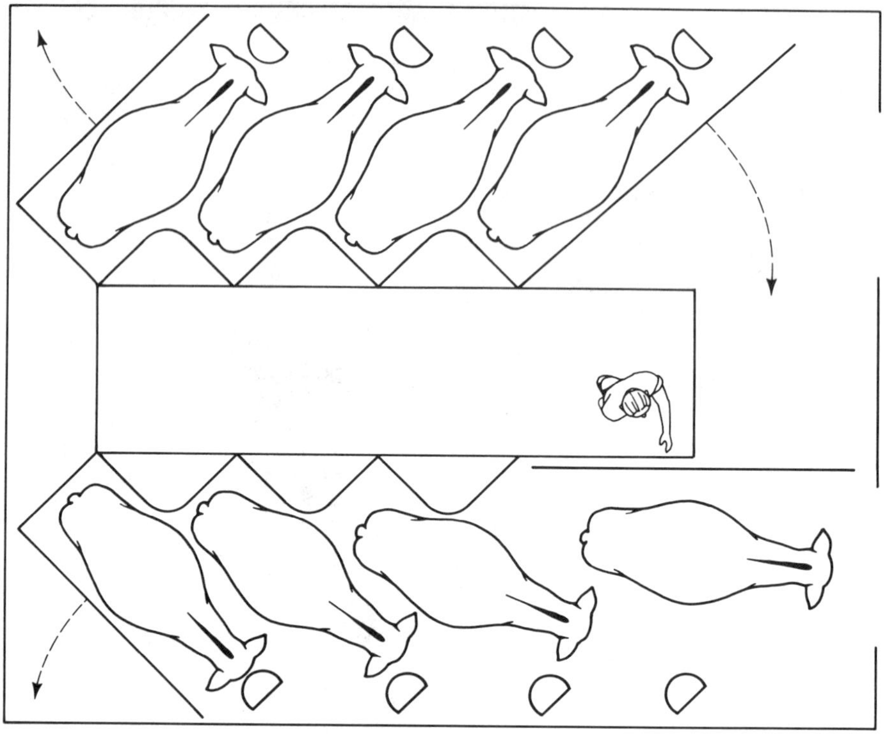

Figure 7.3 Herringbone milking parlor: (a) arrangement of cows in a double-4 herringbone parlor; (b) photograph of cows on one side of a large herringbone parlor. Photo courtesy of Alfa-Laval Agri, Inc., Kansas City, MO.

Polygon Parlors

The polygon parlor was developed at Michigan State University and consists of a four-sided milking parlor with herringbone stalls. Polygon parlors have four entrances and exits (Figure 7.4) instead of two each for herringbone parlors. Polygon parlors are most convenient from the standpoint of rapid cow movement and construction costs per milking unit. They are also best for utilizing labor-saving equipment, especially automatic detaching units. Parlors with 4, 5, 6, 8, and 10 stalls per side have been constructed. Expected performance values are shown in Table 7.2.

Trigon Parlors

Trigon parlors are similar to polygon parlors except that they are three-sided instead of having four sides (Figure 7.5). The first one was constructed in Washington in 1977. Trigon parlors with 12, 16, and 18 stalls are the most common. They are best suited for herds with 250 to 500 cows. They have most of the advantages of the polygon parlors and are usually less expensive to construct. Major advantages over herringbone parlors are that slow-milking cows are less apt to hold up as many cows because the trigon parlor has fewer stalls per side. Also, less walking distance per cow is required and the milker can observe problems more easily than in herringbone parlors. Expected performance values are shown in Table 7.2

7.5 AUTOMATION OF MILKING PARLORS

Since the 1960s, considerable attention has been directed toward the manufacture of milking equipment and the construction of milking parlors to increase labor efficiency. Most of the effort has been directed toward installing mechanical equipment to reduce the time required to be spent with individual cows, thus reducing the MRT for each cow. Those that will be discussed are: crowd gates for the holding area, automatic udder washing, automatic takeoff units, and backflushing equipment for mastitis control.

Crowd Gates

Many types of crowd gates are on the market. Some are advanced manually and others advance automatically. Most crowd gates are made of metal and advance toward the milking parlor as the number of cows in the holding area is reduced. The main purpose of the crowd gate is to encourage the cows to enter the parlor without the milker having to chase them. In most herds, the first two-thirds of the cows enter the parlor without encouragement, but the last one-third need some. This is especially true if no grain is fed in the parlor, and is more common with low-yielding cows than with high producers. The gate should provide a gentle

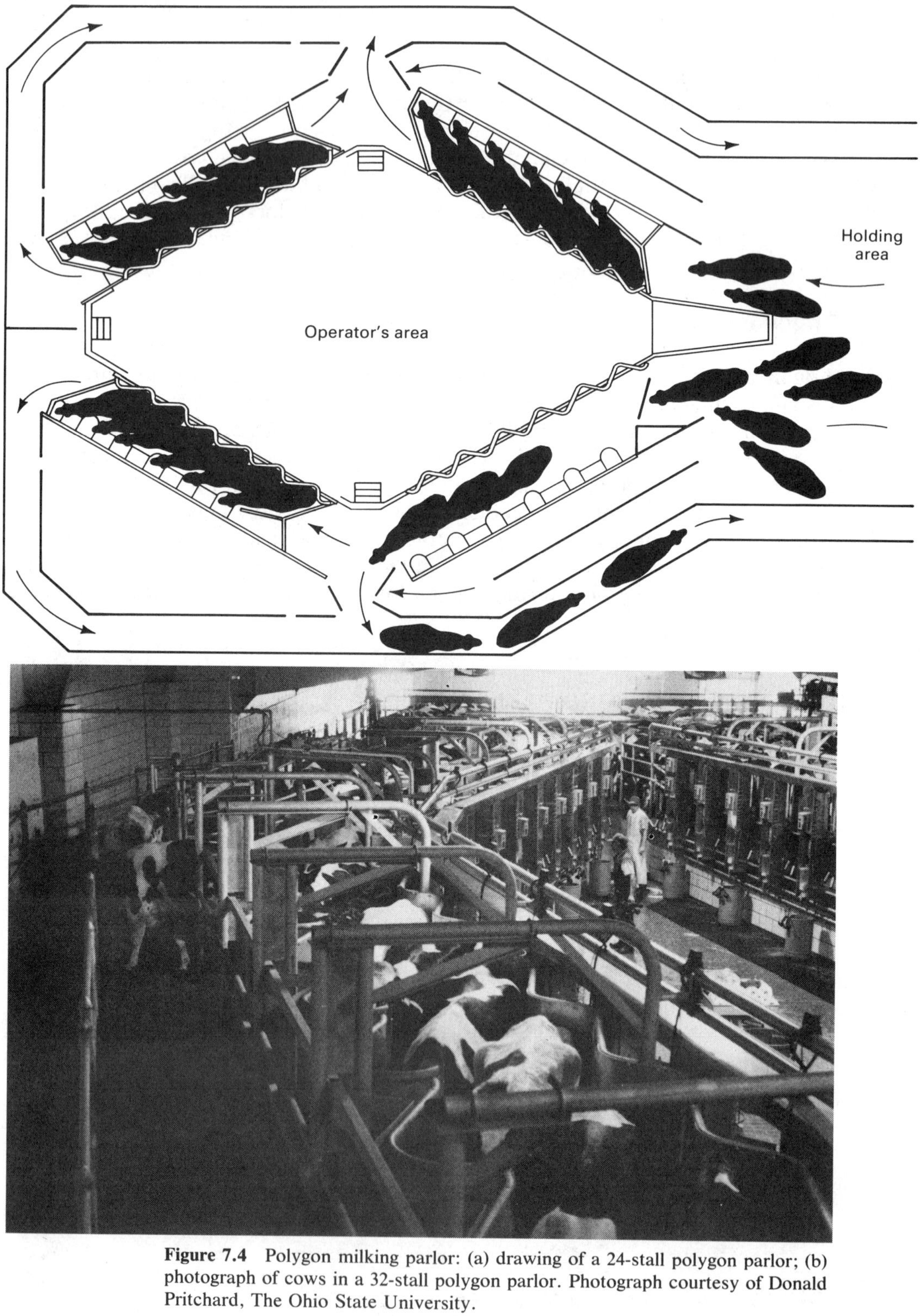

Figure 7.4 Polygon milking parlor: (a) drawing of a 24-stall polygon parlor; (b) photograph of cows in a 32-stall polygon parlor. Photograph courtesy of Donald Pritchard, The Ohio State University.

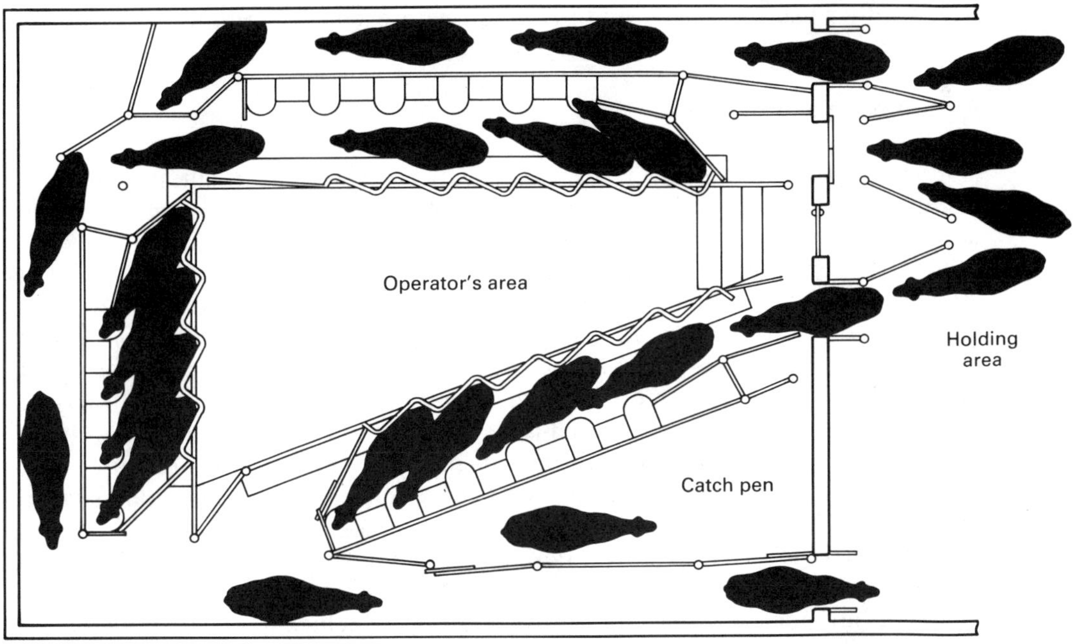

Figure 7.5 Arrangement of cows in a 16-stall trigon milking parlor.

nudge to the cows instead of harsh treatment to prevent the cows from becoming afraid of it.

Automatic Udder Washing

The two most time-consuming jobs in the milking routine are udder washing and machine stripping. Udder washing takes from 15 to 30 seconds per cow in most routines. Automatic udder washing machines have been developed for use outside the parlor, thus eliminating this step in the routine. These are common in the warmer climates, but cannot be used during the winter in the northern parts of the United States. Most automatic udder washing is done in "prep" stalls located in the milking parlor. The number of prep stalls and time interval between udder washing and application of the milking units are critical for high labor efficiency and good milking procedures. These have to be worked out for each system. Another possibility for nearly automated udder washing is to install spray nozzles underneath the cows in the milking stalls. All cows on a side are washed at the same time in order to reduce the milker's time spent on each cow.

Regardless of where the udders are washed, the bottom of the udder and teats should be dried with paper towels. If the udder or body remains wet and contaminated water enters the teat cups during milking, an increase in coliform mastitis infections can occur. In-parlor automatic wash stalls are not viewed with much favor today because of the time required to dry the udder. Most milkers would rather use a spray or wash solution with each cow than to deal with the rather wet cow resulting from automatic udder washing.

Automatic Takeoff Units

Automatic takeoff units (automatic detachers, Figure 7.6) eliminate the necessity of the milker to be present when the cow is milked out. They also eliminate the necessity of machine stripping—one of the most time-consuming jobs in the milking routine. From limited research results it appears that high production can be attained without machine stripping and many dairy farmers have used the automatic takeoff units with good success with high-producing cows. Even though the milker need not be present when the cows are milked out, the milker still needs to dip the teats in an effective teat dip after milking.

Different methods can be used to activate the automatic detachers: (1) milk flow rate sensors, (2) milk temperature sensors, and (3) fixed-time milking. Fixed-time milking is not acceptable to most dairy farmers, and milk temperature sensors are influenced by temperature variations due to season and time of day; therefore, most manufacturers use milk flow rate to determine the endpoint of milking. Most milk flow sensors are set to detect a given flow rate, such as 0.45 kg per minute, at which time the detacher unit is activated. Most detachers have an adjustable time delay between sensing a given rate of flow and activating the detacher.

Automatic detachers should have the following characteristics: (1) vacuum to the teat should be turned off for 2 to 4 seconds before the milker unit is removed; (2) the unit should have good maneuverability with the claw, to follow cow move-

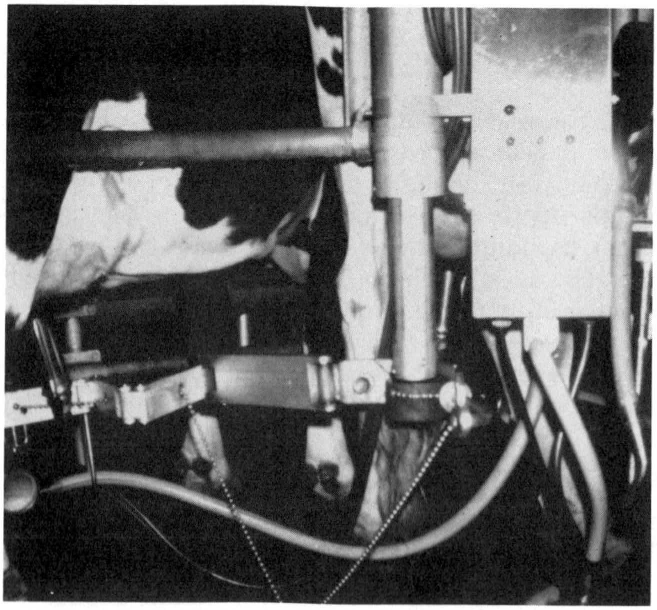

Figure 7.6 Automatic milker detaching units can increase the labor efficiency in larger milking parlors. Courtesy of Donald Pritchard, The Ohio State University.

ment during milking; (3) the milking machine should retract from underneath the cow after the end of milking; and (4) the unit should use low-voltage electricity.

Automatic Backflushing

Since the milking machine cluster can transfer mastitis-causing bacteria from one cow to the next, manufacturers have developed automatic flushing of the teat cup cluster after each cow. The teat cups and claw are first flushed with warm water, followed by a disinfectant solution, and another water rinse. This procedure destroys most of the bacteria located in the teat cups and claw. In research trials where automatic backflushing was beneficial, no teat end dipping was done after milking. Its value in reducing mastitis infections when teat dipping is also used is yet to be determined. Also, the use of iodine as the disinfectant causes an increase in iodine concentration in the milk. When using backflushing, the equipment must be 100% foolproof, since the disinfectant solution should never be added to the milk.

Other Automation Devices

Michigan research workers designed and tested a number of devices to increase the performance of milking parlors. They devised feed bowl covers that automatically cover the feed bowls of cows when the exit gates are activated. These covers prevent cows from stopping and checking or eating from each feed bowl as they leave the parlor. Although feed bowl covers increase the throughput of cows in a parlor, they have not gained widespread acceptance.

Illinois workers, as well as some European workers, have developed methods for the in-line detection of abnormal milk, usually based on the conductivity of the milk. This is a definite aid in detecting cows with mastitis problems so that they can be treated and their milk diverted from sale.

Automatic weighing devices for milk yield have been perfected and the yields recorded automatically with a computer. More interest will be shown in these devices as herds become larger and more farmers use computers. Several companies are trying to develop automatic identification of cows as they enter the milking parlor. When this is perfected, each individual cow's milk yield could be recorded with no effort required by the milker.

Results of Automation

Michigan and Arizona workers have measured the performance in a number of milking parlors. Their results are shown in Table 7.1. Steady-state throughputs in cows per hour do not include the time required to prepare the parlor and equipment for milking or the cleanup time after milking. In general it will be noted that adding a crowd gate and using feed bowl covers and automatic detachers increase the labor efficiency and in most cases allow the milking parlor to be operated by one less person. In some cases it may not pay to automate a parlor because the labor efficiency is not increased enough to pay for the cost of the equipment. This appears to be the case for the double-4 herringbone parlor.

7.6 CHOOSING A MILKING PARLOR

The choice of size and type of milking parlor will depend on a number of factors, which include: number of cows to be milked, number of persons to do the milking, the maximum amount of time that a milker wants to be in the milking parlor and milk house at each milking, cost, and personal preference. New milking parlor construction and remodeling of existing parlors should include the decision on the number of cows to be milked immediately after occupancy and 5 years later. Since the capital investment in a new parlor is high, it should be planned to accommodate an increased herd size in the future. If it is not constructed for a future herd size, it should at least have the capability of being expanded to accommodate the future herd size.

The second consideration centers around two questions: How many persons should be in the parlor? How long should a person stay in the milking parlor and milk house at each milking? The number of persons in the parlor is dependent on the size of the parlor and the degree of automation. Results in Table 7.1 show that automation can reduce by one the number of persons required in the parlor. Having two persons in the parlor makes the training of new personnel easier, reduces the chances of an injury to cow or person to go unnoticed, and multiple-person parlors can be operated by one person for short periods. One-person parlors are somewhat more efficient in cows milked per person per hour, and no personality conflicts arise with just one person. If two or more people work in the parlor, one person should have overall responsibility for the milking.

The length of time to be spent at each milking in the milking parlor, in parlor preparation, and in cleanup depends on the individual's wishes and work habits. Labor efficiency decreases with each hour in the parlor, but this can be restored with a 30-minute break every 3 or 4 hours. Most people prefer an 8-hour straight shift over two 4-hour split shifts. Longer shifts are more practical for the larger dairy herds. In most cases considerable cost in constructing and operating a milking parlor can be saved if the parlor is used for longer periods. The estimated number of cows that can be milked per milking in various types of parlors and milking times is shown in Table 7.2. The available times include 1 hour for parlor preparation and cleanup as well as 15 minutes to change each 100 cows.

The final decision on the type and size of parlor depends on the construction cost of the parlor and the annual costs to operate it. The initial cost is critical for borrowing the money, but the annual costs give a much better estimate of the real costs of the parlor. Figure 7.7 shows the annual costs for operating different types of parlors with different herd sizes. These results are from Washington State data obtained in 1981 and show the four most economical milking parlors for each herd size in terms of annual cost per cow. They should not be used as current

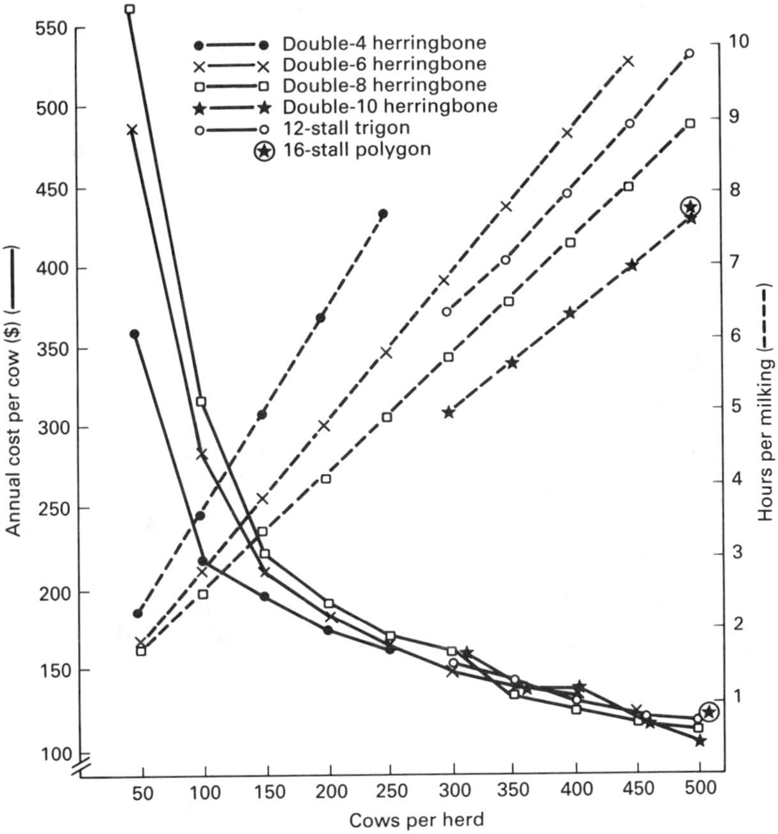

Figure 7.7 Annual operating costs and hours required per milking in various milking parlors. Double-4 herringbone parlor is without mechanization; the others include milking equipment and parlor mechanization. Data from G. S. Willett and D. C. Grusenmeyer, 1982, *Dairy Herd Management*, Jan., p. 41.

figures; however, the relative values should be helpful in choosing a parlor. Note that the annual cost for milking cows in a milking parlor for a 50-cow herd is very high. At a herd size of 100 cows, the costs become more reasonable. The annual cost per cow continues to decrease with increasing herd size; however, the differences among parlors greatly diminish. The double-4 herringbone parlor without automation appears to be best for the smaller herds and continues to be the least expensive through 250 cows; however, the number of hours per milking is great. The larger herringbone parlors with automation are most economical when herd size reaches 300 cows. The trigon parlors do not become competitive until 300 cows are reached and the polygon parlors when 500 cows are reached. Similar costs should be estimated for each type of parlor that is considered for construction.

REFERENCES AND FURTHER READINGS

APPLEMAN, R. D., and R. J. GUSTAFSON. 1985. Source of stray voltage and effect on cow health and performance. *J. Dairy Sci.*, 68:1554.

ARMSTRONG, D. V. 1979. Trigon milking parlor. *West. Reg. Ext. Pub.* 27.

BICKERT, W. G. 1975. A close look at milking machine detachers. *Hoard's Dairyman*, 120:277.

BICKERT, W. G. 1983. Milking parlor selection. *Am. Soc. Agric. Eng. Conf.*

BICKERT, W. G., and D. V. ARMSTRONG. 1976. Herringbone and side-opening milking parlors. *Mich. Ext. Bull.* E-1034.

BICKERT, W. G., and D. V. ARMSTRONG. 1976. Polygon milking parlors. *Mich. Ext. Bull.* E-1035.

CLOUGH, P. A. 1977. Machine milking in cowsheds and milking parlors. In *Machine Milking*, National Institute for Research in Dairying, Shinfield, Reading, England.

SCHMIDT, G. H., and D. E. PRITCHARD. 1983. Types of milking parlors. *Ohio Coop. Ext. Dairy Guide Leaflet* 500.

SMITH, T. W., ET AL. 1985. Effect of automatic backflushing on number of new intramammary infections, bacteria on teatcup liners, and milk iodine. *J. Dairy Sci.*, 68:424

WILLETT, G. S., and D. C. GRUSENMEYER. 1982. Comparing parlor costs. *Dairy Herd Manage.* Jan., p. 41.

8

Udder Diseases

Two major diseases that affect the bovine mammary gland are mastitis and udder edema. Edema is a physiological condition that is not caused by an infectious agent, whereas almost all mastitis is caused by bacteria. Cancer is a common disease of the mammary glands of humans, rats, dogs, and mice; however, it is rarely, if ever, found in cattle. Mastitis and udder edema are discussed in detail in this chapter.

8.1 MASTITIS

Mastitis is an inflammation of the mammary gland and may be the result of chemical or mechanical injury, but in most cases it is caused by bacteria (Figure 8.1). It is the most prevalent and costly cattle disease in the United States. It has been estimated that the average cost per cow is equivalent to approximately 700 kg (1540 lb) of milk. The milking cow population in the United States in 1985 was approximately 10.8 million cows. When milk is worth $12 per 45.5 kg (100 lb), the annual loss to U.S. dairy farmers is slightly over $2 billion. This is equal to about $185 per cow, which is much higher than the average profit per cow on most farms. A quarter that is infected with bacteria for one lactation will have a 10 to 12% decrease in milk production. This loss of milk production, along with

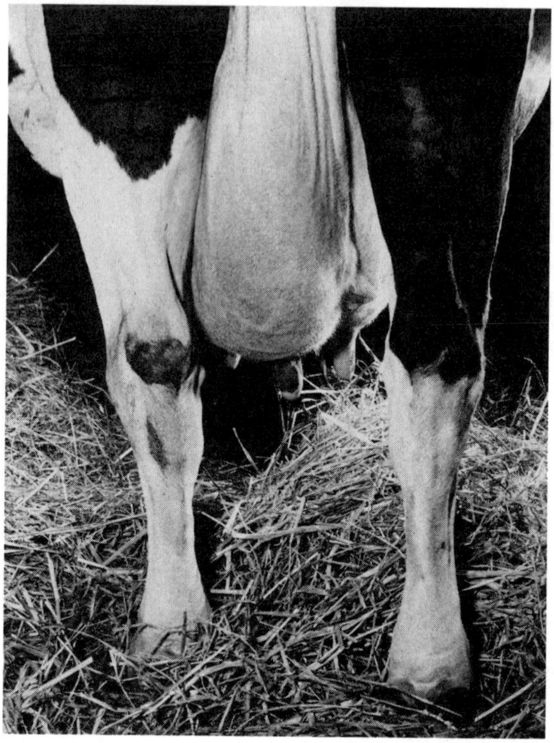

Figure 8.1 Blind quarter resulting from a severe case of clinical mastitis. From G. H. Schmidt, 1971, *Biology of Lactation*, W. H. Freeman and Company.

the discarded milk, costs of antibiotics and veterinary treatment, additional culling, and loss of value of cows with mastitis make up this large loss due to the disease. When a quarter becomes infected or inflammed, an increase in the somatic cell count of the milk occurs. The number of somatic cells (SCC) is related to the severity of inflammation. The SCC of milk is now routinely determined in many milk testing laboratories. The relationship between the somatic cell count of milk of a cow and the loss of milk production is shown in Table 8.1. It will be noted that SCC counts as low as 50,000 per milliliter of milk cause a reduction in milk yield. Raubertas and Shook (1982) averaged the SCC of cows on DHI tests in Wisconsin. They found that for each unit increase in log cell count (such as the DHI scores, Table 8.1), there was a decrease in lactation milk production of 135 kg (297 lb) for first lactation cows and 270 kg (594 lb) for all other lactations.

In herds where a mastitis control program is not carried out or an ineffective one used, approximately 50% of the cows are infected with pathogenic bacteria in an average of two quarters per cow. This means that approximately 25% of all quarters in these herds are infected. These figures are comparable to those found in surveys conducted in the 1930s. In herds where an effective control program has been carried out, the percent of quarters infected can be reduced to 7 to 8%. Oliver and Mitchell (1984) compared the quarter and cow infection rates in 17 herds that were free from *Streptococcus agalactiae* and 17 herds infected with the organism. Herds free from the organism were using mastitis control proce-

TABLE 8.1 RELATIONSHIP BETWEEN PREDICTED DAILY MILK PRODUCTION AND SOMATIC CELL COUNT

Somatic cell count (1000/ml)	DHI score	First lactation Milk (kg)	First lactation Loss (kg)	First lactation Loss (lb)	Older lactations Milk (kg)	Older lactations Loss (kg)	Older lactations Loss (lb)
12.5	0	23.1	—	—	29.2	—	—
25	1	22.9	0.2	0.4	28.6	0.6	1.3
50	2	22.6	0.5	1.1	28.0	1.2	2.6
100	3	22.4	0.7	1.5	27.4	1.8	4.0
200	4	22.1	1.0	2.2	26.9	2.3	5.1
400	5	21.8	1.3	2.9	26.2	3.0	6.6
800	6	21.4	1.7	3.7	25.4	3.8	8.4
1600	7	20.7	2.4	5.3	24.6	4.6	10.1
3200	8	20.0	3.1	6.8	23.6	5.6	12.3
6400	9	19.0	4.1	9.0	22.5	6.7	14.7

Source: G. M. Jones et al., 1984, *J. Dairy Sci.*, 67:1823.

dures of postmilking teat dip and antibiotic dry cow therapy. Herds with the organisms were not using the procedures. In herds with *Strep. agalactiae*, 58.5% of the cows and 37.0% of the quarters were infected, with 87% of the infections due to *Strep. agalactiae* and *Staphylococcus aureus*. In herds free from *Strep. agalactiae*, only 25.8% of the cows and 10.3% of the quarters were infected. Of the infected quarters, 25.2% were due to *S. aureus*, 47.9% due to streptococci other than *Strep. agalactiae*, and 10.5% due to coliforms, and the remainder had other and nonspecific organisms. Thus the number and types of infections change with the institution of a mastitis control program.

In addition to the loss of milk production caused by mastitis, milk from infected and clinical quarters has decreased amounts of total solids, fat, lactose, and casein. Milk from cows with mastitis has a much lower value for cheese making.

Types and Causes of Mastitis

At least 20 different bacteria and other organisms can cause mastitis. They are generally grouped into two categories: common pathogens and environmental pathogens. Common pathogens are primarily staphylococcal organisms (*S. aureus* and *S. epidermidis*) and *Strep. agalactiae*. The primary reservoir of these bacteria is the infected mammary glands of cows. These organisms can be either eliminated or effectively controlled by management. *S. aureus* and *Strep. agalactiae*, along with *Strep. uberis* and *Strep. dysgalactiae*, cause about 95% of mastitis infections and clinical cases in herds where an effective mastitis control program is not carried out.

The environmental pathogens are *Strep. uberis, Strep. dysgalactiae*, and coliform bacteria. Coliform bacteria include a group of bacteria that are fermentative gram-negative bacilli and live in the digestive tract of humans and animals without producing disease. The more common coliform bacteria that cause mastitis are *Escherichia coli, Enterobacter aerogenes* and *cloacae, Klebsiella*, and *Citrobacter*. These are known as environmental bacteria because of their presence in the environment, especially bedding material, poorly sanitized teat cup liners, damp walkways, manure-covered barnyards, and contaminated water in ponds and barnyards. No effective control program has been developed for the environmental pathogens; however, the control program of teat end dipping and dry cow therapy is effective against *Strep. uberis* and *Strep. dysgalactiae*. Two other environmental bacteria are commonly isolated from mammary glands; however, research workers do not agree on their importance in causing reduced milk yields and clinical mastitis. The two organisms are *Cornybacterium bovis* and *S. epideridis*. These are usually considered minor pathogens; however, it is becoming increasingly clear that they do cause udder injury when they are present in the mammary glands.

In addition to bacteria, other organisms can cause mastitis. Among these are yeast, mycoplasma, and *Prototheca*. Mycoplasma are the smallest free-living organisms and lack a cell wall. *Prototheca* are achloric algae that are oval to spherical in shape and vary in size. Although the incidence of infection and clinical mastitis caused by these organisms is low, they can be very troublesome problems on individual farms.

In herds not using an effective mastitis control program or those having an environmental bacteria problem, 1 to 3% of cows will show clinical mastitis at any one time. About 40% of quarters infected with *Strep. agalactiae* and *S. aureus* become clinical during the lactation. These infections also have a tendency to stay in the quarters for relatively long periods of time. Quarters infected with the environmental pathogens show a much higher incidence of clinical mastitis—up to 70%. These infections are of much shorter duration than the other infections. For this reason, survey data of infected quarters or clinical quarters on a given day are not very good indications of the real infection status in a herd.

Considerable confusion still exists between clinical mastitis and udder infection. The sequence of events in the development of an infection and clinical mastitis is shown in Figure 8.2. Udder infection is defined as the presence of pathogenic bacteria in the mammary gland. An infection can be diagnosed with certainty only by laboratory examination of milk samples that have been plated out on a growth media and that have been carefully taken from quarters. Clinical mastitis is an obvious inflammation of the udder and is evidenced by gross changes in the behavior of the cow, the appearance of the udder, and/or the composition of milk. Clinical mastitis usually has the following symptoms: gross swelling of the udder or quarter, heat, redness, pain, decreased milk production, and change in milk composition (Figure 8.3). When the clinical signs are accompanied by fever and mild depression, it is called acute. A peracute condition occurs when the symptoms are accompanied by high fever, depression, shivering, loss of weight, and loss of appetite. In severe cases of peracute mastitis, death may result.

Subacute or mild clinical symptoms include the loss in milk production, changes in milk composition, increased pH of milk, and decreases in milk fat and milk casein. Usually, no abnormal signs of the udder are seen. The normal pH of milk is about 6.6, whereas that of blood is about 7.4. When injury or infection

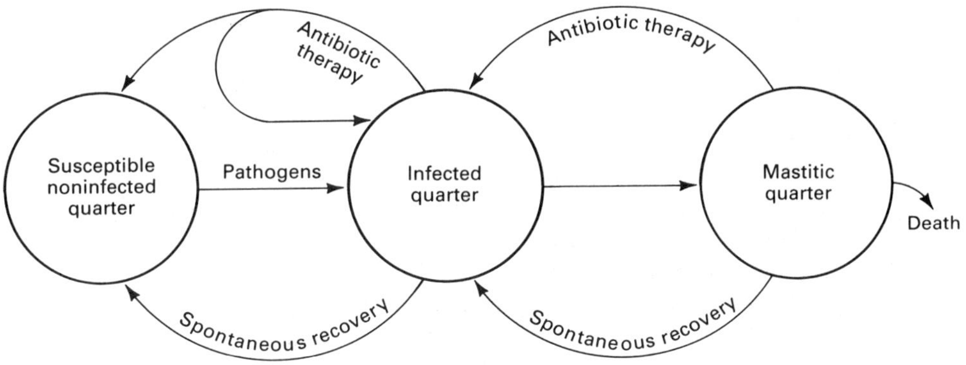

Figure 8.2 Sequence of events in the development of an infection and clinical mastitis in a dairy cow. Redrawn from F. H. Dodd et al., 1964, *J. Dairy Sci.*, 47:1109.

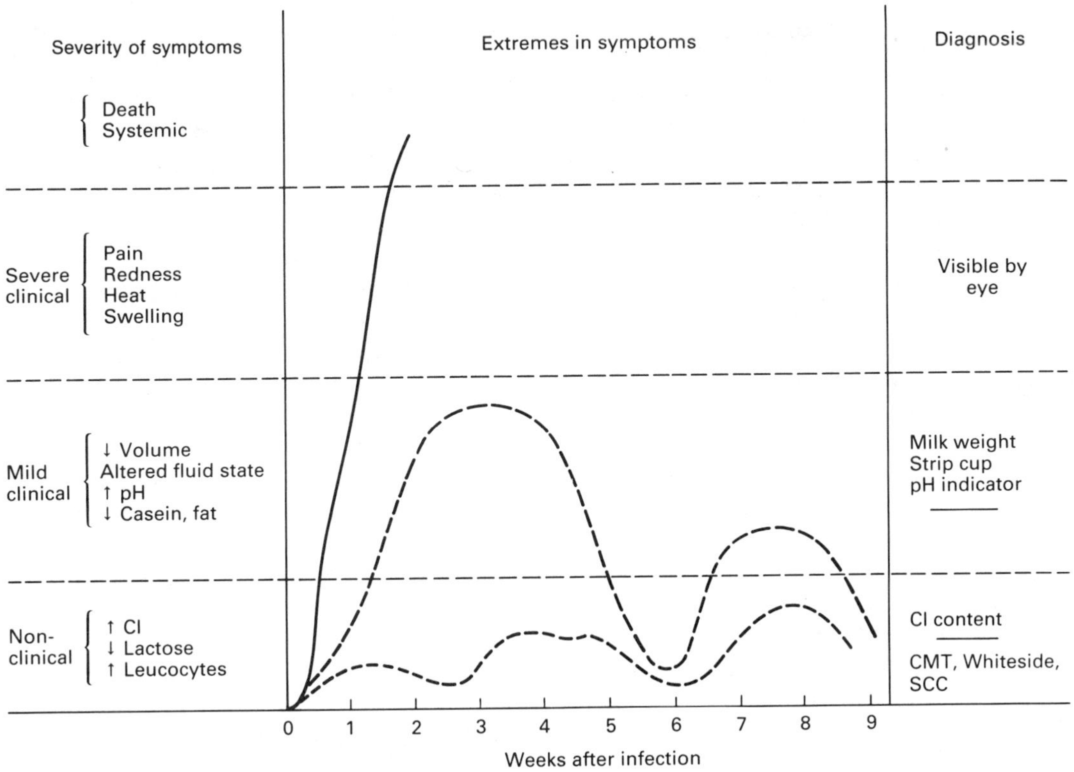

Figure 8.3 Symptoms and diagnosis of the various forms of mastitis. From G. H. Schmidt, 1971, *Biology of Lactation*, W. H. Freeman and Company.

occurs in the udder, the milk has a tendency toward the composition of blood, thus the increase in pH.

Subclinical mastitis refers to an udder or quarter being infected, but the dairy farmer cannot detect any changes in the composition of milk or the appearance of the mammary gland. Even so, it has been shown that an infected quarter without showing clinical mastitis can cause a 10 to 12% reduction in milk production of that cow for the lactation. Milk from a quarter showing subclinical mastitis will have an elevated SCC, an increase in chlorine content, and a decrease in lactose content (Figure 8.3).

The extremes in subclinical infections and clinical mastitis are shown in Figure 8.3. At one extreme a quarter may become infected and become severe clinical in a matter of 48 to 72 hours. In some cases death may result. At the other extreme a quarter may be infected for the entire lactation without ever showing any clinical signs of mastitis. In other cases an infected quarter may show clinical mastitis, be treated with antibiotics, and the abnormal signs disappear. Several weeks or months later the quarter will show clinical signs again. In this case the bacteria were not eliminated and the subclinical form of the disease continued. If several cases of clinical mastitis (flare-ups) occur in the same quarter during one lactation, it is referred to as chronic mastitis.

Diagnosis of Mastitis

The diagnosis of mastitis is of three forms: the identification of the causative agent, the identification of abnormal milk, and the detection of udder and body symptoms. The diagnosis of the causative agent requires that the milk to be analyzed must be taken under aseptic conditions (Figure 8.4a). The end of the teat must be cleaned and then sterilized with alcohol. The milk must be collected in a sterile tube without contamination from the environment. After collection, the milk is streaked onto a culture medium in a laboratory and incubated for 24 to 48 hours. Different ingredients are used in the medium to separate the various types of bacteria. Some identification is based on the morphology of the growth of organisms on the medium. In most cases, a number of different chemical and bacterial tests are required to identify the various species of organisms. The minimum time for a tentative diagnosis of the organism is 24 hours, with 72 hours being much more common. A blood agar medium plate with *S. aureus* organisms is shown in Figure 8.4b. A quarter is usually considered infected if a specific bacterium is isolated from the quarter on two successive samplings or if the identification of the bacterium is associated with an elevated SCC or abnormal milk. In many cases, abnormal milk will be seen from a quarter without isolating a bacterium. This does not necessarily mean that the quarter is not infected; only that the methods were not adequate or that the bacteria would not grow on the specific medium.

The most common method of detecting abnormal milk is the strip cup or plate. Two or three streams of foremilk from each quarter are squirted on the cup or plate and the milk examined for abnormal color or consistency (see Section 6.2). Approximately 50% of clinical mastitis can be detected by this method. The

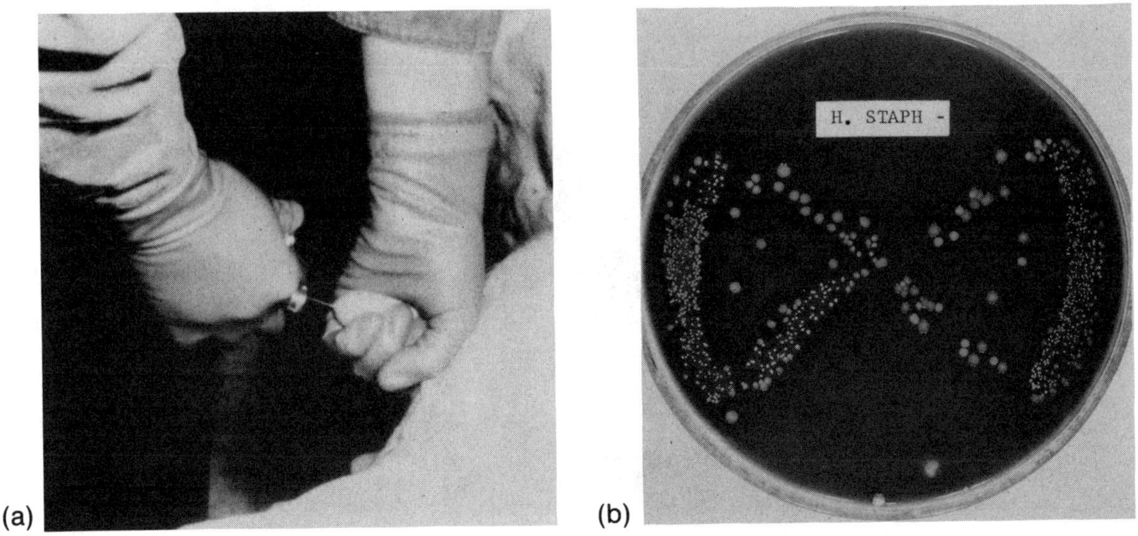

Figure 8.4 To identify the causative bacteria causing mastitis, (a) an aseptically drawn milk sample must be taken and (b) streaked onto a blood agar plate, incubated, and bacteria identified by their morphological characteristics. From *Principles of Dairy Science* by G. H. Schmidt and L. D. Van Vleck. Copyright © 1974 W. H. Freeman and Company. Used by permission.

SCC count is used to measure the number of somatic cells in the milk. The SCC is made up of sloughed-off epithelial cells and leucocytes, with polymorphonuclear (PMN) leucocytes being the most prevalent. The number of somatic cells is related to the infection status of the quarter as well as the degree of inflammation. Central DHI milk testing laboratories now measure this on the milk sample taken for milk composition determination. This gives a measure of the SCC of the milk from the cow, not each quarter. It can serve as an effective screening tool for problem cows as well as a monitoring method of the mastitis status in the herd. DHI laboratories generally use an electronic cell counter for this purpose.

Three screening tests have been developed to measure the SCC of milk. The Modified Whiteside Test (MWT) is usually used by milk plants and the California Mastitis Test (CMT) is a cow-side test in which the milk from individual quarters can be tested (Figure 8.5). In both cases the test is based on the amount of gel formation when milk is mixed with either sodium hydroxide (MWT) or a detergent (CMT). The CMT also has a pH indicator to help in the diagnosis. The CMT can be very helpful in identifying quarters with abnormal milk or to monitor the mastitis status of the herd. The MWT has been largely replaced by either the Wisconsin Mastitis Test (WMT) or the electronic cell counting in laboratories. The WMT uses the same principle as the CMT; however, the amount of gel formation is quantified, which makes it a more accurate test. Most dairy plants use either the WMT or electronic SCC on each dairy farmer's bulk tank milk on a monthly or biweekly basis to screen for milk with high SCC. Changes in milk composition, such as decreases in fat, casein, and lactose concentration and increases in chlo-

Figure 8.5 The California Mastitis Test (CMT) can be used as a cow-side test to determine quarters with a high leucocyte count. Milk from a quarter and a chemical reagent are placed in one section of the paddle and gently rotated. The amount of gel formation is related to the leucocyte count of the milk. Photo courtesy of the Ohio Cooperative Extension Service.

rine amounts require laboratory tests and are usually not used for routine mastitis diagnosis. A drop in milk production can be noted if a dramatic decrease occurs or if milk weights are obtained daily.

Visual observations still play a very important role in mastitis detection. The clinical symptoms of abnormal milk; hard, red, and swollen udders; and a cow being sick or off feed can all be seen by the naked eye.

Transmission of Mastitis-Causing Bacteria

Strep. agalactiae is found only in the mammary glands of animals; therefore, eliminating quarters infected with this organism will eliminate its source of contamination. Although *S. aureus* is found in the environment, most of the organisms come from infected mammary glands. Again, elimination of the infected quarters will remove the major source of contamination. For this reason, these two bacteria can be controlled by a dry cow therapy–teat dip control procedure. These organisms are transferred primarily by hands and equipment that come in contact with infected and noninfected quarters. The milking machine is the most common method of transfer, especially from an infected quarter to a noninfected quarter of the same cow. In fact, about 40% of new infections caused by these organisms are cross-infections in the same cow. Other methods of transfer are wash towels, clothes, and sponges.

Although the other streptococcus organisms live in the udder, they also live in the environment; consequently, they are not as well controlled by dry cow therapy–teat end dip control programs as are *S. aureus* and *Strep. agalactiae*. Most of the other organisms, especially the coliform organisms, are widespread in the environment. These are spread by methods in which the contaminated material comes in contact with the teat ends. Most common among these are: unclean bedding, poorly sanitized milking machines, contaminated water (either excess wash water during udder preparation or ponds), or dirty barnyards.

Factors That Affect Infection Rate and Clinical Mastitis

A number of factors influence whether a quarter becomes infected. Among these are: type of bacteria, strain of bacteria, number of bacteria on teat end, heredity, condition of teat ends, SCC of quarter, and the condition of the milking machine. Cows with teat end erosions are more susceptible to infection than are cows with normal teats. The keratin material lining the teat meatus also plays a role in preventing infection. Removal of the keratin, either intentionally or with an infusion tube, increases the chances of infection. Older cows have a higher incidence of infection than do younger cows, but this is caused by reinfections of previously infected quarters rather than a higher infection rate of noninfected quarters. Fast-milking cows also seem to be more susceptible to mastitis infections.

In most cases the establishment of an infection in a mammary gland depends on the balance between the virulence of the invading organisms and the defense system of the mammary gland. If the bacteria penetrate the streak canal (the primary defense mechanism), the secondary defense mechanisms then come into play. These come from within the mammary gland and form a complex of immunological, chemical, and cellular mechanisms. The cellular defense mechanisms center around phagocytosis, which is the process of recognizing, ingesting, and digesting foreign material. The phagocytic cells are mainly PMNs and macrophages and account for 80 to 90% of the SCC. Milk from noninfected glands contains between 50,000 and 200,000 SCC/ml of milk, although some research indicates that the level of "normal" milk is below 50,000. When foreign material, such as bacteria, enter the mammary gland, the PMNs and macrophages attempt to destroy the bacteria; however, a number below 500,000 per milliliter is not very effective in protecting the gland. Usually, 24 hours elapse before the SCC reaches a number to be very effective. Also, for some reason, the PMNs in milk are less phagocytic than those in blood. In some studies it has been shown that the infection rate is lower when bacteria are introduced into quarters with a high SCC count. In fact, some authorities advocate increasing the number of quarters infected with minor infections in order to reduce the infection rate of major pathogens. This is generally not accepted by most experts in the field.

Chemical defense mechanisms include compounds found in the milk, of which lactoferrin is the most common. Lactoferrin is a protein in milk that inhibits the growth of bacteria that require iron by binding the iron and thus making it unavailable for the bacteria. Lactoferrin increases in concentration about 30 times in infected quarters and 100 times in dry-period secretion.

The immunological defense systems include the humoral and the cell-mediated immune systems. Both play a role in destroying invading foreign material. The humoral immunity is brought about by antibodies that direct their activity toward killing of bacteria and neutralizing their toxins. Cell-mediated immunity is mediated by T-lymphocytes, which kill foreign cells by the elaboration of a cytotoxin.

The milking machine has been shown to be involved in the infection rate. This is especially true with machines that produce cyclic vacuum fluctuations with the opening and closing of the teat cup liners and irregular vacuum fluctuations brought about by large amounts of air admission into the system. These conditions have been shown to increase the infection rate by propelling small droplets of milk contaminated with pathogens against the tip of the teat. These "impacts" are more common toward the end of milking, during machine cluster removal, and when the liners slip on the teat during milking. These findings have resulted in the design of milking machines with larger volume claw pieces, better airbleeds in the claw, larger-diameter short air tubes, and greater vacuum reserves.

In addition to influencing whether an infection takes place in the mammary gland, the defense mechanisms also play a role in preventing the infection from developing into clinical mastitis (Figure 8.4). Again it appears to be a balance between the numbers and virulence of the bacteria and the defense mechanisms in the gland. Although a large number of factors have been incriminated as causing an infected quarter to become clinical, no direct relationship between a specific factor and clinical mastitis has ever been demonstrated under controlled experimental conditions. Among factors incriminated over the years have been: overmilking, high vacuum levels, high production, high grain feeding, high protein feeding, and incomplete milking.

Clinical mastitis can revert to an infected state or to a noninfected state (Figure 8.2). In most cases, antibiotics are infused in quarters with clinical mastitis. If the antibiotic is completely effective in killing the bacteria, a noninfected status develops. Usually, the antibiotic causes a regression of the clinical signs, but the bacteria remain in the gland. The gland thus reverts to the infected state and this cycle may repeat itself several times during a lactation. In severe cases, death may result.

Times of New Infections

New intramammary infection rates are highest in early lactation and in the early dry period. This is especially true for infections caused by *S. aureus* and *Strep. agalactiae*. The infection rate during the first 3 weeks of the dry period is very high, with the number of new infections during the total dry period almost equal to the number occurring during the entire lactation period (Figure 8.6). In an early study by Neave et al. (1970) the infection rate was followed in an 89-cow dairy herd. The rate was 7.0 new infections per week during the first 3 weeks of the dry period, 0.18 per week during the remaining dry period, and 1.12 during the previous lactation. The situation may be different for herds where coliform organisms are the major problem. Smith et al. (1985) studied the coliform infection rate during the dry period in one herd. They found the rate to be much higher during the first and last quarters of the dry period. Coliform infections that originated during the first half of the dry period were other than *E. coli*. Those that were first detected at calving appeared to be predominately *E. coli*.

Mastitis Control Programs

An effective mastitis control program must be simple to carry out and the cost of the program must be much less than the disease itself. Also, there must be

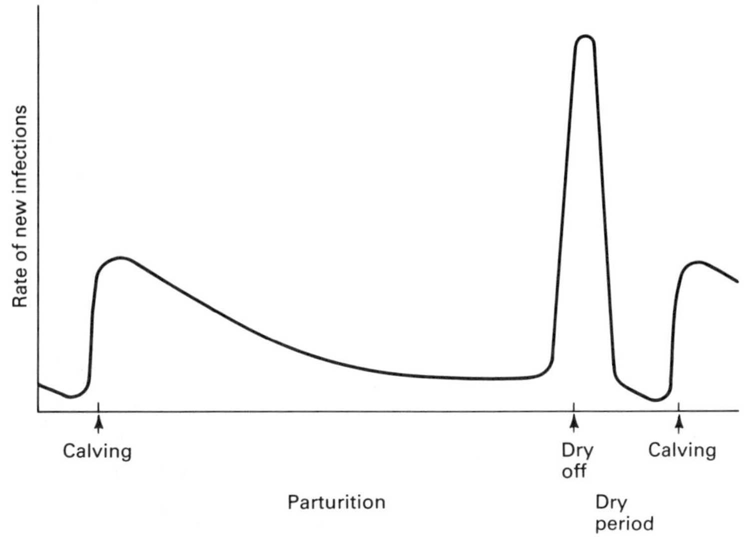

Figure 8.6 New infection rates of dairy cows during the lactation and the dry period. From R. P. Natzke, 1981, *J. Dairy Sci.*, 64:1431.

good experimental evidence that the program is effective under a wide range of conditions. It is also helpful if the dairy farmer can see results in a relatively short period.

A control program has two basic components: preventing new infections and reducing the duration of existing infections. This assumes that all clinical mastitis is caused by bacteria. In cases where trauma causes the problem, that trauma should be removed immediately.

In the absence of an effective mastitis control program, about 50% of the cows are infected when measured on any given day. To keep this value constant, the number of new infections is about equal to the number of infections that disappear. Infections disappear by spontaneous recovery, effective lactation and dry period therapy, and sale of infected animals. Reducing the new infection rate and reducing the duration of the existing infections will reduce the percentage of quarters or cows infected at any given time. For example, a 25% quarter infection rate (at a point in time) can be the result of 25% of quarters infected 100% of the time (1.00×0.25), or 50% of the quarters infected for 50% of the time (0.50×0.50).

If the number of quarters with new infections could be reduced by half ($\frac{1}{2} \times 50\%$) and the duration of the infections also reduced by half ($\frac{1}{2} \times 50\%$), the infection rate at any given time would then be $25\% \times 25\%$ or 6.25%. This is the goal of a control program and can be attained when dealing with nonenvironmental pathogens.

Preventing new infections. Since most of the mastitis pathogens are found in the environment, attempts to produce milk under sterile conditions are impractical. Since the udders of infected cows are the major source of nonenvironmental pathogens, considerable progress has been made in reducing the spread of these organisms from infected to noninfected glands or by killing the organisms soon after they are spread. The major ways of spreading these organisms are the hands of the milker, the washcloth or towel, and the teat cup cluster. Workers at the National Institute for Research in Dairying (NIRD) in England conducted a series of trials to determine the role of sanitation in preventing new infections. In their first field trial they compared a complete hygiene program of sanitation to no hygiene. Dairy farmers using the complete hygiene program used separate paper towels dipped in a sanitizing solution to wash cows' udders, dipped their gloved hands in a sanitizing solution between cows, dipped the teats of cows in a strong teat dip of 5000 ppm Chlorhexidine after milking, and sterilized the teat cup cluster between cows with nearly boiling water. The new infection rate was reduced by 42%. When cross-infections were removed, the prevention rate increased to 54%. In a second field trial they compared the same two treatments but included a third, a partial hygiene program, which was similar to the complete hygiene program with the exception of the sterilization of the teat cup cluster. The new infection rate of the partial hygiene program was 52% lower than that of the control herds, and the complete hygiene program resulted in a 64% reduction in new infections. This work has been confirmed by a number of studies and now points to the practical possibility of reducing new infections. It now appears that the dipping of the teat ends after milking is the most important component of this part of the control program.

Eliminating existing infections. The second component of an effective control program is the elimination of existing infections. A control program based solely on treatment of cows with clinical mastitis does not prevent new infections or clinical mastitis. The antibiotic therapy can reduce the clinical signs and thus reduce the destruction of secretory tissue, in contrast to not treating the animal. In some cases, the antibiotic may destroy the pathogenic bacteria in the udder. This is especially true when treating *Strep. agalactiae*. Lactation therapy cannot be a major component of the control program, since only 30 to 40% of the infected quarters become clinical during lactation. This means that 60 to 70% of quarters would never be treated.

The most logical time to eliminate existing infections is during the early dry period. High levels of antibiotic can be infused into each quarter at drying off since milk does not have to be withheld from market because of antibiotic contamination as it does during lactation therapy. The higher levels are more effective against the pathogenic bacteria. In addition, the presence of the antibiotic in the udder after drying off greatly reduces the infection rate, which is very high during the first 3 weeks of the dry period.

The effectiveness of treating cows during the dry period was shown in a study by Smith et al. (1967). Two amounts of cloxacillin (synthetic penicillin) were infused into quarters of cows at drying off (Table 8.2). The 1-g dose of cloxacillin eliminated 86% of the existing infections. Both cloxacillin preparations reduced the number of existing infections and prevented most of the new infections found

TABLE 8.2 EFFECT OF DRY COW THERAPY ON EXISTING INFECTIONS AND PREVENTING NEW INFECTIONS

			Experimental treatment			
	Control		0.2 g cloxacillin		1.0 g cloxacillin	
Infections	Cows	Quarters	Cows	Quarters	Cows	Quarters
Number infected	286	1144	303	1212	299	1196
Percent infected						
At drying off	53.8	24.2	52.0	27.4	49.9	24.6
At next calving	62.0	30.9	26.0	11.3	15.6	5.4
Percent of quarters with new dry period infections	—	9.5	—	3.3	—	1.7
Percent of infections persisting through dry period	—	88.8	—	28.8	—	14.6

Source: A. Smith et al., 1967, *Vet. Rec.*, 81:504.

in the control animals. Several other products have subsequently been tested and found effective as a dry period drug.

Combination of hygiene and dry period therapy. Since the sanitation procedure does not eliminate existing infections and the dry period therapy does not prevent new lactation infections, a combination of the two procedures was tested in two research trials. A field trial involving 30 commercial dairy herds in England was used to measure the effectiveness of the two components. Half of the herds followed the partial hygiene program in combination with treating all quarters of all cows at drying off with cloxacillin. The other herds used teat dipping only as the hygiene component in conjunction with dry period treatment. At the end of 3 years, the infection rate had decreased from 27% of quarters infected to 8%. Most of the progress was made during the first year.

A similar study was conducted in New York State involving 27 commercial dairy herds (Figure 8.7). The procedures were similar to the English trial except that the dry period drug consisted of 1 million units of penicillin and 1 g of streptomycin. At the beginning of the study, 28.1% of the quarters were infected, which was reduced to 7.1% after 3 years. In both of these studies, the infection rate was reduced by about 75% with a combination of hygiene and dry cow therapy. An effective teat dip appeared to be the important part of the hygiene component. The herds in the New York study were continued on the program for another 3 years to determine if antibiotic-resistant bacteria might develop. No evidence was found that continuous use of antibiotics produces antibiotic-resistant bacteria.

Control of environmental pathogens. An effective control program for mastitis caused by environmental pathogens has not yet been developed. The dry period therapy–teat dip procedure for controlling other pathogens is not effective

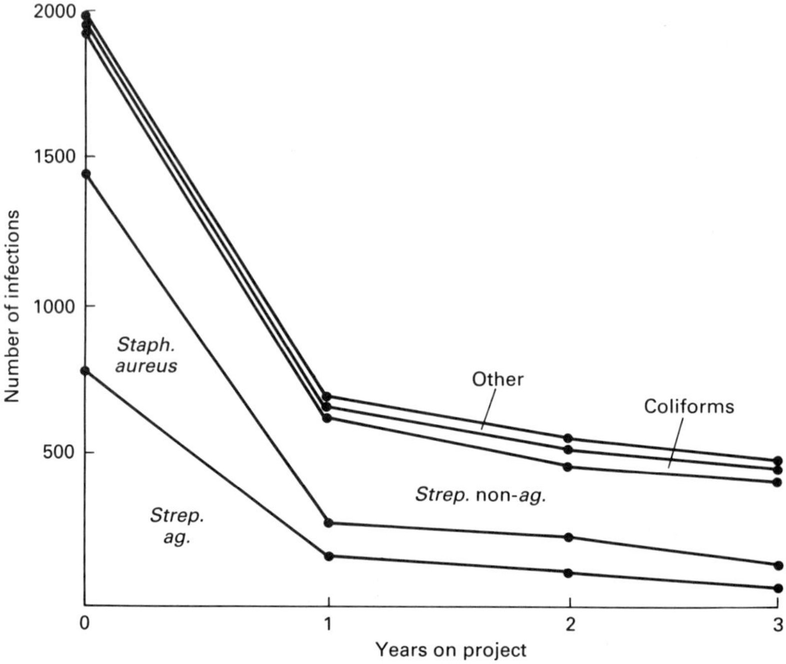

Figure 8.7 Number of mastitis infections in 27 New York dairy farms before and at the end of 3 years of using a mastitis control program consisting of a combination of teat end dipping and dry period therapy. Data from R. P. Natzke, 1978, *Large Herd Management*, University of Florida Press, Gainesville, FL.

in controlling environmental pathogens. However, results from the New York and English studies indicated that the incidence of mastitis caused by environmental pathogens was not increased by the program. Since no effective control program has been developed, the best available recommendations come from empirical observations and from field attempts to control the disease. The first recommendation centers around the bedding material. Since sawdust and recycled manure may be sources of coliform bacteria, a common recommendation is to switch to bedding material such as straw. The exposure to coliform bacteria will also be reduced if the space allocated to each cow is increased, the stalls kept clean, and reducing the time that cows are in the stalls or barn. Also, the milking of wet udders should be avoided. The udders should be dried before the milking machines are applied. In addition, the flanks of the cows should not be so wet during milking that the contaminated water might enter the teat cups. The milking machine should also be checked so that it does not cause damage to the teat, which might aid the passage of bacteria through the teat meatus. The milking machine should also be cleaned so that it is not a reservoir for bacteria. Finally, adequate vacuum reserve should be present so that the machine does not assist the passage of bacteria through the teat meatus by the impacts on the teat end.

Other control methods. A number of other control methods have been advocated, but none have been found to be very effective. Decreasing the susceptibility of cows to infection and clinical mastitis is the desirable long-term objective; however, attempts at breeding programs and vaccinations to accomplish this have not been successful. Regardless of the role of the milking machine and milking procedures in the prevention of infection and clinical mastitis, the best operating machine and procedures should be used for high machine and labor efficiency. Sanitation of all aspects of the milking and housing operations appears to be important and should be emphasized, especially for high-quality milk.

8.2 UDDER EDEMA

Udder edema is the swelling of the udder that occurs either shortly before, at the time of, or shortly after parturition (Figure 8.8). Udder edema, which is more prevalent in first-lactation cows and in cows with pendulous udders, is caused by an accumulation of tissue fluid between the skin and the secretory tissue of the udder, especially in the bottom portion of the udder (Figure 8.8). In severe cases, the edema is as high as the vulva and extends as far forward as the umbilicus or even to the brisket. The accumulation of the tissue fluid makes the udder very hard, causing the teats to be shorter for milking, which makes the milking machine more difficult to attach and to keep it attached to the teats. The cow also experiences some pain and discomfort. All of these conditions make complete removal of milk difficult.

Tissue fluid accumulates because of an upset in the balance between lymph formation and removal of lymph from the udder. Most of the tissue fluid is drained by the veins in the udder. A portion is also drained by the lymph vessels, which empty into the venous system near the heart. Since lymphatic vessels are more permeable than the walls of the capillaries, the larger particles, such as the proteins and lipids, tend to move in the lymph vessels. The lymph vessels contain a series of one-way valves that allow the lymph to flow in the direction toward the heart. The movement of fluid in lymph vessels is brought about by the pressure differential between the tissue fluid and the fluid in the lymph vessels as well as the mechanical action of the visceral organs, muscles, and skin on the lymph vessels.

A single factor or a combination of factors may cause an upset in the balance between the formation and drainage of tissue fluid from the mammary gland. These factors usually cause a change in the hydrostatic and osmotic pressures of the tissue fluids and of the blood. An increase in the capillary pressure or a congestion of the venous flow causes an increased accumulation of tissue fluid. A decrease in the osmotic pressure of the blood and/or damage to the capillary walls causes an increased movement of tissue fluid out of the capillaries. Obstruction of the lymphatic vessels prevents tissue fluid from being absorbed by the lymphatic vessels. Other factors that may contribute to edema are low tissue pressure, a high salt intake (which helps to retain body fluids), a high fluid intake, and a disrupted innervation of the capillaries.

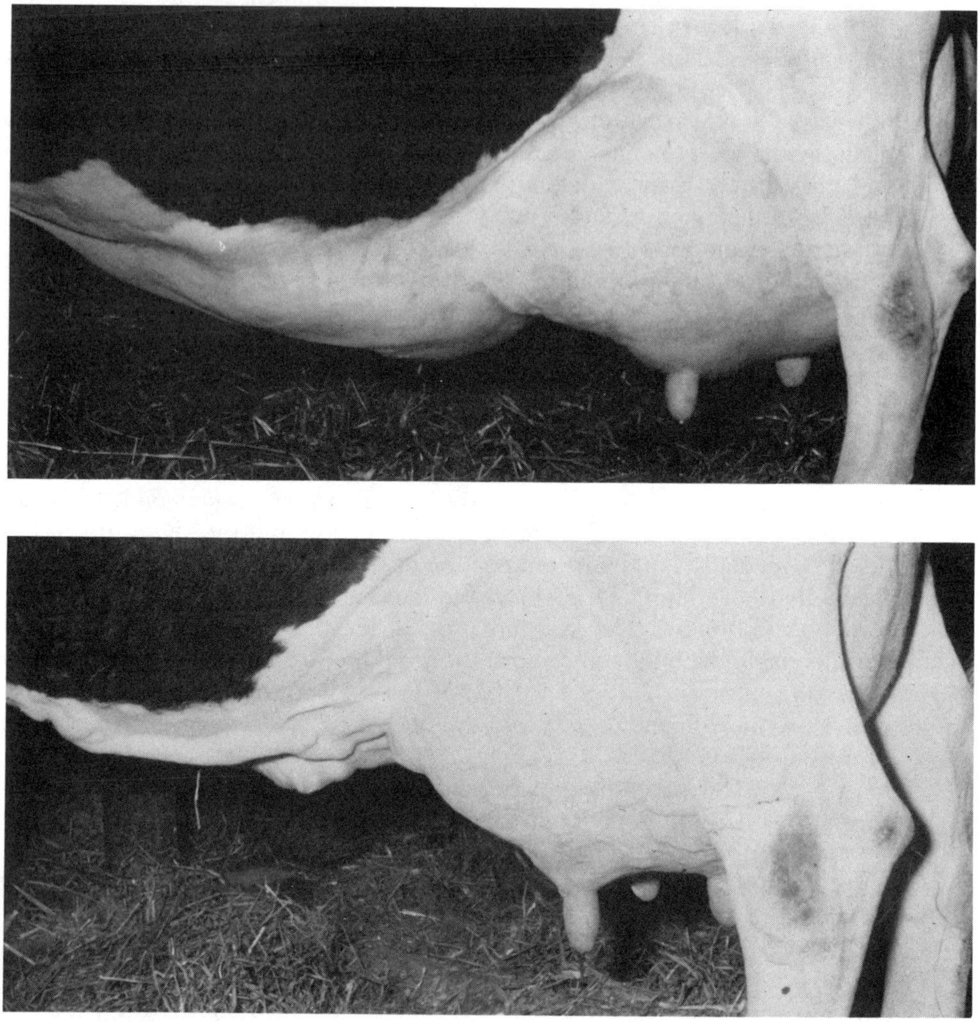

Figure 8.8 (a) Cow showing severe edema at calving; (b) the same cow after recovery 3 weeks later. From D. A. Morrow and G. H. Schmidt, 1964, *CIBA Veterinary Monograph Series/One*.

Milk produced from cows with udder edema is normal in composition. In most cases the edema regresses within 2 to 4 days and has completely disappeared 2 or 3 weeks after parturition. The regression occurs after the lymphatics become adjusted to handling large quantities of tissue fluid produced during milk secretion. In some instances, the edema becomes pathological and persists throughout the entire lactation period. There is some speculation that accumulation of tissue fluid causes damage to the suspensory apparatus of the udder, thus causing the cow's udder to become pendulous. The converse argument claims that a pendulous udder contributes to edema by impairing flow in the blood and lymph vessels that lead to the udder. Dentine and McDaniel (1984) found that udder type was not a good predictor of edema score at the next parturition; however, the edema score at parturition was a better predictor of udder support during the next lactation period. This gives some support to the argument that edema may be a contributing cause to pendulous udders.

Considerable attention has been focused on the effect of feed, especially during the dry period, on the development and severity of the edema. Controlled studies indicate that neither the percent corn or protein in the ration before parturition affects the incidence or severity of edema. Emery and co-workers (Emery et al., 1969) found that first-lactation cows had more udder edema than older cows if they received large amounts of grain before calving. Other research workers found significant increases in severity of udder edema when cows were fed varying amounts of concentrates prepartum. The heritability of udder edema is between 0.10 and 0.13 (Dentine and McDaniel, 1983).

Diuretics offer the most effective treatment for the problem. They are given to cows suffering from severe cases or cows with previous history of severe edema. Most diuretics cause a decreased resorption of sodium and potassium by the renal tubule of the kidney, which causes an increased fluid output. The addition of glucocorticoid hormones to the diuretic increases its effectiveness. Massage of the udder is helpful in some cases; however, the procedure is not practical in large herds. There is no known effective prevention program.

REFERENCES AND FURTHER READINGS

DENTINE, M. R., and B. T. MCDANIEL. 1983. Variation of edema scores for herd-year, age, calving month, and sire. *J. Dairy Sci.*, 66:2391.

DENTINE, M. R., and B. T. MCDANIEL. 1984. Associations of subjective udder edema scores and descriptive trait codes for udder types. *J. Dairy Sci.*, 67:208.

DODD, F. H. 1983. Mastitis—Progress on control. *J. Dairy Sci.*, 66:1773.

DODD, F. H., ET AL. 1964. Control of udder infection by management. *J. Dairy Sci.*, 47:1109.

EBERHART, R. J., ET AL. 1979. Coliform mastitis—A review. *J. Dairy Sci.*, 62:1.

EMERY, R. S., ET AL. 1969. Prepartum grain feeding effects on milk production, mammary edema, and incidence of diseases. *J. Dairy Sci.*, 52:345.

GUIDRY, A. J. 1985. Mastitis and the immune system of the mammary gland. In *Lactation*, B. L. Larson, ed., Iowa State University Press, Ames. IA.

MORROW, D. A., and G. H. SCHMIDT. 1964. Udder edema. *CIBA Veterinary Monograph Series/One*.

NATZKE, R. P. 1981. Elements of mastitis control. *J. Dairy Sci.*, 64:1431.

NATZKE, R. P., ET AL. Mastitis control program: Effects on milk production. *J. Dairy Sci.* 55:1256.

NEAVE, F. K., ET AL., 1970. A practical method of mastitis control. *Proc. 18th Internat. Dairy Cong.*, Sydney, Australia, iE:627.

OLIVER, S. P., and B. A. MITCHELL. 1984. Prevalence of mastitis pathogens in herds participating in a mastitis control program. *J. Dairy Sci.*, 67:2436.

RAUBERTAS, R. F., and G. E. SHOOK. 1982. Relationship between lactation measures of somatic cell concentration and milk yield. *J. Dairy Sci.*, 65:419.

SMITH, A., ET AL. 1967. Methods of reducing the incidence of udder infection in dry cows. *Vet. Rec.*, 81:504.

SMITH, K. L. 1983. Mastitis control: A discussion. *J. Dairy Sci.*, 66:1790.

SMITH, K. L., ET AL. 1985. Environmental pathogens and intramammary infection during the dry period. *J. Dairy Sci.*, 68:402.

TARGOWSKI, S. P. 1983. Role of immune factors in protection of mammary gland. *J. Dairy Sci.*, 66:1781.

9

Review and Application of Mendelian Genetics

What is genetics? William Bateson, who named the field of study in 1906, wrote: "Genetics is the science dealing with heredity and variation seeking to discover laws governing similarities and differences in individuals related by descent." The key words are *heredity*, the transmission of genetic material from one generation to the next, and *variation*, the wide range of values seen in such traits as milk production.

In this chapter the transmission of genetic material from parent to progeny is reviewed. All living matter consists of cells, and each cell has in its nucleus a complete copy of the genetic material of the animal. The genetic material is contained on *chromosomes*, which are long, slender, threadlike structures found in the nucleus of each cell. These chromosomes are paired, although the pairs are not exact duplicates of the genetic material. Each cow has 30 pairs of chromosomes.

Along the chromosomes are located the *genes*. The abstract idea of how the gene system functions was discovered by the Austrian monk, Gregor Mendel, who reported his experiments in 1866. The genes correspond to chemical structures along the chromosome. Since the chromosomes are paired, the genes are also paired, but the pairs may or may not be identical. Nonidentical genes located at the same place along the chromosomes are called *alleles*.

9.1 TRANSMISSION OF GENETIC MATERIAL

Knowledge of how genes are transmitted from parent to offspring is necessary for an understanding of how genetics can be applied to dairy cattle breeding. Germ cells (sperm or ova) are formed that contain only one of the pair of each chromosome and gene. To illustrate, assume that an animal carries the gene for polled (no horns), P; and the gene for horns, p. The pair of genes at the location for horns are not identical for this animal. Each cell of the animal contains both P and p alleles: P on one chromosome and p on the paired chromosome (Figure 9.1). With respect to P or p, two types of sperm or ova can be formed in equal numbers. The same pattern holds true for all other gene pairs that control other traits. The other parent provides germ cells in the same way. When fertilization (union of a sperm and an ovum) occurs, the genes are again paired. For example, suppose that an animal that is PP is mated to another that is pp (Figure 9.2).

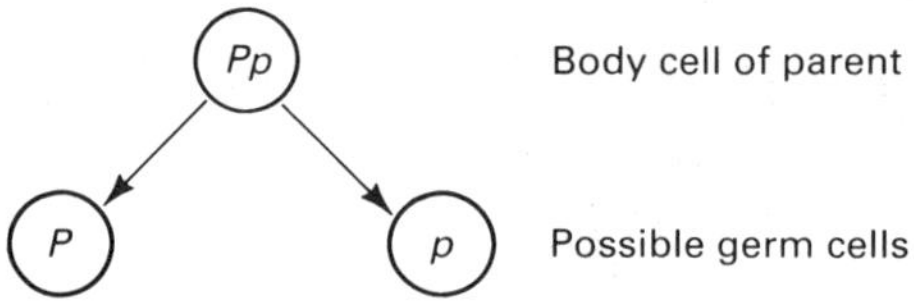

Figure 9.1 From *Principles of Dairy Science* by G. H. Schmidt and L. D. Van Vleck. Copyright © 1974 W. H. Freeman and Company. Used by permission.

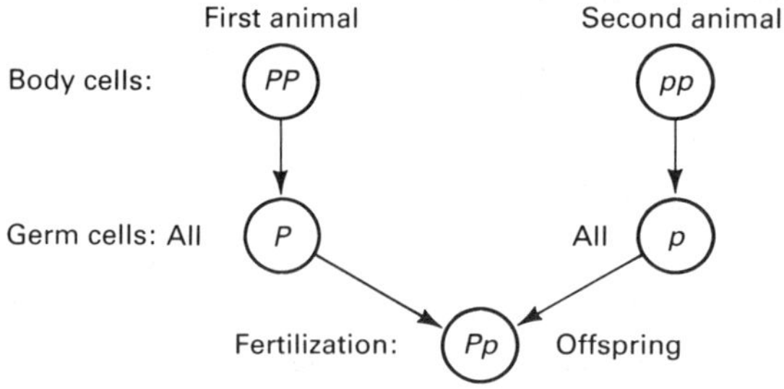

Figure 9.2 From *Principles of Dairy Science* by G. H. Schmidt and L. D. Van Vleck. Copyright © 1974 W. H. Freeman and Company. Used by permission.

Genotypic Frequencies

If the parents can form different types of germ cells, the gene pair of the offspring is determined by chance union of one germ cell from each parent. A simple method for determining the expected or average fraction of the different types of offspring is to write the alleles from each parent, together with their frequency, in the following form. Let $(f_{S1}p_1 + f_{S2}p_2)$ stand for the gametes of the male (sire) and let $(f_{D1}p_1 + f_{D2}p_2)$ stand for the gametes of the female (dam) where the f's are the frequencies of the alleles p_1 and p_2. The f's must be 0, $\frac{1}{2}$, or 1. Multiply these two gametic arrays to obtain the expected offspring genotypes and frequencies:

$$(f_{S1}p_1 + f_{S2}p_2) \times (f_{D1}p_1 + f_{D2}p_2)$$

$$= f_{S1}f_{D1}p_1p_1 + f_{S1}f_{D2}p_1p_2 + f_{S2}f_{D1}p_2p_1 + f_{S2}f_{D2}p_2p_2$$

(This formula is not as complicated as the symbols make it appear.) The mating of animals that are Pp can be diagrammed as in Figure 9.3. The expected fre-

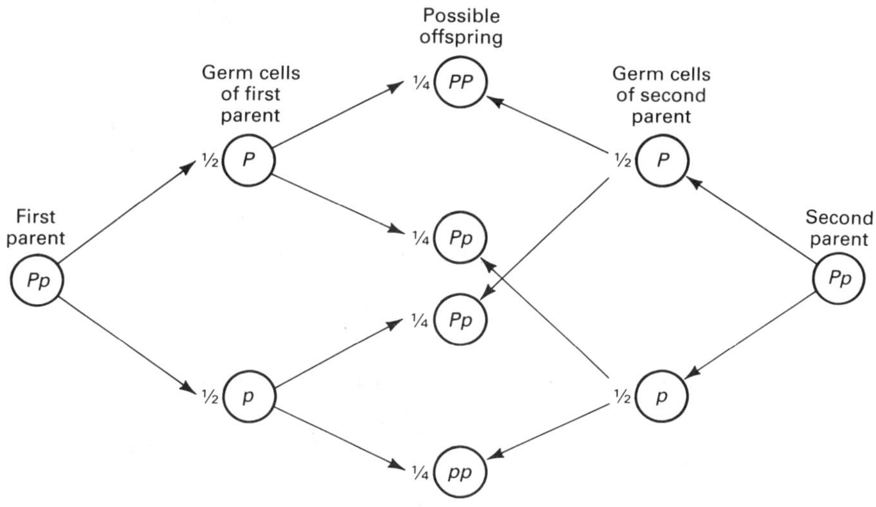

Figure 9.3 From *Principles of Dairy Science* by G. H. Schmidt and L. D. Van Vleck. Copyright © 1974 W. H. Freeman and Company. Used by permission.

quencies of the offspring from mating *Pp* and *Pp* are $\frac{1}{4}$, *PP*; $\frac{1}{2}$, *Pp*; and $\frac{1}{4}$, *pp*. In this example, $f_{S1} = \frac{1}{2}$, $p_1 = P$; $f_{S2} = \frac{1}{2}$, $p_2 = p$; $f_{D1} = \frac{1}{2}$, and $f_{D2} = \frac{1}{2}$. Thus $(\frac{1}{2}P + \frac{1}{2}p) \times (\frac{1}{2}P + \frac{1}{2}p) = \frac{1}{4}PP + \frac{1}{4}Pp + \frac{1}{4}pP + \frac{1}{4}pp$ with *Pp* and *pP* being the same.

When there are two possible alleles at a locus there are six possible types of mating, as summarized in Table 9.1. The expected proportions of offspring from the various matings are the expected fractions of the possible types of offspring if many such matings are made. Any particular mating might give considerably different proportions. For example, suppose that a mating is *Pp* × *Pp*. If only two offspring are obtained, some two-offspring families will be all *PP*, some will be all *pp*, some will be all *Pp*, and others will have various proportions of the three types, all by chance. The average proportions for a large number of offspring will, however, be close to the expected proportions, $\frac{1}{4}PP + \frac{1}{2}Pp + \frac{1}{4}pp$.

TABLE 9.1 SIX POSSIBLE MATINGS FOR TWO ALLELES AT ONE LOCUS

First parent		Second parent	Expected frequencies of offspring
PP	×	*PP*	All *PP*
PP	×	*Pp*	$\frac{1}{2}PP + \frac{1}{2}Pp$
PP	×	*pp*	All *Pp*
Pp	×	*Pp*	$\frac{1}{4}PP + \frac{1}{2}Pp + \frac{1}{4}pp$
Pp	×	*pp*	$\frac{1}{2}Pp + \frac{1}{2}pp$
pp	×	*pp*	All *pp*

Phenotypic Expression

Thus far only the genotypes (the gene pair) of the animals have been discussed. What the phenotypes are (what they look like) in relation to the genotypes has not been described. How is the phenotype related to the genotype? This question is best answered by giving examples of dominance and codominance.

Dominance. The first example is that of a dominant gene—the polled gene. As mentioned earlier, there are three possible genotypes. Table 9.2 shows the corresponding genotypes and phenotypes.

Two polled genes give polled cattle. Two horned genes give horned cattle. When the genes of the pair are identical, the animal is said to be *homozygous*. But one polled gene and one horned gene also give a polled animal. Thus the polled gene is *dominant* over the horned gene and masks any effect of the horned gene. The horned gene is said to be *recessive* to the polled gene. When the genes of a pair are not identical, the animal is said to be *heterozygous* at that locus.

Another dominant gene is the one for black in Holstein and Angus cattle, which is dominant to red. The curly coat in Ayrshires is dominant to a slick coat.

TABLE 9.2 EXAMPLE OF DOMINANT GENE ACTION

Gene pair (genotype)	What is seen (phenotype)
P and *P*	Polled (no horns)
P and *p* or *p* and *P*	Polled
p and *p*	Horned

Codominance. In the second situation, the heterozygote is different from either of the homozygotes. The classic example of this is the roan color of Shorthorns. Let *W* be the gene for white and let *w* be the gene for color. Shorthorns do not have the gene for black, so all colored shorthorns are red unless the gene for white at another locus is present. The phenotypes are described as in Table 9.3. This type of gene action is called codominance. Neither the white nor the colored gene is dominant over the other. The colored and white pair of genes result in a mixing of colored and white hairs.

TABLE 9.3 EXAMPLE OF CODOM'NANCE

Genotype	Phenotype
WW	White
Ww	Roan
ww	Red

If there are more than two alleles, many more genotypes and phenotypes are possible. For example, the B system of blood antigens contains many alleles. Such systems with a large number of alleles are useful in cases of disputed parentage, as will be discussed in Section 9.3. When allelic genes are codominant, the phenotype corresponds exactly to the genotype; that is, the number of different phenotypes is the same as the number of different genotypes. When allelic genes are dominant and recessive, the number of different genotypes is greater than the number of phenotypes because each heterozygous genotype has the same phenotype as the homozygous dominant genotype.

9.2 COMBINATIONS OF TRAITS

Independent Loci

What are the expected results for more than one trait? If the phenotype of each trait is determined separately by allelic genes (as described previously) and if the genes determining one trait are independent of the genes for the other traits, another simple rule holds: To be independent, the genotype at one locus must not affect the composition of the genotype at another locus. According to this rule, which is based on independence of loci, the expected results at one locus are multiplied by the expected results at the other locus. If more than two loci are involved, the multiplication continues. Consider the example of mating a roan shorthorn that is heterozygous for polled to a roan horned shorthorn; that is, $WwPp \times Wwpp$. The expected results for the "color" locus are

$$(\tfrac{1}{2}W + \tfrac{1}{2}w) \times (\tfrac{1}{2}W \times \tfrac{1}{2}w) = \tfrac{1}{4}WW + \tfrac{1}{2}Ww + \tfrac{1}{4}ww$$

The expected results for the polled locus are

$$(\tfrac{1}{2}P + \tfrac{1}{2}p) \times (1p) = \tfrac{1}{2}Pp + \tfrac{1}{2}pp$$

The joint results are

$$(\tfrac{1}{4}WW + \tfrac{1}{2}Ww + \tfrac{1}{4}ww) \times (\tfrac{1}{2}Pp + \tfrac{1}{2}pp)$$
$$= \tfrac{1}{8}WWPp + \tfrac{1}{8}WWpp + \tfrac{1}{4}WwPp + \tfrac{1}{4}Wwpp + \tfrac{1}{8}wwPp + \tfrac{1}{8}wwpp$$

These are the expected frequencies of the six different joint genotypes. The expected frequencies of the joint phenotypes can be found by adding together the frequencies of the genotypes that look alike. In this example the phenotypic frequencies are the same as the genotypic frequencies. However, this relationship might not always exist, if, for example, Holsteins heterozygous for the black-and-white color pattern replace the roan shorthorns.

Epistasis

The joint phenotypic frequencies may also be different from the joint genotypic frequencies if the genes at one locus affect the expression of the genes at another locus. An example is the recessive gene for albino, which, when homozygous, prevents any color expression. Suppose that C allows color and that c is the albino gene. If two Holsteins heterozygous both for black and white and for the albino gene are crossed, the expected results at the black-and-white locus are $\tfrac{1}{4}BB$ +

$\frac{1}{2}Bb + \frac{1}{4}bb$ and at the albino locus, the results are $\frac{1}{4}CC + \frac{1}{2}Cc + \frac{1}{4}cc$. The joint genotypic frequencies (by multiplication) would be

$$\tfrac{1}{16}BBCC + \tfrac{1}{8}BBCc + \tfrac{1}{16}BBcc + \tfrac{1}{8}BbCC + \tfrac{1}{4}BbCc + \tfrac{1}{8}Bbcc + \tfrac{1}{16}bbCC + \tfrac{1}{8}bbCc + \tfrac{1}{16}bbcc$$

The phenotypic frequencies would be

$$(\tfrac{1}{16}BBCC + \tfrac{1}{8}BBCc + \tfrac{1}{8}BbCC + \tfrac{1}{4}BbCc) = \tfrac{9}{16} \quad \text{black and white}$$

$$(\tfrac{1}{16}bbCC + \tfrac{1}{8}bbCc) = \tfrac{3}{16} \quad \text{red and white}$$

$$(\tfrac{1}{16}BBcc + \tfrac{1}{8}Bbcc + \tfrac{1}{16}bbcc) = \tfrac{4}{16} \quad \text{albino}$$

A gene that masks the effect of a gene at a different locus is said to be *epistatic* to the masked gene. (The albino gene is epistatic to the genes for black and red.)

Linked Genes

The simple formula for finding genotypic results for several loci will not work if the loci are linked. Linkage, a tendency for genes at certain loci to be inherited together, occurs when the loci are near each other on the same chromosome. The phenomenon of crossing over between the pair of chromosomes in production of germ cells results in recombination of genes at different linked loci. The frequency of recombination types depends on the map distance between the loci. The number of map units can be determined by genetic experiments. For example, suppose that two loci have alleles (*A*, *a* and *B*, *b*) and that the loci are 20 map units apart on the same chromosome. Figure 9.4 shows the chromosome structure for a double heterozygote that received the *A* and *B* genes from one parent and the *a* and *b* genes from the other.

If the genes were not linked, the possible gametes would be *AB, Ab, aB*, and *ab* with equal frequency. But since they are linked, the *A* and *B* gametes and the *a* and *b* gametes will occur more frequently than the *Ab* and *aB* gametes. The *AB* and *ab* types are called nonrecombination gametes because the genes stay together (see Figure 9.5). They comprise 80% of all the gametes when the map distance is 20 units: 0.40 *AB* and 0.40 *ab*. As shown in Figure 9.6, the recombination or crossover types, *Ab* and *aB*, make up 20% of all gametes: 0.10 *Ab* and 0.10 *aB*.

The map distance corresponds to the frequency of recombination gametes. For example, if the map distance is 40 units, the recombination gametes will be 40% of the total and the nonrecombination gametes will compose the remainder. If the distance between loci on the same chromosome is 50 or more map units or if the loci are on different chromosomes, the genes at those loci are inherited

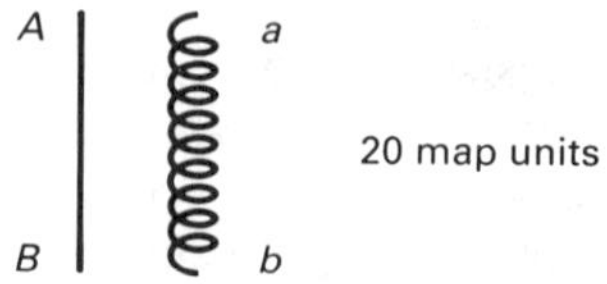

Figure 9.4 Loci separated by 20 map units. From *Principles of Dairy Science* by G. H. Schmidt and L. D. Van Vleck. Copyright © 1974 W. H. Freeman and Company. Used by permission.

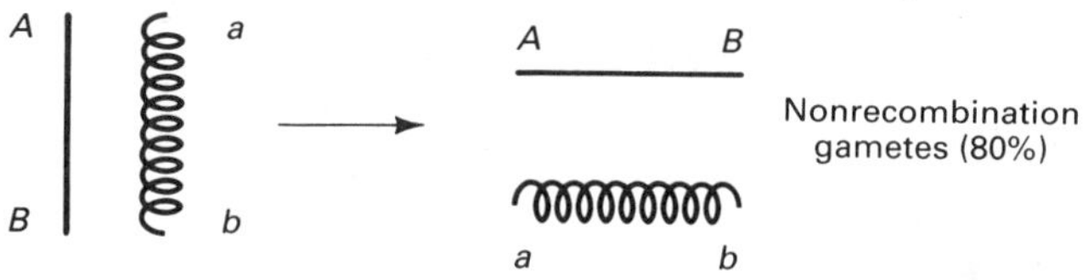

Figure 9.5 Nonrecombination gametes. From *Principles of Dairy Science* by G. H. Schmidt and L. D. Van Vleck. Copyright © 1974 W. H. Freeman and Company. Used by permission.

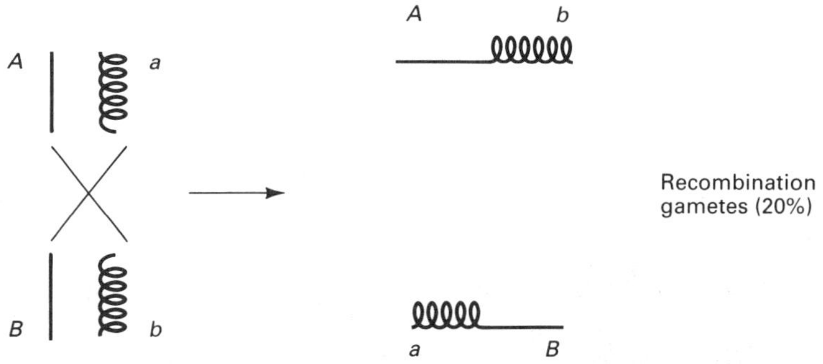

Figure 9.6 Recombination gametes. From *Principles of Dairy Science* by G. H. Schmidt and L. D. Van Vleck. Copyright © 1974 W. H. Freeman and Company. Used by permission.

independently. Since cattle have only 30 pairs of chromosomes, many genes are located on the same chromosomes.

9.3 GENETIC TESTS FOR PARENTAGE

The most practical use for simply inherited traits is in deciding cases of disputed or doubtful parentage. For example, an animal of known *AA* genotype cannot be the parent of an animal with the *aa* genotype.

Variation in many physiological characteristics of blood is known to be caused by alleles at particular loci. Many loci are known to control blood antigens. At some of these loci, an allele may control several antigenic factors. The details are complicated, but the important point is that standard procedures are available to identify the antigenic factors that an animal carries and thus, in many cases, the genotype of the animal at the controlling loci. Table 9.4 lists the standard nomenclature of 10 blood group systems which are associated with a particular locus.

Genetically controlled variation also occurs in several blood proteins and in several proteins of milk (Table 9.4). The casein loci have been shown to be very closely linked on the same chromosome.

Many studies have attempted to find relationships between certain of these relatively easily identified genes and productive traits such as milk yield or fat test. The results have been variable and generally inconclusive, although appar-

TABLE 9.4 GENETIC SYSTEMS CONTROLLING BLOOD ANTIGEN SYSTEMS, BLOOD PROTEINS, AND MILK PROTEINS

	Blood proteins		Milk proteins	
Blood antigen systems locus symbol	Protein	Locus symbol	Protein	Locus symbol
A	Hemoglobin	Hb	β-Lactoglobulin	Lg
B	Albumin	Al	α-Lactalbumin	α-La
C	Postalbumin	Pa	β-Casein	β-C_n
FV	Transferrin	Tf	α_{s1}-Casein	α_{s1}-C_n
J	α_2-Globulin	S_α	κ-Casein	κ-C_n
L	Alkaline phosphatase	F	γ-Casein	γ-C_n
M	Amylase	Am		
S				
R′-S′				
Z				

ently any difference in production resulting from a particular gene is very small compared with differences resulting from other genetic and environmental causes.

All bulls used artificially in the United States are tested for several blood systems and for some blood protein types. These systems can be used in cases of disputed parentage.

The alleles in these gene systems are mostly codominant; therefore, if an animal is heterozygous, both alleles can be determined. Codominance is a feature of most protein systems and of some antigen systems. If the blood type of only one parent is known, only codominant genes are useful in determining parentage because of the difficulty in determining whether an animal with a dominant phenotype is homozygous or heterozygous for the dominant gene.

Exclusion Principle

The basic law of parentage tests based on these genetic factors must be emphasized. An animal cannot be proved to be the parent of another animal. An animal can only be proved not to be the parent of another animal. Generally, however, if an animal is not excluded as a parent in a parentage test, the animal may be accepted as the parent for reasons of opportunity.

An example will explain how parentage tests work. The alleles will not be

TABLE 9.5 EXAMPLE OF A PARENTAGE TEST

	Blood system			Protein system	
Animal	B	FV	J	Hb	Tf
Sire I	B_1/B_5	F_1/F_1	J_1/J_2	H_1/H_1	T_1/T_3
Sire II	B_5/B_5	F_1/F_2	J_2/J_4	H_1/H_1	T_2/T_3
Calf	B_1/B_5	F_1/F_3	J_1/J_3	H_1/H_1	T_1/T_2

labeled by their usual lengthy superscripts; rather, numbers will be used to indicate different codominant alleles at five different loci for three animals as shown in Table 9.5.

At the B, FV, Hb, and Tf loci, either sire could be the sire of the calf. But at the J locus, sire II can contribute only a J_2 or a J_4 gene to its offspring, but the calf does not have either. Therefore, based on the *exclusion principle*, sire II is excluded as a parent of the calf. Sire I is not excluded and may be the parent.

Twins

Determination of one-egg (identical) twins may also be attempted with blood typing. All twins with different blood types are two-egg twins. However, a situation may arise in which two-egg twins may have the same apparent blood types. Early

TABLE 9.6 PARTIAL LIST OF SIMPLY INHERITED LETHALS AND SEMILETHALS AFFECTING DAIRY CATTLE

Characteristic	Comments
Acroteriasis	Amputated, recessive
Adenohypophysial aplasia	Prolonged gestation; small size, recessive
Arthrogryposis	Permanent joint contractures, recessive
Atresia ilei	Blockage of small intestine, possibly genetic
Cerebellar hypoplasia	Recessive
Cryptorchidism	Recessive
Deficiency of uridine monophosphate synthase (DUMPS)	Embryonic death when homozygous, recessive
Dropsy	Fatal fluid retention in fetus, recessive
Dwarfism	Rare, recessive
Epileptic fits	Dominant
Epithelial defects	Defective skin formation, recessive
Fused teats	Recessive
Hairlessness	Recessive
Hydrocephalus	Recessive
Hypotrichosis and anodontia	Lack of hair and teeth, sex-linked recessive
Ichthyosis cogenita	Fishskin, recessive
Limber leg	Nonfunctional legs, recessive
Porphyria	Pink tooth, recessive
Prolonged gestation	Oversize, recessive
Recto-vaginal constriction (RVC)	Difficult calving, recessive
Short spine	Recessive
Spastic (crampy)	Contracture of muscles, recessive
Strabismus	Crossed eyes by 12 months, recessive
Streaked hairlessness	Sex-linked dominant
Syndactylism	Mule foot, recessive
Umbilical hernia	Polygenic
Weaver	Lack of control of legs, recessive
White heifer disease	Abnormal reproductive tract, recessive
Wry tail	Recessive

in the fetal stage there is exchange of blood between the two fetuses in approximately 90% of all twin pregnancies. This exchange often results in a mixture in the blood of one twin of its blood type and also of its twin, so that the true blood type cannot be determined. This condition is called *erythrocyte mosaicism*, that is, evidence for a mixture of blood types, and indicates that the twins are from two eggs. In some cases the shift to the blood type of one twin is so great that both appear to have the same blood type, when in fact they are genetically different. There are, however, some loci that are not affected by the exchange between the two fetuses (vascular anastomosis). Thus the J blood antigen system and the transferrin, postalbumin, and phosphatase systems can be used as further tests for *monozygotic* (one-egg) twins.

This phenomenon of erythrocyte mosaicism can be used to predict which heifers born twin to a bull will be fertile. Erythrocyte mosaicism indicates vascular anastomosis, which also allows movement of male hormones to the female, resulting in an "intersex" and infertile female, commonly called a free martin. If the blood antigens of the heifer twin are different from those of her male twin, she will probably be of normal fertility.

Blood and protein typing has economic value in cases of disputed parentage and in the prediction of fertile heifer twins. Recessive genes that are lethal or semilethal in the homozygous state are also economically important. Table 9.6 lists some of these. Although most have not been serious problems, elimination of such genes through detection of heterozygous carrier males may be important.

9.4 DETECTING CARRIERS OF GENES FOR ABNORMAL TRAITS

Many dairy managers using artificial insemination (AI) have probably had the following nightmare. A well-promoted and heavily used bull has been found, much too late of course, to be a carrier of a seriously undesirable genetic factor. The question that should be resolved for all dairy managers, stud managers, and the AI industry is: Can the spread of undesirable genetic factors be prevented by detection of carriers before they are used heavily? The following discussion shows that carriers can be detected by several mating methods having predetermined chances of nondetection of carriers. These results suggest that AI, if properly used, can help eliminate problems caused by hidden undesirable genes.

The problem of detecting carriers is obvious. There is no way to tell the difference between a "normal" animal with two desirable genes and a "normal" animal that carries one undesirable gene in addition to one desirable gene. The carrier, however, will provide half its offspring with the undesirable gene. In general, if a carrier male is mated randomly to a group of cows, the expected proportion of normal and undesirable calves can be determined if the proportions of eggs carrying desirable and undesirable genes are known.

Mating to Homozygous Recessive Cows

A carrier will be detected if at least one undesirable calf appears because an undesirable calf can occur only if an undesirable gene is received from the father as well as from the mother. In mating a carrier bull to cows homozygous for the undesirable gene, the chance of getting all normal calves out of n calves is $(\frac{1}{2})^n$, because the type of offspring at each birth is independent of the type of offspring at other births. In other words, the chance that the first calf is normal is $\frac{1}{2}$ and the chance that the second calf if normal is also $\frac{1}{2}$, so the chance that both the first and second calves are normal is $(\frac{1}{2})(\frac{1}{2}) = (\frac{1}{2})^2$. Thus the chance of *not* detecting a carrier out of n calves is $(\frac{1}{2})^n$. Because the chance of detection plus the chance of nondetection must be 1, the chance of detection is $1 - (\frac{1}{2})^n$ (Table 9.7).

Mating to Carrier Cows

Another method is to mate the male to known carrier cows—of the type *Bb*, as in the example of genetic recombination. Known carriers have produced at least one affected calf. The chance of obtaining all normal calves out of n calves will be $(\frac{3}{4})^n$, which is the chance of nondetection. The chance of detection will be $1 - (\frac{3}{4})^n$ (Table 9.7).

Mating to Own Daughters

A third method is to mate a bull back to his own daughters. In Figure 9.7, both genes of each of the original mates are assumed to be normal genes. The fraction of eggs of the daughters carrying *B* will be $\frac{3}{4}$ and $\frac{1}{4}$ will be *b*. Thus mating the sire to his own daughters gives the pattern shown in Figure 9.8. The chance of obtaining all (n) normal calves from mating a carrier to his own daughters is $(\frac{7}{8})^n$ and the chance of detecting a carrier is $1 - (\frac{7}{8})^n$ (Table 9.7).

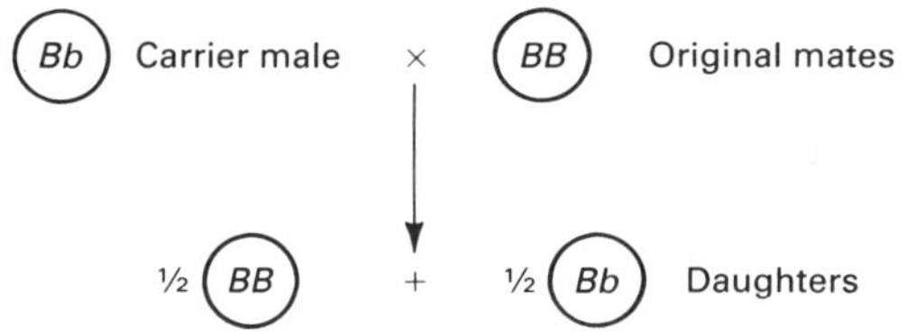

Figure 9.7 Mating of a carrier bull to homozygous cows. From *Principles of Dairy Science* by G. H. Schmidt and L. D. Van Vleck. Copyright © 1974 W. H. Freeman and Company. Used by permission.

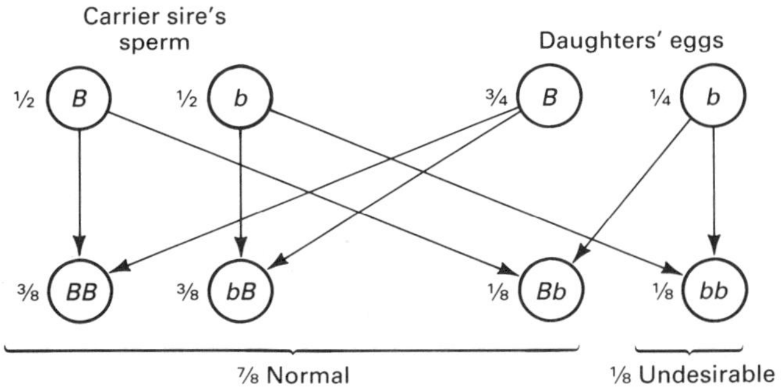

Figure 9.8 Mating of a carrier bull to his daughters. From *Principles of Dairy Science* by G. H. Schmidt and L. D. Van Vleck. Copyright © 1974 W. H. Freeman and Company. Used by permission.

TABLE 9.7 CHANCES OF DETECTING A CARRIER BULL IN VARIOUS MATING COMBINATIONS

	Detects only one recessive		Detects all recessives carried
Number of calves, n	Mating to homozygous recessive cows $1-(\frac{1}{2})^n$	Mating to known carrier cows $1-(\frac{3}{4})^n$	Mating to own daughters $1-(\frac{7}{8})^n$
1	0.50	0.25	0.12
2	0.75	0.44	0.23
3	0.88	0.58	0.33
4	0.94	0.68	0.41
5	0.97	0.76	0.49
6	0.98	0.82	0.55
7	0.99	0.87	0.61
8	1.00	0.90	0.66
9		0.92	0.70
10		0.94	0.74
15		0.99	0.87
20		1.00	0.93
50			1.00

Mating to the Whole Population

The fourth method is to mate the bull to a random group of cows. Some cows in the population will be carriers of the undesirable gene, and therefore a certain fraction of the eggs from the whole group of cows will carry the undesirable gene. If q is the fraction of undesirable genes and $1 - q$ is the fraction of normal genes, then, for the mating system shown in Figure 9.9, the chance of all normal will be $(1 - \frac{1}{2}q)^n$ and the chance of detecting a carrier by obtaining one or more calves with the undesirable trait out of n calves will be $1 - (1 - \frac{1}{2}q)^n$, following the same procedure used previously. The chance of detection obviously depends on the abundance of the undesirable gene in the female population (Table 9.8).

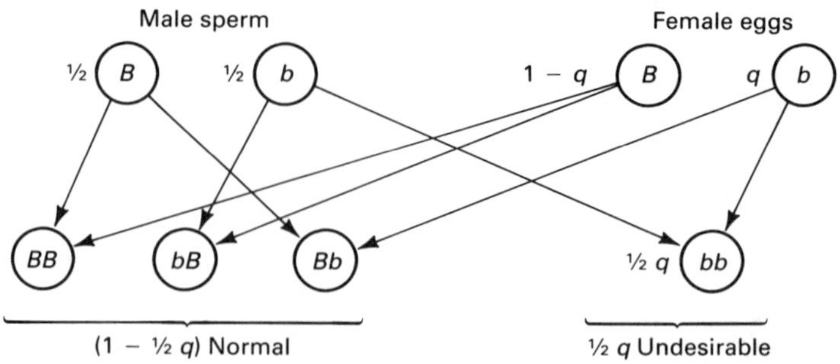

Figure 9.9 Mating a carrier bull to a random group of cows. From *Principles of Dairy Science* by G. H. Schmidt and L. D. Van Vleck. Copyright © 1974 W. H. Freeman and Company. Used by permission.

TABLE 9.8 CHANCES OF DETECTING A CARRIER BULL BY RANDOM MATING IN THE WHOLE POPULATION

Number of calves, n	Detects all recessives depending on frequency $1 - (1 - \frac{1}{2}q)^n$ Recessive gene frequency $= q$				
	0.2	0.1	0.05	0.01	0.001
1	0.08	0.05	0.02	0.00	0.00
2	0.16	0.09	0.05	0.01	0.00
3	0.23	0.13	0.07	0.01	0.00
4	0.29	0.19	0.09	0.02	0.00
5	0.35	0.21	0.11	0.02	0.00
6	0.41	0.24	0.13	0.03	0.00
7	0.46	0.28	0.16	0.03	0.00
8	0.50	0.31	0.18	0.04	0.00
9	0.54	0.34	0.20	0.04	0.00
10	0.58	0.37	0.21	0.05	0.00
15	0.73	0.50	0.30	0.07	0.01
20	0.82	0.61	0.38	0.09	0.01
50	0.99	0.90	0.70	0.22	0.02
100	1.00	0.99	0.91	0.39	0.05
200		1.00	0.99	0.63	0.10
300			1.00	0.77	0.14
400				0.86	0.18
500				0.92	0.22

Impossibility of Proving Noncarrier

A noncarrier bull can never have a calf with the undesirable trait. Thus any bull with an undesirable calf is immediately labeled a carrier. On the other hand, a bull can never be completely proved to be a noncarrier; there will always be a small chance of nondetection, no matter how many normal calves a bull sires. The advantages and disadvantages of the methods described here are obvious; the fourth method merits additional discussion.

Consequences of Testing for Carriers in AI

Random mating to a population of cows detects all undesirable genes, but the chance of detection depends on the abundance of the undesirable gene; that is, the method is effective when the population has many "bad" genes and is less effective when the population has mostly "good" genes. For example, when one of every 10 genes from cows is undesirable, a bull having 100 calves has a chance of detection of 99%. But if only 1 in 1000 genes is undesirable, the chance of detection is only 5%. Thus bulls carrying genes that have a low frequency in the cow population will not be readily detected. This disadvantage, however, is not really important. For example, suppose that a nondetected carrier bull is mated to a population of cows producing only 1 out of 1000 eggs carrying the undesirable gene (Figure 9.10). Only 5 out of 10,000 calves will be expected to show the undesirable trait, a proportion that, economically, is not very important. The frequency of the undesirable gene in the next generation of cows, however, will be increased because one-half the calves of the carrier will carry the undesirable gene (0.4995 + 0.0005 are *Bb*). These new carriers will cause no damage because they are normal in appearance and performance. In the next generation, carrier bulls mated to this group of carrier cows will be detected with a chance of nearly 100% during the sampling period. Thus no carrier bull will be used heavily and none or only a very few calves with the undesirable trait will be born. The net effect of AI and proper testing by mating to 200 or more cows will, through this alternation of high chances of detection and buildup of carrier cows, reduce the number of calves having simply inherited undesirable traits to almost zero in a few generations, as shown in Table 9.9.

Two examples of ways in which genes for serious problems can be widely spread will illustrate the importance of the AI testing procedure.

In the 1920s the Ayrshire bull Dunlap Talisman brought with him to Finland the gene for cogenital dropsy, a lethal condition. By 1937 he had 189 registered grandsons, of which 82 came from sons known to carry the gene. The gene still persists in the United States, although AI testing is reducing the prevalence of the condition.

In the 1960s the Eastern A.I. Cooperative (EAIC) had a Holstein bull, UNH Burke Graduate, which had about 50,000 calves before he was proved to be a carrier of mulefoot, a type of syndactylism. The stud immediately culled the bull, despite numerous objections of some breeders because of the bull's popularity. Actually, EAIC had been more than careful in using Graduate. When they discovered that his sire was a carrier, they test-mated Graduate to daughters of known carrier bulls. Out of 27 calves, none were mulefoot (97% chance of detection), so they continued to use him until one mulefoot calf was born. The reasons for not detecting him earlier were the low frequency of the gene in the population and the bad luck of not detecting him in the test matings.

The infrequency of mulefoot calves illustrates that even if a carrier bull

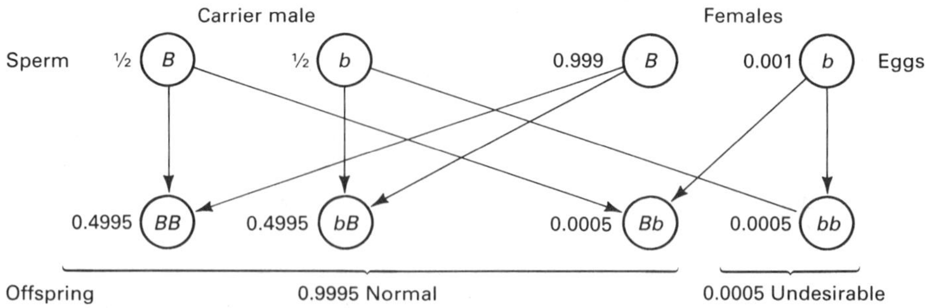

Figure 9.10 Mating a carrier bull to cows furnishing the undesirable trait to 1 egg in 1000. From *Principles of Dairy Science* by G. H. Schmidt and L. D. Van Vleck. Copyright © 1974 W. H. Freeman and Company. Used by permission.

TABLE 9.9 NUMBER OF HOMOZYGOUS RECESSIVE CALVES PER 1000 CALVES BORN AS A RESULT OF NO TESTING AND AI TESTING IF ALL HOMOZYGOUS RECESSIVES AND ALL DETECTED BULLS ARE CULLED

Initial percentage of carrier cows	Type of testing	Generation							
		Initial	1st	2nd	3rd	4th	5th	···	10th
100	None	250	111	63	40	28	20		7
	AI	250	0	0	0	0	0		0
40	None	40	28	20	16	12	10		4
	AI	40	0	0	0	0	0		0
20	None	10	8	7	6	5	4		3
	AI	10	0	0	0	0	0		0
10	None	$2\frac{1}{2}$	2	2	2	2	2		1
	AI	$2\frac{1}{2}$	0	0	0	0	0		0

escapes detection, not much harm results because in the next generation, carriers can be detected more readily and therefore not used heavily. If, however, a known carrier bull had been used to breed heifers sired by Graduate, approximately one-fourth of the calves would have the mulefoot condition. The only solution is to rely on the test results and not take any chances.

Of the four methods of testing for carriers of undesirable genes, the method that adapts easily to AI and production testing of young bulls through restricted mating to a random group of cows is also best from an economic point of view and will eliminate most undesirable traits caused by recessive genes if such undesirable traits are quickly and completely reported. Failure to report may considerably decrease chances of detection.

FURTHER READINGS

Auerbach, C. 1961. *The Science of Genetics*. Harper & Row, Publishers, Inc., New York.

Hines, H. C., G. F. W. Haenlein, J. P. Zikakis, and H. C. Dickey. 1977. Blood antigen, serum protein and milk protein gene frequencies and genetic interrelationships in Holstein cattle. *J. Dairy Sci.*, 60:1143.

Hines, H. C., J. P. Zikakis, G. F. W. Haenlein, C. A. Kiddy, and C. L. Trowbridge. 1981. Linkage relationships among loci of polymorphisms in blood and milk of cattle. *J. Dairy Sci.*, 64:71.

Hutt, F. B. 1964. *Animal Genetics*. Ronald Press, New York.

Johansson, I. 1961. *Genetic Aspects of Dairy Cattle Breeding*. University of Illinois Press, Urbana, IL.

Johansson, I., and J. Rendel. 1968. *Genetics and Animal Breeding*. W. H. Freeman and Company, Publishers, San Francisco.

Leipold, H. W., and S. M. Dennis. 1986. Congential defects affecting bovine reproduction. *Theriogenology*. W. B. Saunders Company, Philadelphia, p. 177.

STORMONT, C. 1958. Genetics and disease. *Adv. Vet. Sci.*, 4:137.

STORMONT, C. 1967. Contribution of blood typing to dairy science progress. *J. Dairy Sci.*, 50:253.

WOMACK, J. E., and Y. D. MOLL. 1986. Gene map of the cow: Conservation of linkage with mouse and man. *J. Hered.*, 77:2.

10

Keys to Genetic Progress

10.1 THE KEYS

Genetic improvement requires variation; otherwise, selection would be impossible. Some of the variation must result from genetic differences that result in *genetic variation*. In addition, only the best must be selected. The extent to which selection will result in genetic improvement depends on the *intensity of selection*, that is, whether the best 1 of 10, the best 5 of 10, the best 9 of 10 or some other fraction is selected.

The herd owner must also decide which cows are best. Usually, the record of the cow is of major importance, although several records provide more accuracy. The knowledge that close relatives of the cow have high records would slightly increase confidence in the cow. Thus the number and type of records on the cow and its relatives determine the *accuracy of selection*.

These three factors determine genetic progress per generation. A fourth factor is quite important in determining the rate of genetic improvement in some situations because herd owners are usually interested in genetic progress per year. Genetic progress per generation divided by the number of years per generation gives genetic progress per year. Thus *generation interval* is important. The generation interval is the average time between birth of an animal and birth of its replacement.

The preceding description can be summarized by a simple equation:

$$\text{genetic gain/year} = \frac{\text{accuracy} \times \text{intensity} \times \text{genetic variation}}{\text{years per generation}}$$

A more detailed equation is given in Section 10.4.

Since breeders want to make genetic gain as large as possible, the three numerator parts should be as large as possible and the generation interval should be as small as possible. An understanding of the parts of this equation will contribute to an understanding of what can or cannot be done to increase genetic progress.

Accuracy

Accuracy of predicting true genetic value of a cow or bull is defined as the correlation between the prediction of genetic value and the *true* genetic value of the animal, which ranges from 0% by guessing to a perfect 100% when genetic value is known exactly. Perfection in predicting the genetic value of dairy cows is impossible to obtain. Fortunately, if enough daughters of a bull are considered, his true daughter superiority, a measure of half his genetic value, can be predicted nearly perfectly.

Accuracy also depends on the fraction of the variation among cows caused by genetic effects and the fraction caused by other effects. The fraction caused by genetic differences is called *heritability*. If all differences are genetically determined, heritability is 100%; thus an animal's appearance or records would perfectly measure its genetic value. If none of the differences are genetically determined, heritability is zero, which means that genetic progress is impossible because all animals have the same genetic value.

In general, as heritability increases, the accuracy of predicting genetic value of a cow also increases. Estimates of heritability for milk production vary from 20 to 40%. A value of 25% is commonly used. Heritability of final type score is approximately the same. Heritability of fertility is, however, much lower—0 to 10%. Thus progress by selection is much more likely for milk yield or type than fertility. Heritability is discussed in more detail in Section 10.3.

Use of more than one record on a cow and records of its relatives improves accuracy if the records are combined in a way that maximizes the accuracy of predicting genetic value. The selection index, commonly called the EBV (estimated breeding value), provides a "best" estimate of the genetic or breeding value of a cow.

The procedure is somewhat complicated if records on many relatives are used. Table 10.1 shows that a cow's records are more important than records of many relatives. (Heritability of 25% and repeatability of 50% have been assumed.) Records of daughters or dam add little accuracy unless the cow has no records. If the cow has no record, the proof of her sire is an important indication of her genetic value. If records of more relatives are used, accuracy is increased only slightly.

The situation is different when evaluating bulls. Nearly perfect accuracy of

TABLE 10.1 ACCURACY OF PREDICTING GENETIC VALUE FOR MILK PRODUCTION FROM A COW'S RECORDS AND RECORDS OF COW'S RELATIVES

Number of records on cow	No other relatives	Records on cow plus a record on dam or daughter	Records on cow plus her sire's proof: Number of paternal half sisters 50	Records on cow plus her sire's proof: Number of paternal half sisters 200
0	0.00	0.25	0.44	0.48
1	0.50	0.53	0.60	0.63
2	0.57	0.60	0.64	0.66
3	0.61	0.63	0.67	0.68
6	0.65	0.67	0.70	0.71

predicting a bull's genetic value can be achieved if enough daughters in many herds are analyzed. USDA sire proofs include a value called repeatability, which is the square of accuracy.

Table 10.2 presents the accuracy of sire proofs, assuming heritability of 25% and environmental correlation of 6% among daughters in the same herd. An environmental correlation exists when daughters of a sire are treated more alike than unrelated cows.

Intensity of Selection

Intensity of selection can have a substantial effect on genetic progress. A breeder cannot be as selective with cows as with bulls because not enough heifers are available for replacements. But there can be intense selection of dams to be mothers of young bulls, as well as intense selection of bulls based on daughter records. The selection intensity factor is a relative measure of how much the average of a selected fraction will exceed the average of the population before selection. Table

TABLE 10.2 ACCURACY OF SIRE PROOFS FOR MILK YIELD

Number of daughters, n	Accuracy when all cows in different herds, $\left(\frac{n}{n+15}\right)^{0.5}$	Accuracy when all cows in one herd, $\left(\frac{n}{2n+14}\right)^{0.5}$
1	0.25	0.25
5	0.50	0.46
10	0.63	0.53
20	0.76	0.60
40	0.85	0.65
50	0.88	0.66
70	0.91	0.67
100	0.93	0.68
200	0.96	0.69
1000	0.99	0.70

10.3 gives the intensity factors for various percentages of selection. If the intensity factors are to apply, selection must be exactly according to the estimated genetic value. In other words, selection of any 10 out of the top 50 of 100 animals is not 10% selection but 50% selection. The intensity factor increases at a faster rate as the percentage selected declines.

Because genetic progress is proportional to the product of accuracy of predicting breeding value and the selection intensity factor, an estimate can be made of the relative importance of four selection paths for genetic improvement as shown in Figure 10.1. As a simple example, suppose that the accuracies of prediction and selection intensity factors are as shown in the figure. The selection intensity factors demonstrate how much more intense selection can be for bulls than for cows except for selection of dams to produce sons. Although in practice other factors may make some of the assumptions invalid, relative superiority of the selected sires and dams of bulls and cows is shown in Figure 10.2 for the accuracy and intensity of selection given in Figure 10.1. Certainly, the potential importance of sire selection is very great compared with selection of dams to have replacement heifers.

Genetic Standard Deviation

Genetic variation is very important. If there are no genetic differences among animals, there will be no genetic progress no matter how intense the selection. Some traits have little genetic variation; others have considerable variation. For-

TABLE 10.3 FACTORS ASSOCIATED WITH INTENSITY OF SELECTION

Select top percentage	Intensity factor	Comments
100	0.00	
90	0.20 ←	Usual level for selecting cows in a herd
85	0.27	
75	0.42 ←	Maximum level for selecting cows in a herd
70	0.50	
60	0.64	
50	0.80	Usual range for selecting dams of natural service and AI young bulls; also, usual range for selection of bulls out of young bulls sampled
40	0.97	
30	1.16	
20	1.40	
10	1.75	
5	2.06	Possible range of selection for dams of young bulls to sample in AI; also, possible range of selection of sires of young bulls to sample in AI
4	2.15	
3	2.27	
2	2.42	
1	2.67	

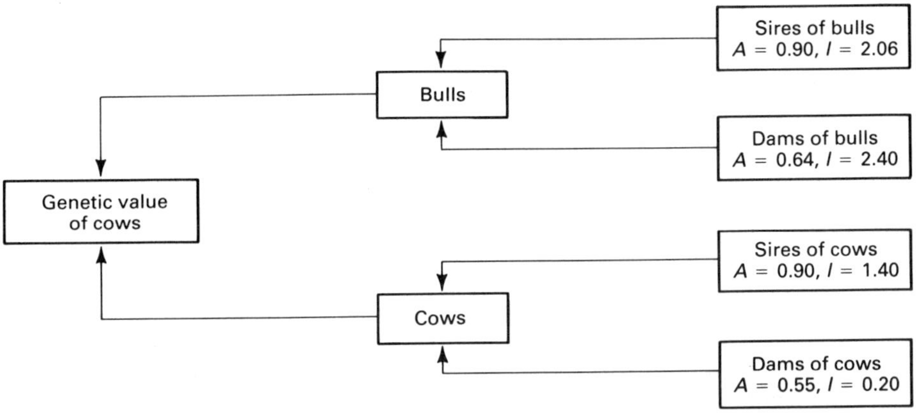

Figure 10.1 Some typical accuracy and selection intensity values under AI conditions for each of the four paths of genetic superiority. From *Principles of Dairy Science* by G. H. Schmidt and L. D. Van Vleck. Copyright © 1974 W. H. Freeman and Company. Used by permission.

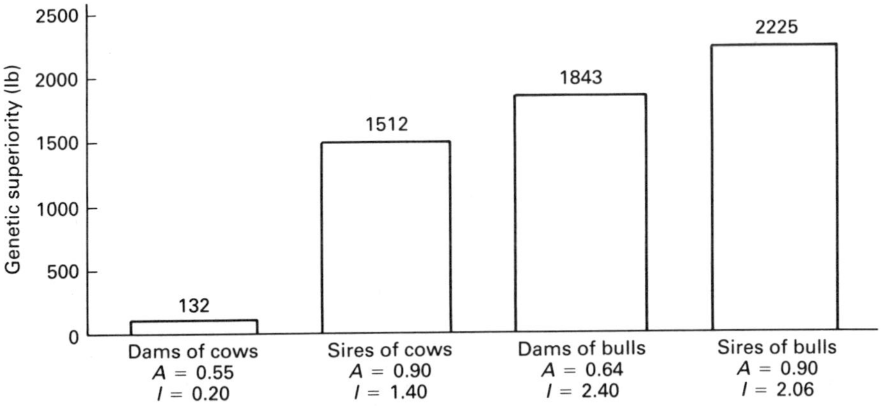

Figure 10.2 Possible genetic superiority for milk yield of selected sires and dams, assuming that the genetic standard deviation of milk is 1200 lb. From *Principles of Dairy Science* by G. H. Schmidt and L. D. Van Vleck. Copyright © 1974 W. H. Freeman and Company. Used by permission.

tunately, there are relatively large genetic differences among cows with regard to milk production. The measure commonly used, the genetic standard deviation, is approximately 1500 lb of milk for Holstein, Brown Swiss, and Ayrshire and somewhat less (1000 to 1200 lb) for Jersey and Guernsey first-lactation records in the United States. Where average production is lower, the genetic standard deviation is likely to be proportionately less. The genetic standard deviation is related to heritability and overall variation as discussed in Section 10.4.

Generation Interval

Generation interval influences the amount of progress per year. The shorter the generation interval, the more the progress per year. The generation interval is usually different for bulls and cows. For cows it averages approximately 5 to 6 years in the United States.

The generation interval of 8 to 10 years for bulls often is longer than it should be. One problem is waiting for a progeny proof. This wait adds 4 years to the one year between the time the bull is born and the time he is used. Thus a practical minimum is 5 to $5\frac{1}{2}$ years. Partial lactation records may reduce this minimum to 5 years. The minimum of 5 to $5\frac{1}{2}$ years assumes that as soon as a proof is obtained, the best bulls are used to sire the next generation of young bulls. If records on the second crop of daughters are required before selection, another 4 years must be added to the generation interval, making a total interval of 8 to 10 years.

The length of the generation interval is also a major problem when naturally proved (NS) bulls are used in AI. The original NS proof is approximately as accurate as a pedigree index for a young bull based on a dam with several records, an AI proof on the dam's sire, and an AI proof on the sire of the young bull. However, the natural service proof usually is not available until the bull is 6 to 7 years old. If an AI proof is then required, the bull is 10 to 11 years old when he enters heavy service. The generation interval of 10 or more years would impede progress even without waiting for a second set of AI daughters.

Balancing the Keys

Some of the keys to genetic progress affect other keys. For example, decreasing generation interval probably decreases accuracy since fewer records are available for use in selection. Accuracy may need to be balanced against intensity of selection, especially if only a fixed number of cows are available for progeny testing of young bulls. Increasing the number of records of animals before selection decisions are made will reduce the intensity of selection because some animals will have become unavailable for selection.

10.2 VARIATION: THE KEY TO THE UNEXPECTED

An understanding of "heritability" and how it can be used in a breeding program is important to cattle breeders. An accurate definition of heritability requires an understanding of variation.

Jay L. Lush of Iowa State University stated that variation is the raw material with which the breeder works. An example of variation is the differences among animals for characteristics such as milk production or type score. A simple measure of variation, but one that is difficult to use, is the range, or the difference between the extremes. For example, mature equivalent (ME) 305-day milk production within a herd may be from a high of 30,000 lb of milk to a low of 10,000 lb. The difference of 20,000 lb is the range for that herd.

The usual measure of variation is the *variance,* which, in technical terminology, is the average squared deviation of individual records from their population average. The variance within a herd is the average of the squared deviations of all individual records from the herd average. A statistical discussion of variance, as well as a procedure for computing it, is presented in Section 10.4. A more empirical discussion of variation follows.

Cattle breeders would be fooled if they thought all the variation within a herd resulted from genetic differences among the cows. Usually, only part of the total variation can be traced to genetic causes. The variation that is caused by genetic differences among individuals is called the *genetic variance.*

As an example, assume that the average production of daughters of sire A is reported to be 1000 lb above their herdmates' average. The daughters of sire B average the same as their herdmates. Without really thinking, many herd managers would say that the records of nearly all the daughters of sire A would be better than those of sire B. However, a herd averaging 18,000 to 19,000 lb of milk may have some cows that produce 25,000 to 28,000 lb of milk and other cows that produce 12,000 to 13,000 lb. Would not the same be true for daughters of a sire? Some daughters would be much higher producers and others would be much lower producers than the average daughter of a sire. Half would be expected to be above the average and half would be below the average of the sire group.

How much below the average are the records of some of these daughters expected to be? To find an empirical answer to this question, first-lactation records of the first 500 daughters of 117 AI Holstein sires, expressed as deviations from their herdmates, were grouped by 1000-lb intervals above and below the average for the sire. The results averaged for all the sires are shown in Figure 10.3. The interval with the largest fraction of daughter records is from 500 lb below the sire average to 500 lb above the sire average. Approximately 17% of the daughters

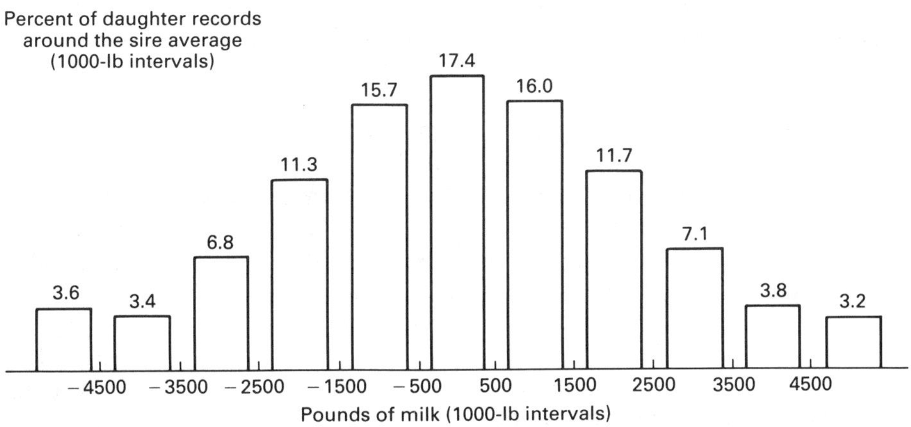

Figure 10.3 Distribution of daughter records composited for 117 sires, each having 500 daughters. From *Principles of Dairy Science* by G. H. Schmidt and L. D. Van Vleck. Copyright © 1974 W. H. Freeman and Company. Used by permission.

were within 500 lb of their sire group average. Roughly half of the daughters were above average and half were below average, again as expected.

Figure 10.3 shows that each interval has a different fraction and that the fraction decreases as the amount from the average increases or decreases. Not many cows are expected to be 5000 to 6000 lb above the herd average, especially after records are adjusted for age differences, but many cows are expected to be near the average.

Distribution of Records

Figure 10.3 shows that records of daughters of a sire follow a normal distribution. The term "normal" refers to the so-called bell-shaped distribution of records. If a smooth, curved line is drawn connecting the tops of the groups, the typical bell shape appears, which is characteristic of many types of biological effects. The normal distribution is such that approximately 68% of the records of daughters of the average sire would be expected to be within one standard deviation, in this example 2500 lb, of the sire's average. Addition of the fractions in the middle five groups, between −2500 and +2500, shows that 72% or slightly more than 68% were in that interval.

To illustrate the similarity of the distributions of daughter records for all 117 sires, three sires (A, B, and C) were selected whose daughters' production averaged approximately 1000 lb more than, equal to, or 1000 lb less than the herd averages. The shapes of the distributions in Figure 10.4 are similar and approximate the bell shape. The major difference in the curves is that they are offset from one another by either 1000 or 2000 lb—the differences in the sire averages. For example, the charts for sires A and C look alike except that the peak interval for sire A is +500 to +1500, whereas the largest interval for sire C is −1500 to −500.

Even though the daughter averages are markedly different, several daughters of sire C outproduced some daughters of sire A. For example, 61% of daughters of sire C were above −1500, and 13% of the daughters of sire A were lower than −1500. The important point is more daughters of the high-average sire were higher than those of the low-average sire and fewer daughters of the high-average sire were as low as those of the low-average sire, as shown by Table 10.4, which is based on Figure 10.4.

The preceding discussion shows the extent to which variation can be seen within a genetically similar group (comprised of daughters of the same sire), but the variation in genetic values is also important. The average of 500 daughters of a sire should indicate his daughter superiority; that is, there should be a rough balance between his poorer daughters and his better-than-average daughters. If only 10 daughters are averaged, the average could be boosted if, by luck, all daughters were above average, or lowered if, by chance, all daughters were below average. With enough daughters, usually 50 to 100 or more, the odds even out, so that sire averages of 500 daughters provide a good measure of genetic differences.

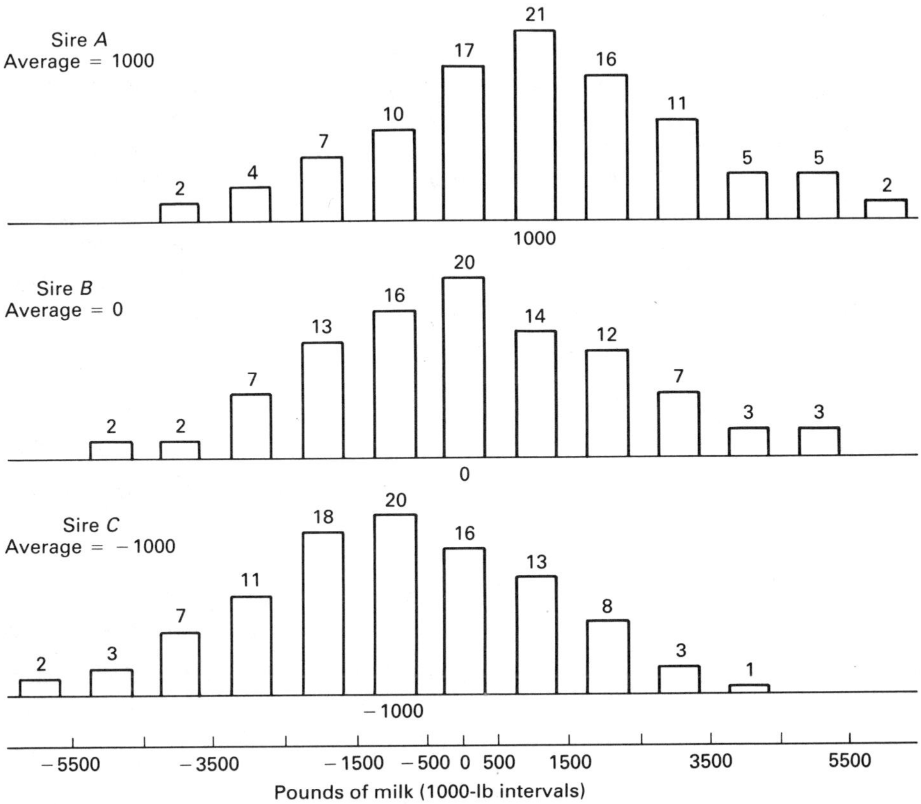

Figure 10.4 Percentage of daughter records for high, average, and low sires (in 1000-lb intervals). From *Principles of Dairy Science* by G. H. Schmidt and L. D. Van Vleck. Copyright © 1974 W. H. Freeman and Company. Used by permission.

TABLE 10.4 PERCENTAGE OF DAUGHTERS OF A HIGH-AVERAGE BULL (+1000 POUNDS), A MIDDLE-AVERAGE BULL (+0 POUNDS), AND A LOW-AVERAGE BULL (−1000 POUNDS) IN THE SAME GENERAL PRODUCTION LEVELS

Sire	Percentage of daughters above +1500	Percentage of daughters above +500	Percentage of daughters below −500
A	37	60	23
B	25	39	40
C	12	25	61

Genetic Variation

How does the distribution of sire averages compare with the distribution of daughter records? The averages of 117 sire groups of 500 daughters are plotted in Figure 10.5 for the same 1000-lb intervals as their daughter records. The daughter distribution at the bottom is for comparison with the distribution of sire averages. Sire averages peak mostly between −500 and +500. All sires except one were included in the range −1500 to +1500. Since the daughter averages are almost exact measures of the sires' daughter superiorities (or one-half the genetic values of the sires), this figure represents the distribution of genetic values. The range in genetic values shown in Figure 10.5 is much less than for the range of records in a sire group.

One last question should be answered. Do genetic values follow the same bell-shaped distribution as daughter records? The distribution of sire values in Figure 10.5 is not clearly bell-shaped. The sire (genetic) values in that figure were replotted into 400-lb intervals and recorded in Figure 10.6. The bell shape is now quite evident, although because there were only 117 sires, the curve is not very smooth.

The large amount of variation in milk records shows that unusually high or low records should be expected to occur no matter what is predicted from the

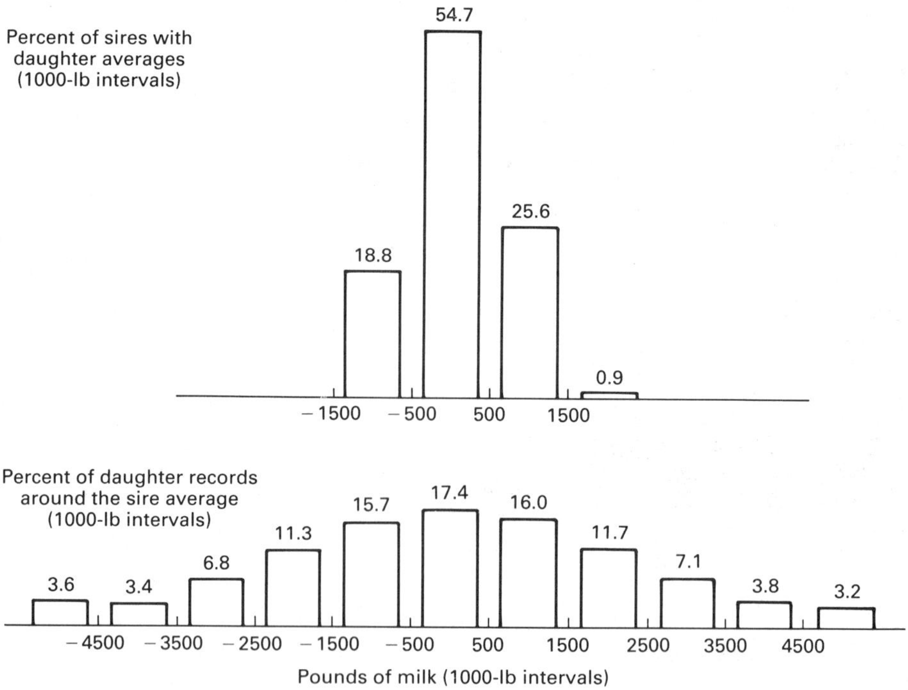

Figure 10.5 Comparison of distribution of sire averages and distribution of daughter records. From *Principles of Dairy Science* by G. H. Schmidt and L. D. Van Vleck. Copyright © 1974 W. H. Freeman and Company. Used by permission.

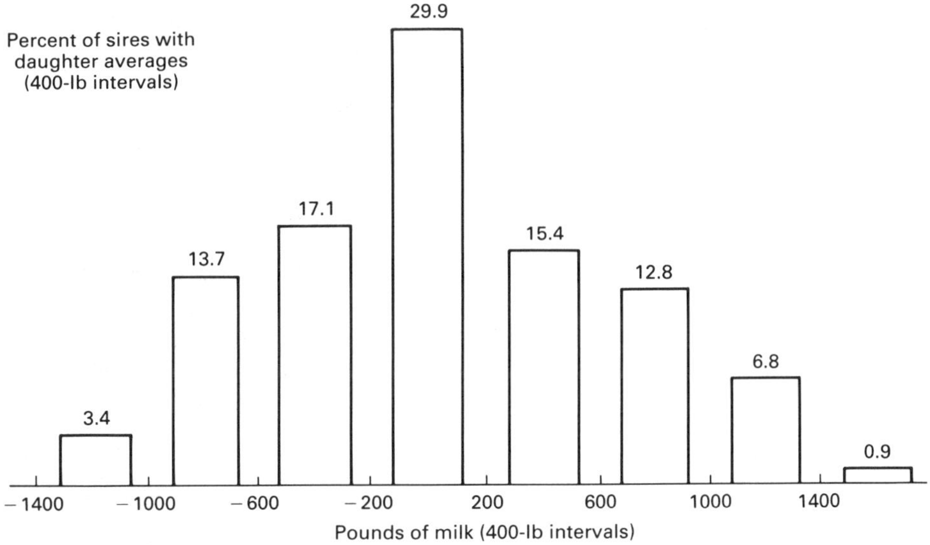

Figure 10.6 Distribution of sire averages. From *Principles of Dairy Science* by G. H. Schmidt and L. D. Van Vleck. Copyright © 1974 W. H. Freeman and Company. Used by permission.

sires' estimated daughter superiority. The presence of genetic differences is important because genetic improvement by selection is then possible. The relative importance of genetic differences and environmental differences determines heritability for a trait.

10.3 HERITABILITY

Heritability is defined as the fraction of the total variance that is genetic variance. The remainder of the variance (total variance minus genetic variance) is defined as environmental variance. The genetic standard deviation is the square root of the genetic variance.

A large heritability (0.50 or greater) suggests that a record on an individual is a good indicator of that individual's genetic value. If heritability is relatively small (0.15 or less), one record has some meaning in estimating genetic value, yet much is unknown about the individual's true genetic value. A progeny test and pedigree analysis will then help in predicting genetic value. For example, if two cows have the same production records, why will one have daughters that are superior to the daughters of the other, even when the sires are the same? If heritability were near 100%, this would not happen. But heritability of milk production is only approximately 25%.

On the average, daughters of cows that produce, say, 1000 lb of milk above the herd average will produce 125 lb ($\frac{1}{2} \times 250$—relationship of one-half times estimated genetic value) above the herd average if sired by average bulls. Some daughters will, however, produce much more and some will produce much less.

Although the expected genetic superiority of cows with 1000 lb of actual production superiority is +250 lb, some will exceed +250 and some will be below +250 in true genetic merit. Heritability times the production record, however, is the best estimate of genetic superiority. Figure 10.7 shows this relationship for a set of records that had an average heritability of 0.40. The dams were divided into 20 equal groups according to their herdmate deviation. Those averages were then plotted against the corresponding daughter averages.

Heritability values for some typical traits are given in Table 10.5. Many characteristics of dairy cattle have relatively low heritability values. Selection for such traits based on individual records is not likely to lead to rapid genetic progress, particularly since many traits cannot be measured directly on the bull. Progeny tests of sires and the resulting frequent use of the few found superior are necessary to increase genetic merit more than mass selection for most productive traits of dairy cattle.

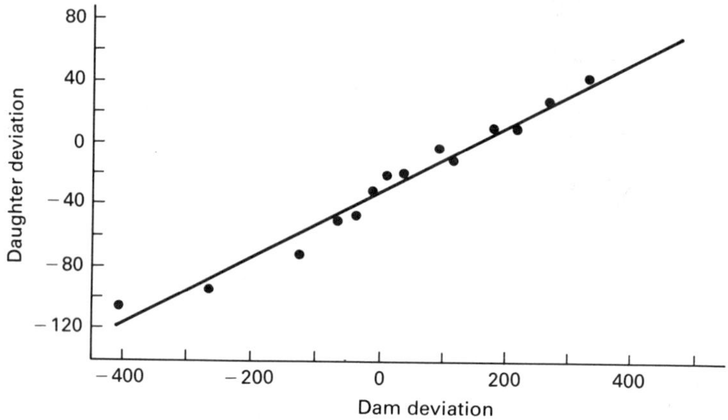

Figure 10.7 Regression of daughter milk deviation from herdmate average on dam deviation in pounds × 10. Each point represents the means for 2370 pairs, or 5% of the total. From L. D. Van Vleck and C. L. Hart, 1965, *J. Dairy Sci.*, 48:1376.

TABLE 10.5 AVERAGE HERITABILITY ESTIMATES FOR SOME TRAITS

Trait	Heritability
Milk or fat yield (one record)	0.25
Final type classification score	0.25
Longevity	0.05
Rate of maturity	0.05
Measure of fertility	0.05
Mastitis incidence	0.20
Milking qualities	0.25
Mature size	0.40
Fat, protein, nonfat solids test	0.45–0.55

10.4 APPENDIX

Variability

Symbols. The standard deviation is usually assigned the symbol σ (lowercase Greek sigma) if the value is the true standard deviation. If the standard deviation is estimated from data, the usual symbol is s. Often the symbol is subscripted to indicate which trait is involved. For example, σ_x is the standard deviation of trait x, where x may be the symbol for a trait such as milk production.

Although reasonably easy to interpret, the standard deviation is difficult to estimate directly. The usual method is to estimate the variance and then to estimate the standard deviation as the square root of the variance. The variance is defined as the average squared difference from the average. The symbols for variance are σ_x^2 (sigma squared) if the true variance is known or s_x^2 (s squared) if estimated from data. The symbols $V(x)$ and $\mathrm{Var}(x)$ are also sometimes used. The lowercase Greek mu, μ, is the usual symbol for the overall average, which is often called the mean.

The standard deviation. The most intuitive measure of variation is the standard deviation. For records that follow the normal (bell-shaped) distribution, tables are available that describe what fraction of those records are expected to be between the overall average and other values measured in terms of standard deviation units. For example, 68% of all records will be between the average minus one standard deviation and the average plus one standard deviation. One-half of these or 34% will be less than the average and 34% will be greater than the average, because the normal distribution is symmetrical around the overall average. Table 10.6 lists the fraction between the average and the average plus a given fraction (n) of standard deviation units.

TABLE 10.6 FRACTION OF RECORDS BETWEEN THE TRUE AVERAGE AND THE AVERAGE PLUS A STATED NUMBER OF STANDARD DEVIATIONS

Standard deviations, n	Fraction of records between μ and $\mu + n\sigma$ or between μ and $\mu - n\sigma$
0.1	0.04
0.2	0.08
0.3	0.12
0.4	0.16
0.5	0.19
0.6	0.23
0.7	0.26
0.8	0.29
0.9	0.32
1.0	0.34
1.5	0.43
2.05	0.48
2.33	0.49

The fraction of records between $\mu + n\sigma$ and $\mu - n\sigma$ is obtained by doubling the table values, as illustrated in Figure 10.8. The fraction less than $\mu + n\sigma$ is obtained by adding the fraction below the average (0.50) to the table value, that is, the fraction below $\mu + n\sigma = 0.5$ plus the fraction between μ and $\mu + n\sigma$, and similarly the fraction below $\mu - n\sigma = 0.50$ minus the fraction between μ and $\mu - n\sigma$. These relationships are shown in Figures 10.9 and 10.10.

Computing the variance. When the true average is known, the variance is estimated from the sum of the squared differences between the records and the true average divided by the number of records. If x_i is the symbol for the record

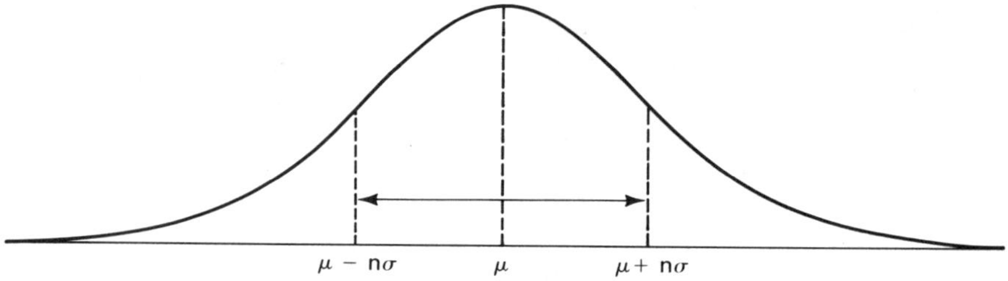

Figure 10.8 Fraction of records between $\mu - n\sigma$ and $\mu + n\sigma$. From *Principles of Dairy Science* by G. H. Schmidt and L. D. Van Vleck. Copyright © 1974 W. H. Freeman and Company. Used by permission.

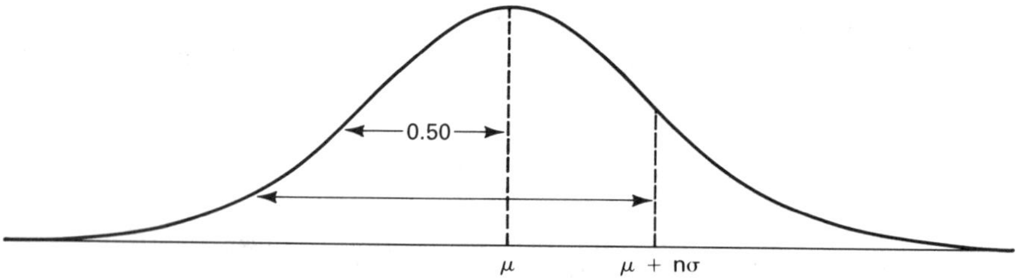

Figure 10.9 Fraction of records less than $\mu + n\sigma$. From *Principles of Dairy Science* by G. H. Schmidt and L. D. Van Vleck. Copyright © 1974 W. H. Freeman and Company. Used by permission.

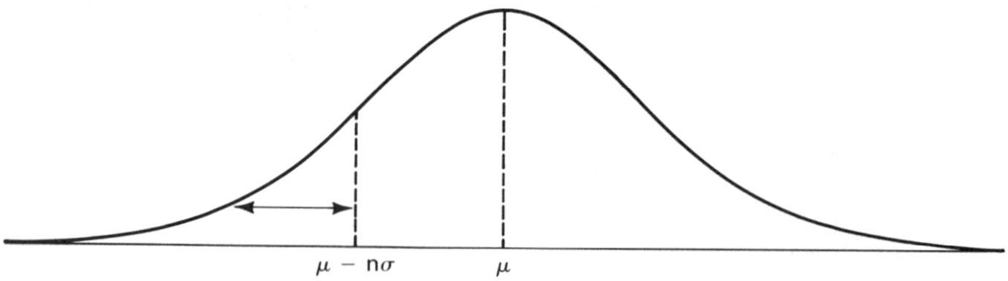

Figure 10.10 Fraction of records less than $\mu - n\sigma$. From *Principles of Dairy Science* by G. H. Schmidt and L. D. Van Vleck. Copyright © 1974 W. H. Freeman and Company. Used by permission.

for trait x for animal i and N is the number of animals, then

$$\sigma_x^2 = \frac{\sum_{i=1}^{N} (x_i - \mu)^2}{N} = \frac{(x_1 - \mu)^2 + (x_2 - \mu)^2 + \cdots + (x_N - \mu)^2}{N}$$

The symbol $\sum_{i=1}^{N}$ is the statistical notation indicating summation ($\sum$) of what follows the $\sum$ from a subscript of $i = 1$ (the first record) up to $i = N$ (the last record).

If the variance is estimated from a sample of data with the true average unknown, then

$$s_x^2 = \frac{\sum_{i=1}^{N} (x_i - \bar{x})^2}{N - 1} = \frac{(x_1 - \bar{x})^2 + (x_2 - \bar{x})^2 + \cdots + (x_N - \bar{x})^2}{N - 1}$$

The symbol $\bar{x}$ (x-bar) represents the average of all the records used, that is,

$$\bar{x} = \frac{\sum_{i=1}^{N} x_i}{N}$$

The major difference in the formulas is that when $\bar{x}$ is used, the division is by $N - 1$ rather than N because of the statistical definition of average. $N - 1$ is often referred to as degrees of freedom. The standard deviation is the square root of the variance, $\sqrt{\sigma_x^2} = \sigma_x$, in the first instance and the square root of the estimate of the variance, $\sqrt{s_x^2} = s_x$, in the second case. The units of measurements are always a part of the computation of averages, variances, and standard deviations. If milk is recorded in pounds, the average is in terms of pounds, the variance is in terms of pounds squared since each difference is squared, and the standard deviation is then returned to pounds.

Heritability

Heritability can be defined as the ratio of variance resulting from genetic differences among cows to the total variance resulting from all differences among records. A usual shorthand way of describing this definition is to let $P_i = G_i + E_i$, where P_i is the phenotypic (actual) measurement of the trait for animal i, G_i is the contribution to P_i by the total *genotype* of the animal, and E_i is the contribution to P_i by all other factors, which collectively are called *environmental* effects. Usually, the assumption is made that the genotypic value and environmental effects do not influence each other. If the G's for each cow could be measured, then a variance of G's could be calculated, that is, σ_G^2, as well as the variance of P's, that is, σ_P^2. If the E's could be measured, σ_E^2 could be calculated. With some statistical reasoning, the relationship $\sigma_P^2 = \sigma_G^2 + \sigma_E^2$ can be shown. Thus the definition for heritability, which has the usual symbol h^2, is

$$h^2 = \frac{\sigma_G^2}{\sigma_G^2 + \sigma_E^2} = \frac{\sigma_G^2}{\sigma_P^2}$$

The square indicates that heritability cannot be negative. As stated earlier, if $\sigma_G^2 = 0$, then $h^2 = 0$, which happens if all animals have the same genetic value for a trait. On the other hand, $h^2 = 1$ if $\sigma_E^2 = 0$ and $\sigma_G^2 \neq 0$, which is true if the environmental effects make the same contribution to each record (that is, if the environmental effects are all equal).

The genetic standard deviation used in calculating genetic progress is σ_G, the square root of σ_G^2. By the previously described relation, $\sigma_G^2 = h^2\sigma_P^2$ and thus $\sigma_G = \sqrt{h^2\sigma_P^2}$, the square root of heritability times the phenotypic or overall standard deviation.

Estimating heritability. Since the G's cannot be measured, heritability must be estimated indirectly, which is possible because extension of the genetic principles stated in Chapter 9 shows that the relationship between records of relatives is proportional to the relationship between the relatives. The measure of relationship between records is the *covariance*. The symbol used is σ_{xy}, where x and y may be two traits or where x and y are two relatives (daughter $= x$ and dam $= y$).

For example, the relationship between noninbred daughters and dams is $\frac{1}{2}$ (as will be shown in Chapter 11) and the daughter will receive one-half her genotype and genetic value from her dam. The environment, of course, is not inherited. Thus the covariance between records of daughter and dam can be shown to equal one-half the genetic variance. A similar result holds for other pairs of relatives. Although certain complications may occasionally invalidate that procedure, σ_G^2 can generally be estimated rather easily from the covariance between records of relatives. Computing a covariance is no more difficult than computing a variance. The only difference is that instead of squaring the deviation from the average for one trait, the product of the deviations from their respective averages is computed:

$$\sigma_{xy} = \frac{\sum_{i=1}^{N} (x_i - \mu_x)(y_i - \mu_y)}{N}$$

$$= \frac{(x_1 - \mu_x)(y_1 - \mu_y) + (x_2 - \mu_x)(y_2 - \mu_y) + \cdots + (x_N - \mu_x)(y_n - \mu_y)}{N}$$

or

$$s_{xy} = \frac{\sum_{i=1}^{N} (x_i - \bar{x})(y_i - \bar{y})}{N - 1}$$

In this case x_i represents a record on relative x and y_i represents the record on the corresponding relative y; i refers to the specific pair of relatives.

Two statistical terms are sometimes encountered in estimating heritability. The *regression* of daughter record on dam record refers to fitting a straight line to a plot of daughter records against dam records, as was shown in Figure 10.7.

The slope of this line is the *regression coefficient*, which is the covariance between daughter and dam records divided by the variance of dam records. In the absence of selection, the covariance between daughter and dam records equals $\frac{1}{2}\sigma_G^2$ and the variance of dam records is $\sigma_P^2 = \sigma_G^2 + \sigma_E^2$. Thus the regression coefficient of daughter records on dam records equals $\frac{1}{2}\sigma_G^2/(\sigma_G^2 + \sigma_E^2)$. Obviously, twice the regression coefficient is the estimate of heritability. This procedure would also be appropriate if there was selection on dam records to decide which dams had daughters.

The correlation coefficient between daughter and dam records is defined as the covariance divided by the square root of the product of the variance of dam records and the variance of daughter records. Again, because the variance of daughter records equals $\sigma_G^2 + \sigma_E^2$, the correlation between daughter records and dam records equals $\frac{1}{2}\sigma_G^2/\sqrt{(\sigma_G^2 + \sigma_E^2)(\sigma_G^2 + \sigma_E^2)} = \frac{1}{2}\sigma_G^2/(\sigma_G^2 + \sigma_E^2)$. Doubling the correlation provides an estimate of heritability. This procedure works only if the records are unselected, because selection results in a nonproportional reduction in the variances and covariance.

Other methods of estimation are also used but they generally rely on the same basic principle: The relationship between records of relatives is proportional to the relationship between the relatives. Computations of relationships and inbreeding are described in Chapter 11.

Calculation of Genetic Progress

The formula of Section 10.1 ignored the fact that genetic superiority, ΔG, is usually different for selected males and females because the accuracy (r_{TI}), selection intensity factor (D), and generation interval (L) may all be different for males and females. Total expected genetic response per year depends on both as will be seen, although the expected genetic superiority of offspring in any generation is the average of superiorities of selected males and females.

Genetic improvement per year. Let $\Delta S = r_{G\hat{G}_S} D_S \sigma_G$, where ΔS is the genetic superiority of selected sires, $r_{G\hat{G}_S}$ is the accuracy of prediction for sires, and D_S is the selection intensity factor for sire selection. Similarly, let $\Delta D = r_{G\hat{G}_D} D_D \sigma_G$, the genetic superiority of selected dams. Let ΔG be genetic improvement per year, L_S be the generation interval in years for sires, and L_D be the generation interval for dams. Then $\Delta G = (\Delta S + \Delta D)/(L_S + L_D)$, which is not $[(\Delta S/L_S) + (\Delta D/L_D)]/2$.

Proof. Let S be the genetic value of sires selected to produce the next generation and D be the value of selected dams. These selected sires are born L_S years before they produce replacement progeny which have genetic value P. The genetic average of bulls born L_S years ago is $P - L_S\,\Delta G$. The superiority of the selected bulls over that average is ΔS. Thus $S = P - L_S\,\Delta G + \Delta S$. Similarly, $D = P - L_D\,\Delta G + \Delta D$. $P = (S + D)/2$, so that by substitution $P = (S + D)/2 = \frac{1}{2}(P - L_S\,\Delta G + \Delta S + P - L_D\,\Delta G + \Delta \mathrm{D})$. Then, subtracting P from both sides yields $0 = -L_S\,\Delta G - L_D\,\Delta G + \Delta \mathrm{S} + \Delta \mathrm{D}$. Rearranging gives $\Delta G(L_S +$

$L_D) = \Delta S + \Delta D$, and finally, $\Delta G = (\Delta S + \Delta D)/(L_S + L_D)$. [This result is due to Dickerson and Hazel (1944).]

Rendel and Robertson (1950) have extended this procedure to consider selection paths for sires of sires (SS), dams of sires (DS), sires of dams (SD), and dams of dams (DD), each with different generation intervals (L_{SS}, L_{DS}, L_{SD}, and L_{DD}, respectively).

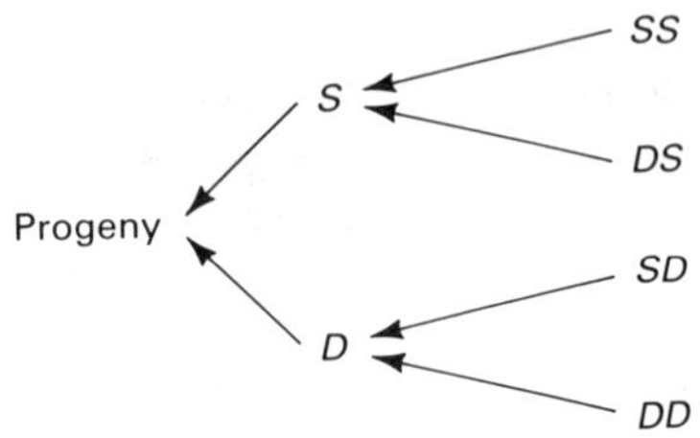

Let ΔSS, ΔDS, ΔSD, and ΔDD be the respective genetic superiorities of the selected grandparents above their generation average. For example, $\Delta SS = r_{G\hat{G}_{SS}} D_{SS} \sigma_G$. By reasoning as before,

$$SS = S - L_{SS}\,\Delta G + \Delta SS,$$

$$DS = D - L_{DS}\,\Delta G + \Delta DS,$$

$$SD = S - L_{SD}\,\Delta G + \Delta SD$$

$$DD = D - L_{DD}\,\Delta G + \Delta DD$$

Because $S = (SS + DS)/2$, $D = (SD + DD)/2$, and $G_{\text{progeny}} = (S + D)/2 = (SS + DS + SD + DD)/4$, then by substitution,

$$(S + D)/2 = \tfrac{1}{4}(S - L_{SS}\,\Delta G + \Delta SS + D - L_{DS}\,\Delta G + \Delta DS + S - L_{SD}\,\Delta G + \Delta SD + D - L_{DD}\,\Delta G + \Delta DD)$$

After rearranging and subtracting $(S + D)/2$ from both sides,

$$\Delta G(L_{SS} + L_{DS} + L_{SD} + L_{DD}) = \Delta SS + \Delta DS + \Delta SD + \Delta DD$$

and

$$\Delta G = (\Delta SS + \Delta DS + \Delta SD + \Delta DD)/(L_{SS} + L_{DS} + L_{SD} + L_{DD})$$

Genetic progress per year is equivalent to the average superiority of the four selection paths divided by the average generation interval of the four paths. This expression or the preceding one involving just sires and dams can be used to compare the expected genetic progress for different selection programs considering differences in generation intervals, selection intensities, and accuracies of prediction.

REFERENCES AND FURTHER READINGS

DICKERSON, G. E., and L. N. HAZEL. 1944. Effectiveness of selection on progeny performance as a supplement to earlier culling in livestock. *J. Agric. Res.*, 69:459.

RENDEL, J. N., and A. ROBERTSON. 1950. Estimation of genetic gain in milk yield by selection in a closed herd of dairy cattle. *J. Genet.*, 50:1.

ROBERTSON, A., and J. M. RENDEL. 1950. The use of progeny testing with artificial insemination in dairy cattle. *J. Genet.*, 50:21.

VAN VLECK, L. D. 1985. Notes on the theory and application of selection principles for the genetic improvement of animals. Dept. Anim. Sci., Cornell University, Ithaca, NY.

11

Relationships and Inbreeding

An animal is frequently said to have 50 or 25% of the blood of another. This, of course, cannot really be true, but these percentages do apply to the average fraction of genes in common from a common ancestor.

As an example, a calf receives a sample half of the genes of its mother. The relationship is said to be 50%. The relationship of an animal to itself or to its identical twin is, of course, 100%; its relationship to its offspring is ordinarily 50%.

What is an animal's relationship to its grandprogeny? As might be expected, it is 25%. A grandprogeny receives, on the average, a sample quarter of its genes from each of its four grandparents. This concept of halving in each generation continues with the great grandprogeny, $12\frac{1}{2}$%; and great, great grandprogeny, $6\frac{1}{4}$%, related to the great grand- and great, great grandparents, respectively.

Thus after two or three generations any particular ancestor is likely to supply only a very few of its genes to its descendants. The closer the relationship, the more genes that the ancestor and descendant have in common.

Similarly, relationships between individuals with a common ancestor can be determined. For example, daughters by the same bull and different dams (paternal half sisters) have an average relationship of 25%. Maternal half sisters also are 25% related to one another. Fraternal twins have the same sire and dam, as do

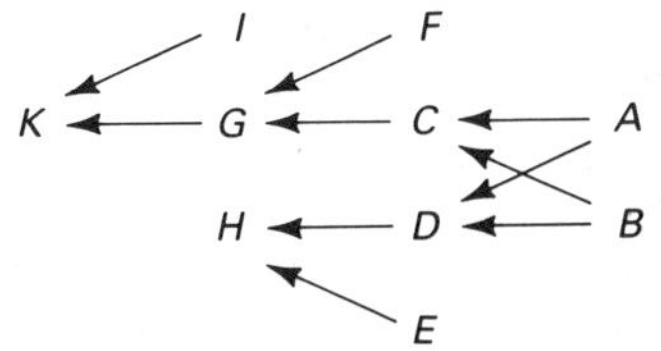

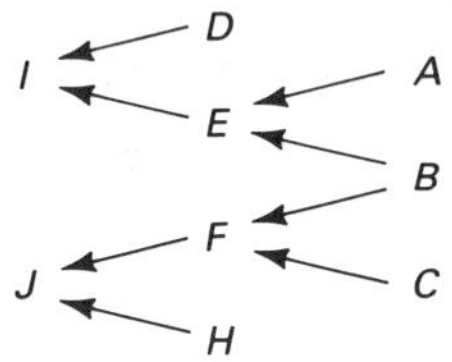

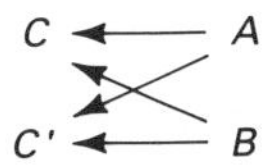

Pair	Relatives	Relationship (%)
B,C	Parent-progeny	50
C,D	Full brothers or sisters	50
A,G	Grandparent-grandprogeny	25
A,K	Great grandparent-great grandprogeny	12½
C,H	Full aunt-niece	25
G,H	Full cousins	12½
E,F	Half brothers or sisters	25
E,J	Half aunt-niece	12½
I,J	Half cousins	6¼
C,C'	If identical twins	100
C,C	Noninbred animal to itself	100

Figure 11.1 Examples of common relationships. From *Principles of Dairy Science* by G. H. Schmidt and L. D. Van Vleck. Copyright © 1974 W. H. Freeman and Company. Used by permission.

full brothers or sisters, but are born at the same time. Some common relationships are given in Figure 11.1; arrows indicate pedigrees to avoid misunderstanding the nomenclature of relationships.

11.1 IMPORTANCE OF KNOWING RELATIONSHIPS

Likeness between Relatives

What is the importance of knowing these relationships? First, there is a direct connection between the similarity in genetic value of related individuals and their relationships. Suppose that the genetic value of a cow is +500 lb of milk. The genetic value may be estimated as heritability times the cow's record. Her daughter would be expected to have a genetic value of $\frac{1}{2}$ times 500 or 250 lb of milk (relationship times genetic value of related individual).

This procedure does not always work so simply when the genetic value of more than one relative is known. For example, suppose that the genetic value of a cow is +500 and the genetic value of her dam is +1000; the genetic value of a daughter of the cow is to be predicted. The predicted genetic value of the daughter is not quite $\frac{1}{2}$ times 500 plus $\frac{1}{4}$ times 1000. Since the cow and her dam are related, the weights for the records are somewhat lower.

Inbreeding

Mating of related individuals produces an inbred offspring. Usually, a large degree of inbreeding is not desirable, because vigor, reproduction, and even production usually are lowered by inbreeding. The inbreeding coefficient of an animal is one-half the relationship between its parents. For example, if a bull is mated to his own daughters, what will be the inbreeding of the offspring? The relationship between father and daughter is 50%. Therefore, the offspring are $\frac{1}{2}$ times 50% or 25% inbred.

Knowledge of relationships can be helpful in selecting animals to keep in the herd or in making matings to avoid high levels of inbreeding. Line breeding is a form of inbreeding in which an attempt is made to concentrate the genes from a single ancestor so that a line of related animals is developed.

Even if relationships are not used for the purposes described here, herd owners can at least tell their neighbor that a particularly good cow of theirs is, say, 25% related to the heifer they want to sell that neighbor, rather than simply saying that the cow is from the same cow family.

This could wear off some of the glamor from the idea of cow families, since members of the same cow family may be practically unrelated. For example, that great brood cow back three generations is related by only $12\frac{1}{2}$% to her great granddaughter. Actually, the great granddaughter is more closely related to her paternal half sister than to her great grandam (25%, compared with $12\frac{1}{2}$%).

11.2 COMPUTING RELATIONSHIPS

Unfortunately, there is no easy way to compute the relationships and inbreeding of cows in a dairy herd, but the computations can usually be performed by following a few simple rules. The next few paragraphs describe a procedure for calculating the relationships among all animals in a herd. The procedure is time-consuming and requires a large sheet (or sheets) of paper but is relatively easy.

The method is based on the fact that if two animals are related, one or both of the parents of the younger of the two must also be related to the older animal of the pair. In fact, if C and D are the two animals and A and B are the parents of D, the relationship between animals C and D is one-half the relationship between A and C plus one-half the relationship between B and C. ($a_{CD} = \frac{1}{2}a_{AC} + \frac{1}{2}a_{BC}$, where a stands for the relationship and the subscripts refer to the animals that are related.) The basis for this method is illustrated in Figure 11.2. Similarly, the inbreeding coefficient of an animal is calculated as one-half the relationship between its parents, as shown in Figure 11.3.

Tabular Method

The procedure is called the tabular method because a table is computed, which, when finished, will give the relationship of any animal to any other in the herd, as in Table 11.1.

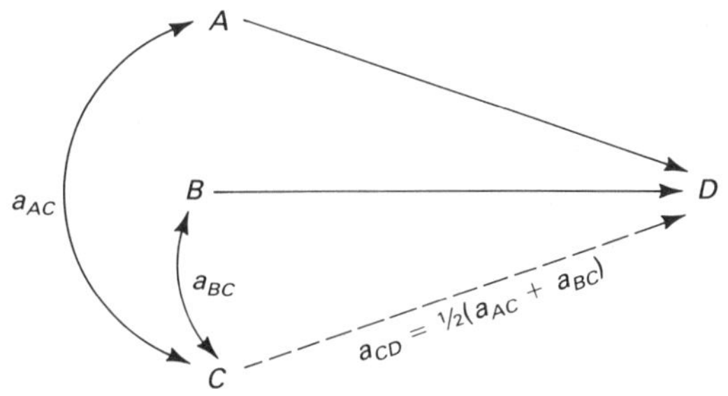

Figure 11.2 Basis for tabular method of computing relationships: The relationship between C and D is $\frac{1}{2}$ the relationship between C and the sire of D plus $\frac{1}{2}$ the relationship between C and the dam of D. From *Principles of Dairy Science* by G. H. Schmidt and L. D. Van Vleck. Copyright © 1974 W. H. Freeman and Company. Used by permission.

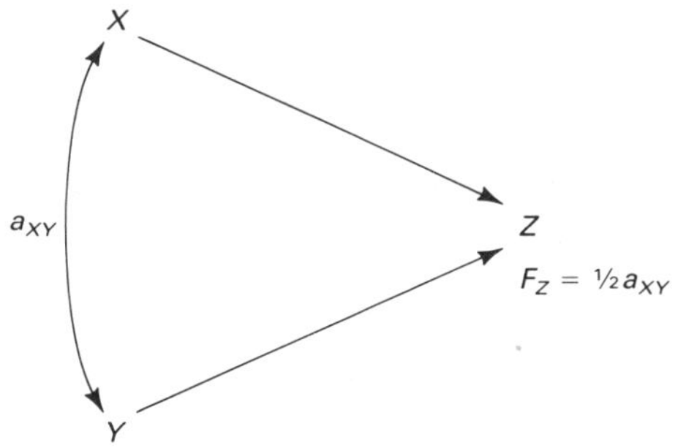

Figure 11.3 The inbreeding coefficient for an animal (F_Z for animal Z) is $\frac{1}{2}$ the relationship between the parents of Z, a_{XY}. From *Principles of Dairy Science* by G. H. Schmidt and L. D. Van Vleck. Copyright © 1974 W. H. Freeman and Company. Used by permission.

In Table 11.1 A and B are parents of D; C has as one parent animal A; and the other parent is not related to A or B, so it is ignored. The table gives the relationships among any of the four animals. To find the relationship between C and D, go to row C and then across to column D. The relationship is $\frac{1}{4}$ or 25%.

The rules for computing relationships and inbreeding are as follows.

1. First, determine which animals are to be included. Put them in order by date of birth, oldest first. This step is very important so that a progeny does not appear before its parents in the table.
2. Write the names or numbers of the animals in order of birth across the top of the table to identify the columns and down the left side of the table to identify the rows.
3. Write above the numbers of the animals the names or numbers of their parents, if known. If unknown, put a dash or a question mark. No entries should be made in the table for unknown parents.

TABLE 11.1 RELATIONSHIP TABLE

Parents: Animal:	A	B	A–? C	A–B D
A	1	0	$\frac{1}{2}$	$\frac{1}{2}$
B	0	1	0	$\frac{1}{2}$
C	$\frac{1}{2}$	0	1	$\frac{1}{4}$
D	$\frac{1}{2}$	$\frac{1}{2}$	$\frac{1}{4}$	1

4. Put a 1 in each of the diagonal cells of the table, such as row 1, column 1; row 2, column 2. The 1 is the animal's basic relationship to itself unless the animal is inbred. If the inbreeding of any of the first or base animals is known, add that to the diagonal for that animal. All other inbreeding coefficients will be added to the diagonals as they are computed according to rule 6. For the base generation animals, enter their relationships if known or, if not known, assume the relationships are zero.
5. Compute entries for each off-diagonal cell of row 1 according to the rule of one-half the entry for the first parent in this row corresponding to the column of the first parent plus one-half the entry for the second parent in this row corresponding to the column of the second parent. When the first row is finished, write the same values down the first column. If this step is not followed correctly, difficulty will almost always result.
6. Go to the next row and begin at the diagonal, which now has a 1 in it. To that 1, either add one-half of the relationship between the animal's parents, which can be found from an earlier entry in the table at the intersection of the row and column of the two parents, or add zero if the parents are or are assumed to be unrelated. One-half the relationship between the parents is the inbreeding coefficient. Often, the inbreeding coefficient is 0. Continue across the row as before, computing the off-diagonal entries by rule 5. Put the values for this row down the corresponding column.

Repeat rules 5 and 6 in this manner until the table is computed, always remembering to do a row at a time, and to put the same values in the corresponding column before going to the next row.

In summary, the first basic step is to add the inbreeding coefficient, which is one-half the relationship between the animal's parents (the row and column of the parents) to the 1 already in the diagonal, as shown in Figure 11.3. The second basic step is to compute the off-diagonal relationships as illustrated by Figure 11.2, where the relationship is one-half the sum of two relationships from columns corresponding to the parents that appear to the left in the same row of the table. These rules may seem complicated but an example will clarify the procedure.

Suppose that sire A is mated back to his own daughter, C. This pedigree is represented by the following diagram, in which the numbers refer to month, day, and year of birth. Note that the only part of an animal's pedigree that must be recorded is that of its parents.

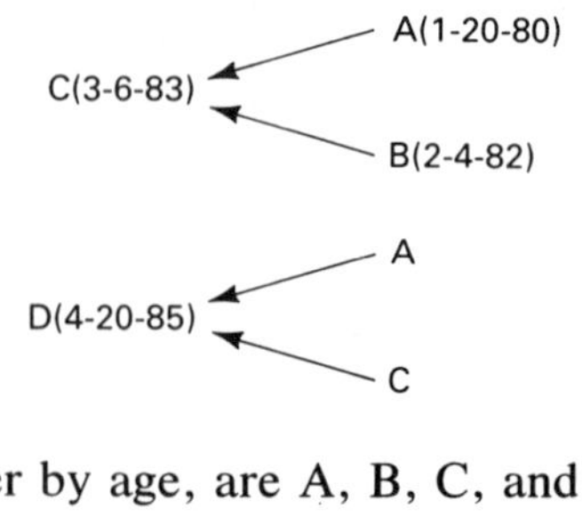

The animals, in order by age, are A, B, C, and D. These letters should be written as described in step 1. Write the parents, if known, above the letters as shown in Table 11.2. The parents of A and B can be ignored unless they are known to be related to each other. Put 1's in all four diagonal cells [where the (A, A), (B, B), (C, C), and (D, D) rows and columns meet].

Now begin on row 1 or the row for A. B is assumed not to be related to A, so enter 0 under the column labeled B. (If the relationship of A and B is known and it is not 0, enter that instead.)

The entry for row A, animal C, is determined according to rule 5 ($\frac{1}{2}$ the entry for row A, animal A, plus $\frac{1}{2}$ the entry for row A, animal B, because A and B are the parents of C). Thus $\frac{1}{2}(1) + \frac{1}{2}(0) = \frac{1}{2}$. The entry for row A, animal D, will be $\frac{1}{2}$ the (A, A) entry plus $\frac{1}{2}$ the (A, C) entry, which is $\frac{1}{2}(1) + (\frac{1}{2})(\frac{1}{2}) = \frac{3}{4}$. The relationships between animal A and the other three animals have now been computed.

The second row, row B, is done similarly, beginning at row B, animal B. The parents of B were not related, so B is not inbred. Therefore, do not add anything to the 1 already in the (B, B) cell.

The entry for row B, animal C, is $\frac{1}{2}(0) + \frac{1}{2}(1) = \frac{1}{2}$. The value for row B, animal D, is $\frac{1}{2}(0) + \frac{1}{2}(\frac{1}{2}) = \frac{1}{4}$. Now write the values for row B down column B: 0, 1, $\frac{1}{2}$, and $\frac{1}{4}$. The 0 and 1 were already there.

Next, row C: animal C entry in that row is the 1 already there plus $\frac{1}{2}$ the relationship between the parents of C, which are A and B. Look at the entry for row A, animal B, and find zero for the relationship between A and B. The entry is then $1 + \frac{1}{2}(0) = 1$.

The animal D entry in row C is: $\frac{1}{2}(\frac{1}{2}) + \frac{1}{2}(1) = \frac{3}{4}$. Put this value into row D, column C, according to step 5.

The last entry is row D, animal D. Add to the 1 already there, $\frac{1}{2}$ the relationship between the parents, A and C. A and C are found to be related by $\frac{1}{2}$ from the animal C entry in row A. The animal D entry in row D is then $1 + \frac{1}{2}(\frac{1}{2}) = 1 + \frac{1}{4}$. Animal D is thus $\frac{1}{4}$ or 25% inbred.

Note from this special kind of mating that animal D is related to A, her father, by 75% and also to her mother, C, by 75%, but is related to her grandmother, B, by only 25%. The only way that two animals other than identical twins can be related by more than 50% is if one or both the animals is (are) inbred. This was true with D and C and also with D and A. D is inbred.

For a herd of any size, the amount of paper and the amount of time required to complete the table will be large. The labor involved, however, is simple and

TABLE 11.2 RELATIONSHIP TABLE FOR THE SAMPLE PEDIGREE[a]

Birth dates: Parents: Animal:	1-20-80 — A	2-4-82 — B	3-6-84 A–B C	4-20-85 A–C D
A	1	0	$\frac{1}{2}$	$\frac{3}{4}$
B	0	1	$\frac{1}{2}$	$\frac{1}{4}$
C	$\frac{1}{2}$	$\frac{1}{2}$	$1 + \frac{1}{2}(0) = 1$	$\frac{3}{4}$
D	$\frac{3}{4}$	$\frac{1}{4}$	$\frac{3}{4}$	$1 + \frac{1}{2}(\frac{1}{2}) = 1 + \frac{1}{4}$

[a] Relationships are given in fractions; to obtain percentages, multiply by 100.

straightforward. Before starting a table for a large herd, it would be a good idea to practice on a small segment of the herd and to work through the example given here. Alternatively, the relationships among the members of a particular cow family could be found. Many dairy managers would be surprised by the results.

Reading the Table

The relationship of any animal to any other is obtained by finding the row of the first animal and then going over to the column of the second animal. The value at the intersection of the row and column is the relationship between the first animal and the second animal.

The entries in the diagonal (e.g., row A and column A) give the relationship of an animal to itself. If the value is greater than 1, the excess is the inbreeding coefficient for that animal. For example, the value for animal D is found in row D, column D, to be $1 + \frac{1}{4}$. The excess over 1 is $\frac{1}{4}$, so the inbreeding coefficient of D is $\frac{1}{4}$ or 25%.

11.3 INBREEDING DEPRESSION

Most research has shown that each percentage of inbreeding leads to an approximate 50-lb decrease in milk production, as well as increased expression of lethal factors since inbreeding increases the fraction of homozygous loci. Inbreeding

TABLE 11.3 AVERAGE CHANGE RESULTING FROM A 1% INCREASE IN INBREEDING COEFFICIENT

Trait	Change
Milk yield	−50 lb
Fat yield	−1.5 lb
Fat (%)	+0.005
Weight	
At birth	−0.24 lb
At 1 year	−1.5 lb
At 2 years	−3.0 lb
At 4 years	−5.0 lb
At maturity	?
Height	Slight decrease
Girth	Slight decrease
Mortality to first calf (% above noninbred)	+2
Conception	
Number of services	+0.05
Days from first service	+3
Failed to conceive (%)	0.5
Age at puberty	Delayed

Source: Adapted from C. W. Young et al., 1969, *North Central Reg. Res. Pub.* 191.

also tends to depress vitality early in life but has only small effects later in life. A summary of the results of a regional breeding project is given in Table 11.3.

Inbreeding appears to have little value in dairy cattle breeding programs because of its numerous detrimental effects, especially the increase in mortality and fertility problems that result from increased inbreeding.

The use of inbreeding might be justified in a situation where a bull is greatly superior to any other bulls available but is related to sires of cows in the herd. Then a decline in production of 300 or 600 lb of milk, caused by inbreeding depression for 6 or 12% inbreeding, might not be as great as the superiority transmitted by the superior sire.

Inbreeding may have some role in breeding programs where sires are mated to daughters to test for lethal recessives (see Section 9.4). A large-scale breeding program could make use of this effect of inbreeding to find and eliminate deleterious genes by rigorous selection. The small breeder, however, stands a good chance of ruin by doing too much inbreeding.

11.4 LEVELS OF INBREEDING

Nearly all cows in all breeds are inbred to some extent because if pedigrees are traced far enough back, one or more ancestors will be found in common on the sire side and the dam side. The magnitude of the inbreeding coefficients, however, is surprisingly small as shown in Table 11.4. The table shows that since the 1950s

TABLE 11.4 ESTIMATES OF AVERAGE INBREEDING IN U.S. DAIRY BREEDS AT VARIOUS TIMES[a]

Breed	Year	Estimate of average inbreeding (%)	Source
Holstein	1889	2.4 ± 0.4	Lush et al. (1936)
	1899	2.1 ± 0.3	Lush et al. (1936)
	1909	4.7 ± 0.5	Lush et al. (1936)
	1919	3.4 ± 0.4	Lush et al. (1936)
	1928	4.7 ± 0.5	Lush et al. (1936)
	1931	4.0 ± 0.5	Lush et al. (1936)
	1970	5.0 ± 0.4	Young (1973)
	1976	3.8	Bonczek and Young (1980)
Brown Swiss	1909	5.0 ± 0.7	Yoder and Lush (1937)
	1919	4.0 ± 0.6	Yoder and Lush (1937)
	1929	3.8 ± 0.6	Yoder and Lush (1937)
Ayrshire	1960–1980	<3.0	Hudson and Van Vleck (1982)
Guernsey	1960–1980	<1.0	Hudson and Van Vleck (1983)
Holstein	1960–1980	<0.5	Hudson and Van Vleck (1983)
Jersey	1960–1980	<1.0	Hudson and Van Vleck (1983)
Brown Swiss	1960–1980	<1.0	Hudson and Van Vleck (1983)

[a] Estimates are to a base prior to 1900 except for those by Hudson and Van Vleck, which are to an approximate 1954 base.

Source: C. W. Young, 1984, *Proc. Natl. Invit. Workshop Genet. Improv. Dairy Cattle,* Dept. Anim. Sci., Cornell University, Ithaca, NY.

inbreeding levels have increased very little, a period when artificial insemination was extensively used. Total inbreeding, since herdbooks were started in the late nineteenth century, has averaged only about 5%. With such a slow increase in inbreeding levels, a small amount of selection can easily compensate for the detrimental effects of inbreeding (Table 11.3). Breeders have evidently avoided matings between close relatives. The effects of high levels of inbreeding which are likely from brother–sister matings and father–daughter matings ($F = 0.25$) can be substantial and should be avoided. With widespread artificial insemination, the average relationship between bulls and cows will increase, so that random mating will result in more inbred cows than in the past, but the level of the inbreeding will not necessarily be high. Matings between bulls and dams related by $12\frac{1}{2}\%$ or more should be avoided. Generally, if no grandparents are the same between a bull and a cow, the inbreeding of the progeny is not likely to be important.

11.5 APPENDIX

Practice Problem in Computing Relationships and Inbreeding

Following is an example of systematic inbreeding where sire A is mated back to his daughters in each succeeding generation.

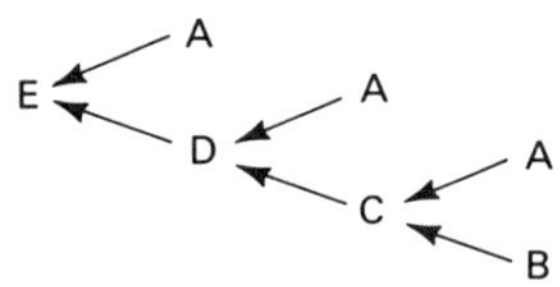

Compute the relationships among the five animals as well as the inbreeding coefficients of D and E. Assume that the parents of A and B are not related to each other (see Tables 11.5 and 11.6).

TABLE 11.5 PRACTICE PROBLEM: COMPUTE THE RELATIONSHIPS

Parents:	—	—	A–B	A–C	A–D
Animal:	A	B	C	D	E
A	1	0			
B	0	1			
C					
D					
E					

TABLE 11.6 SOLUTION TO PRACTICE PROBLEM

Parents:	—	—	A–B	A–C	A–D
Animal:	A	B	C	D	E
A	1	0	$\frac{1}{2}$	$\frac{3}{4}$	$\frac{7}{8}$
B	0	1	$\frac{1}{2}$	$\frac{1}{4}$	$\frac{1}{8}$
C	$\frac{1}{2}$	$\frac{1}{2}$	1	$\frac{3}{4}$	$\frac{5}{8}$
D	$\frac{3}{4}$	$\frac{1}{4}$	$\frac{3}{4}$	$1 + \frac{1}{4}$	1
E	$\frac{7}{8}$	$\frac{1}{8}$	$\frac{5}{8}$	1	$1 + \frac{3}{8}$

Computations:

Row, column	*Row, column*
$(A, C) = \frac{1}{2}(1) + \frac{1}{2}(0) = \frac{1}{2}$	$(C, A) = \frac{1}{2}$
$(A, D) = \frac{1}{2}(1) + \frac{1}{2}(\frac{1}{2}) = \frac{3}{4}$	$(D, A) = \frac{3}{4}$
$(A, E) = \frac{1}{2}(1) + \frac{1}{2}(\frac{3}{4}) = \frac{7}{8}$	$(E, A) = \frac{7}{8}$
$(B, C) = \frac{1}{2}(0) + \frac{1}{2}(1) = \frac{1}{2}$	$(C, B) = \frac{1}{2}$
$(B, D) = \frac{1}{2}(0) + \frac{1}{2}(\frac{1}{2}) = \frac{1}{4}$	$(D, B) = \frac{1}{4}$
$(B, E) = \frac{1}{2}(0) + \frac{1}{2}(\frac{1}{4}) = \frac{1}{8}$	$(E, B) = \frac{1}{8}$
$(C, C) = 1 + \frac{1}{2}(A, B) = 1 + \frac{1}{2}(0) = 1$	
$(C, D) = \frac{1}{2}(\frac{1}{2}) + \frac{1}{2}(1) = \frac{3}{4}$	$(D, C) = \frac{3}{4}$
$(C, E) = \frac{1}{2}(\frac{1}{2}) + \frac{1}{2}(\frac{3}{4}) = \frac{5}{8}$	$(E, C) = \frac{5}{8}$
$(D, D) = 1 + \frac{1}{2}(A, C) = 1 + \frac{1}{2}(\frac{1}{2}) = 1 + \frac{1}{4}$	
$(D, E) = \frac{1}{2}(\frac{3}{4}) + \frac{1}{2}(1 + \frac{1}{4}) = 1$	$(E, D) = 1$
$(E, E) = 1 + \frac{1}{2}(A, D) = 1 + \frac{1}{2}(\frac{3}{4}) = 1 + \frac{3}{8}$	

REFERENCES AND FURTHER READINGS

BONCZEK, R. R., and C. W. YOUNG. 1980. Comparison of the ancestry and inbreeding of two lines of Holstein cattle and the Holstein breed. *J. Dairy Sci.,* 63(Suppl. 1):107.

CRUDEN, D. 1949. The computation of inbreeding coefficients. *J. Hered.,* 40:248.

HUDSON, G. F. S., and L. D. VAN VLECK. 1984. Inbreeding of artificially bred dairy cattle in the northeastern United States. *J. Dairy Sci.,* 67:161.

HUDSON, G. F. S., and L. D. VAN VLECK. 1984. Effects of inbreeding on milk and fat production, stayability, and calving interval of registered Ayrshire cattle in the northeastern United States. *J. Dairy Sci.,* 67:171.

LUSH, J. L., J. C. HOLBERT, and O. S. WILLHAM. 1936. Genetic history of the Holstein–Friesian cattle in the United States. *J. Hered.,* 27:61.

VAN VLECK, L. D. 1985. Notes on the theory and application of selection principles for the genetic improvement of animals. Dept. Anim. Sci., Cornell University, Ithaca, NY.

WRIGHT, S. 1922. Coefficients of inbreeding and relationship. *Am. Nat.,* 56:330.

YODER, D. M., and J. L. LUSH. 1937. A genetic history of the Brown Swiss cattle in the United States. *J. Hered.,* 28:154.

Young, C. W. 1973. Inbreeding and ancestry of registered Holstein–Friesian females born in 1970. *J. Dairy Sci.,* 56:659.

Young, C. W. 1984. Systems of mating—Problems and opportunities. In *Proc. Natl. Invit. Workshop Genet. Improv. Dairy Cattle*. Dept. Anim. Sci., Cornell University, Ithaca, NY.

Young, C. W., W. J. Tyler, A. E. Freeman, H. H. Voelker, L. D. McGilliard, and T. M. Ludwick. 1969. Inbreeding investigations with dairy cattle in the north central region of the United States. *N. Cent. Reg. Res. Pub.* 191.

12

Correction of Records for Nongenetic Factors

Selection usually implies genetic selection or selection for an animal's true ability. Many environmental factors, however, may mask the animal's true and genetic abilities, as indicated in Figure 12.1. For convenience these nongenetic factors can be grouped as (1) environmental effects attributable to known environmental factors, and (2) other effects not attributable to any specific cause or causes but that are nevertheless real.

Genetic evaluation procedures generally require that records be adjusted to a common basis for known environmental effects (e.g., age at freshening, season of the year, days dry and days open in the previous lactation). The nonattributable environmental effects cannot be adjusted for but are considered in another way in genetic evaluation.

Adjustment factors for effects such as age at freshening may be different for different breeds, regions, and even seasons of the year. Thus specific sets of adjustment factors will not be listed here. In the following section evidence is presented showing the possible need for adjustment of records.

12.1 ADJUSTMENT FOR AGE OF COW

Need for Adjustment

The purpose of mature equivalent age adjustment is to adjust a cow's record to what the record might have been had the cow been mature when the record was

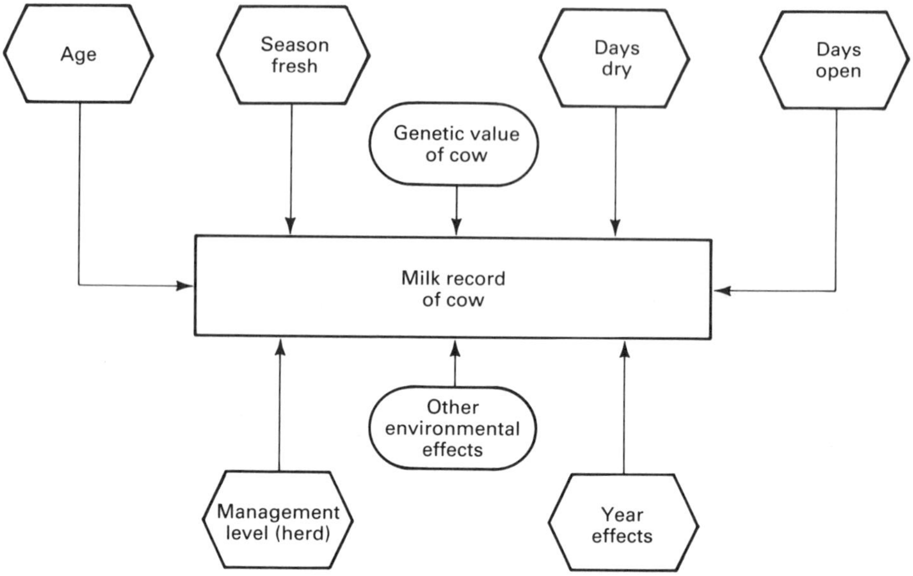

Figure 12.1 Diagram of some factors that mask the genetic value of a cow to produce milk. From *Principles of Dairy Science* by G. H. Schmidt and L. D. Van Vleck. Copyright © 1974 W. H. Freeman and Company. Used by permission.

made. Such factors are thus called mature equivalent (ME) factors; however, they do not predict what a cow will produce when mature. The adjustment can be made to other standard ages, such as 2 years (yr) or 42 months (mo), but traditionally the base for adjustment has been production of mature cows.

The basic reason for age adjustment is that differences in age at freshening may introduce sizable biases into genetic evaluation of cows or sires unless some adjustment is made. For example, which of two cows should be selected on the basis of their actual records if one was 24 mo of age and the other was 6 yr of age at freshening? If the younger cow actually produced more, there would be no question. But in most cases the older of the pair would produce more, since for most breeds 6-year-olds produce 20 to 50% more than 2-year-olds. The only way to compare such cows is to adjust their records to a common age equivalent.

The need for age adjustment is nearly as great when comparing first-lactation cows as when comparing cows in different lactations. A simple example illustrates this need. First-lactation records can begin earlier than 22 mo of age or even later than 36 mo of age, although most first-lactation records in the United States begin at 24 to 29 mo of age. A typical age adjustment factor for a Holstein freshening at 23 mo is approximately 1.38. The ME factor for a 36-month-old Holstein is 1.21. Most herd owners probably would not compensate that much when comparing first-lactation cows of such different ages.

Adjusting for age of daughters for first proofs of bulls may not be as important because the average age at first calving usually is approximately the same for daughters of different sires. The extremes in average age at first freshening between two sires, each having at least 20 daughters, are likely to be between 25 and 30 mo with corresponding age adjustment factors of 1.33 and 1.25 for Holsteins. Suppose that two such groups of 20 daughters produced an average of 15,000 lb of milk. If the herd levels were the same, there would be no apparent difference between the sires. On a mature equivalent basis, however, those cows that freshened at 25 mo would average 20,000 ME (1.33 × 15,000) and those calving at 30 mo would average 18,750 ME (1.25 × 15,000). On an ME basis, the bull with the younger daughters would appear to be considerably superior to the other bull.

Such large differences in average calving age seldom occur between groups of daughters. Sometimes, however, young bulls may be sampled in AI very quickly so that all daughters would be born at approximately the same time. If one bull is sampled 4 or 5 mo later than another bull, both groups of daughters might freshen in September, October, or November, since that is a popular time to freshen heifers. Thus the age of the two groups of daughters at freshening would differ by 5 mo.

Determining Age Correction Factors

One way to find the adjustment factor for a particular age is to divide an estimate of the production of mature cows by an estimate of the production of cows of the particular age:

$$\text{ME factor for 24 mo} = \frac{\text{estimate of production for mature cows}}{\begin{array}{c}\text{estimate of production for cows}\\ \text{freshening at 24 mo of age}\end{array}}$$

When a cow has a record beginning at 24 mo of age, her actual record is multiplied by the age factor for 24 mo to compute the mature equivalent record.

Gross comparison method. Estimates of production for different age groups are obtained in many ways. The *gross comparison* method uses the simple average for all cows of each age, although in some cases the herd levels may be taken into account. If herds of different production levels have different age compositions, the herd and age effects will be partially confounded (a statistician's jargon word) if the simple averages are used. The gross comparison method is subject to bias because older cows usually are a selected group. Cows with the smallest first-lactation records may not survive to have records at older ages. Thus averages of older cows are biased upward and consequently, ME factors for young cows are too large.

Paired comparison method. A bias in the opposite direction may occur with the *paired comparison* method. The basic principle of the method is to compare records of the same cow at different ages. These comparisons are then averaged for all cows by statistical procedures to give a complete set of age adjustment factors. The bias results because the repeatability of lactation yield is not 100% (only approximately half the superiority of a cow is expected to be repeated). Thus the increase in production caused by increased age will be underestimated, and consequently the mature equivalent factors will be too small for young cows.

Because the biases of the gross and paired comparison methods are theoretically in different directions, some researchers have thought that averaging the two sets of age adjustment factors would be more appropriate than using either the gross or paired comparison factors.

Mixed model method. A theoretically best method is commonly called the *mixed model* method. This statistical procedure accounts for imperfect repeatability of milk records from lactation to lactation and also for effects of selection on records of older cows. Estimates of age effects are then converted into multiplicative age adjustment factors, as described previously. Plots of a set of such factors for Holsteins are shown in Figure 12.2. These cows freshened from July to October. The plot for cows freshening from November to June produced a similar figure.

Paired comparison factors are larger than expected because the environment was improving so rapidly that the environmentally caused production increase at older ages more than balanced the incomplete repeatability in milk yield. The

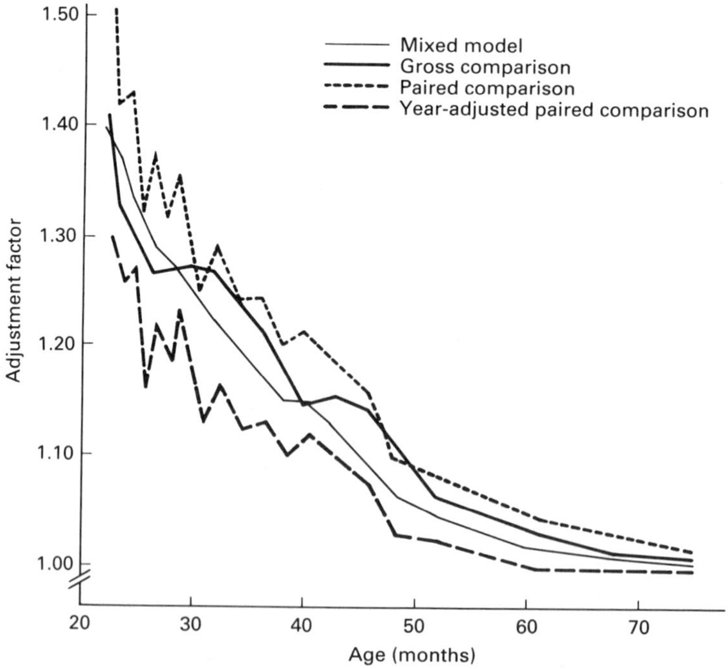

Figure 12.2 Age adjustment factors for cows freshening from July to October in New York State, computed by different methods. From *Principles of Dairy Science* by G. H. Schmidt and L. D. Van Vleck. Copyright © 1974 W. H. Freeman and Company. Used by permission.

bottom line in Figure 12.2 represents the expected result for paired comparison age adjustment factors after correction for production increase caused by environmental improvement. As a result the mixed model factors are generally intermediate.

Breed-age-average method. An age adjustment method that is commonly used for milk records in Canada and that is often used with overall type scores is the *breed-age-average* (BAA) percentage. The average for all cows of a certain age and breed is computed to serve as a base. Then the record of each cow of that age is expressed as a percentage of the breed-age-average. One avoidable difficulty with this procedure is that the breed-age-averages used as the bases for computing the percentages often are not recomputed for several years. If a general increase in production occurs, there will be many more cows over 100% than below 100%. Another danger is making the age interval too long—such as 3 to 6 mo rather than 1 mo.

A method commonly used in Great Britain and in some European countries is similar to the BAA method. In sire evaluation, daughters are compared only with other cows of the same age having records in the same year. The problem is that the age group may include all first-lactation animals, so that cows freshening at 24 mo of age may be compared with herdmates averaging 30 or more mo of age at freshening. Such long age intervals are necessary to have enough cows for comparison. There are other ways of finding age factors, but most are similar to the methods described here.

Differences in Age Adjustment Factors

Genetic and herd differences. A few minor problems may be encountered in using age adjustment factors. Some bulls may have daughters that mature at a slower rate than other bulls' daughters. Thus the argument can be made that records of daughters of such bulls should have larger age adjustment than daughters of other bulls. The nonscientific rebuttal is that those cows should actually have less age adjustment because they are not the types of cows that will make money for their owners. The younger a cow enters production at a high level, the more money she will make. There is no evidence that slower-maturing cows stay in the herd longer than their earlier-maturing herdmates. Actually, late-maturing cows often are culled sooner than early-maturing cows because they are not producing enough milk and are not making enough money to justify maintaining them until they are mature.

A herd owner might also ask: "These age factors are fine for the average herd but my herd is not average. Shouldn't I have a different set of factors for my high-producing herd?" Some research has shown that separate age adjustment factors for each herd are not worth the effort needed to develop them. The decrease in accuracy that results from using the overall age adjustment factors rather than factors for specific herds is not very important. Usually, the number of records in any herd is too small to provide good estimates of adjustment factors.

Breed differences. Some other problems with age adjustment are more important than herd specific factors. Different age adjustment factors are needed for most breeds. For example, 2-yr-old Guernseys require the least adjustment to a mature basis and 2-yr-old Brown Swiss require the most. These differences can be seen in Table 12.1, which is a partial table of age adjustment factors. Different age adjustment factors should be expected for different breeds. But when only a few records are available for a breed, accurate estimation of the age adjustment factors is difficult.

TABLE 12.1 EXAMPLE OF AGE ADJUSTMENT FACTORS FOR DIFFERENT BREEDS[a]

	Breed				
Age (months)	Ayrshire	Guernsey	Holstein	Jersey	Brown Swiss
23	1.30	1.24	1.38	1.34	1.54
24	1.28	1.22	1.35	1.31	1.51
25	1.25	1.20	1.33	1.29	1.48
26	1.22	1.18	1.31	1.28	1.45
27	1.20	1.17	1.29	1.27	1.43
28	1.19	1.16	1.27	1.27	1.40
29	1.19	1.15	1.26	1.27	1.38
30	1.18	1.15	1.25	1.27	1.36
31	1.17	1.15	1.24	1.26	1.34
32	1.17	1.14	1.23	1.24	1.32
33	1.16	1.13	1.23	1.23	1.31
34	1.14	1.13	1.22	1.21	1.30
35	1.14	1.13	1.22	1.20	1.23
36	1.14	1.12	1.21	1.18	1.26
38	1.15	1.10	1.19	1.16	1.22
40	1.14	1.08	1.16	1.13	1.19
42	1.11	1.07	1.14	1.11	1.17
44	1.10	1.06	1.12	1.10	1.15
46	1.10	1.06	1.11	1.08	1.14
48	1.10	1.06	1.10	1.07	1.12
50	1.08	1.04	1.08	1.06	1.11
52	1.06	1.03	1.06	1.05	1.10
54	1.05	1.02	1.05	1.04	1.08
56	1.04	1.01	1.04	1.04	1.07
58	1.03	1.01	1.04	1.03	1.06
60	1.02	1.01	1.02	1.02	1.05
66	1.00	1.00	1.00	1.00	1.02
72	1.00	1.00	1.00	1.00	1.00
78	1.00	1.00	1.00	1.00	1.00
84	1.00	1.00	1.00	1.00	1.00
90	1.03	1.00	1.01	1.02	1.00
96	1.03	1.00	1.01	1.02	1.00
102	1.03	1.01	1.02	1.03	1.01
108	1.03	1.01	1.02	1.03	1.01
114	1.03	1.01	1.04	1.05	1.03
120	1.03	1.01	1.04	1.05	1.03

[a] Partial set of age adjustment factors used in New York State before different age adjustment factors were developed for different seasons of freshening.

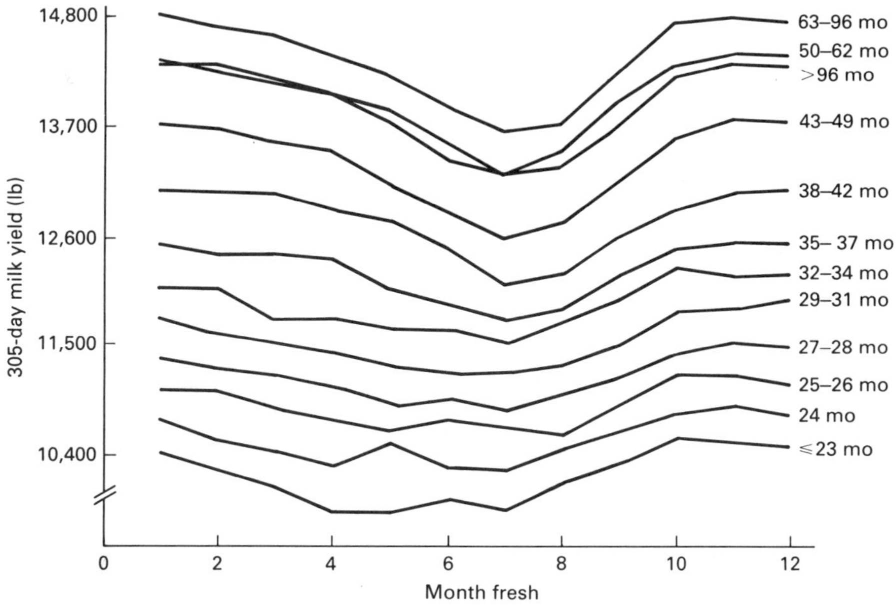

Figure 12.3 Seasonal change in milk yield for each age at calving. From P. D. Miller, W. E. Lentz, and C. R. Henderson, 1970, *J. Dairy Sci.*, 53:351.

Differences in milk production and milk fat and protein. There is also evidence that age adjustment factors for milk production may differ from those for fat or for protein production. Similarly, different sets of factors also may be needed for fat test or protein test.

Regional differences. Different areas of a country may require different age factors. ME factors for Holsteins in the northeastern United States are larger for 2-year-old records than ME factors for the Midwest. The reason is that mature cows outproduce their younger herdmates less in the Midwest than in the Northeast. Such differences may be caused by different methods of growing and feeding heifers and cows. Regional age adjustment factors are now commonly used for Holsteins. Regional factors for breeds other than Holstein may be needed but not enough records are available for accurate determination of regional factors.

Seasonal differences. Different age adjustment factors are needed for different seasons of freshening even in the same region. Figure 12.3 shows how month of freshening affects milk production of Holsteins calving at various ages and illustrates the need for simultaneous adjustment for age at freshening and month of freshening. A partial set of joint age and seasonal factors for Holsteins is given in Table 12.2. The average yield of all mature cows was used as the base, although any other base could be used. This set of factors increases or decreases records of mature cows depending on whether they freshened in a favorable or

TABLE 12.2 MULTIPLICATIVE FACTORS FOR JOINT ADJUSTMENT OF MILK YIELD FOR MONTH AND AGE OF CALVING

Age	Month of calving											
(months)	Jan.	Feb.	Mar.	Apr.	May	June	July	Aug.	Sept.	Oct.	Nov.	Dec.
24	1.34	1.36	1.37	1.39	1.40	1.40	1.40	1.37	1.35	1.33	1.32	1.33
25	1.31	1.32	1.34	1.35	1.37	1.36	1.37	1.36	1.33	1.30	1.30	1.31
26	1.29	1.30	1.31	1.32	1.35	1.33	1.34	1.34	1.31	1.28	1.28	1.28
27	1.27	1.20	1.29	1.31	1.33	1.32	1.33	1.32	1.30	1.26	1.26	1.26
28	1.25	1.27	1.28	1.29	1.31	1.30	1.32	1.30	1.28	1.25	1.24	1.24
29	1.23	1.25	1.26	1.27	1.29	1.29	1.30	1.29	1.26	1.23	1.23	1.22
30	1.22	1.24	1.24	1.26	1.27	1.28	1.28	1.27	1.25	1.22	1.21	1.20
31	1.21	1.22	1.23	1.24	1.26	1.26	1.27	1.26	1.23	1.20	1.20	1.19
32	1.20	1.21	1.22	1.23	1.24	1.25	1.26	1.25	1.22	1.19	1.19	1.19
85	0.96	0.97	0.97	0.99	1.01	1.03	1.04	1.04	1.01	0.97	0.96	0.97

Source: P. D. Miller et al., 1970, *J. Dairy Sci.*, 53:351.

an unfavorable month. Unless the proper seasonal factors are used, bull proofs may be affected if most of one bull's daughters freshen in one season and most of another bull's daughters freshen in another season.

In general, age factors should be developed specifically for trait, breed, region, and season unless evidence shows that some of these factors are not important. Having accurate age factors and using them properly is a valuable and basic tool in selecting the best cows and bulls.

Traits Other than Milk Production

Traits other than production traits are also influenced by the age of the cow. Body size and stature obviously continue to increase for some years after lactation begins. The frequency of defects of feet and legs, crampiness, and various joint weaknesses increase with age. Type scores, although they are less influenced by age than milk yield, also need adjustment, as shown in Table 12.3.

Susceptibility to some production-related diseases is greater in later lactations. The results shown in Table 12.4 are from a mastitis survey. One sample

TABLE 12.3 AGE ADJUSTMENT FACTORS FOR HOLSTEIN FINAL CLASSIFICATION SCORES

Age (months)	Factor
Less than 24	1.07
24	1.06
24–27	1.05
28–31	1.04
32–38	1.03
39–46	1.02
47–52	1.01
53–107	1.00
108–143	0.98
144–167	0.96
More than 167	0.95

Source: R. H. Kliewer, 1971, *Proc. NAAB Natl. Meet.*, p. 54.

TABLE 12.4 DEGREE OF UDDER INFECTION IN HOLSTEIN COWS OF DIFFERENT AGES

	Age group		
	<36 months	36–48 months	>48 months
Number of cows	12,392	10,697	25,469
Fraction clinical	0.026	0.058	0.129
Fraction with organisms	0.276	0.410	0.614
Number of quarters per cow with organisms	0.41	0.68	1.2
Number of kinds of infections/cow	0.32	0.51	0.83

Source: J. W. Wilton et al., 1972, *J. Dairy Sci.*, 55:183.

was taken from each quarter of the cow during each lactation. Bacterial cultures were made to determine which of nine mastitis organisms were present, even if the quarter was not clinically diagnosed with mastitis.

Similarly, ketosis and mik fever practically never appear during the first lactation but may occur quite frequently in older cows. Norman estimated the fraction of New York Holsteins affected in different lactations. His data, shown in Table 12.5, consisted of herd owners' reports of any difficulty during the lactation prior to the time when the cow was appraised for type. The mastitis trend

TABLE 12.5 FRACTION OF HOLSTEIN COWS REPORTED TO HAVE KETOSIS, MILK FEVER, MASTITIS, AND CYSTIC OVARIES

Disorder	Number of lactations					
	1	2	3–4	5–6	7–8	9–10
Ketosis	0.02	0.04	0.09	0.15	0.20	0.23
Milk fever	0.00	0.01	0.02	0.07	0.15	0.25
Mastitis	0.10	0.19	0.30	0.39	0.48	0.60
Cystic ovaries	0.03	0.06	0.09	0.10	0.15	0.17

Source: H. D. Norman, 1970, Ph.D. thesis, Cornell University.

TABLE 12.6 MATURE EQUIVALENT 305-DAY MILK PRODUCTION ADJUSTMENT FACTORS FOR DAYS OPEN, BY BREED[a]

Days open	Breed				
	Ayrshire	Guernsey	Holstein	Jersey	Brown Swiss
20–29	1.20	1.17	1.20	1.22	1.20
30–39	1.15	1.15	1.17	1.16	1.15
40–49	1.12	1.12	1.12	1.14	1.14
50–59	1.10	1.07	1.08	1.09	1.08
60–69	1.06	1.06	1.05	1.06	1.05
70–79	1.03	1.02	1.03	1.03	1.02
80–89	1.00	1.00	1.01	1.01	1.00
90–99	1.00	1.00	1.00	1.00	1.00
100–109	0.98	0.99	0.99	0.99	1.00
110–119	0.97	0.99	0.98	0.99	0.99
120–129	0.96	0.99	0.98	0.98	0.99
130–139	0.96	0.97	0.97	0.97	0.99
140–149	0.96	0.95	0.97	0.97	0.97
150–159	0.96	0.95	0.96	0.97	0.96
160–169	0.94	0.95	0.96	0.96	0.95
170–179	0.94	0.95	0.96	0.95	0.95
180–189	0.94	0.95	0.95	0.95	0.95
190–199	0.94	0.94	0.95	0.94	0.94
200–209	0.93	0.94	0.95	0.94	0.94
210–219	0.93	0.93	0.94	0.94	0.93

[a] Base equals the ME 305-day milk yield of a cow open between 90 and 99 days.

Source: L. R. Schaeffer, R. W. Everett, and C. R. Henderson, 1973, *J. Dairy Sci.*, 56:602.

is slightly different from that shown in Table 12.4. Table 12.5 also shows that reported cases of cystic ovaries increase with age.

12.2 ADJUSTMENT FOR BODY WEIGHT

Some researchers have suggested adjusting milk yield for body weight of the cow as an alternative to age adjustment, reasoning that body weight may be more indicative of physiological maturity than chronological age. Body weight at a given age, however, is largely hereditary. In addition, many of the genes affecting milk yield also affect body weight or growth rate. Thus, adjusting for body weight simultaneously adjusts genetic value for milk yield. Most research suggests that adjustment should be for age rather than body weight.

12.3 ADJUSTMENT FOR DAYS DRY

The length of the resting period of the udder prior to lactation has an effect on yield, which may be important in cow selection. Generally, lactation yield increases nearly linearly as the previous dry period increases up to 50 to 60 days. A dry period longer than 60 days does not improve yield in the next lactation. Actually, cows that have long dry periods but average calving intervals tend to have low average production for actual or 305-day lactations because the number of days milked has a high correlation with lactation yield. As with body weight, the genetic relationship of days dry to yield complicates adjustment of yield for the previous dry period. Adjustment of days dry to a common basis is not usually done. A dry period of 30 to about 60 days is attainable by proper management and is economically optimum.

12.4 ADJUSTMENT FOR CALVING INTERVAL

The interval between the start of one lactation and the next is determined primarily by the number of days open (the time from parturition to the next successful breeding). Milk yield is influenced both by preceding calving interval and by days pregnant during the current lactation. If 60 days are allowed to elapse before first breeding after calving, average days open will be 75 to 90 days after allowance for repeat breedings and time of estrus. As shown in Figure 12.4, an open period between 60 and 90 days appears to be ideal. The influence of pregnancy becomes apparent in the fourth month of gestation. Thus, if lactation records of 200 days or less are used for genetic evaluation, correction for calving interval is not necessary. Differences in calving interval may result in production differences of 5 to 10%, which are important in comparing 305-day production of cows. Because the genetic influence on calving interval is very small, adjustment for calving interval will improve genetic evaluation of yield. A set of multiplicative adjustment factors is given in Table 12.6. The random, environmental nature of effects that

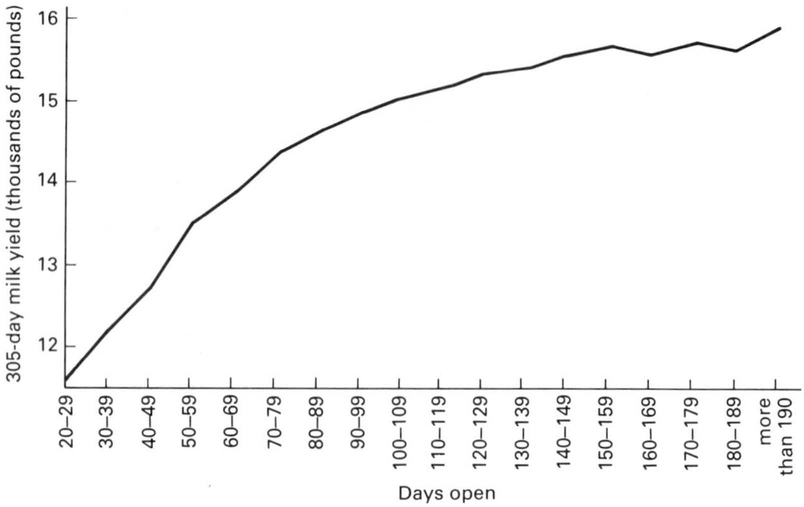

Figure 12.4 Average adjusted 305-day milk production for open periods (all lactations). From L. R. Schaeffer and C. R. Henderson, 1972, *J. Dairy Sci.*, 55:107.

influence calving interval makes adjustment for calving interval less important for sire evaluation than for cow selection if herd management effects are taken into account.

From an economic point of view, the optimal calving interval is thought to be approximately 13 mo between the first and second calvings and 12 mo between other calvings, since that interval tends to maximize the yield per day in the herd and thus lifetime production. The recommended practice of a 305-day lactation with a 60-day dry period and a wait of 60 days before breeding seems to be well accepted.

The reason why increased calving interval accompanies higher production has not been determined, but many explanations are feasible. High producers may have more breeding problems. A high producer may be deliberately not bred back as soon as a low producer. A high producer may be given more chances to conceive, whereas a low producer may be culled with the same number of returns to service. A high producer may show less evidence of estrus than a low producer. The conception rate among second and third services is nearly the same as for first services.

12.5 ADJUSTMENT FOR HERD MANAGEMENT LEVEL

The most important effect on yield and on some other traits is called the herd effect. Actually, many effects combine to form the herd effect for such traits as milk yield. Feeding and management differences are undoubtedly the major herd effects on milk yield. The apportionment of importance to various feeding or management methods is a nongenetic problem—the genetic problem is to adjust for the overall effect of herds, which is usually done by comparing cows with others in the same herd. Genetic comparison of cows in the same herd is not a problem unless effects of season of freshening are important. Cows calving in the same herd, year, and season of freshening whose records are adjusted for age and calving interval can be compared relatively safely. Comparison of cows with records started in different seasons is essentially done by deviating the records from the herd and season average. Most systems of sire evaluation consider herd and season effects in one way or another.

12.6 ADJUSTMENT FOR TIMES MILKED PER DAY

Before labor costs became prohibitive, milking three and four times a day was frequent. An increase in total production of 10 to 20%, according to age of cows and increase in feeding, was common. Corrections were then necessary to convert to a twice-daily standard. Now only a negligible fraction of cows are milked more than twice a day. If these cows are compared with cows milked only twice a day, corrections are necessary.

12.7 STAGE OF LACTATION

Milk yield varies with stage of lactation, but milk yield is usually recorded for a specified length of lactation. Type classification is recorded only once during a lactation. The stage of lactation has a major influence on the four components of final type score as well as on final type score. Adjustments in type scores should be made for stage of lactation as well as for age at classification.

12.8 EXAMPLE OF USE OF ADJUSTMENT FACTORS

Two Holstein cows both produce 14,000 lb of milk in 305 days in lactations that started in the same herd–year–season.

Cow	Actual 305-day yield	Age at freshening (mo)	Days to next freshening	Times milked per day
A	14,000	24	455	3
B	14,000	60	375	2

The two cows did not have the same opportunity to produce even though they were producing at the same time and under the same management. Younger cows produce less than older cows (advantage to cow B). A shorter calving interval means more days carried calf and fewer days open (advantage to cow A). Cows milked three times a day produce more than if milked twice a day (advantage to cow A).

The mature equivalent factors (Table 12.2) are 1.35 and 1.02.

Calving interval can be converted to days open to use Table 12.6 and subtracting the average gestation length of 280 days:

$$\text{days open for cow A} = 455 - 280 = 175 \text{ with factor, } 0.96$$

$$\text{days open for cow B} = 375 - 280 = 95 \text{ with factor, } 1.00$$

Note that the base for days open in Table 12.6 is 90 to 99 days. Factors for times milked are not well estimated, but a standard correction is 0.85 to convert three times a day to two times a day.

These factors are multiplicative, so which adjustment is made first is not important.

Standard record for cow A:

$$(14{,}000 \text{ lb})(1.35)(0.96)(0.85) = 15{,}422 \text{ lb}$$

Standard record for cow B:

$$(14{,}000 \text{ lb})(1.02)(1.00)(1.00) = 14{,}280 \text{ lb}$$

If cow A and cow B had made their records under standard conditions—mature age, days open 90 to 99 days and milked twice a day—then the estimates of what they would have produced are 15,422 lb and 14,280 lb, over 1000 lb more for cow A. The standard records do not estimate what the cows will produce when they are mature. New random temporary environmental effects will affect those records so that the adjusted records are not the same each time a cow makes a record.

If a mixture of additive and multiplicative factors are used, the order of use makes a difference. Additive factors involve adding or subtracting an appropriate factor. Suppose, for example, each day carried calf reduces milk yield by 2 lb. Then cow A would have for her 305-day lactation, days carried calf of $305 - 175 = 130$. For cow B, days carried calf is $305 - 95 = 210$. Days carried calf is the standard lactation length (305 days) minus days open. The additive factor for cow A would be $-(2)(130) = -260$ lb and for cow B would be $-2(210) = -420$ lb. In this case the base is zero days carried calf. Now suppose this adjustment is done first and then multiplicative factors for age and times milked are applied.

Adjusted record for cow A:

$$(14{,}000 - 260)(1.35)(0.85) = 15{,}767$$

Adjusted record for cow B:

$$(14{,}000 - 420)(1.02)(1.00) = 13{,}852$$

If the days carried calf adjustment was done last:

Adjusted record for cow A:

$$(14{,}000)(1.35)(0.85) - 260 = 15{,}805$$

Adjusted record for cow B:

$$(14{,}000)(1.02)(1.00) - 420 = 13{,}860$$

For practical purposes in this example, the order of adjustment is not of much importance. What is important to remember is that the total adjustment is different depending on the order of the adjustments. Fortunately, multiplicative factors are usually the only ones used, and in fact such adjustments are routinely made by Dairy Records Processing Centers before records are used for genetic evaluations.

12.9 SUMMARY OF ADJUSTMENTS

Records used for predicting genetic value for milk yield should be adjusted for age at freshening and calving interval, using appropriate adjustment factors. The more important adjustment, for herd, year, and season of freshening, is usually made by deviating records from herd–year–season averages or by similar statistical procedures. Correction for other factors is not generally recommended, although each situation must be evaluated separately. The methods of adjusting other traits for the major environmental factors are the same. Age is often a major consideration. After these adjustments have been made, the important problem of cow and bull evaluation and selection can be tackled with fewer worries about potential biases.

REFERENCES AND FURTHER READINGS

HANSEN, L. R., G. R. BARR, and D. A. WIECKERT. 1969. Effects of age and stage of lactation on type classification. *J. Dairy Sci.*, 52:646.

HYATT, G., JR., and W. J. TYLER. 1948. Variations in type ratings of individual Ayrshire cows. *J. Dairy Sci.*, 31:71.

LOUCA, A., and J. E. LEGATES. 1968. Production losses in dairy cattle due to days open. *J. Dairy Sci.*, 51:573.

MCDANIEL, B. T., R. H. MILLER, E. L. CORLEY, and R. D. PLOWMAN. 1967. DHIA age adjustment factors for standardizing lactations to a mature basis. *DHI Newsl.*, ARS-44-188.

MILLER, P. D., W. E. LENTZ, and C. R. HENDERSON. 1970. Joint influence of month and age of calving on milk yield of Holstein cows in the northeastern United States. *J. Dairy Sci.*, 53:351.

SCHAEFFER, L. R., and C. R. HENDERSON. 1972. Effects of days dry and days open on Holstein milk production. *J. Dairy Sci.*, 55:107.

FREEMAN, A. E., C. G. HICKMAN, P. MILLER, and B. T. MCDANIEL. 1973. Symposium on age adjustment. *J. Dairy Sci.*, 56:941.

13

Cow Evaluation and Selection

Evaluation of cows can be for several purposes. Three are of primary importance: (1) selection of cows to remain in the herd, (2) selection of cows to breed to obtain replacement heifers, and (3) selection of cows to breed to obtain sons for herd use or use in artificial insemination. In addition, several traits may be considered important. For example, selection might be to increase milk yield, decrease milking time, increase mastitis resistance, and maintain fat test. The relative economic value of milk yield as compared with other traits, however, dictates that primary emphasis should be placed on milk yield or the value of the milk. Consequently, discussion of cow and sire selection will first include methods of evaluation for a single trait—milk yield. The principles, advantages, and disadvantages of multiple-trait selection are discussed in Chapter 15.

Discussion of genetic evaluation and selection in this and subsequent chapters assumes that the records have been adjusted for such important fixed environmental factors as age. The exception is adjustment for herd and year and season of freshening, which is important at least for milk records. Adjustment for herd–year–season effects can be accomplished by taking differences between the records and the average of herdmates where herdmates are defined as cows freshening in the same herd and in the same season of the same year. Most statistical procedures adjust for opportunity in a similar way.

Dairy cattle breeders have a host of jargon words associated with genetic value. Many of these have the same meaning. To reduce confusion, these terms will be described now.

Genetic value is the effect the genes of the animal have on her production. *Breeding value* has the same meaning except that it is more likely to be used to describe the effects of genes the cow can pass on to her progeny. Because an animal can pass on only a sample one-half of its genes (with corresponding gene effects) to its offspring, the term *transmitting ability* is often used as a substitute for one-half breeding value. Thus

$$\text{transmitting ability} = \text{breeding value}/2$$

$$\text{breeding value} = \text{genetic value}$$

Genetic value is never known exactly and must be estimated. Two terms used widely are estimated breeding value (EBV) and estimated transmitting ability (ETA). As expected,

$$\text{ETA} = \text{EBV}/2$$

Consequently, ETA and EBV can be used interchangeably by remembering that ETA is exactly one-half of EBV. EBV indicates the cow's genetic ability to produce based on her genetic makeup, and ETA indicates the part of a cow's ability to produce that is passed to her daughter.

The accuracies of estimating breeding value or transmitting ability are the same because ranking does not change because ETA is exactly one-half EBV. Confidence ranges are a measure of how close the true breeding value will be to estimated breeding value. Because ETA = EBV/2, the confidence range for transmitting ability is exactly one-half of the confidence range for breeding value.

"Cow index" (CI) is a term used by the USDA which is synonymous with ETA. At least one region of the country uses the term "estimated average transmitting ability" (EATA). The extra word "average" emphasizes that on the average one-half of the genetic value is transmitted but that individual progeny may receive a better-than-average sample of genetic effects and others a poorer-than-average set of genetic effects. Thus equivalent terms for cow evaluations are ETA, EBV/2, CI, and EATA.

Although the genetic principles are the same for bulls, evaluations of bulls have received different jargon words. Values on the USDA sire summary are referred to as predicted differences (PDs) and sometimes as modified contemporary comparisons (MCCs). The values on the Northeast sire summary are referred to as sire comparisons (SCs). All are estimates of transmitting ability, although the computations are done differently and may have different properties. Thus equivalent terms for bull evaluations are PD, MCC, SC, ETA, and EBV/2.

13.1 EVALUATION AND SELECTION OF COWS FOR REAL PRODUCING ABILITY

Selection of cows that are to be retained in the herd can be made either on the basis of anticipated milk yield in the next lactation (cow's real producing ability) or estimation of genetic superiority for milk yield (cow's genetic or breeding value), which is important for selection of dams of replacement heifers. An expression equivalent to "real producing ability" is "breeding value." There is a definite relationship between real producing ability and breeding value, as Figure 13.1 illustrates for milk yield. For a given record of a cow (a record corrected for age and expressed as a difference from the herdmate, year-season average), the real producing ability includes the genetic value in addition to a permanent environmental effect that remains with the cow throughout her lifetime. Real producing ability of a cow for milk yield, on the average, accounts for approximately one-half of the variation from herd average. (This fraction is called "repeatability.") The chance, temporary environmental factors account for the other half. Variance caused by differences in genetic values makes up about one-half of the variance of real producing ability or about one-fourth of the total variance for milk records. The genetic fraction of the total variance is defined as heritability.

Figure 13.1 shows that selecting cows for real producing ability may not be the same as selecting cows for breeding value. In most cases ranking of cows within a herd on the basis of estimated breeding value (EBV) or estimated real producing ability (ERPA) will be similar. In fact, if only the cow's own records are available, the ranking will be the same, although the estimates will be different. The rankings on EBV and ERPA can be different only if records on relatives are also used in the evaluation.

Whether to select for producing ability or for genetic value is a difficult decision to make. Each herd owner must evaluate the question in terms of either short-term gain (selecting for real producing ability) or long-term gain (selecting for breeding value). For example, a dam can contribute a sample half of her breeding value to her daughters. (The other half comes from the sire.) The permanent environmental effect associated with the dam cannot be transmitted and dies with her but does contribute to her productivity while she lives.

In some cases selection for a combination of real producing ability and breeding value may be desirable to maximize profits over a given period of time. The

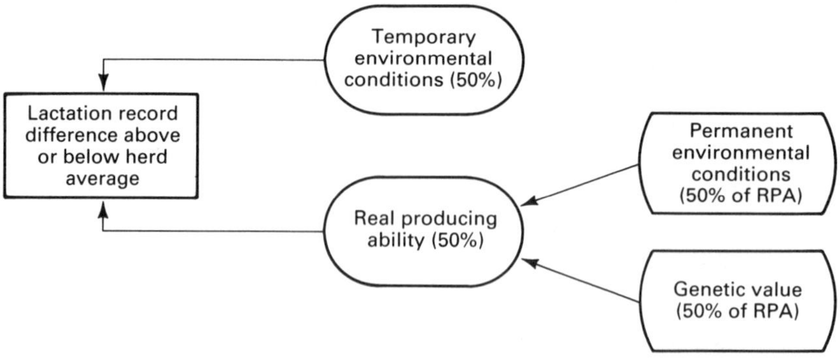

Figure 13.1 Breakdown of variation in differences above and below herd average for milk yield after adjustment for age effects and days carried calf. From *Principles of Dairy Science* by G. H. Schmidt and L. D. Van Vleck. Copyright © 1974 W. H. Freeman and Company. Used by permission.

procedure to find the best combination is complicated and is outside the scope of this book.

Accuracy of Estimating Real Producing Ability

The herd owner who wants to select cows for maximum productivity in their next lactations needs a rule to rank cows for their producing ability. Genetic research has provided the *selection index method* and similar statistical procedures for ranking cows within a herd, which maximize the probability of correctly ranking the animals. Of course, no method is perfect, and some errors in ranking will be made even with the best statistical procedure.

On the average, a first milk record (adjusted for age) is correlated with a future lactation record by approximately 50% (i.e., repeatability is 50%). If a heifer produces 1000 lb of milk on a mature equivalent basis above the average of her herdmates in her first lactation, she will be expected to average approximately 500 lb of milk ($\frac{1}{2} \times 1000$) more than the herd average in her next lactation. Similarly, a first-calf heifer that yields 1000 lb (corrected for age) below the herd average is expected to average 500 lb less than the herd average in the remainder of her records. Not all cows will produce as expected, but on the average the previous description will be true.

In her next lactation a cow will have the same real ability to produce but will receive a new set of environmental conditions. If sufficient records on the cow can be obtained, her real producing ability can be accurately determined because the temporary environmental conditions for her records will balance out to an average of near zero. Estimating future production from two lactations is more accurate than from one record. Similarly, three records are better than two; but two records are not twice as good as one. In fact, five records are only slightly better than four, as shown in Figure 13.2, where the accuracy of estimation is given as the percentage correlation between the average difference of a cow's

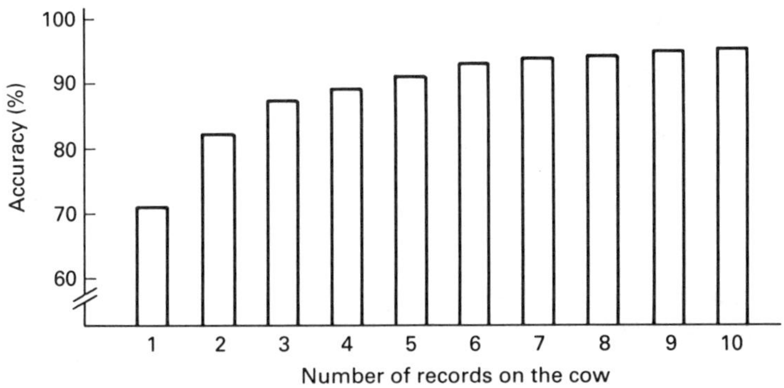

Figure 13.2 Accuracy of estimating real producing ability for milk yield, assuming that repeatability is 50%. From *Principles of Dairy Science* by G. H. Schmidt and L. D. Van Vleck. Copyright © 1974 W. H. Freeman and Company. Used by permission.

records from her herd average and her true ability. The accuracy of estimating real producing ability from a cow's records can be determined by the following formula:

$$\text{accuracy} = \sqrt{\frac{nr}{1 + (n - 1)r}}$$

In this equation accuracy is the correlation between the cow's true producing ability and the average of n of her records and r is repeatability of the trait. Repeatability is the correlation between two records for the same cow and measures the fraction of variation in records which is due to differences in real producing abilities.

Estimating Real Producing Ability

Many dairy records processing centers provide estimates of real producing ability on a routine basis. The estimates may have different names in different parts of the country: estimated real producing ability (ERPA), estimated producing ability (EPA), and most probable producing ability (MPPA).

A simple way to estimate real producing ability follows, which also illustrates some of the principles used in more complicated methods. Records of the cow and her herdmates will be used. Records of other relatives and the proof of her sire could be used but are of little importance in estimating producing ability.

1. Compute the herd-season average corresponding to each record of the cow. Such an average should include age-corrected records of cows that freshened at approximately the same time, but not the record of the cow. A records program may supply such information. If not, take, for example, cow A that freshened in June. A corresponding herd average might include records of other cows freshening between April and July of that year.
2. Find the difference between each record of the cow and the corresponding herd average.
3. Compute the average difference for each cow by dividing the cow's total difference by the number of its records.
4. Multiply by the weighting factor:

$$\frac{nr}{1 + (n - 1)r}$$

The result is the amount of milk above or below the herd average her future lactation is estimated to be—the ERPA. In summary,

$$\text{ERPA} = \frac{nr}{1 + (n - 1)r} \text{ (average of differences from herdmates)}$$

As an example, suppose that cows A, B, and C are to be ranked according to their estimated real producing ability. Cow A has one record of +1000 above herd average. Cow B has three records of −200, +500, and +300, averaging +200. Cow C has two records of +800 and +900, averaging +850. The estimates are:

$$\text{ERPA for A} = 0.5 \times 1000 = +500$$

$$\text{ERPA for B} = 0.75 \times 200 = +150$$

$$\text{ERPA for C} = 0.67 \times 850 = +570$$

The rank is then C > A > B.

Ranking of cows for culling can be on the basis of estimated real producing ability. In some situations, however, this ranking system would need to be modified. If few replacement heifers were forthcoming, keeping a young cow might be preferable to keeping a higher-ranking older cow, so that herd size could be maintained more easily.

Furthermore, most of the lower-ranking animals may be first-lactation heifers because more younger cows are available to be culled than any other age group, and older cows, having survived one or more cullings, may be a selected and superior group compared with the average younger cows. If genetic improvement is the major goal, the genetic value of the offspring should be considered; that is, selection should be based on the estimated transmitting ability of the dam rather than on ERPA.

13.2 EVALUATION AND SELECTION FOR TRANSMITTING ABILITY

Many dairymen select a cow for its breeding value rather than for its real producing ability, that is, for its genetic ability to produce, of which a sample half (its transmitting ability) can be passed to its offspring.

Accuracy of Estimating Transmitting Ability

The best information about a cow's transmitting ability is her records. Each additional record adds to the accuracy of the estimate, but the added accuracy becomes less with each extra record. The accuracies of ETAs from a cow's own records, from her mother's records, or from a grandmother's records are shown for milk records in Figure 13.3. A cow's records are twice as valuable as those of its mother and four times as accurate as those of its grandmother because accuracy depends on the relationships to the cow being evaluated. Figure 13.3 also shows that adding more records increases the accuracy slowly after the first one or two records.

The accuracy values that result when records on the cow, as well as records on either its dam or granddam or its daughters, are used, are shown in Figure 13.4. The figure illustrates clearly the importance of using records of closely related animals in estimating genetic value. Records of grandparents increase accuracy much less than records of progeny or parents. Since a cow is completely

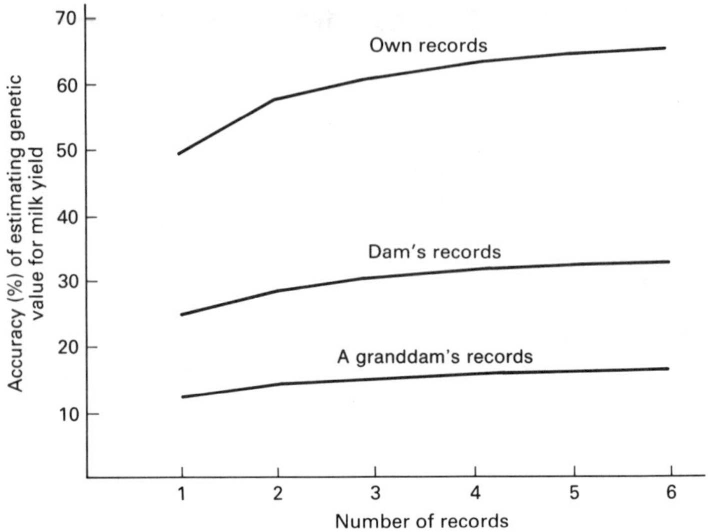

Figure 13.3 Comparison of the accuracy of estimating a cow's genetic value from the cow's own records, her dam's records, and her granddam's records, assuming that heritability is 25% and repeatability is 50%. From *Principles of Dairy Science* by G. H. Schmidt and L. D. Van Vleck. Copyright © 1974 W. H. Freeman and Company. Used by permission.

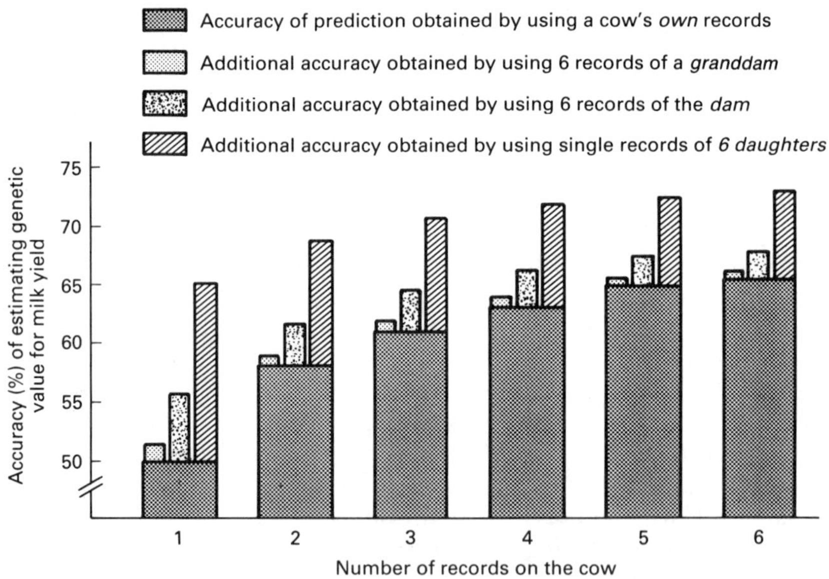

Figure 13.4 Comparison of the accuracy of estimating a cow's genetic value from the cow's own records, her own plus six records of a granddam, six records of the dam, and single records of six daughters, assuming that heritability is 25% and repeatability is 50%. From *Principles of Dairy Science* by G. H. Schmidt and L. D. Van Vleck. Copyright © 1974 W. H. Freeman and Company. Used by permission.

related to herself, her records are the best source of information about her breeding value.

The proof of her sire can add more to the accuracy of the estimate of a cow's breeding value than her dam's records, her granddam's records, or her daughter's records because a sire can have many daughters, thus making his proof quite accurate. Using the sire proof as well as the cow's own records increases the accuracy of estimating her breeding value, as shown in Figure 13.5. A within-herd natural service proof may contain as many as 10 or more daughters, whereas AI proofs may contain as many as 50 or more. Using a proof with 10 records results in a sizable increase in accuracy, whereas one with 50 records increases accuracy less than twice as much. More than 50 records in the sire proof does not significantly increase the accuracy of prediction. The next 150 daughter records (up to 200) increase the accuracy by approximately 2% when one record on the cow is known; accuracy is increased 1% when six of the cow's records are available.

Records on relatives can be ranked for their importance in predicting a cow's genetic value in this order:

1. A cow's own records
2. A proof on her sire
3. Her daughter's records
4. Her dam's records
5. A proof on her grandsire
6. Her granddam's records

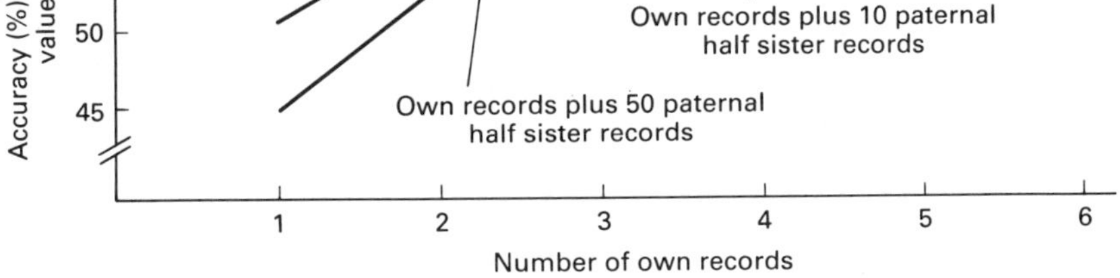

Figure 13.5 Comparison of the accuracy of estimating a cow's genetic value from the cow's own records and her own plus a sire proof based on 10, 50, and 200 half sister records, assuming that heritability is 25% and repeatability is 50%. From *Principles of Dairy Science* by G. H. Schmidt and L. D. Van Vleck. Copyright © 1974 W. H. Freeman and Company. Used by permission.

Estimated transmitting abilities are routinely provided to herds on DHI testing. Many centers use contemporary or herdmate comparisons and records on the cow and her dam and the proof of her sire. The name for the USDA estimate of transmitting ability is the cow index (CI). Some centers provide the same estimate but name it estimated average transmitting ability (EATA). Others use the name ETA. The Northeast method is based on a more sophisticated statistical method and uses all relationships in a herd to estimate ETAs.

In general, heifers cannot be evaluated very accurately before they have completed some of their first record, but an ETA for a heifer can be computed from the ETAs of the mother and father: (ETA of her dam + ETA of her sire)/2. If no sire proof is known, the heifer's ETA is (ETA of her dam)/2. If no records are known on her mother, the heifer's ETA is (ETA of her sire)/2. If the average or base ETA for potential sires and dams is not zero, that base can be substituted for zero when the parent ETA is unknown.

The ETA of the sire is the sire's predicted difference or daughter superiority, as described in Chapter 14.

FURTHER READINGS

AGRICULTURAL RESEARCH SERVICE. (Annual). DHIA cow performance index list. U.S. Department of Agriculture, Beltsville, MD.

AGRICULTURAL RESEARCH SERVICE. 1976. USDA–DHIA modified contemporary comparison sire summary and cow index procedures. *USDA Prod. Res. Rep.* 165, U.S. Department of Agriculture, Beltsville, MD.

BATH, D. L., F. N. DICKINSON, H. A. TUCKER, and R. D. APPLEMAN. 1985. *Dairy Cattle: Principles, Practices, Profits*, 2nd ed. Lea & Febiger, Philadelphia.

DICKERSON, G. E. 1941. Estimates of producing ability in dairy cattle. *J. Agric. Res.*, 61:561.

HENDERSON, C. R. 1975. Use of all relatives in intraherd prediction of breeding value and producing abilities. *J. Dairy Sci.*, 58:1910.

LEGATES, J. E., and J. L. LUSH. 1954. A selection index for fat production in dairy cattle utilizing the fat yields of the cow and her close relatives. *J. Dairy Sci.*, 37:744.

POWELL, R. L. 1980. Shedding a little light on cow indexes. *Holstein World*, Feb. 10.

ROBERTSON, A. 1959. A simple method of pedigree evaluation in dairy cattle. *Anim. Prod.*, 1:167.

14

Bull Evaluation and Selection

Most breeders know that the phrase "the herd sire is half the herd" means that half the future herd inheritance comes from the bull and half from his mates. Actually, bulls usually contribute more than half to the genetic improvement in most herds since herd sires can be more intensely selected than cows—perhaps out of the top cows in the herd or in the nation. In the past, the problem was that the accuracy of evaluation of a bull was small because frequently all that was known about the bull was the performance of his dam.

The user of artificial insemination (AI), however, expects available sires to have been evaluated relatively accurately from many daughter records in many herds. Selection of AI bulls can be very intense, so that truly herd-improving sires are available in AI. The percentage of the total of genetic improvement resulting from such superior sires may be 90% or more.

The discussion of bull evaluation and selection in this chapter will assume that the daughter records have been adjusted for important known effects and will emphasize selection of bulls for milk production or product value. Selection for more than one trait, such as for product value or for type-production indexes, is discussed in Chapter 15.

General methods of bull evaluation from progeny records are described first, followed by an outline of a recommended two-stage selection process for obtaining genetically superior sires for use in AI. Special difficulties with progeny proofs in one herd—the so-called natural service proof—are also described.

Bases

Genetic evaluations such as the USDA's predicted difference, PD, or the Northeast's AI sire comparison, NEAISC, can only predict differences between bulls. The magnitude of the proofs is not important—only differences have meaning. The proofs, however, can be standardized to what is called a fixed base. One way that is often used is first to obtain statistical solutions for the proofs. Then, if, for example, the base is to be 1974, the weighted average of proofs for bulls with daughters first freshening in 1974 is calculated. Weighting is by number of daughters freshening for the first time in 1974. This weighted average or base is then subtracted from all proofs. After this adjustment, the weighted average for 1974 freshenings will be zero.

If genetic progress is made, bull proofs will increase with time. Soon most bulls will have proofs of 1000 lb or higher when expressed from the 1974 base. People have a tendency to forget that only differences between proofs are meaningful. All bulls with proofs over some magic number such as 1000 lb might wrongly be considered to be equally good, so that selection standards are relaxed with a corresponding decrease in rate of genetic gain.

Thus a new base might be established to decrease the average proof to more nearly zero. For example, the USDA in 1984 changed the base to that for cows first freshening in 1982. The proofs were then labeled PD_{82} rather than PD_{74}. The change in base decreased proofs for Holstein bulls by about 900 lb of milk. Differences between bulls did not change except for new records included in later evaluations.

14.1 MODIFIED CONTEMPORARY COMPARISON

The chief dairy bull evaluation report in the United States is the Sire Summary List, which is calculated and published twice each year from DHI records by the USDA Animal Improvement Programs Laboratory. Nearly all other reports are taken from this one with the exception of the Northeast AI Sire Report, based on the NEAISC, which is calculated separately and published for the northeastern states twice each year. The procedures were similar for both reports until 1970, when the Northeast report utilized advanced computer speeds to take into account additional factors involved in sire evaluation with a mixed model procedure called the sire comparison. In 1974, the USDA replaced the herdmate procedure with a modification called the modified contemporary comparison (MCC) in an attempt to account for those factors.

Estimated Daughter Superiority

The goal of all sire evaluation procedures is to estimate differences in daughter performance between a bull and another bull, which is equivalent to estimating the difference between one-half the genetic values of the two bulls. The usual names given to the USDA evaluation are *predicted difference* (PD) and *modified contemporary comparison* (MCC). The evaluation is equivalent to estimated transmitting ability (ETA) or one-half estimated breeding value (EBV/2).

Fixed Adjustments

All records are standardized to a lactation length of 305 days, twice-a-day milking, and a mature age equivalent. The age adjustment factors are different for the various geographical regions. Records that are incomplete because of abortion and culling are extended to a 305-day length. These adjustments are important to reduce possible biases in sire evaluation.

An essential step in the evaluation is to adjust each record for the herd where the record was made by expressing the record as a deviation from the average of the contemporaries (cows of the same age in the same herd that started records at approximately the same time of year). All herdmates, regardless of age, are used since the records are adjusted for age, but records of cows of the same lactation number receive more weight in the comparison. The difference from contemporary average is the basic unit for the contemporary comparison method of sire evaluation. The number of contemporaries partially determines the value of the contemporary difference. Therefore, each contemporary comparison is adjusted according to the number of contemporaries. The contemporary average is also adjusted for the merit of the sires of the contemporaries. The modified contemporary average is computed counting the noncontemporary average (herdmates of a different lactation group) as one contemporary. Thus if a first-lactation cow has five first-lactation herdmates with an average of 12,000 lb and 15 later-lactation herdmates with an average of 15,000 lb, her modified contemporary average is

$$\text{MCA} = \frac{(5 \times 12{,}000) + (1 \times 15{,}000)}{5 + 1} = 12{,}500$$

If the cow were in a later lactation, the later contemporaries receive the most weight:

$$\text{MCA} = \frac{(1 \times 12{,}000) + (15 \times 15{,}000)}{1 + 15} = 14{,}812$$

Modified Contemporary Comparison Method of Evaluation

This procedure is designed to reduce problems due to the possibility of older herdmates being more highly selected in some herds than in others and also problems due to age factors not being appropriate for a particular herd.

The MCA is then adjusted for the average value of the sires of the contemporaries by subtracting an appropriately weighted average of the predicted merits of the contemporary sires, $\overline{PD}_{\text{SMC}}$.

The basic unit of the modified contemporary comparison is the modified contemporary deviation:

$$\text{MCD} = D - (MCA - \overline{PD}_{\text{SMC}})$$

where D is the cow's record, MCA is the modified contemporary average for her herdmates, and $\overline{PD}_{\text{SMC}}$ is the adjustment for the average transmitting ability of the sires of the contemporaries. This adjustment attempts to account for genetic differences in the competition that daughters of various bulls may have in different herds.

Numbers and heritability adjustment. Another adjustment of the average daughter contemporary difference is a weighting factor, b, which takes into consideration the number of daughters, heritability, number of herds represented, environmental correlation between paternal sisters having records in the same herd, lengths of lactations, number of modified contemporaries, and the weights used for the predicted differences of the sires of the contemporaries. For example, a complete record will receive slightly more weight than an incomplete or in-progress record. A daughter with more than one record will be slightly more important for her sire's evaluation than a daughter with only one record.

The final adjustment factor can be approximated as

$$b = \frac{Nh^2}{4 + (N - 1)h^2 + (4c^2/N)(\sum n_i^2 - N)}$$

where N is the total number of effective daughters of the bull, $\sum n_i^2$ is the number of daughters in each herd squared and summed for all herds, $h^2 = 0.20$ is the USDA estimate of heritability of milk yield, and $c^2 = 0.14$ is the estimate of the environmental correlation between paternal sisters in the same herd.

The effective number of daughters is somewhat different from the number of daughters because of adjustment for the number of records per daughter, which means that a daughter with more than one record is slightly more important than a daughter with only one record, and for lengths of lactations.

Summary of predicted difference. The summary of the computations for the PD of a bull is

$$\text{PD} = b[\overline{D - (MCA - \overline{PD}_{\text{SMC}})}] + (1 - b)G$$

where b is the weighting factor (also called repeatability by the USDA), and $[\overline{D - (MCA - \overline{PD}_{\text{SMC}})}]$ is the average of daughter differences from their modified contemporary averages after adjusting for the merit of sires of the contemporaries. The last part of the PD is the adjustment toward the genetic average, G, of the group or population to which the bull belongs. As the amount of information on a bull increases, b approaches 1 and the influence of the group average on the PD becomes less and less. The PD is labeled with the year used as a base. When the base of comparison is changed, the PDs receive another label (e.g., PD_{74}, PD_{82}).

Repeatability of Predicted Difference

The correlation between the true daughter difference for a bull and his PD is the square root of the final adjustment factor, b. The USDA has defined repeatability to be the squared correlation, which is b. As the number of daughters becomes large and with only one daughter per herd ($n_i = 1$ for all herds), the value of b approaches 1, perfect repeatability. Figure 14.1 shows the effect of number of daughters per herd on repeatability, using USDA values for heritability and environmental correlation between daughters in the same herd.

The most important factor in deciding which bull to use is the PD. Selection theory says that the probability of choosing the better bull is always higher when the bull with the higher PD is chosen, no matter what the repeatability values are. Therefore, repeatability should not be used to rank bulls. Repeatability can be used to put confidence ranges on bulls' true transmitting abilities as discussed in Section 14.7.

However, repeatability values of less than 50 to 60% may indicate the possibility of biases because the daughters will be in only a few herds or even in a single herd. Biases may result if some daughters receive preferential treatment in feeding and management. There are no methods to adjust for preferential treatment.

Sires Included in Sire Summaries

Both AI and natural service sires are included in the USDA sire summary. Natural service sires generally have low repeatabilities because most of their daughters are in a single herd, where the possibility of environmental correlations exists.

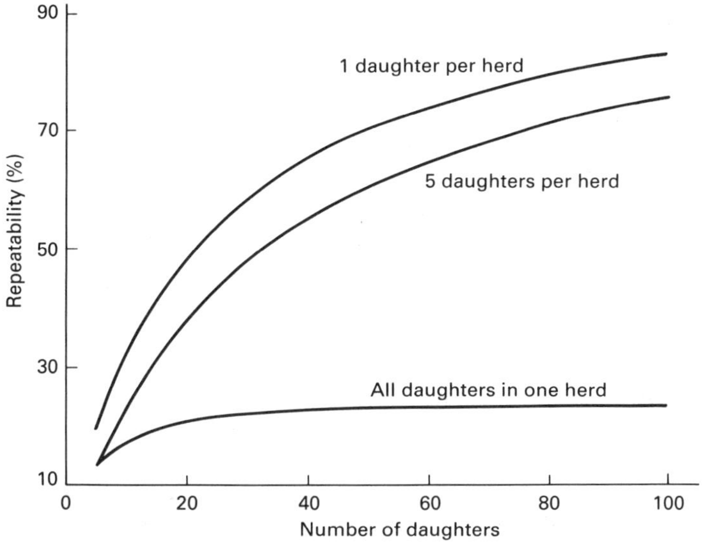

Figure 14.1 Repeatabilities for USDA sire evaluations for different numbers of daughters per herd. From *Sire Summary List*, U.S. Department of Agriculture, May, 1969.

14.2 THE SIRE COMPARISON METHOD

The Northeast sire evaluation procedure is a mixed model method that takes into account several factors that have not yet been possible to implement on a national basis.

Basic Unit of Comparison

The basic idea of the sire comparison method is quite simple. Daughters of one sire are compared with daughters of another sire that freshen in the same herd, year, and season. This comparison is an estimate of the difference between transmitting abilities of the sires and is the basic unit for the sire comparison method. For example, if three sires, A, B, and C, have daughters freshening in a herd during the same year-season, the differences between A and B, between A and C, and between B and C can be estimated. More sires would give more combinations. These estimated differences can then be combined for all herds and year-seasons to obtain a final set of differences among sires. Heritability, the number of comparisons, and relationships among the bulls are considered in combining the differences.

The Need for Sire Comparison

The herdmate method was developed in 1954 for conditions existing at that time. Because computers in the early 1950s were relatively small and slow, simplifying assumptions had to be made. The herdmate method depended on the validity of certain of these assumptions, the most important of which are that there is no genetic improvement over time and that sires are used randomly in all herds. In the early 1950s, bulls in AI appeared to average no better than those used naturally, and new AI bulls appeared to be no better than older ones. Technicians carried liquid semen from only a few bulls each day, with a bull being available at most twice a week, resulting in nearly random use of available bulls. The herdmate method gave excellent results under those conditions. However, as proofs became available, poor sires were eliminated, whereas better ones were used more heavily, resulting in both genetic improvement and nonrandom use of sires. Testing of sons of the best sires and heavy use of only the best sires resulted in additional genetic gain.

Effects of sires of herdmates on the herdmate method. The herdmate method compares records of a sire's daughters with records of their herdmates, that is, cows out of other sires that freshen in the same herd, year, and season. The herdmate method did not consider how good the sires of the herdmates were and therefore gave biased evaluations if the average value of sires of the herdmates was not the same for all sires (Figure 14.2). Pricing semen of bulls differently may, for example, result in nonrandom use of sires in some herds.

In the northeastern United States, the Ayrshire bull Selwood Betty's Com-

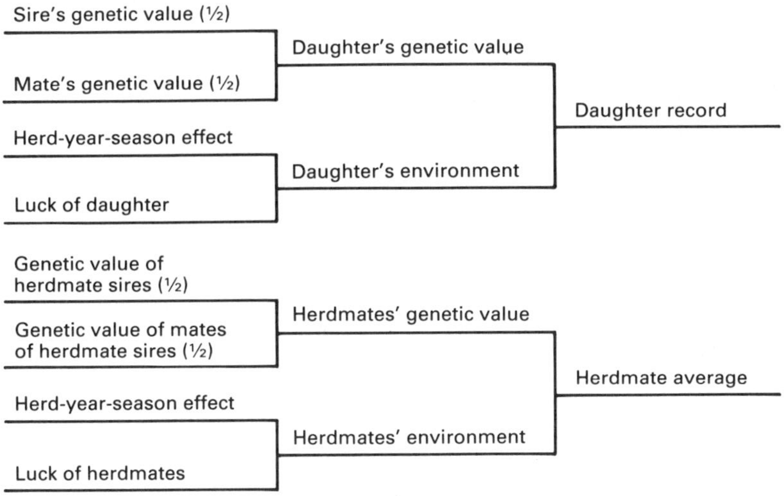

Daughter
herdmate difference = ½(sire's genetic value − average of sires of herdmates)
+ ½(mate's genetic value − average of mates of herdmates' sires)
+ (luck of daughter − average luck of herdmates)

Necessary assumptions for unbiased sire evaluation:

1. Average of sires of herdmates is the population average or the same for all sire groups.
2. Average of mates of sire is the same as average of mates of herdmate sires.

Figure 14.2 Model of the daughter–herdmate comparison, in which records of daughters are compared with the average of records of cows freshening in the same herd, year, and season.

mander had an evaluation far superior to any other Ayrshire bull. Consequently, a majority of services were to him. Thus other bulls' daughters were being compared with herdmates that were primarily "Betty" daughters. Those bulls were at a disadvantage when compared with bulls having daughters that did not have Betty daughters as herdmates.

Effect of genetic improvement on the herdmate method. Genetic improvement can introduce substantial bias against young bulls if the herdmate method is used because daughters of young sires are compared with the latest, best herdmates, whereas daughters of older bulls are compared with poorer herdmates of earlier years. The effect of genetic improvement in productivity of 200 lb per year is illustrated in Figure 14.3. Daughter superiority drops at half that rate because mates of the bull that contribute half of the genes of the daughters are improving as rapidly as their herdmates. The sire comparison method overcomes these biases by using as the basic comparison the difference between the daughter of one sire and the daughter of another sire in the same herd, as shown in Figure 14.4. Thus random use of sires is not required. Similarly, the situation where sires of herdmates of a young bull's daughters are superior to the sires of herdmates of an older bull's daughters is taken into account by comparing sires whose daughters freshen in the same herds and year-seasons. The USDA pro-

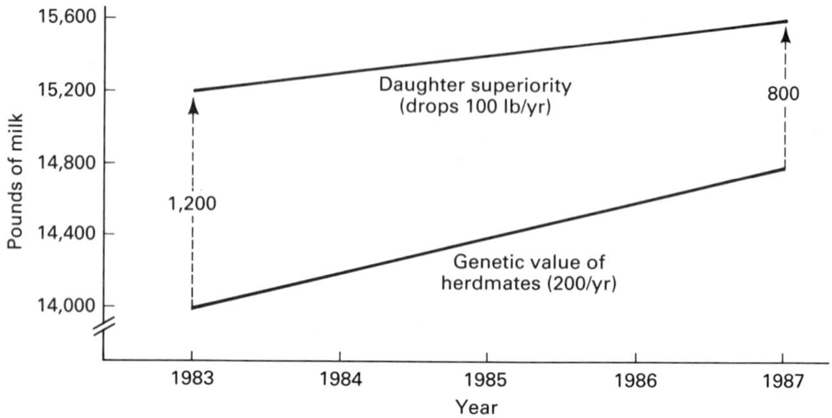

Figure 14.3 Effect of genetic trend on apparent daughter superiority assuming genetic improvement of 200 lb per year. From *Principles of Dairy Science* by G. H. Schmidt and L. D. Van Vleck. Copyright © 1974 W. H. Freeman and Company. Used by permission.

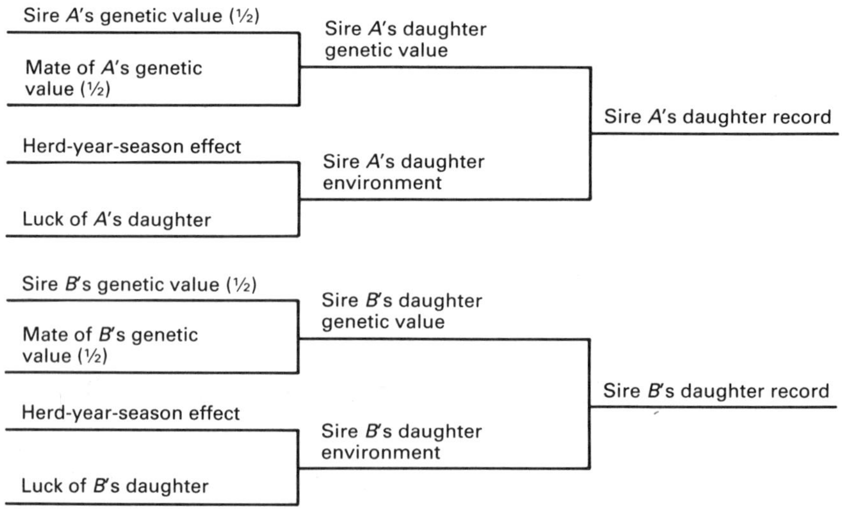

Daughter of Sire *A*
− daughter of Sire *B* = ½(*A*'s genetic value − *B*'s genetic value)
+ ½(genetic value of *A*'s mate − genetic value of *B*'s mate)
+ (luck of *A*'s daughter − luck of *B*'s daughter)

Necessary assumption for unbiased sire evaluation: *A*'s mates' average is the same as *B*'s mates' average.

Figure 14.4 Model for the sire comparison where daughters of one sire (A) are compared with daughters of another sire (B), all freshening in the same herd, year, and season.

cedure accounts for sires of herdmates by adjusting the herdmate average (MCA) for the PDs of sires of the herdmates.

Limitations of both methods. A limitation of both the modified contemporary comparison and sire comparison methods is that any special treatment given to daughters of some sires to increase their productivity cannot be measured from DHI records.

Neither method fully considers the possibility that some sires might be mated to better cows in a herd than other sires. Most research shows that the differences between mates of the sires usually are not significant enough to cause much problem, but either method of sire evaluation can be modified to take into account the records of the mates. The sire comparison uses the proof of the sire of the mate (the maternal grandsire of the daughter) to adjust for the genetic contribution of the mate to a bull's daughter.

Records used. The sire comparison method, as used in the Northeast AI Sire Comparison, compares only first-lactation daughters of AI sires. Any bias from selective culling of cows because of low production during or at the end of their first lactation is eliminated by using only first records. Only AI sires are evaluated because inclusion of natural service sires increases the computational difficulties. In areas where herds are relatively large and include many artificially sired cows, these restrictions cause no problems. The sire comparison method can be modified so that all records can be included for daughters of both AI and natural service sires, but the computational problems would increase. Such problems, however, will be reduced or eliminated by new generations of computers.

The Northeast sire comparison now provides joint estimates for milk, fat, and protein yield. Joint estimation takes into account genetic correlations among the three yield traits. Accuracy of evaluation for protein yield is improved for bulls having daughters with few protein records and many milk and fat records. For bulls with all daughters having all three traits measured, the increase in accuracy of evaluation is slight as compared to single trait evaluation.

14.3 USE OF DAM–HERDMATE DIFFERENCES IN SIRE EVALUATION

There are two possible reasons for using the dam–herdmate difference in sire evaluation. The first and more important reason is to eliminate bias resulting from the mating of some bulls to genetically better cows. Bulls with better-than-average mates would have higher daughter average differences from herdmates than they really deserve. This is a positive bias. Similarly, bulls mated primarily to poorer cows would receive a negative bias in evaluation.

Correction for Selection of Mates

Possible biases. Table 14.1 shows how much positive bias can be expected depending on how highly selected a bull's mates are. For example, if a

TABLE 14.1 BIAS IN DAUGHTER–HERDMATE DIFFERENCE CAUSED BY SELECTED MATES[a]

Top percentage from which sire's mates are selected	Bias (pounds of milk) in daughter–herdmate average $\frac{1}{8}$ (dam–herdmate deviation)
10	439
20	350
30	290
40	242
50	200
60	161
70	124
80	88
90	49

[a] A within-herd standard deviation of 2000 pounds and heritability of 25% were assumed.

bull is mated to cows in the top 10% of the herd, the average daughter difference from herdmates would be biased by approximately 440 lb of milk. If the matings are to the top 50% of cows, the bias would be 200 lb.

Bias in sire evaluation after adjustment for numbers of daughters will be less than bias in average daughter–herdmate difference because the multiplicative adjustment, b, is always less than 1. The weight given to the dam's herdmate difference when one record is available on both the daughter and dam is one-eighth of the weight for the daughter–herdmate difference. Table 14.2 shows the bias in the estimated daughter superiority for different numbers of daughters and levels of selection of mates, assuming a heritability of 0.25. When many AI daughters are scattered throughout many herds, there is only a slight chance of having much

TABLE 14.2 BIAS IN DAUGHTER–HERDMATE SIRE EVALUATION CAUSED BY SELECTED MATES[a]

Top percentage from which sire's mates are selected	Number of daughters in evaluation						
	5	10	20	40	70	100	200
10	+50	+80	+116	+145	+165	+173	+185
20	40	64	92	116	132	138	148
30	33	53	76	96	109	114	123
40	27	44	64	80	91	95	102
50	23	36	53	66	75	79	84
60	18	29	42	54	61	64	68
70	14	23	33	41	47	49	52
80	10	16	23	29	33	35	37
90	5	9	13	16	19	20	21

[a] Bias in sire evaluation in pounds of milk, numbers adjusted.

selection for mates of various bulls. Selected mates are more likely for only a very few daughters. Then the bias in the bull evaluation will not be as great because the adjustment for number of daughters is a number less than 1.

Wrong use of dams' records. One misuse of the daughter–dam comparison method that preceded the herdmate contemporary comparison method was to subtract the dam's record from the daughter's record. Any changes in the overall herd average resulting from better feeding and management were included in the daughter–dam difference, which made daughters look better than they really were, as illustrated in Figure 14.5. If both daughter and dam records are expressed as differences from herdmate averages, the change in herd average does not introduce such a bias.

A method called joint sire and cow evaluation based on mixed model procedures automatically accounts for genetic value of mates of a bull in his evaluation very much as described in this section. The method requires an equation for each cow with a record and for each bull. Consequently, the method is computationally demanding and has not yet been widely implemented. The Northeast AI Sire Comparison includes a maternal grandsire effect (sire of the mate) in the model to account partially for the genetic value of the mate.

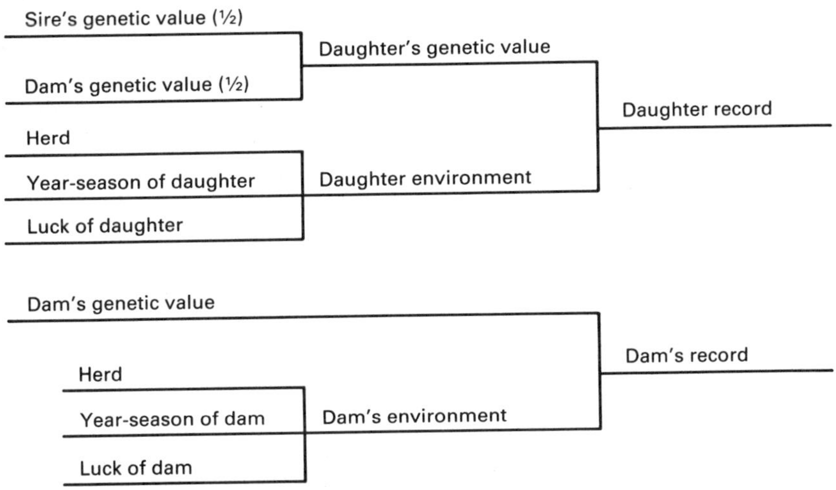

Daughter-dam = ½(sire's genetic value − dam's genetic value)
+ (year-season of daughter − year-season of dam)
+ (luck of daughter − luck of dam)

Necessary assumptions for unbiased sire evaluation:

1. Dam's genetic value averages are the same for all sire groups.
2. Average year-season effect of daughters equals the average year-season effect of dams. (This is not likely to occur with changing levels of production.)

Figure 14.5 Model of the daughter–dam comparison. From *Principles of Dairy Science* by G. H. Schmidt and L. D. Van Vleck. Copyright © 1974 W. H. Freeman and Company. Used by permission.

14.4 NATURAL SERVICE SIRE EVALUATION

A natural service sire evaluation is defined as a proof based on daughter records in only one or very few herds. Such evaluations have been used successfully in some countries to find bulls for use in AI. Their success resulted primarily from the very intense selection applied—for example, purchase of only the top 1 of 100 or 1 of 200 naturally proved bulls and only after careful examination of the conditions under which daughter records were made. Experience with NS proofs in most countries has not been very satisfactory unless selection was done carefully and intensely.

Disadvantages of Natural Service Sire Evaluation

The first two disadvantages of natural service sire evaluations are obvious. The natural service evaluation usually consists of only a few daughter records, which reduces the accuracy of evaluation in comparison with AI proofs with many daughters in many herds. In addition, daughters of the same sire tend to be treated somewhat more alike in a herd than unrelated cows. Research shows that this environmental correlation occurs for both naturally and artificially sired sisters in a herd. Most AI proofs have only one or two daughters per herd, and thus the environmental correlation is much less important for AI proofs than for natural service proofs where most daughters are in a single herd. The result, as shown in Figure 14.1, is to reduce accuracy of evaluation and consequently genetic progress. Natural service evaluations can be adjusted for small numbers and environmental correlations to make AI and natural service evaluations more comparable. The third disadvantage is most important.

The tarnished reputation of the natural service proof resulted from the questionable but understandable practice of making a bull "look good" by giving special care to his daughters. To what extent this was done in the past and to what extent it is done is not clear. Many people think that such preferential treatment is the main reason why some "can't miss" bulls on the basis of their natural service proof "did miss" when put into AI service.

No Correction for Preferential Treatment

There is no way of correcting either an AI or a natural service proof for preferential treatment, but AI daughters exist in many herds where there are less likely to be special reasons to make a bull look good. Thus AI proofs based on one or two daughters per herd in 50 or more herds seem unlikely to be affected very much by preferential treatment.

Research can tell approximately how much preferential treatment has occurred, but only by comparing what the bull did later or what his sons did in AI with what he was expected to do based on a first proof made under what turned out to be preferential conditions. Some studies indicate that preferential treatment has been important in evaluations of some bulls used in AI. Those bulls were generally available only for special fees and before a random AI proof was available. When such bulls were later used in random herds and on a random set of cows, their proofs declined because the preferential treatment given to their first AI daughters was not repeated on the larger number of later daughters. Later daughters of bulls made popular by their first production and type proofs may also receive substantial preferential treatment so that later proofs are biased.

14.5 YOUNG BULL TESTING

Two-stage selection of bulls is a recommended practice for AI studs. Most bull studs now follow this plan. In the future, most of the best AI-proved bulls will have survived these two stages of selection. First, cows with the largest cow indexes (CI) are mated to bulls with the highest predicted differences to produce young bulls for further testing. This first stage of selection should produce young bulls that average 1000 lb of milk or more in estimated daughter superiority (one-half the genetic value of the young bull) above a bull produced from an average mating. Second, these selected young bulls are tested by mating to a limited number of cows to produce 50 to 100 daughters. The best one of 5 or 6 of young bulls that survive this testing procedure should average 750 lb or more of milk above the 1000 lb from first-stage selection.

Matings to Produce Young Bulls

The effectiveness of stage 1 selection depends on the fact that an average bull calf receives one-half of his genetic value from his sire and the other half from his dam. Thus the predicted difference or pedigree index (PI) of a young bull is

$$\text{PI} = (\text{ETA of dam} + \text{PD of sire})/2$$

Similarly, the average daughter superiority of all young bulls will be half the average daughter superiority of the dams plus half the average daughter superiority of the sires that are mated to produce the young bulls. The problem, simply put, is to find the best cows and mate them to the best proved sires.

Cows should not be excluded from consideration if they have only one record. By the time the cow is bred to the best available AI sire and a bull calf is obtained, she will probably have at least two more records. The cow could then be reevaluated before the bull calf is purchased, although evidence shows that once a cow is considered for such a special mating, her records improve considerably, probably as a result of preferential treatment, which makes those records unreliable. The proof of a cow's sire (based on her paternal sisters) increases the accuracy of predicting her transmitting ability (Table 14.3). Records of paternal sisters add approximately 15% to the accuracy of evaluating the cow's genetic value. The values in the table are estimated daughter superiority for different levels of selection for Holstein records.

Table 14.4 also shows the importance of selecting prospective dams from the top of the list. The top 1 or 2% of cows should be enough to produce young

TABLE 14.3 COMPARISON OF DAUGHTER SUPERIORITY OF DAMS SELECTED ON THE BASIS OF THEIR OWN RECORDS WITH THAT OF DAMS SELECTED ON THE BASIS OF THEIR OWN RECORDS AND RECORDS OF PATERNAL HALF SISTERS[a] (POUNDS OF MILK)

Dams' ETA (select top percentage)	Own records (2)	Own records (2) + half sisters' (100)
1	980	1170
2	840	960
5	720	820
10	615	700
20	490	560
50	280	320
100	0	0

[a] Based on heritability of 25% and repeatability of 50%.

TABLE 14.4 SUPERIORITY OF DAUGHTERS OF YOUNG BULLS RESULTING FROM MATING TOP AI SIRES TO TOP DAMS

Dams' ETA		AI sires' average predicted difference (lb)					
		1000	800	600	400	200	0
Select top percentage	Average (lb)	Young bull pedigree index (lb)					
1	1170	1085	985	885	785	685	585
2	960	980	880	780	680	580	480
5	820	910	810	710	610	510	410
10	700	850	750	650	550	450	350
20	560	780	680	580	480	380	280
50	320	660	560	460	360	260	160
100	0	500	400	300	200	100	0

bulls unless selection for many other characteristics decreases the selection intensity.

In the United States, sires chosen to sire the young bulls should average over 1000 lb in daughter superiority above cows born the same year. Table 14.4 shows the expected results for different intensities of selection of both bull dams and bull sires. If the bull sires average 1000 lb and the top 1 or 2% of bull dams are used, the average predicted difference of the young bulls should average approximately 1000 lb. This high expected average is possible because of the recognition of the top cows through publication of the USDA cow index list and of similar lists by universities that use the cow's own records and the records of her relatives for the cow's evaluation.

Although the averages of these young bulls should be very high, the accuracy of the estimates is only approximately 60%, compared with 90 to 95% for proved bulls. Table 14.5 shows the fraction of young bulls that should have true daughter superiority above 500 lb of milk for the various average ETAs and PDs given in

TABLE 14.5 FRACTION OF YOUNG BULLS HAVING TRUE DAUGHTER SUPERIORITY ABOVE +500 POUNDS FOR VARYING AMOUNTS OF SELECTION OF THEIR DAMS AND SIRES

Dams' ETA		AI sires' average predicted difference (lb)					
		1000	800	600	400	200	0
Top percentage	Average (lb)	Fraction of young bulls with true daughter superiority above +500					
1	1170	0.89	0.84	0.79	0.72	0.65	0.57
2	960	0.84	0.79	0.72	0.65	0.57	0.48
5	820	0.80	0.74	0.67	0.59	0.51	0.42
10	700	0.77	0.70	0.62	0.54	0.46	0.38
20	560	0.72	0.65	0.57	0.48	0.40	0.32
50	320	0.63	0.55	0.47	0.39	0.31	0.24
100	0	0.50	0.42	0.34	0.27	0.20	0.15

Table 14.4. With intense selection of bull sires and bull dams, the chances are that 84 to 89 of 100 young bulls will have daughter superiority greater than +500 lb of milk. If only average bulls and average cows are mated (both with an ETA of 0 lb), their sons will average 0 lb PD; only 15 of 100 would be expected to exceed +500 lb in true daughter superiority.

These results show that a dairy cow population receives a major genetic boost if bull studs test young bulls out of the best bull dams and bull sires. The size of this increase depends on how high the standards are set to qualify as "best" bull dams and bull sires. Enough superior cows and bulls are available so that the standards should be set very high. If these young bulls are then tested and again carefully selected in the second stage of selection, the resulting doubly selected sires could average up to 1700 lb in daughter superiority compared to cows born when the bulls were.

These results are theoretically possible, but success may make the results look less impressive. If the system works according to theory, after two stages of selection young bulls will have a large crop of daughters that average less than the 1700 lb of expected superiority above their herdmates. The selection program will have raised the genetic merit of the whole population so that the latest group of daughters will be compared with a group of herdmates that are much superior to those existing when the mates were first selected to produce the young bulls. The PDs of the bulls may actually be 1700 lb better than the ETAs of the original population of cows but not that much better than the herdmates some 9 to 10 years later. The sire comparison and modified contemporary comparison methods take into account improvement in herdmates, so if the bulls are evaluated by those methods, their average superiority should be close to the expected superiority. Any biases in first-stage selection as a result of preferential treatment of potential bull dams to make them look good will result in less gain than expected. There is some evidence that the first-stage selection is only about 60% effective, possibly because of preferential treatment of dams of young bulls and biases in proofs of sires of young bulls.

The fundamental principles of sampling the highly selected but unproven young bulls are discussed in the remainder of this section.

Sampling Young Bulls in AI

The second stage of selection consists of sampling pedigree-selected young bulls by mating them at random to obtain 50 or more tested daughters for more accurate estimation of daughter superiority based on records of their own daughters. The optimal number of tested daughters for breeds with small population size may be less—between 20 and 30.

The entire group that is sampled could average +1000 lb in daughter superiority but how much higher the best ones will be depends on two things under the breeders' control and one thing over which there is little control. Accuracy can be increased by obtaining more daughters for each bull, although there is not usually much advantage to having more than 50. The intensity of selection factor, I, can be increased by selecting fewer bulls out of all bulls sampled. For example, for selecting the top 1 of 2, $I = 0.80$; for selecting the top 1 of 5, $I = 1.40$; for selecting the top 1 of 10, $I = 1.76$. Thus daughter superiority of the group of bulls selected after sampling depends a great deal on the selection intensity. The genetic deviation is a constant value that cannot be changed.

Table 14.6 gives the accuracy value for different numbers of daughters, all by AI sires in different herds, for a heritability of 0.25. Accuracy will be less if several daughters are in the same herd, as was shown in Figure 14.1.

The increase in daughter superiority as a result of stage 2 selection would be about 750 lb by selection of the top 1 of 5 sampled bulls and would be in addition to the increase resulting from stage 1 selection. Increasing intensity of

TABLE 14.6 THEORETICAL DAUGHTER SUPERIORITY OF SELECTED SAMPLED BULLS[a,b]

		Fraction of bulls selected/bulls sampled									
		4/5	3/4	2/3	1/2	1/3	1/4	1/5	1/8	1/10	1/20
		Intensity of selection factor									
Number of daughters	Accuracy	0.35	0.42	0.55	0.80	1.10	1.27	1.40	1.65	1.76	2.06
1	0.25	1050	1060	1080	1120	1160	1190	1210	1240	1260	1300
5	0.50	1100	1120	1160	1240	1330	1380	1420	1490	1520	1610
10	0.63	1130	1150	1200	1300	1410	1480	1520	1620	1660	1770
20	0.76	1160	1190	1250	1360	1500	1570	1630	1750	1800	1930
40	0.85	1170	1210	1280	1400	1560	1640	1710	1840	1890	2050
50	0.88	1180	1220	1290	1420	1580	1670	1730	1870	1920	2080
70	0.91	1190	1220	1300	1430	1600	1690	1760	1900	1960	2125
100	0.93	1190	1230	1300	1440	1610	1700	1780	1920	1980	2140
200	0.96	1200	1240	1310	1460	1630	1730	1800	1950	2010	2180
1000	0.99	1200	1240	1320	1470	1650	1750	1820	1980	2040	2220

[a] Superiority in pounds; includes +1000 lb from stage 1 selection.

[b] Genetic standard deviation of 1200 lb and heritability of 25% are assumed.

selection to 1 of 8 or 1 of 20 would give an even greater boost to daughter superiority of the group of bulls finally selected for heavy service. However, the cost of sampling so many bulls prevents many studs from attempting such a large sampling program. For example, suppose that a stud needs 5 new proved bulls each year. To select 1 of 20, 100 bulls would have to be proved and be alive after proving each year. Many young bulls will be lost from the time they are purchased until they are proved because of such factors as disease, semen problems, and functional defects.

In spite of the high cost of sampling bulls, the genetic returns will more than pay for the cost. During each year of the program, on the average after the equilibrium rate of gain is reached, new cows would genetically be able to produce 200 lb of milk more than they did the year before. Thus even if the cost is great, the expected increased production is also very great.

The initial problem is to finance the program since several years will elapse before any benefits can be seen. The sampling procedure takes a minimum of 4 to 5 years and the first calves of the best of the sampled bulls do not produce milk until 3 years later.

Balancing Accuracy and Selection Intensity

Another question that might be asked is: Why not have 1000 daughters or more for each young bull? The accuracy value would be nearly 100%, which should result in increased daughter superiority. Table 14.7 shows that increasing accuracy to 0.99 with 1000 daughters does not raise the daughter superiority substantially over that for 40 to 70 daughters. Also, if 1000 daughters are required for each young bull and if 40 bulls are sampled each year, approximately 200,000 to 400,000 services will be required to the young bulls because only 10 to 20% of services in tested herds result in a DHI-tested daughter. If the total breeding population were 1 million cows, 40% would have to be bred to young bulls, which would be an impossibility because in many areas the proportion of DHI-tested cows is less than 40%.

There is an optimal fraction of services that should be made to the proved

TABLE 14.7 COMPARISON OF DAUGHTER SUPERIORITY FOR DIFFERENT COMBINATIONS OF SELECTION AND ACCURACY OF EVALUATION FOR 1000 TESTED DAUGHTERS WHEN FIVE BULLS ARE NEEDED EACH YEAR

Number of bulls tested	Number of daughters per bull	Bull selection intensity	Accuracy	×	Selection intensity factor	×	½ Genetic standard deviation (in pounds)	=	Average daughter superiority (in pounds)
5	200	1/1	0.96		0.00		600		0
10	100	1/2	0.93		0.80		600		440
20	50	1/4	0.88		1.27		600		670
40	25	1/8	0.85		1.65		600		840
50	20	1/10	0.76		1.76		600		800
100	10	1/20	0.63		2.06		600		770

bulls. In a 400,000-cow population this fraction is 90% or higher, so that only 40,000 cows would be available for testing young sires. If 1 of 40 services result in a tested record (only approximately one-fourth of cows are on testing programs), then 1000 daughter records would be available each year for testing. Suppose that 5 bulls are to be selected each year. Would it be best to obtain 200 daughters for each of 5 bulls, 100 daughters for each of 10 bulls, 50 for each of 20 bulls, 25 for each of 40 bulls, or 10 for each of 100 bulls?

Table 14.7 shows how such a question can be answered. The optimal daughter superiority comes at the intermediate accuracy of 25 daughters per bull and selection of the top 1 of 8. The essential point is that accuracy has to be balanced against selection intensity. Both accuracy and selection intensity have to be weighed against cost of the program and increased income for the dairy owners.

14.6 SUCCESS OF SELECTION

Most of the preceding discussion was based on theoretical grounds. The real success of any breeding program is determined by the results. The average improvement has been about 100 lb of milk per lactation per year during the 1970s and 1980s. This improvement is less than theoretically possible, but no breeding program has yet followed theory closely enough for actual and expected progress to be in agreement.

Improvement in yield has occurred in programs that have attempted to follow

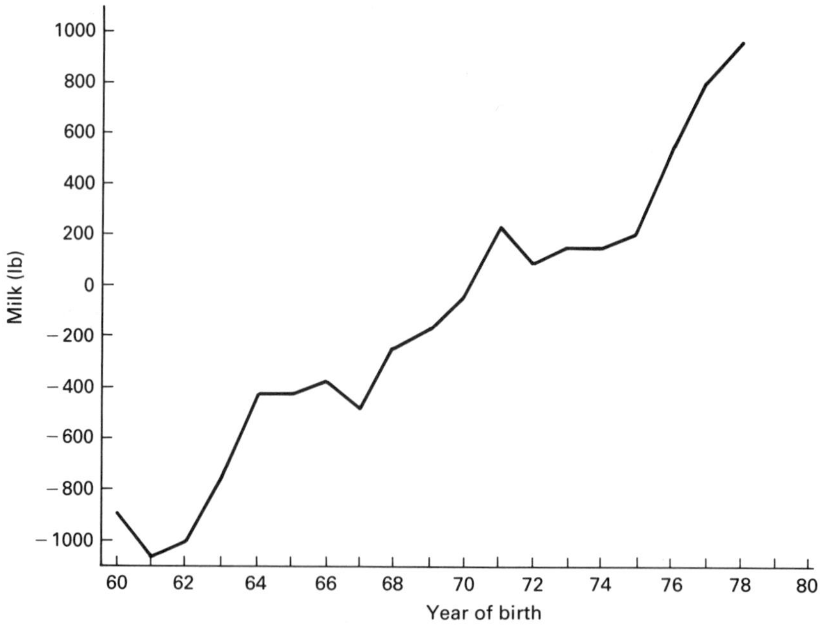

Figure 14.6 Average genetic value for milk yield by year of birth of Holstein bulls that were used in artificial insemination in the northeastern United States.

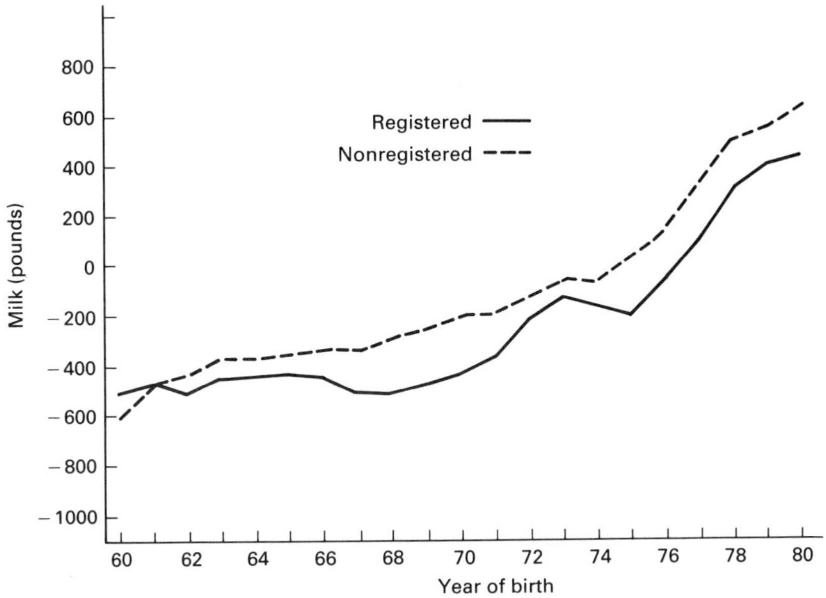

Figure 14.7 Average genetic value for milk yield by year of birth for registered and nonregistered Holstein cows in the northeastern United States.

recommended practices. Figure 14.6 shows the change in genetic value of Holstein sires born from 1960 through 1978 and entering AI service in the Northeast. The increase has not been steady. For three periods totaling one-half of the years there was essentially no gain. For the other 9 years, gain was about 220 lb per year.

Figure 14.7 shows genetic gain in artificially sired Holstein cows. After 1961 the genetic value of nonregistered cows exceeded that of registered cows. Gain was slow until about 1970, especially for registered cows. After 1970 gain in both groups averaged about 90 lb of milk per year.

14.7 ACCURACY AND CONFIDENCE RANGES FOR A SIRE'S TRANSMITTING ABILITY OR FOR A DAUGHTER'S GENETIC VALUE OR RECORD

Repeatability or squared accuracy of estimating a bull's transmitting ability (TA) can also be used to put confidence ranges on a new daughter's genetic value or on a new daughter's milk record.

Figure 14.8 shows the relationship between the three things that are often predicted: (1) the transmitting ability of a bull (half his genetic value), (2) the genetic value of a new daughter not in the original proof, and (3) the record of such a daughter. Records of a bull's daughters can be used to predict any of the three. The daughter records are used to obtain the predicted difference, which, in fact, also predicts a future daughter's genetic value (if from a random mate)

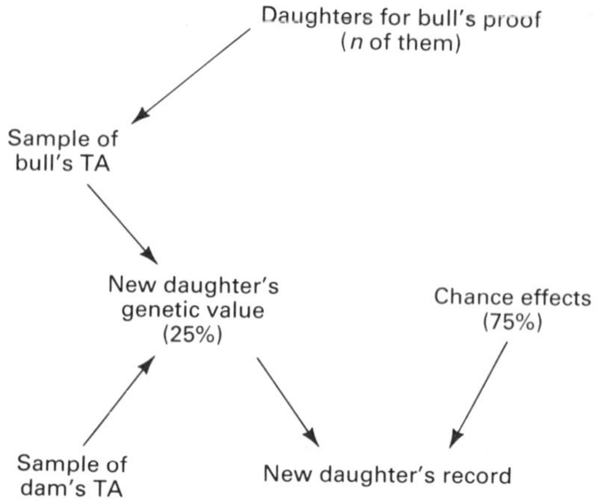

Figure 14.8 Diagram of the three things that are predicted from a sire's daughters: (1) his transmitting ability, (2) a new daughter's genetic value, and (3) a record of a new daughter. Notice the factors that make the accuracy of prediction less for a daughter's genetic value and a daughter's record. From *Principles of Dairy Science* by G. H. Schmidt and L. D. Van Vleck. Copyright © 1974 W. H. Freeman and Company. Used by permission.

as well as her record. The prediction is exactly the same for all three. Accuracy is, however, not the same for all three. A bull's TA can be predicted quite accurately if enough daughters are available, as shown in the left-hand columns of Table 14.8. Although a bull's TA may be predicted nearly perfectly by the PD, a new daughter also receives half her genes from her mother and only a random half of the genes of her sire. Thus the accuracy squared for a daughter's genetic value is only one-fourth as large as the accuracy squared for the sire.

The accuracy for predicting a milk record is even less than for predicting

TABLE 14.8 ACCURACY SQUARED OR REPEATABILITY FOR PREDICTING A BULL'S TRUE TRANSMITTING ABILITY, THE GENETIC VALUE OF A NEW DAUGHTER, OR A RECORD OF A NEW DAUGHTER[a]

Number of daughters in bull's proof (PD), n	Bull's PD, $\frac{n}{n + 15}$	New daughter's genetic value, $\frac{1}{4}\left(\frac{n}{n + 15}\right)$	New daughter's record, $\frac{1}{16}\left(\frac{n}{n + 15}\right)$
0	0	0	0
5	0.25	0.06	0.016
10	0.40	0.10	0.025
20	0.57	0.14	0.036
50	0.77	0.19	0.048
100	0.87	0.22	0.054
1000	0.99	0.25	0.062

[a] 1.00 would be perfect accuracy.

genetic value (right-hand column of Table 14.8). Two parts other than management level contribute to the record—the daughter's genetic value and the chance, luck, or environmental factors which account for 75% of the variation in records. The chance effects cannot be predicted at all. Only the genetic part can be predicted, so that accuracy squared for predicting a record of a new daughter of a bull is only one-sixteenth of the squared accuracy for predicting the sire's transmitting ability when heritability is .25.

These squared accuracy values can be used to determine the confidence ranges for a new record as well as for the transmitting ability of the bull. As seen in Figure 14.9, as the number of daughters in the bull's proof increases, the 68% confidence range for his true transmitting ability becomes smaller and with enough daughters becomes nearly zero.

The 68% confidence range for a bull extends equally about his PD, so that there is a 16% chance that his true value is above the upper line and a 16% chance that his true transmitting ability is below the lower line of Figure 14.9.

The ranges for a new daughter's genetic value can be interpreted similarly. The range is quite a lot wider and does not become much narrower as the accuracy of predicting the sire increases. The range again extends about the best prediction of the daughter's genetic value, which is her sire's PD.

The lack of accuracy in predicting a record of a new daughter is shown by the wide and nearly constant confidence ranges for the new record. The 68% range is approximately from the PD +2500 lb to the PD −2500 lb. Notice that despite the wide range, the range is about the sire's PD. This wide range explains the frequent occurrence of daughters that produce much more or less than the prediction for the sire.

For example, suppose that the PD is +1500 lb. Records of a daughter will be as mature equivalent difference from herdmate average. There is about a 16% chance that a daughter will have a record 2500 lb less than +1500 lb; that is, a 16% chance of being more than 1000 lb or more (1500 − 2500) below herd average. The 68% confidence range is for a record to be between 4000 lb (1500 + 2500) and −1000 lb (1500 − 2500) as a difference from herdmate average. Nevertheless, the average of records of all his daughters would be expected to be +1500 lb, of which half would be greater than +1500 lb and half would be less than +1500 lb.

For the bull with a PD of −300 lb, the chance of a daughter having a record 2500 lb greater than −300 lb is 16%, that is, a 16% chance of being +2200 lb or more. The chance of being that much below the PD is also 16%—a 16% chance of the record being −2800 lb or less. Yet the average of all his daughters' records would be expected to be −300 lb.

Wider confidence ranges can also be determined. For example, the 96% confidence range can be found by multiplying the ranges in Figure 14.9 by 2. Then only 2% of the records would be expected to exceed the upper range and 2% to be less than the lower range. The 90, 80, 70, 60, and 50% ranges are calculated by multiplying the ranges in Figure 14.9 by 1.65, 1.28, 1.04, 0.84, and 0.67, respectively.

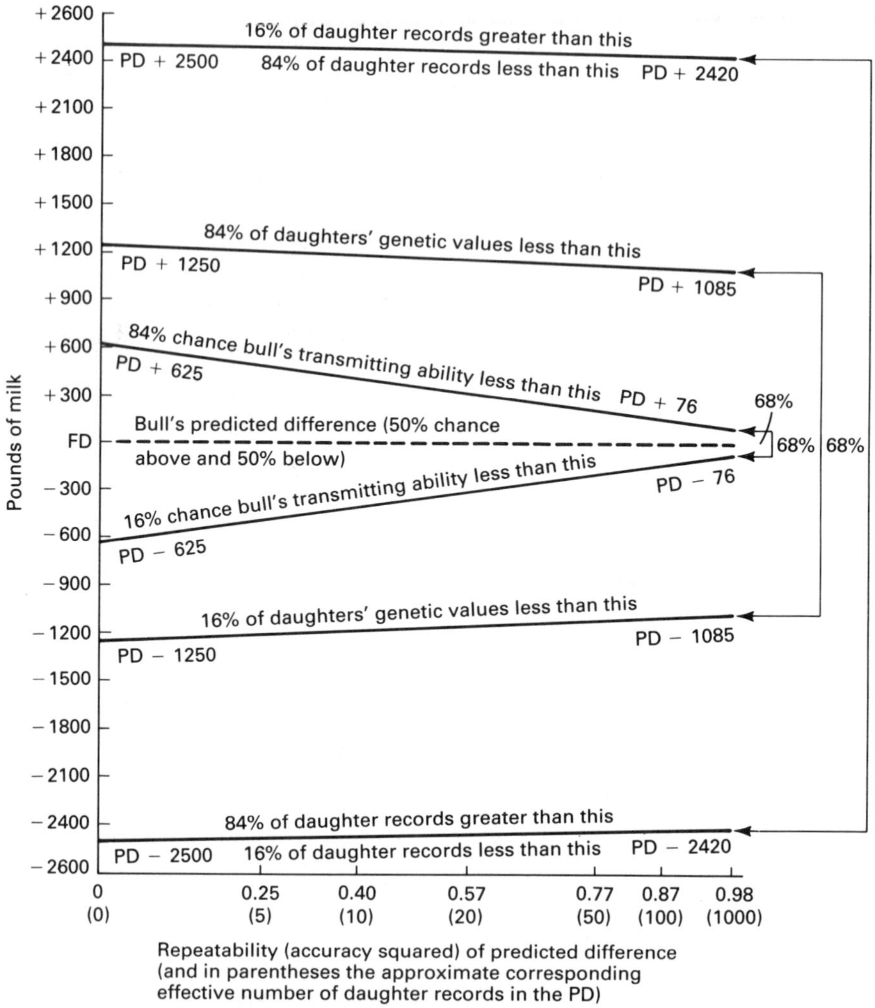

Figure 14.9 Confidence ranges (68%) for bull's true transmitting ability, a new daughter's genetic value, and a new daughter's record as a difference from herd-mate average for different repeatabilities of the bull's predicted difference.

The important point is that although much variation can be expected in the records of daughters of a sire, the average of his future daughters is likely to be close to the predicted difference for the sire, particularly if the number of daughters in the initial proof is large. This discussion has shown that repeatability or accuracy can be used as a guide to how much records of individual daughters are likely to vary from their sire's PD. It should be equally clear that repeatability of a sire proof does *not* show what fraction of daughters are going to exceed their sire's PD or how much of a sire's PD is going to show up in his future daughters.

14.8 RISK

The confidence range on a bull's proof is an indication of a risk that the bull is not as good as his proof (pessimistic viewpoint) and an indication of the chance that the bull is better than his proof (optimistic viewpoint). For example, there is a 16% chance that his true transmitting ability is less than the lower limit of his 68% confidence range. Similarly, there is about a 2% chance that his true transmitting ability is less than twice the lower limit of the 68% range, the 96% range. True transmitting ability is never known exactly. Nevertheless, a proof based on a large second crop of daughters or on proofs of his sons is a good indication of his true transmitting ability. In practice, what is likely to happen is that a bull has a proof based on an effective number of 40 or so sampling daughters. If he is ranked high, he will be used heavily enough so that beginning 4 years later his proof will be based on hundreds and possibly thousands of later daughters. About 6 years after his first proof, his first sons will also have proofs.

In theory, about 1 of 6 such first-proof bulls will have a final proof that drops below the lower limit of the original 68% CR and 1 of 40 bulls will have a proof that drops twice that much or more. When that happens, the bull is seen to be, at best, an average and not an outstanding bull. Thus some risk is involved in using the bull. The basic rule is that the highest-ranked bulls should always be used, but with the knowledge that some will drop in rank within the same group by the time a final proof is published.

What can be done to protect against the chance a bull's proof will drop by an amount that would decrease the herd average by an economically important amount? The obvious answer is to use more than one bull because the chance that the average of several high-ranked bulls will drop by an important amount is not very large. The idea of a confidence range works with the average of bulls as well as for individual bulls. The standard deviation which determines the upper and lower limits of the confidence range is only $1/\sqrt{\text{number of sires}}$ as large as for a single bull. The midpoint of the range is the average of the bulls' proofs. This average obviously will be smaller if several bulls are used rather than the top bull. Thus the expected daughter performance will be somewhat less. This decrease can be looked at as the cost of insurance against the top bull being overevaluated by an amount that is economically important. Table 14.9 gives the standard deviations of the average of bull proofs for different effective daughter numbers.

For example, suppose that the top bulls, A, B, C, D, and E have proofs of 1100, 1000, 900, 800, and 700 lb based on 40 effective daughters. If A and B are used equally, the 68% confidence range on their average transmitting ability is (their average is 1050)

$$1050 \text{ lb} \pm 231 \text{ lb}$$

that is, from 819 lb to 1281 lb. If A, B, C, D, and E are used equally, the 68% confidence range is (their average is 900 lb)

$$900 \text{ lb} \pm 146 \text{ lb}$$

TABLE 14.9 STANDARD DEVIATION (POUNDS) OF AVERAGE TRUE TRANSMITTING ABILITY GIVEN AVERAGE PREDICTED TRANSMITTING ABILITY BY NUMBER OF SIRES IN THE AVERAGE AND EFFECTIVE NUMBER OF DAUGHTERS IN THE PROOFS[a]

Effective number of daughters	Number of sires used				
	1	2	5	10	20
10	474	342	217	153	108
40	326	231	146	103	73
100	226	160	101	71	50
1000	76	54	34	24	17

[a] The standard deviations can be used as upper and lower limits of 68% confidence ranges by adding and subtracting the standard deviations from the average proof.

that is, from 754 lb to 1046 lb. The expected cost of a somewhat smaller confidence range due to using the top 5 rather than the top 2 bulls is the difference in the averages:

$$900 \text{ lb} - 1050 \text{ lb} = -150 \text{ lb}$$

In this example, because of the considerably smaller average, the lower limit on the confidence range is actually smaller when the five bulls are used than when only two are used. The upper limit for the 5 bulls is slightly below the average of the top 2 bulls. If the five top bulls were more similar, the average for five would be close to that for the two, so that the lower limit for five bulls would be greater than for two bulls. If risk is defined as a drop of more than two standard deviations (one chance in 40), the lower limit for the two bulls is slightly smaller, 589 lb, than that for the five bulls, 608 lb. The decision to use several bulls must consider the average as well as the desired confidence range (risk) for the group of bulls used.

Risk of decreasing the herd average may be affected more by other factors than by a bull being overevaluated. The sample of genes received from the bull as well as from the dams may result in a much larger confidence range on the average genetic ability of cows entering the herd in any year than is caused by the over- or underevaluation of a bull. Herd size, in this case the number of fresh heifers entering the herd each year, also affects the confidence range.

Table 14.10 gives the standard deviation of average cow genetic value given the average proof of their sires for number of bulls used each year of 1 to 20 and for number of heifers freshening each year of 1 to 100. For relatively few heifers, even with many sires, the confidence range for genetic value is much larger than the confidence range attributed to evaluation of the sires. For example, assume that bulls have proofs based on 40 daughters and a herd with 10 fresh heifers divided among 5 sires (2 daughters each). The confidence range is +372 lb for average genetic value of the 10 heifers. The confidence range on the bulls' average TA was only ±146 lb.

TABLE 14.10 STANDARD DEVIATION (POUNDS) OF AVERAGE GENETIC VALUE OF A CROP OF HEIFERS GIVEN THE AVERAGE PROOF OF THE SIRES

Number of heifers per year	Number of sires				
	1	2	5	10	20
1	1131	—	—	—	—
2	832	800	—	—	—
5	584	—	506	—	—
10	473	413	372	358	—
20	406	334	283	263	253
40	369	287	225	200	186
100	344	255	182	150	131

The confidence ranges on records (differences from herd average) become even larger because temporary environmental factors have a much larger effect (about six times) on daughter performance than does the sire's genes. This confidence range represents the overall risk (1 in 6) of a yearly crop of heifers having an average that much below the herd management level. For example, the confidence range for 10 heifers is 832 lb if sired by only one bull, 800 lb if from two bulls, and 772 lb if from 10 bulls (one daughter each) (see Table 14.11). These confidence ranges for what are essentially herd averages in comparison with the confidence ranges for average transmitting ability of the sires show that the bulls used have a smaller effect ($\frac{1}{3}$ to $\frac{1}{10}$ as much) on risk than other random factors affecting records.

In practice a major factor in the confidence range for yearly herd average production is change in management conditions. Such changes can easily be 750 lb or more.

An important risk factor in choosing bulls cannot be covered by theory. If an initial bull proof is inflated because of special treatment of daughters or better than average mates, a second crop of normally managed daughters with average mothers cannot be expected to match the performance predicted by the biased proof. Risks from using such a bull are likely to be much greater than random

TABLE 14.11 STANDARD DEVIATION (POUNDS) OF AVERAGE DIFFERENCE FROM MANAGEMENT LEVEL FOR A CROP OF HEIFERS GIVEN THE AVERAGE PROOF OF THE SIRES

Number of heifers per year	Number of sires				
	1	2	5	10	20
1	2443	—	—	—	—
2	1742	1727	—	—	—
5	1131	—	1092	—	—
10	832	800	779	772	—
20	632	588	561	551	546
40	503	447	410	396	390
100	406	334	283	263	253

risks described in this chapter. The possibility of biased proofs is a good reason to use three to six bulls each year to diminish the average risk.

REFERENCES AND FURTHER READINGS

AGRICULTURAL RESEARCH SERVICE. (Semiannual). USDA–DHIA *Sire Summary List.* U.S. Department of Agriculture, Beltsville, MD.

AGRICULTURAL RESEARCH SERVICE. 1976. USDA–DHIA modified contemporary comparison sire summary and cow index procedures. *USDA Prod. Res. Rep.* 165, U.S. Department of Agriculture, Beltsville, MD.

BATH, D. L., F. N. DICKINSON, H. A. TUCKER, and R. D. APPLEMAN. 1985. *Dairy Cattle: Principles, Practices, Profits*, 2nd ed. Lea & Febiger, Philadelphia.

CORNELL UNIVERSITY. (Semiannual). *Northeast AI Sire Comparison.* Dept. Anim. Sci., Cornell University, Ithaca, NY.

DEMPFLE, L. 1977. Comparison of several sire evaluation methods in dairy cattle breeding. *Livest. Prod. Sci.*, 4:129.

HENDERSON, C. R. 1954. Selecting and sampling young bulls. *Proc. 7th Annu. Conv. Natl. Assoc. Artif. Breed.*

HENDERSON, C. R. 1964. Selecting the young sire to sample in artificial insemination. *J. Dairy Sci.*, 47:439.

HENDERSON, C. R. 1970. A new sire evaluation method. *Proc. 23rd Annu. Conv. Natl. Assoc. Artif. Breed.*

HENDERSON, C. R. 1972. Sire evaluation and genetic trend. *Proc. Symp. Honor Dr. Jay L. Lush.* American Society of Animal Science, Champaign, IL.

HENDERSON, C. R. 1975. Use of relationships among sires to increase accuracy of sire evaluation. *J. Dairy Sci.*, 58:1731.

HENDERSON, C. R. 1976. Multiple trait sire evaluation using the relationship matrix. *J. Dairy Sci.*, 59:769.

QUAAS, R. L., R. W. EVERETT, and A. E. MCCLINTOCK. 1979. Maternal grandsire model for dairy sire evaluation. *J. Dairy Sci.*, 62:1648.

ROBERTSON, A., and J. M. RENDEL. 1950. The use of progeny testing with artificial insemination in dairy cattle. *J. Genet.*, 50:21.

SKJERVOLD, H., and J. J. LANGHOLZ. 1963. The optimum size of progeny groups and optimum use of young bulls in AI breeding. *Acta Agric. Scand.*, 13:131–140.

THOMPSON, R. 1979. Sire evaluation. *Biometrics*, 35:339.

VAN VLECK, L. D. 1964. Sampling the young sire in artificial insemination. *J. Dairy Sci.*, 47:441.

VAN VLECK, L. D. 1985. Notes on the theory and application of selection principles for the genetic improvement of animals. Dept. Animal Sci., Cornell University, Ithaca, NY.

15

Selection for Economic Value and Economics of Selection

Discussion of cow and bull selection thus far has been concerned primarily with selection for high genetic value for milk yield. Such selection implies that milk yield is the only trait that has economic value. The value of a cow, of course, is not determined solely by her milk yield. Fat or protein content of the milk may influence income. The incidence and severity of mastitis affects yearly income, as do other disease problems. Speed of milking may affect labor costs. Calving interval, if lengthened, reduces income per day of life. Some traits may affect herd life and ease of handling. Temperament may be difficult to evaluate in monetary terms, but there is often some cost value in terms of patience and tolerance. All these and other traits may have economic value. Two widely used indexes of economic value are (1) product or dollar value based on milk and its components, and (2) type-production indexes based on final type score and production traits.

15.1 TOTAL ECONOMIC VALUE

The problem of estimating total economic genetic value of a cow is complicated for several reasons. First, the genetic value for each trait can only be estimated. Second, economic values are difficult to determine for such traits as temperament and are complex in that a unit increase in the trait may have a different value depending on the level of the trait. For example, an increase in type score from 89 to 90 may be more valuable for selling the cow or her embryos than an increase from 83 to 84. Third, interrelationships of genetic values and environmental effects for the various traits influence estimation of total economic value and response to selection, but these genetic and environmental correlations are difficult to estimate.

Methods of Selection for Total Economic Value

The theory of selection for more than one trait results in an index value for economic value made up of records on measured traits of the animal and its relatives. Each record is weighted according to relationships of the relatives, economic values, genetic and environmental correlations among the traits, and heritabilities of the traits (see, e.g., Van Vleck, 1985).

Some research has suggested that when the genetic and environmental correlations are not likely to be estimated accurately, a simpler procedure may be appropriate. The genetic value for each trait is estimated only from records for that trait. These estimated genetic values are then weighted according to the economic value for each trait. This simpler procedure is identical with the more complicated one when the genetic and environmental correlations are zero.

As an example, suppose that the predicted difference for milk yield has been estimated to be +1000 lb of milk and that for milking speed has been estimated to be −0.5 minute per milking or −5.08 hours in a 305-day lactation.

Next, suppose that the net value of a pound of milk above feed cost is \$0.06 and that the labor cost per milking unit is \$1.50 per hour. Then the index equals (\$0.06/lb)(1000 lb) + (−\$1.50/hour)(−5.08 hour) = \$67.62. Note that a decrease in milking time results in a higher economic value since increased milking time is a negative labor cost.

The general equation is $I = v_1I_1 + v_2I_2 + \cdots + v_mI_m$, where I is the overall index of economic value, I_1 is the index estimate of genetic value for trait 1, I_2 is the estimate for trait 2, . . . , I_m is the estimate for trait m, and the v's are the corresponding net economic values. The indexes, I_j, can be PDs for bulls or ETAs for cows.

Relative Progress

Often, when selection is for one trait or for total economic value, knowledge of expected correlated responses for other traits can be useful in deciding whether selection should be for more than one trait. The equation for determining the expected correlated response is deceptively simple but requires numerous computations, as can be seen in Van Vleck (1985).

The usual rule of thumb is that if selection is jointly for n traits, the relative progress for any one of them is $1/\sqrt{n}$ of the progress possible by selection for the single trait only. The approximation applies exactly when the traits have equal heritabilities, the genetic and phenotypic correlations are zero, selection emphasis is equal for all traits, and the economic value of a standard deviation increase is the same for all traits. Such assumptions are not likely to be really correct, although the rule does provide a rough guideline for most cases. This rule is illustrated in Figure 15.1, in which milk is the trait for which relative progress is being compared.

Because traits have different scales of measurement (milk yield in pounds, fat content in percent, milking speed in minutes per milking), care must be taken in assigning the economic selection emphasis to various traits. According to the rule of thumb, equal selection emphasis assumes that a standard deviation increase in one trait has the same value as a standard deviation increase in any other trait.

Economic Values

If a trait has no economic value, there is no economic incentive to select for that trait. A trait can, however, have a personal economic value even though there is no market value for the trait. These traits are commonly called glamour traits. The so-called workability and wearability type traits have more obvious economic value.

The determination of economic values for important traits is a difficult problem in establishing a selection program. The value of milk production is the most easily determined of all characteristics of the dairy cow. The price of milk should be reduced by the cost of feed for increased production of 1 lb of milk. Values for type classification score can be obtained if the value of each increase in score is known. Otherwise, an increase in type score may have only an aesthetic value. Fat test helps to determine the value of each pound of milk but in itself has no value. Values of management traits are even more difficult to determine. Milking speed may be economically important because each additional minute per milking

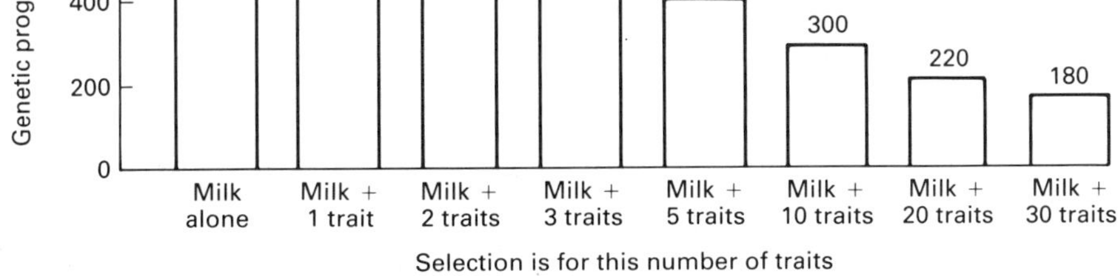

Figure 15.1 Relative progress in milk yield if selection emphasis is equal for several traits. (Assumes that selection for milk alone results in progress of 1000 lb.) From *Principles of Dairy Science* by G. H. Schmidt and L. D. Van Vleck. Copyright © 1974 W. H. Freeman and Company. Used by permission.

per cow adds up to more than 10 hours during a 305-day lactation. On the other hand, if rapid milking speeds are associated with milk leakage and increased udder disease, the value of faster milking time is less.

Such traits as temperament are especially difficult to assign economic values. Often they become all-or-none traits—either the cow's behavior is so bad that she is sold or is moderate enough that the cow is tolerated. If production is high and the calving record is good, bad temperament will be tolerated longer than if the cow is a poor producer or has fertility problems. Thus some economic value is usually placed on behavior, but the exact value often cannot be clearly stated.

Next to production, lack of fertility is probably the trait that has the most economic significance. Failure to conceive will curtail production after 300 to 400 days of lactation. Fortunately, cows that do not conceive on first service frequently conceive later before the calving interval is substantially lengthened. In practice, a cow that fails to conceive within a reasonable time is culled without

TABLE 15.1 HERITABILITY ESTIMATES FOR SOME TRAITS OF THE DAIRY COW

Trait	Heritability	Economic value
Milk production		
Milk yield	0.20–0.35	Direct
Fat yield	0.20–0.35	Direct
Solids yield	0.20–0.35	Direct
Fat test	0.45–0.60	Price differential
Protein test	0.45–0.60	Price differential
Solids test	0.45–0.60	Price differential
Dairy management		
Milking speed	0.20–0.35	Labor costs
Milk leak	0.05–0.10	{Sales loss, Disease}
Udder edema	0.05–0.15	{Vet costs, Sales loss}
Feeding speed	0.10–0.20	Labor costs
Temperament	0.05–0.20	Labor, ulcers, etc.
Mastitis resistance	0.20–0.35	{Vet costs
Ketosis resistance	0.20–0.35	Sales loss}
Breeding trouble	0.00–0.10	{Replacement loss, Sales loss}
Udder characteristic		
Depth	0.15–0.25	?
Floor levelness	0.15–0.25	?
Strength of attachment	0.05–0.20	?
Shape	0.05–0.15	?
Teat position	0.05–0.15	?
General		
Overall type	0.15–0.30	Breeding sales
Dairy character	0.15–0.30	Breeding sales
Herd life	0.00–0.10	Replacement loss

consideration of any superior characteristics because a sterile cow is worthless for production. This trait tends to be culled independently of other traits.

Freedom from mastitis is certainly economically important because of the cost of treatment and the value of decreased production and discarded milk. The heritability of mastitis resistance may be large enough for selection to be effective. The problem of selection for mastitis resistance is that mastitis rarely occurs in the first lactation. Thus selection of sires based on first-lactation daughter performance does not reflect mastitis resistance. Waiting until many daughters have several lactations increases the generation interval and reduces genetic progress for milk yield.

Milk constituents, fraction fat and protein, have some economic value in pricing milk. Selection, however, for product value automatically takes into account the fraction of fat and protein in the milk. These constituents probably need to be monitored only to make sure they stay above minimum levels because more feed is needed to produce the milk constituents than to produce the constituentless milk.

Some traits that might be considered in a breeding program have already been discussed. Several times as many could be added. Table 15.1 lists some of them and includes heritability estimates and comments about economic values.

Milk yield is the most important trait in a breeding program, but the number of combinations of other traits and relative economic values is very large. Discussion of all such combinations is outside the scope of this book. In the following sections, selection for milk and one other trait will be discussed.

15.2 SELECTION FOR PRODUCTION AND TYPE

Nothing in dairy cattle breeding stirs more controversy than mention of type. No dairy manager is likely to select completely for either type or milk production. The relative economic values assigned to type and production, however, are likely to be different from one breeder to the next. For example, if all economic emphasis is put on type, what is expected to happen to progress for milk yield? The answer to this or the reverse question may help in deciding how much emphasis a dairy manager may want to put on type and how much on production.

Heritability is approximately 25% for both type and production. Assume the following: a genetic correlation of 10%, a phenotypic correlation of 10%, a standard deviation of milk yield of 2500 lb, and a standard deviation of 3.7% for first type score. What must be determined is the effect of the relative economic values of type and production on progress for type or production. First the units of type and production should be standardized to equate percentage type scores to pounds of milk.

Relative Economic Values

A relatively easy way to determine the ratio of economic values per standard deviation is as follows. (1) Find the net value of a pound of milk and multiply by its standard deviation, 2500 lb. (2) Determine the value of a 1% change in type score or perhaps find the value of one grade change of 5% in type score and convert that value back to 1%. Then multiply by 3.7%. For example, suppose that the net price of milk is \$0.06 per pound; then a standard deviation increase in milk yield is worth \$0.06/lb × 2500 lb = \$150. Assume that five points in final type score is worth \$200; therefore, each point is worth \$40.00 and a standard unit of type is worth \$40/% × 3.7% = \$148. Type and production would appear to be nearly equally valuable.

Research has suggested that a ratio of 4 to 1 would be appropriate for breeder herds that sell many animals. Commercial herds would be expected to place more economic emphasis on production. Figure 15.2 answers the questions of what progress to expect for milk yield or type if selection is for both. If all emphasis is on milk yield, progress is 100% of maximum possible for milk. But if all emphasis is on type, progress for milk yield is only 10% as great as if selection were for milk yield alone.

If three (or more) times as much emphasis is given to milk yield, progress for milk yield is nearly 100% (90% for 2:1 up to 98% for 5:1). Note that ratios of 10:1 or 20:1 would result in nearly 100% progress for milk yield. Thus some selection for type would not significantly decrease progress for milk yield. But if type receives equal or greater emphasis than milk yield, progress for milk yield is markedly decreased.

If a dairy manager is interested in what happens to type, a similar figure can be drawn by using the units marked in terms of (type: milk) and (relative progress for type). The situation is by symmetry merely reversed. Thus if all emphasis were on milk yield, type would improve only 10% as fast as it would if all emphasis were on type.

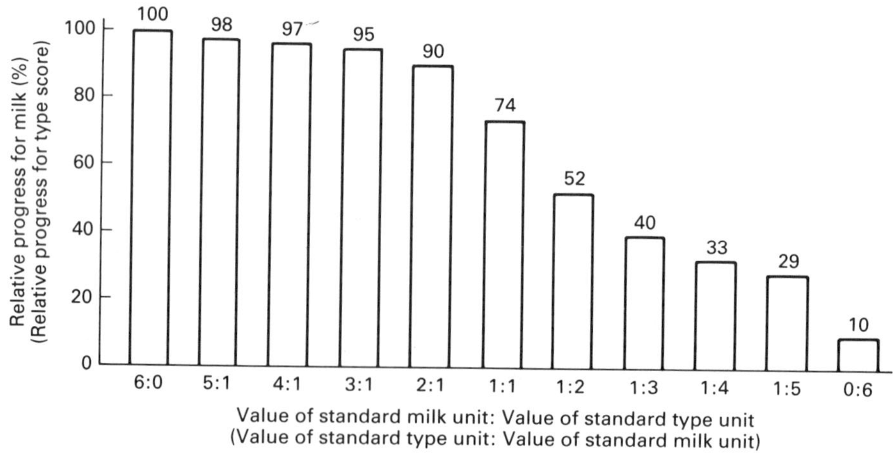

Figure 15.2 Comparison of progress in milk yield when selecting for both milk yield and type and progress in milk yield when selecting for milk alone. Comparison of progress for type when selecting for milk yield and type is found by using the values and progress given in parentheses. From *Principles of Dairy Science* by G. H. Schmidt and L. D. Van Vleck. Copyright © 1974 W. H. Freeman and Company. Used by permission.

A major problem in evaluating improvement in type is the measuring system. Milk yield can be recorded in pounds up to 20,000, 30,000, 40,000, or more. But type scores have an automatic top limit of 100 points. This limit makes the average type score remain approximately the same year after year, even though there may actually be considerable improvement. Certainly, the type of a modern cow is by many standards better on the average than it was 20 or 30 years ago, although the average score probably has not changed much. If improvement in type is to be evaluated in a meaningful way, another method of measuring type is needed.

If selection is for several components of type, progress for milk production might be slowed considerably unless milk production receives by far the most emphasis. For example, if the relative weights for five breakdown categories are 1:1:1:1:1 with relative weight to milk yield of 5, the situation is approximately the same as equal weight to type and milk production—a 26% reduction in progress for milk production as compared with selection for milk production alone. But giving each of five breakdown traits the same relative value as production is equivalent to assigning five times as much weight to type as to production. The importance of assigning reasonable values to the various type traits relative to production cannot be overemphasized.

Type–Production Indexes

Many breed organizations promote both type and production. As an aid to members, they often publish indexes to combine production and type evaluations into one value—an index. As economic conditions change, the economic weights for type and production are changed. For example, in the 1970s and until the mid-1980s, the Holstein Association of America assigned relative economic weights of 3:1:1 to milk yield, fat percentage, and final type score in what was then called the total performance index (TPI). As the milk market changed in the mid-1980s, they changed the traits and economic weights to emphasize the nutrient content of milk—the fat and protein yields in pounds—with less emphasis on type. The formula is

$$\text{TPI} = \left[\frac{2\text{PD(fat)}}{22.5} + \frac{2\text{PD(protein)}}{19} + \frac{\text{PD(type)}}{0.7}\right] \times 50 + 234$$

The first three terms in TPI make up the index. The last two terms, the multiplier of 50 and the 234, scale the TPI values to be in the range of about 300 to 800. Therefore, the economic weights assigned are 2/22.5, 2/19, and 1/0.7 or 0.08888, 0.10526, and 1.42857. The denominator terms are the standard deviations of sire proofs for fat yield, protein yield, and type score.

In the section on relative economic values, relative economic values were defined as the value per phenotypic standard deviation, not as the value per standard deviation of predicted differences. The weights for TPI can be multiplied by their corresponding phenotypic standard deviations to obtain relative economic values. If the phenotypic standard deviations for fat yield, protein yield, and type are 100 lb, 80 lb, and 3.7%, the economic weights become:

- For fat 8.89
- For protein 8.42
- For type 5.29

If these are divided by 5.29, the result is relative economic values of 1.68, 1.59, and 1.00 for fat (lb), protein (lb) and type (%), which are somewhat different from 2:2:1. In fact, because of the relatively large correlation between fat and protein, the relative economic weight for production to type is probably more similar to 3:1 than 4:1. Such a small difference in emphasis as shown previously in this section is not very important. Other breed associations are assigning relative economic weights of 5 or 6 for production and 1 for type.

15.3 INDEX FOR PRODUCT OR DOLLAR VALUE

In the 1970s, sire summaries began to combine evaluations for milk yield and fat yield into a single product value evaluation by weighting the milk and fat by gross prices for fat and milk. These evaluations were called dollar value or product value proofs. For example, the USDA evaluation is commonly listed as predicted difference dollars, PD$. In the 1980s, as protein testing became more nearly universal, sire evaluations for protein yield or cheese yield were included in some product value summaries.

Product value evaluations incorporate the change in value of the milk when it contains more or less fat and protein. The relationships, for example, among milk yield, fat yield, and fat test are biologically not completely understood. The difficulty arises because fat test = fat yield/milk yield. Correlations between a ratio and its components are not easy to explain.

Milk is priced in most markets depending on fat test and sometimes on protein test. Product value evaluations, however, are based on fat and protein yield, as well as milk yield, and not on fat and protein content. Consequently, dairy managers are sometimes confused and often select bulls based both on product value and on fat test even though fat test is already included in product value. The result is that fat test receives more economic weight than is appropriate. Only in extreme cases, when herd averages for milk and fat test are greatly different from breed averages, is the approximation to product value based on milk yield and fat yield much different from the direct approach which incorporates proofs for milk and fat test rather than milk yield and fat yield. What is usually not realized is that most sire summaries calculate the proof for fat test indirectly using proofs for milk and fat yields. For example, with the indirect evaluation

$$\text{PD(test)} = \frac{\text{breed average fat yield} + \text{PD(fat)}}{\text{breed average milk} + \text{PD(milk)}}$$

so that the choice of breed averages becomes important. The reason for the indirect approach to evaluate fat test is that similar computing procedures can be used for milk and fat yield because they have similar heritabilities. The heritability for fat test, however, is about twice the magnitude of heritability for milk yield. The usual way of calculating product value evaluations using yields only and a way to include fat and protein tests for unusual situations will be described.

The Milk-Price Equation

The way milk markets describe the pricing of milk tends to lead to confusion among people who do not use the terminology regularly. The value of milk is quoted as price per hundredweight (cwt = 100 lb) with a fat differential per point (0.1%). For example, price per cwt is $12 with a fat differential of $0.16. The fat differential is the change in price per 100 lb of milk for each unit of 0.1% difference from a base fat test usually set at 3.5%.

The pricing system can be unraveled in several ways to arrive at a value for milk per pound with an addition or subtraction due to the percentage of fat as a difference from the base fat percentage. For example:

$$\text{Milk: } \$12.00/100 \text{ lb} \xrightarrow{\div\, 100} \$0.12/\text{lb}$$

Fat differential: $0.16/100 lb/0.1%

to put on lb basis ↓ ÷ 100

$0.0016/lb/0.1%

to put on full % ↓ × 10

$0.016/lb/%

The milk price equation becomes

$$\text{milk(lb)}[\$0.12/\text{lb} + (0.016/\text{lb}/\%)(\text{fat }\% - 3.5\%)] \tag{1}$$

Equation (1) can be factored into an expression invoving milk(lb) and fat(lb) by multiplying through by milk(lb):

$$\text{milk(lb)}[\$0.12/\text{lb} + (\$0.016/\text{lb}/\%)(-3.5\%)] + \text{milk(lb)}(\$0.016/\text{lb}/\%)(\text{fat }\%)$$

In the second term, fat % = [fat(lb)/milk(lb)]100. Thus the second term becomes fat(lb)($1.60)/lb. In the first term ($0.016/lb/%)(−3.5%) = −$0.0560/lb, so that $0.12/lb −$0.0560/lb = $0.0640/lb, the price of milk containing no fat. The milk price equation thus becomes

$$(\$0.0640/\text{lb})(\text{milk, lb}) + (\$1.60/\text{lb})(\text{fat, lb}) \tag{2}$$

Equation (2) is commonly used in sire summaries by substituting PD(milk) for milk and PD(fat) for fat:

$$\text{PD\$} = (\$0.0640) \times \text{PD(milk)} + (\$1.60) \times \text{PD(fat)} \tag{3}$$

Milk-Fat Product Value

This product value is equivalent to the difference from a zero–zero bull, PD(milk) = 0 and PD(fat) = 0. Technically, breed average (milk) + PD(milk) and breed average (fat) + PD(fat) are substituted into equation (2) to calculate expected product value for a daughter's record in an average herd. Then breed average (milk) + 0 and breed average (fat) + 0 are substituted into equation (2) to calculate the expected product value for the record of a daughter of a zero-zero bull. The difference from a zero-zero bull is the same as equation (3).

A slightly more accurate procedure for individual herds is to substitute herd average (milk, lb) + PD(milk, lb) and herd average (fat, %) + PD(fat, %) for (milk, lb) and (fat, %) in equation (1). The resulting equation predicts the product value of a daughter in that herd. Herd average (milk, lb) + 0 and herd average (fat %) + 0 are substituted into equation (1) to establish a base corresponding to a zero-zero bull.

In most cases bulls will be ranked the same with either equation (1) or (2) procedures. When the herd average is considerably below breed average for milk and considerably above herd average for fat test or if the reverse is true, bulls with high milk and low fat test proofs or with low milk and high fat test proofs may rank differently with the two procedures.

Milk–Fat–Protein Product Value

If milk is purchased with a protein differential as well as a fat differential, product value (milk–fat–protein) should be used for selection of bulls and cows. The equations in the preceding section can be extended to include protein.

The previous example will illustrate the extension. Assume that the protein differential is \$0.12 per cwt per 0.1% and the base protein percentage is 3.1%. Equation (1) becomes

$$\text{milk(lb)}[\$0.12/\text{lb} + (\$0.016/\text{lb}/\%)(\text{fat }\% - 0.35\%) + (\$0.012/\text{lb}/\%)(\text{protein }\% - 3.1\%)] \quad (4)$$

The decimals in the differentials are moved one position to the right to account for the differentials being expressed per 0.1% rather than per 1%. Equation (2) becomes

$$\text{milk(lb)}[\$0.12 + \$0.016(-3.5) + \$0.012(-3.1)] + \$1.60[\text{fat(lb)}] + \$1.20[\text{protein(lb)}]$$

so that equation (3) becomes

$$\$0.0268\ \text{PD(milk)} + \$1.60\ \text{PD(fat)} + \$1.20\ \text{PD(protein)} \quad (5)$$

As with milk-fat product value, equation (5) is usually used in sire summaries to predict product value in an average herd, although equation (4) can be used with substitutions of herd average + PD for milk, fat %, and protein % for herds with unique herd averages.

The important point to reemphasize is that product value proofs automatically account for fat and protein content. If fat and protein tests are considered in addition to product value, the effect is to reduce genetic gain for milk income.

15.4 A NET PRODUCT VALUE AND TYPE INDEX

Product value evaluations account for the value of milk yield and its components and thus can be considered to be milk proofs standardized for value of fat and protein. Thus economic weights for milk and type can be used with product value and type. Gross economic value of milk is already built into product value so that product value needs only to be reduced by feed cost. Generally, multiplication by a fraction of 0.5 to 0.6 would match feed costs corresponding to 50 to 40% of the gross value of milk.

If the appropriate relative economic values for production and type are 4:1, the value of one standard deviation of type (e.g., 3.7%) would have a value of one-fourth the net value of one standard deviation of milk (e.g., 2500 lb). If the net price for milk is \$0.06/lb, the value of 2500 lb of milk is \$150. Thus the value of 3.7% of type is \$150/4 = \$37.50. Consequently, each unit (% of type score) has a value of \$37.50/3.7 = \$10.14. In summary, for feed costs of 50% of milk price per pound of milk:

$$\text{net product value and type index} = 0.5\ (\text{PD\$}) + \$10.14\ \text{PD(type)}$$

If a ratio of relative economic values other than 4:1 is wanted, similar calculations can find the appropriate net product value and type index. For example, if a ratio of 1:1 is appropriate, then

$$\text{net product value and type index} = 0.5\ \text{PD(\$)} + \$40.54\ \text{PD(type)}$$

15.5 SELECTION FOR FERTILITY AND PRODUCTION

Until scientists can make a cow produce milk economically without calving, a calf must be born before lactation begins. Thus the economics of fertility and production are tightly intertwined. However, whether there is any relationship between how soon a cow conceives and how much milk she produces if she does calve is another question. The answer to this question will show whether selection can be for both improved fertility and improved production.

There are many ways of measuring fertility. The usual measures are: number of services to conception, days open, and calving interval. Nearly all these measures are closely related because gestation length varies relatively little within a breed. Probably the important measure of fertility is calving interval.

Most research reports suggest that heritability of calving interval is 5% or less. The genetic correlation between milk production and calving interval, however, is apparently relatively high (50%). The phenotypic correlation (the relationship between production in the same lactation and calving interval) is much lower; most reports show a range of 0 to 20%. The genetic relationship, although high, creates a problem because it indicates that if selection is for increased milk production, calving interval will also increase.

The correlations (50% and 20%) and heritabilities (25% for milk and 5% for calving interval) can be used to determine what effect different amounts of economic emphasis on milk yield and calving interval will have on genetic change for milk yield and calving interval. Relative economic emphasis is per standard deviation, that is, 2500 lb for milk yield and 36 days for calving interval. To decrease calving interval while increasing milk yield, calving interval would have negative emphasis and milk yield positive emphasis.

The expected results for selection based on first-lactation production and calving interval (calving interval can be approximated by adding 280 days to days open after first calving) are given in Table 15.2. Because milk yield and calving interval are positively correlated, an increase in milk yield will almost always result in a predicted increase in calving interval even when more negative emphasis is put on calving interval than positive emphasis is put on milk yield. If the value of 2500 lb of milk at $0.05 per pound and the value of a 36-day interval at $0.50 per day are $125 per standard deviation of milk yield and $18 per standard deviation of calving interval, the relative economic weights are about 6: −1.

The practical solution appears to be to ignore calving interval from a genetic standpoint, which would apparently result in an increase of about 3 days in calving interval for every 1000 lb of genetic increase in milk yield. Economically, an increase of 1000 lb of milk is certainly much more valuable than any loss caused by an increase in calving interval of 3 days.

The dairy manager should attempt to reduce calving interval by better management. Since 95% of the variation in calving interval is attributable to nongenetic

TABLE 15.2 EFFECT OF DIFFERENT AMOUNTS OF EMPHASIS ON MILK YIELD AND CALVING INTERVAL ON GENETIC IMPROVEMENT FOR MILK YIELD

Emphasis on milk yield: calving interval	Relative genetic progress for milk yield (%)	Change in calving interval for 1000-lb increase in milk yield (days)
6:0 (all milk)	100	3.3
5: −1	100	3.2
4: −1	100	3.2
3: −1	100	3.1
2: −1	100	3.1
1: −1	98	2.8
1: −2	87	2.0
1: −3	57	0.1
1: −4	15	−11.6
1: −5	−18[a]	−15.6
0: −6 (all calving interval)	−83[a]	−4.8
Ignore calving interval completely	100	3.2

[a] For milk decrease of 1000 lb.

reasons, better management can be used to provide a shorter interval between calves.

15.6 SELECTION FOR MILK YIELD AND HERD LIFE

Most breeders naturally think first of the production of their cows, then type and its components, and finally longevity. A high-producing, long-living cow is ideal. Fortunately, long herd life and a high level of production appear to have a close, positive relationship. There is really one major reason for the desirability of long herd life. The cost of raising or purchasing a replacement heifer is usually twice or more the salvage or meat market value of the cow being replaced. Arithmetic shows that this is a sizable cost in dairy operations.

Two forces that affect longevity require balancing. A faster turnover rate because of selection for production increases genetic improvement for milk yield but decreases herd life, especially if genetic progress attributable to AI bulls is rapid. Genetic improvement, of course, means greater income, but faster turnover rate means increased costs of management.

On a short-term basis (5 to 10 years), a low turnover rate, which is the same as long herd life, is economically sound. On a long-term basis, genetic improvement would more than offset the extra costs of replacements.

What about the relationship between longevity and production? Do high producers leave the herd early? Do low producers last the longest? The answer to the last question is obvious. Low producers are culled quickly and thus do not last, which shows that the ability to live a long life is not necessarily the same as herd life. Longevity in a dairy herd is the same as herd life, which is the summation of general health, fertility, and production.

Although the ability to live a long life is undoubtedly influenced by an individual's genetic makeup, the same is not as true for herd life. Heritability of productive herd life is only approximately 5%. The genetic correlation between herd life and milk yield is relatively high—approximately 60%. Thus many genes affecting herd life also influence milk yield. Environmental effects on herd life and milk yield are not closely related, so that correlation between a cow's first milk record and herd life is only approximately 20%. Table 15.3 uses these values and a value of heritability of 25% for milk yield to show how much relative progress can be made for milk yield or longevity with different amounts of emphasis on milk yield and longevity for five records on the cow if cows were allowed the opportunity for five lactations before being culled. All emphasis on milk yield (6:0) would give 100% progress for milk yield, and 93% as much progress for longevity as when all emphasis is placed on longevity. But even five times as much emphasis on longevity (1:5) still would give 98% progress for milk yield. Progress for both milk yield and longevity would be nearly maximum no matter what the relative economic emphasis on milk yield and longevity.

The hidden factor, however, is the length of time required to measure longevity. If cows must have five lactations before selection of herd replacements, the generation interval is 2 years longer than if cows must have three lactations.

TABLE 15.3 LACK OF IMPORTANCE OF RELATIVE EMPHASIS ON MILK PRODUCTION AND LONGEVITY IN SELECTING FOR MILK YIELD AND LONGEVITY AMONG COWS WITH FIVE LACTATIONS

Relative emphasis on milk yield: longevity	Relative progress for milk yield (%)	Relative progress for longevity (%)
6:0 (all milk)	100	93
5:1	100	94
4:1	100	94
3:1	100	94
2:1	100	95
1:1 (equal)	100	96
1:2	99	97
1:3	99	98
1:4	98	98
1:5	98	99
0:6 (all longevity)	93	100

Genetic progress decreases as the generation interval increases. Thus waiting for five lactations rather than measuring the first-lactation milk production may actually decrease progress for longevity as well as for milk yield.

Table 15.4 gives the complete picture when the generation interval for selection on the first lactation is assumed to be 4 years. Each additional lactation is assumed to add one year to the generation interval. In these calculations all emphasis is put on milk yield.

Maximum progress per year for milk yield would be from selection only for milk yield after the first lactation. Surprisingly, maximum progress per year for longevity also would occur if selection is for milk yield only in the first lactation. The base of 100% for progress in longevity corresponds to all emphasis put on longevity after five lactations.

TABLE 15.4 RELATIVE PROGRESS FOR MILK YIELD AND LONGEVITY PER YEAR, DEPENDING ON NUMBER OF LACTATIONS COMPLETE BEFORE SELECTING REPLACEMENTS

Number of lactations completed	Relative progress for milk yield (%)	Relative progress for longevity (%)
	All emphasis on milk yield	
1	100	147
2	92	136
3	81	120
4	72	106
5	64	95
	All emphasis on longevity	
5	60	100

The reasons why progress is faster for longevity when longevity is essentially ignored are (1) the high genetic relationship between longevity and production, and (2) the short generation interval for first-lactation selection. The low heritability of longevity also has an effect.

The result is a pleasant paradox—the dairy manager can ignore longevity and obtain more genetic progress for longevity than by selecting directly for longevity. Selection, however, must be for milk yield, which would be done anyway.

However, a word of caution must be added. Selection may not increase herd life much, if at all, because replacement rates are not likely to change because of management practices. What will improve are the components of herd life that would yield a longer herd life if management practices would allow. The genetic improvement also would not be very rapid because genetic differences for herd life among cows are small.

15.7 SELECTION FOR MILK YIELD AND COMPONENTS OF MILK

Total solids yield—fat, protein, lactose, and minerals—is an important consideration in determining the total energy value and nutritional value of milk. Yet the price paid for milk to dairy managers in the United States and in many other countries depends mostly on the amount of milk and the fraction of either fat or protein or both in the milk.

Market prices for milk do not suggest selection for increases in any of the components of milk with the possible exception of fat. In spite of the economics, milk with a high content of fat, protein, and lactose is a valuable food source. Thus the question should be answered whether total energy or total protein or total fat yield will be greater if selection is for milk yield alone or if selection is for content or yield of some component of milk.

To answer this question the relationships among these components and milk are needed. There are some differences among scientific studies, but some typical values are given in Table 15.5, which lists heritabilities, phenotypic variances, and genetic and phenotypic correlations.

The variances of the yields of milk components are considerably lower than those for milk yield, but the standard deviation of total solids is much larger than that for protein or fat. Similarly, the variance of total solids content is larger than that for fat or protein contents. Genetic progress for either protein yield or protein content would be considerably slower than that for yield or content of fat or total solids.

Selection Based on Only One Component of Milk

Predicted responses to selection for a single trait were calculated relative to an increase in genetic value of milk of 1000 lb, assuming that selection is based solely on the cow's record for that trait. A reasonable estimate is that selection of cows for milk yield can increase their genetic merit by 25 lb of milk per year. Thus approximately 40 years would be necessary to make the changes shown in Table

TABLE 15.5 ESTIMATES OF PHENOTYPIC[a] AND GENETIC[b] CORRELATIONS BETWEEN MILK, FAT, TOTAL SOLIDS, AND PROTEIN YIELDS AND FAT, TOTAL SOLIDS, AND PROTEIN CONTENT OF MILK, AS WELL AS PHENOTYPIC VARIANCES AND HERITABILITIES USED IN PREDICTING GENETIC CHANGES IN EACH, WITH SELECTION BASED ON MILK AND COMPONENTS OF MILK

	Yield				Content			
	Milk	Fat	Total solids	Protein	Fat (%)	Total solids (%)	Protein (%)	Variance
Milk	—	0.65	0.90	0.80	−0.60	−0.60	−0.45	6,250,000 lb^2
Fat	0.85	—	0.85	0.75	0.20	0.15	0.20	10,000 lb^2
Total solids	0.95	0.95	—	0.90	−0.30	−0.25	−0.10	95,000 lb^2
Protein	0.95	0.90	0.95	—	0.00	0.10	0.15	6,600 lb^2
Fat (%)	−0.15	0.30	0.05	0.00	—	0.95	0.75	$0.09\%^2$
Total solids (%)	−0.20	0.25	0.05	0.10	0.90	—	0.85	$0.20\%^2$
Protein (%)	−0.35	−0.05	−0.15	0.05	0.50	0.65	—	$0.036\%^2$
Heritability	0.25	0.20	0.20	0.20	0.50	0.60	0.50	

[a] Phenotypic correlations are below the diagonal of dashes.

[b] Genetic correlations are above the diagonal of dashes.

Source: Adapted from K. R. Butcher et al., 1967, *J. Dairy Sci.*, 50:185.

15.6. Sire selection based on 50 daughter records gives similar genetic responses to selection but should take only 10 years or less under optimal selection procedures. Only cow selection will be discussed here since the results for sires are similar except that the time required for the improvement is compressed.

How a component of milk is expected to change if selection is for that or another constituent is shown in Table 15.6. Selection for any percentage component increases the percentages of all components. Selection for total solids content increases the yield of all components more than selection for any other

TABLE 15.6 RESPONSE IN COMPONENTS OF MILK IF SELECTION IS FOR ONE COMPONENT, BASED ON A SINGLE RECORD FOR THAT COMPONENT[a]

	Yield (in pounds)				Content (%)		
Basis of selection	Milk	Fat	Total solids	Protein	Fat	Total solids	Protein
Yield							
Milk	1000	23.3	99.2	23.2	−0.10	−0.17	−0.05
Fat	581	32.0	83.8	19.5	0.03	0.04	0.02
Total solids	805	27.2	98.6	23.4	−0.05	−0.06	−0.01
Protein	716	24.0	88.8	26.0	0.00	0.02	0.01
Content							
Fat (%)	−849	10.1	−46.8	0.0	0.24	0.37	0.11
Total solids (%)	−930	8.3	−42.7	4.5	0.25	0.43	0.14
Protein (%)	−636	10.1	−15.6	6.2	0.18	0.33	0.15

[a] Response is relative to 1000-lb increase in milk yield obtained by selecting for milk yield from a single milk record on the cow that corresponds to a selection intensity factor of 1.6.

component because of the higher heritability of total solids content. Selection for either total solids yield or milk yield decreases the percentages of all components slightly. Yet over a 40-year period or even over a 10-year period, such percentage declines would not be very noticeable.

A more important consideration is what would happen to total milk yield or total yield of solids, protein, or fat if selection were for an increased fraction of any of those components. Total energy value and total milk value are determined primarily by the amount of the product; within normal limits, they are determined only slightly by the fractional content. Progress in both milk yield and total solids yield would decline dramatically if selection were made on the basis of the fraction of any constituent. Fat yield would increase somewhat and protein yield slightly less if selection were for the fraction of total solids or protein or fat. Selection for yield of milk or any constituent would give greater responses in all yield traits than would selection for fraction of a constituent.

Naturally, selection for milk yield alone gives the greatest gain in milk yield—the 1000 lb used as a basis of comparison. The surprising result is that selection for milk yield is expected to be optimal for improving total solids yield and nearly optimal for improving protein yield. Progress in fat yield is expected to be somewhat less when selection is for milk yield than if selection is directly for fat yield.

Selecting directly for protein yield would be best for increasing total yield of protein, although selection for milk yield would result in nearly as much correlated response in protein yield. Selecting for protein content would result in a decline in milk yield because of the negative correlation between protein content and milk yield. Selection based on protein yield alone would increase milk yield substantially but not nearly as much as selection for milk yield. Dairy managers in countries where protein is scarce and expensive may want to select for protein yield but not for protein content. Response in protein yield if selection were for milk yield would be so high, however, that selection for milk yield may be preferable because protein yield must be determined by expensive, rather sophisticated methods. Milk can simply be weighed.

Joint Selection for Milk Yield and Components of Milk

Product value evaluations automatically assign appropriate economic weights to milk and its components to maximize economic return from selection (see Section 15.3).

15.8 ECONOMICS OF SELECTION

The criteria for making mating and selection decisions include genetic evaluations for traits of economic importance but obviously also must consider cost and investment value of the decisions. For example, two typical questions are: What can I afford to pay for this cow as compared to that cow? Should I buy semen of this bull or that bull? Answers to such questions naturally depend on many factors, including expected production per lactation and over a lifetime, production of descendants, net price of the production, interest rates, investment period, and costs of the semen or cows. Several assumptions are involved in developing answers to such questions, and consequently the answers are subject to errors in the assumptions. The transmission of genetic effects to descendants is on a probability basis and many factors affecting records involve random occurrences. Therefore, answers are expected "on the average" if the assumptions are true. Individual decisions may turn out much better than or much worse than expected. If averaged over many decisions, however, the results will be close to the expected results.

Expected returns from a management decision can be calculated either forward from a beginning time to a fixed end time or back to a beginning time, including all returns through the fixed end time. The forward calculation is more straightforward, but the backward calculation is typically used by economists to describe the net present value (NPV) of a management decision. The equivalence of the two procedures will be illustrated, and then the NPV method will be used to evaluate mating and selection decisions.

Net Present Value

The forward calculation method is well known. Assume that \$100 is invested for 4 years at an annual interest rate of 0.10. The cumulative values (also illustrated symbolically with an investment of X dollars at an interest rate of d) at the end of each year are:

	End of year			
Start	1	2	3	4
100				
	100 + 0.1(100) = 110			
		110 + 0.1(110) = 121		
			121 + 0.1(121) = 131.10	
				131.10 + 0.1(131.10) = 146.41
Summary				
100	100(1.1) = 110	$100(1.1)^2 = 121$	$100(1.1)^3 = 131.10$	$(100)(1.1)^4 = 146.41$
X	$X(1 + d)$	$X(1 + d)^2$	$X(1 + d)^3$	$X(1 + d)^4$

The last two lines show how the value of an initial investment at any time period can be computed. The value of X dollars becomes at the end of year n, $X(1 + d)^n$ dollars.

The back-calculation to present value now becomes obvious. What is the value at the starting time that would have resulted in a value of $X(1 + d)^n$ at the end of year n? The calculation is to divide the value at year n by $(1 + d)^n$. Thus $X(1 + d)^n/(1 + d)^n = X$. For example, what investment at 10% interest is necessary to have \$131.10 at the end of year 3? The answer is $\$131.10/(1.1)^3 = \$100$.

For a breeding program the expected income may fluctuate from year to year. For example, suppose that the expected income by year is as follows. The problem is to calculate the present value of those returns at some specified interest rate, for example, 10%.

End of year	Income in that year	Present value for that year	Symbolic present value for that year
1	$100 = R_1$	$100/(1.1)^1 = 90.91$	$R_1/(1 + d)^1$
2	$200 = R_2$	$200/(1.1)^2 = 165.29$	$R_2/(1 + d)^2$
3	$140 = R_3$	$140/(1.1)^3 = 105.18$	$R_3/(1 + d)^3$
		Total present value $361.38	

The income for year i is divided by $(1 + d)^i$ to calculate the corresponding value at year 0. Present values can be summed over years to calculate total present value for all three years. The procedure of dividing by $(1 + d)^i$ is sometimes called discounting back to the initial investment.

If in the example the income for each year is calculated to the end of year 3 and credited with an interest rate of 10%, the result would be as follows:

- Income for year 1 will draw interest for 2 years.
- Income for year 2 will draw interest for 1 year.
- Income for year 3 will not draw interest because the income is credited at the end of the year.

The numerical calculations are:

$$(\$100)(1.1)^2 + (\$200)(1.1)^1 + (\$140)(1.1)^0 = \$121 + \$220 + \$140 = \$481$$

The $(1.1)^0$ is calculated from the arithmetical rule that any number N except zero to the power zero equals 1 (i.e., $N^0 = 1$).

The total return through the end of year 3 of $481 can be discounted back to the starting point to find present value by dividing by $(1.1)^3$. The result, $\$481/(1.1)^3 = \$361.38$, is, not surprisingly, the same as if the discounted returns for each year had been summed, that is, $\$90.91 + \$165.29 + \$105.18 = \$361.38$.

These simple calculations demonstrate the equivalence between accumulating income and interest to a specified end time or discounting income for each year back to the beginning of the investment period when the interest rate paid for money invested is the same as the interest rate for borrowed money. One last comment: A close approximation to the so-called "true" interest rate to use in discounting is the rate charged for loans minus the inflation rate.

Comparing net value for two cows. The problem of selection of cows can always be reduced to deciding which one of two cows is to be selected and the other culled. For this example, only expected income of the cow herself will be considered. Production of descendants will not have great importance in making the decision. Other than the investment period and discount rate, the factors to consider are (1) the expected production per lactation, (2) thc chances of surviving to have other lactations, and (3) the net price for production. To simplify the example, only milk yield will be considered.

Expected mature equivalent production is largely determined by the management level of the herd (which would be the same for any pair of cows in a herd) and can be approximated as the mature equivalent herd average plus the predicted deviation for that cow. The predicted deviation is the expected producing ability (EPA). If the EPA is not known, the EPA can be estimated as $PD_{sire} + ETA_{dam}$.

All predictions of transmitting ability are on a mature equivalent basis. Thus

$$\text{predicted ME record} = \text{herd average (ME)} + \text{EPA}$$

A reasonable approximation for the EPA for a milk record is

$$\text{EPA} = \{nr/[1 + (n-1)r]\} \text{ (average difference from herdmate averages)}$$

where n is the number of measured records of the cow and r is repeatability of milk yield, which is about 0.50. When $r = 0.5$, the weighting factor $\{n(0.5)/[1 + (n-1)(0.5)]\} = n/(n+1)$.

The herd manager, however, is paid not for mature equivalent production but for actual yield. The predicted ME record must be adjusted to a predicted actual yield by dividing by the age adjustment factor corresponding to the age of the cow when the records will be made. For purposes of calculation, the age adjustment factors for cows freshening at ages in years of 2, 3, 4, 5, 6, and older are: 1.37, 1.18, 1.05, 1.01, and 1.00, which correspond to factors to adjust ME records to an actual basis of 1/1.37 = 0.73, 1/1.18 = 0.85, 1/1.05 = 0.95, 1/1.01 = 0.99, and 1/1 = 1.00.

The assumption will be made that each cow is just fresh and will complete the just started lactation. The chance of surviving to have other records may depend on how old the cow is. Table 15.7 is based on northeastern data and shows

TABLE 15.7 SURVIVAL PROBABILITIES FOR COWS STARTING A SPECIFIED LACTATION

This and future lactations	Current lactation			
	1	2	3	4 or 5
1	1.00			
2	0.82	1.00		
3	0.68	0.83	1.00	
4	0.52	0.63	0.76	1.00
5	0.34	0.41	0.50	0.65
6	0.25	0.30	0.37	0.48
7	0.16	0.20	0.24	0.31
8	0.11	0.13	0.16	0.21
9	—	0.08	0.11	0.14
10	—	—	0.06	0.10
11	—	—	—	0.05

the chances for a cow to have up to seven later records, depending on her age at the start of the current lactation.

As an example, assume that the value of a pound of milk above feed costs is \$0.06, a 5-year investment period and a discount rate, $d = 0.1$. Two cows have the following information:

Cow	Age (yr)	Starting lactation	Previous ME record − ME herd average	Predicted ME herd average	Predicted ME record
A	5	4	+7000	14,000	17,500
B	3	2	−6000	14,000	11,000

1. Predicted ME record = predicted ME herd average + EPA. In this case, $n = 1$, $r = 0.5$

$$\text{EPA (for A)} = [n/(n + 1)](7000) = 3500$$

$$\text{EPA (for B)} = [n/(n + 1)](-6000) = -3000$$

Predicted ME record (for A) = 14,000 + 3500 = 17,500
Predicted ME record (for B) = 14,000 − 3000 = 11,000

2. For each lactation; the predicted actual net income, discounted to starting time is

$$[(\text{net milk price})(\text{predicted ME record})(\text{survival probability})(\text{appropriate age factor})]/(1 + d)^i$$

where $i = 1$ for the lactation just started.

Cow A:

Started lactation	$(0.06)(17{,}500)(1.00)(0.98)/(1.1)$ +
2nd year	$(0.06)(17{,}500)(0.65)(0.99)/(1.1)^2$ +
3rd year	$(0.06)(17{,}500)(0.48)(1.00)/(1.1)^3$ +
4th year	$(0.06)(17{,}500)(0.31)(1.00)/(1.1)^4$ +
5th year	$\underline{(0.06)(17{,}500)(0.21)(1.00)/(1.1)^5}$
Discounted sum:	(0.06(17,500)(2.1254) = \$2232

Cow B:

Started lactation	$(0.06)(11{,}000)(1.00)(0.85)/(1.1)$ +
2nd year	$(0.06)(11{,}000)(0.83)(0.95)/(1.1)^2$ +
3rd year	$(0.06)(11{,}000)(0.63)(0.98)/(1.1)^3$ +
4th year	$(0.06)(11{,}000)(0.41)(0.99)/(1.1)^4$ +
5th year	$\underline{(0.06)(11{,}000)(0.30)(1.00)/(1.1)^5}$
Discounted sum:	(0.06)(11,000)(2.3518) = \$1552

The example illustrates an important economic principle. The difference in expected producing ability of the two cows is about as large, 6500 lb, as will be found. Yet the difference in net present value is only $680. What that means is that cow A is worth only $680 more at auction than cow B, due to expected production over the next 5 years if survival probabilities are average for the two cows. Most people are likely to believe that a larger difference in purchasing price is justified between what are equivalent to the best and poorest cows in the herd.

The important principle for calculating net present value is that the net milk price and predicted record appear in the calculation for each year. Thus the calculation can be condensed by summing the products of survival probability, age adjustment, and discount factor over years and then multiplying by net price and the predicted record. The sum can be called the discounting factor. The discounting factor for cow B (2.3518) is larger than for cow A (2.1254) because the survival probabilities are larger for the younger cow.

In this example, difference in production of descendants has not been considered. For such a short time period the added advantage for cow A would be relatively small. Production of descendants will be considered in determining the value of semen of a bull.

A modification of this procedure can be used to determine whether a cow should be purchased to fill a vacant spot in the milking string. The only additional factor to consider is the maintenance cost for the cow, although the estimated producing ability of the cow and expected herd average are important. The maintenance cost must be exceeded or the cow would be expected to lose money.

As an example, assume that the nonfixed cost of maintaining a cow is $400, which with a net value of $0.06/lb = 6667 lb of production, which is production needed to break even. Thus the discounted value of production in the ith year is

$$\text{(survival probability)[(net milk price)(predicted record)} \\ \text{(age adjustment to actual basis)} - \text{break-even yield]}/(1 + d)^i$$

The yearly discounted values can be summed over the investment period to obtain net present value adjusted for maintenance cost.

If milk is priced on components of fat and protein, then estimated product value adjusted for cost of added production, typically 40 to 50%, should be substituted for net price of milk times predicted milk record.

Return Over Investment in Semen Cost

Net present value of expected production for an average daughter of a bull is the expected return from the investment in enough semen to result in a daughter freshening at about 2 years of age. The number of services required will be about six.

On a fresh heifer basis:

$$\begin{array}{c}\text{return on}\\ \text{semen investment}\end{array} = \begin{array}{c}\text{discounted}\\ \text{production}\\ \text{of heifer and}\\ \text{descendants}\end{array} - \begin{array}{c}\text{cost of semen}\\ \text{needed to}\\ \text{obtain a 2-year-}\\ \text{old heifer}\end{array}$$

Semen cost depends on number of conceptions needed to obtain a 2-year-old heifer and the number of services per conception. Three conceptions are required if 0.80 of conceptions result in a live birth, 0.50 of live births are heifers, and 5 of 6 survive 2 years to freshen in the herd: $(\frac{4}{5})(\frac{1}{2})(\frac{5}{6}) = \frac{1}{3}$, the fraction of conceptions resulting in a 2-year-old heifer.

The number of services per conception depends on conception rate per service as shown in Table 15.8, assuming the conception rate is the same for first and later services. A typical bull will have a first service conception rate of 50%, which results in an average of two services per conception. Services per conception can be approximated by 1/conception rate. Some inseminators have a labor fee for first services, none for second services, and a fee for all later services. The last column of Table 15.8 can be used to estimate the number of nonsecond services.

The semen cost will be (cost/conception) times 3:

$$\text{cost/conception} = (\text{services/conception})(\text{cost/unit of semen}) + (\text{number paid services/conception})(\text{labor fee})$$

Thus on a fresh heifer basis (2-year-old daughter):

$$\text{semen cost} = 3[(\text{services/conception})(\text{unit cost}) + (\text{paid services})(\text{service fee})]$$

TABLE 15.8 NUMBER OF SERVICES PER CONCEPTION AND NUMBER OF NONSECOND SERVICES FOR VARIOUS CONCEPTION RATES PER SERVICE

Conception rate per service, w	Number services per conception, c	Number of nonsecond services, $c - (1 - w)$
0.70	1.43	1.13
0.68	1.47	1.15
0.66	1.52	1.18
0.64	1.56	1.20
0.62	1.61	1.23
0.60	1.67	1.27
0.58	1.72	1.30
0.56	1.79	1.35
0.54	1.85	1.39
0.52	1.92	1.44
0.50	2.00	1.50
0.48	2.08	1.54
0.46	2.17	1.63
0.44	2.27	1.71
0.42	2.38	1.80
0.40	2.50	1.90
0.38	2.63	2.01
0.36	2.78	2.14
0.34	2.94	2.28
0.32	3.12	2.44

The return on the expected production due to the genetic value of the semen contributed to the daughter is determined by the same factors as those discussed in the preceding section plus the gene flow to descendants of the daughter. Figure 15.3 illustrates the contribution of the daughter and her descendants through several lactations adjusted for survival and adjusted to actual rather than ME production. Each freshening of the daughter has a one-third chance of a granddaughter beginning production 2 years later. Each record of the average daughter expresses the transmitting ability of her sire (half his genetic value). Each average granddaughter expresses one-fourth of his genetic value, a great-granddaughter one-eighth, and so on. In the calculations, income and cost of production will be assumed to occur at the end of the lactation year. Thus, by the end of the sixth year after semen purchase, a daughter may have finished three lactations and a granddaughter, one lactation. The predicted production will be adjusted to an actual basis rather than ME and discounted back to time of semen purchase.

As in the earlier sections, the survival probabilities, age adjustment, discount rate, and in addition the fraction of the sire's superiority expressed in the daughter's descendants can be calculated and summed into a discount factor for the

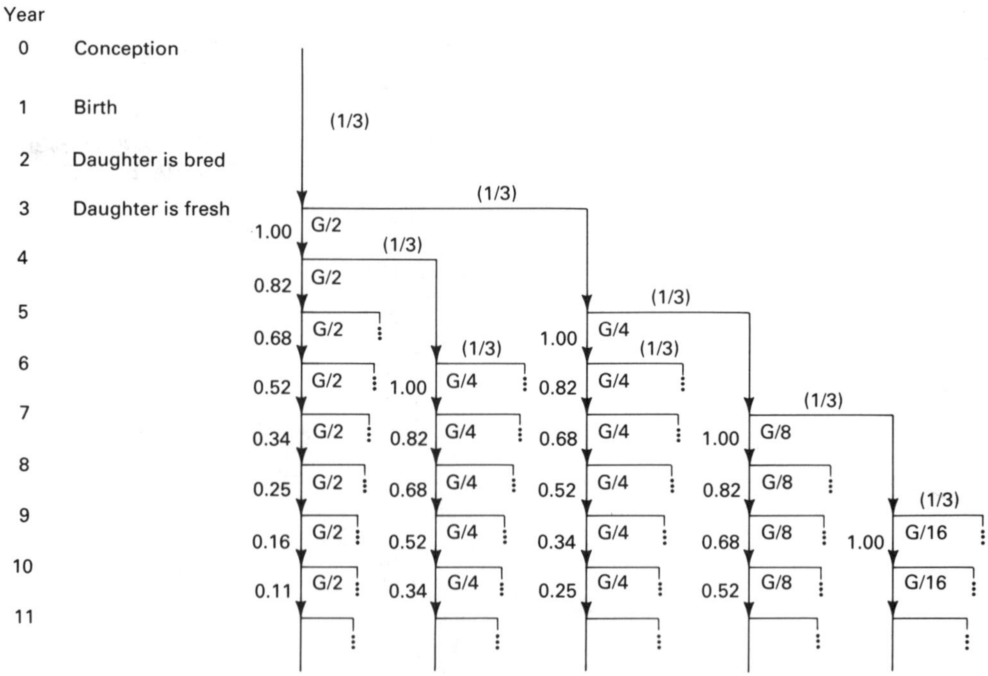

Figure 15.3 Flow of a sire's genes (G) to records of a daughter and her descendants. The ($\frac{1}{3}$) indicates that a conception has only one chance in three of resulting in a 2-year-old fresh heifer. Cow survival probabilities are listed to the left of the genetic expression (G/2 or G/4 or G/8 or G/16). Expected genetic expression on a mature equivalent basis should be adjusted to an actual age basis. The dots indicate continuation of the pattern.

investment period. The discount factor is multiplied by net product value of the bull or net milk price times predicted difference for milk of the bull. Discount factors for various discount rates and investment periods are given in Table 15.9.

The net present value for the bull is

$$\text{NPV} = (\text{discount factor})(\text{net product value}) - (\text{semen cost})$$

Note that this calculation is on the basis of enough semen to obtain a 2-year-old daughter. If six units of semen are required, the formula for the NPV of one unit of semen would be

$$\text{net return/unit of semen} = (\tfrac{1}{6})(\text{discount factor})(\text{net product value}) - (\text{semen and cost of one service})$$

Several computer programs calculate semen value based on similar assumptions, although often the investment period is fixed and production of descendants may be ignored.

TABLE 15.9 FRACTION OF SIRE AND COW SUPERIORITY IN TRANSMITTING ABILITY EXPECTED FROM A MILKING DAUGHTER AND HER DESCENDANTS[a]

Investment period (yr)	True discount rate								
	0.00	0.02	0.04	0.06	0.08	0.10	0.12	0.14	0.16
1	0.00	0.00	0.00	0.00	0.00	0.00	0.00	0.00	0.00
2	0.00	0.00	0.00	0.00	0.00	0.00	0.00	0.00	0.00
3	0.00	0.00	0.00	0.00	0.00	0.00	0.00	0.00	0.00
4	0.73	0.67	0.62	0.58	0.54	0.50	0.46	0.43	0.40
5	1.43	1.31	1.20	1.10	1.01	0.93	0.86	0.79	0.74
6	2.19	1.99	1.80	1.74	1.49	1.36	1.25	1.14	1.05
7	2.92	2.62	2.35	2.12	1.92	1.74	1.58	1.43	1.31
8	3.57	3.17	2.82	2.53	2.27	2.04	1.84	1.66	1.50
9	4.18	3.68	3.26	2.89	2.57	2.30	2.06	1.85	1.67
10	4.74	4.14	3.63	3.20	2.83	2.51	2.24	2.00	1.79
11	5.26	4.56	3.97	3.48	3.05	2.69	2.39	2.12	1.89
12	5.66	4.88	4.22	3.68	3.21	2.82	2.49	2.21	1.96
13	6.05	5.18	4.46	3.86	3.36	2.94	2.58	2.28	2.02
14	6.41	5.45	4.66	4.02	3.48	3.03	2.65	2.33	2.06
15	6.73	5.69	4.84	4.15	3.58	3.11	2.71	2.38	2.10
16	7.02	5.90	5.00	4.27	3.67	3.17	2.76	2.42	2.12
17	7.29	6.09	5.13	4.36	3.74	3.22	2.80	2.44	2.15
18	7.52	6.26	5.25	4.45	3.80	3.27	2.83	2.47	2.16
19	7.74	6.40	5.35	4.52	3.85	3.30	2.85	2.48	2.17
20	7.93	6.53	5.44	4.58	3.89	3.33	2.87	2.50	2.18
21	8.10	6.65	5.52	4.63	3.92	3.35	2.89	2.51	2.19
22	8.26	6.75	5.59	4.67	3.95	3.37	2.90	2.52	2.20
23	8.41	6.84	5.64	4.71	3.98	3.39	2.91	2.52	2.20
24	8.53	6.92	5.69	4.74	4.00	3.40	2.92	2.53	2.21
25	8.65	6.99	5.74	4.77	4.01	3.41	2.93	2.53	2.21

[a] Adjusted for age, survival probability, and discounted back to time of conception for investment periods up to 25 years and for nine discount rates (interest rate minus inflation rate).

As an example, assume two bulls with the following information:

Bull	Product value	Conception rate	Cost of unit of semen
A	$150	0.50	$50
B	$100	0.60	$10

Further assume a 22-year investment period and a discount rate of 0.06 with a return on production over feed cost of 60% with a service fee of $6 per service:

$$\text{net return (bull A)} = (4.67)(0.60)(\$150) - 3[2(\$50) + 2(\$6)]$$
$$= \$420.30 - \$336 = \$84.30$$
$$\text{net return (bull B)} = (4.67)(0.60)(\$100) - 3[1.67(\$10) + 1.67(\$6)]$$
$$= \$280.20 - \$80.16 = \$200.04$$

Obviously, on the basis of product value the return on investment in semen of bull B exceeds the return from semen of bull A. For a shorter investment period, the advantage would be greater. The superiority for production associated with bull A cannot overcome the greater semen cost.

Return on Sexed Semen

The formula for return over investment in semen can be used to calculate the expected return over investment in sexed semen. Sexing of semen has not proved feasible, but if it does, the economics of the cost must be considered. The extra return will be from mating only the top cows in the herd with heifer producing semen and the lower half of the herd with beef bull semen (except in a few cases of importance to bull studs to produce sons from the very top cows to sample for use in AI). The average superiority of the top cows can be taken from Table 15.10.

TABLE 15.10 AVERAGE TRANSMITTING ABILITY OF COWS SELECTED TO BE BRED WITH SEXED SEMEN FOR DIFFERENT PRECISIONS OF SEXING[a]

Precision of sexing: % heifers	Percent cows bred to sexed semen	Select top % of cows to produce replacements	Average transmitting ability for milk of selected cows (lb)
100	50	45	430
90	56	50	390
80	63	57	340
70	71	64	290
60	83	75	200
50	100	90	100

[a] Accuracy of predicting transmitting ability is assumed to be 0.65, and standard deviation of transmitting ability is assumed to be 750 lb.

Source: Adapted from L. D. Van Vleck and R. W. Everett, 1976, *J. Dairy Sci.*, 59:1802.

For example, if sexing semen is 100% accurate, the average superiority in transmitting ability of the top cows will be 430 lb of milk compared to average of all cows. This superiority will be transmitted to daughters the same as that coming from the bull's semen.

Assume that the 430 lb of milk has a net value of $\$0.06(430) = \$25.80$. The genetic value of the average daughter will be the average transmitting ability of the dams plus predicted difference of the bull. Thus the return over investment will be

$$\text{return} = (\text{net dam transmitting ability} + \text{net PD})(\text{discount factor}) - \text{semen cost}$$

As an example, consider perfect sexing, a 22-year investment period, a discount rate of 0.06, and a conception rate of 50%. Only three services with sexed semen are required to obtain a 2-year-old heifer because one-half will not be lost as bull calves. However, the remainder of the herd must be bred so that unless those calves have extra value, the cost of the other breedings must also be charged against the heifers produced as replacements.

Assume that the net product value of the bull is $100 with a regular semen price of $10 and a sexed semen cost of $50, a regular or beef bull semen cost of $2, and service fee of $6. For sexed semen,

$$\text{return} = (4.67)(100 + 25.80) - 3(56) - 3(8) = 587.49 - 192 = \$395.49$$

For regular semen,

$$\text{return} = 4.67(100 + 0) - 6(16) = 467 - 96 = \$371.00$$

For this example, investment in sexed semen if used on the top half of the herd would be profitable but would require perfect sexing. Therefore, the risk may not be acceptable. The expected net profit for the three units of sexed semen is $\$395.49 - \$371 = \$24.49$. Then $8 per unit of sexed semen is the extra price beyond the $40 difference that could be paid to break even for perfectly sexed semen used on the top half of a herd.

Embryo Transfer

Multiple ovulation and embryo transfer (MOET) can result in many heifer replacements from the best cows in the herd. On a population basis, MOET has the greatest potential in selection of dams of sons to test in AI with the result of greater genetic gain per year. On a herd basis, the question is whether increasing genetic value for production by MOET of the top cows is an economical procedure. The equations that estimate return over investment in semen can estimate return over investment in MOET. Superiority in transmitting ability of the top cows is expressed in their daughters, granddaughters, and later descendants just as the effects of genes of a bull. The estimated transmitting ability (ETA) of a cow for production or product value can be substituted in the equation instead of the predicted difference. The assumption will be made that semen costs and return from bull selection will be the same whether MOET is used or not used.

Instead of individual cow ETA, which is subject to a large confidence range,

TABLE 15.11 EXPECTED TRANSMITTING ABILITY FOR MILK YIELD (AND PRODUCT VALUE AT A NET PRICE OF $0.06 PER POUND) AND DISCOUNTED PRESENT VALUE[a]

	Embryos from top fraction				
	0.10	0.20	0.30	0.40	0.50
Milk (lb)	858	682	566	473	390
Milk at $0.06/lb	51.48	40.95	33.93	28.38	23.40
	Present value for discount rate of 0.10 for 5 years (factor = 0.93)				
	$47.88	38.08	31.55	26.39	21.76
	Present value for discount rate of 0.10 for 20 years (factor = 3.33)				
	$171.43	136.36	112.99	94.51	77.92
	Present value for discount rate of 0.04 for 5 years (factor = 1.20)				
	$61.78	49.14	40.72	34.06	28.08
	Present value for discount rate of 0.04 for 20 years (factor = 5.44)				
	$295.50	235.05	194.76	162.90	134.32

[a] Accuracy of 0.65 and standard deviation of transmitting ability of 750 lb are assumed.

the average expected transmitting ability for the top cows can be used in the equation. Embryos from these cows will then be transferred to all cows in the herd to produce replacements. About one in three of the transfers will result in a 2-year-old fresh heifer. Thus the cost will be about three times the cost of a single transfer.

The expected return depends on how few of the best cows furnish the embryos. Table 15.11 lists the expected superiority of top cows for milk and product value (approximated by milk at 0.06/lb) and discounted present value for discount rates of 0.10 and 0.04 and investment periods of 5 and 20 years.

The discounted present values can be used to estimate the break-even cost of an embryo transfer. The tabled values for present value have a cost of about three transfers. For example, for the most optimistic situation of selecting only from the top 10% of the herd with a discount rate of 0.04 and a long investment period of 20 years to gather all the benefits of the superior dams, the present value of the return is $295.50 with a cost of three transfers. The estimated break-even cost is $295.50/3 = $98.50. If the cost is greater than $98.50, a loss is projected. For a short investment period, expected returns from increased production due to MOET is worth less than $100 and in many cases much less.

REFERENCES AND FURTHER READINGS

Agricultural Research Service. 1976. USDA-DHIA modified contemporary comparison sire summary and cow index procedures. *USDA Prod. Res. Rep.* 165, U.S. Department of Agriculture, Beltsville, MD.

Bath, D. L., F. N. Dickinson, H. A. Tucker, and R. D. Appleman. 1985. *Dairy Cattle: Principles, Practices, Profits,* 2nd ed. Lea & Febiger, Philadelphia.

Butcher, K. R., F. D. Sargent, and J. E. Legates. 1967. Estimates of genetic parameters for milk constituents and yields. *J. Dairy Sci.,* 50:185.

Everett, R. W. 1975. Income over investment in semen. *J. Dairy Sci.,* 58:1717.

Everett, R. W., ed. 1984. *Proceedings Natl. Invit. Workshop Genet. Improv. Dairy Cattle,* Cornell University, Ithaca, NY.

Kliewer, R. H. 1972. Advances in the use of type data for evaluation of registered Holstein sires. *Proc. 25th Conv. Natl. Assoc. Artif. Breed.*

McGilliard, M. L., and J. S. Clay. 1983. Selecting groups of sires by computer to maximize herd breeding goals. *J. Dairy Sci.,* 66:647.

Miller, P. D., L. D. Van Vleck, and C. R. Henderson. 1967. Relationships among herd life, milk production, and calving interval. *J. Dairy Sci.,* 50:1283.

Norman, H. D. 1979. An economic index for use in selecting bulls evaluated on protein or solids-not-fat. *USDA Prod. Res. Rep.* 178. U.S. Department of Agriculture, Beltsville, MD.

Politiek, R. D. 1968. Breeding dairy cows for better yields of milk protein. *Proc. 2nd World Conf. Anim. Prod.,* University of Maryland, College Park, MD.

Rendel, J. M., and A. Robertson. 1950. Some aspects of longevity in dairy cows. *Emp. J. Exp. Agric.,* 18:49.

Shanks, R. D., and A. E. Freeman. 1979. Choosing progeny-tested Holstein sires that meet genetic goals at minimum cost. *J. Dairy Sci.,* 62:1429.

Specht, L. W., H. W. Carter, and L. D. Van Vleck. 1967. First classification score and length of herd Life. *J. Dairy Sci.,* 50:1690.

Van Vleck, L. D. 1964. First lactation performance and herd life. *J. Dairy Sci.,* 47:1000.

Van Vleck, L. D. 1985. Notes on the theory and application of selection principles for the genetic improvement of animals. Dept. Anim. Sci., Cornell University, Ithaca, NY.

White, J. M., and J. R. Nichols. 1965. Relationships between first lactation, later performance, and length of herd life in Holstein–Friesian cattle. *J. Dairy Sci.,* 48:468.

Wilcox, C. J., S. N. Gaunt, and B. R. Farthing. 1971. Genetic interrelationship of milk composition and yield. *South. Coop. Ser. Bull.* 155, University of Florida, Gainesville, FL.

Wilcox, M. L., C. R. Shumway, R. W. Blake, and M. A. Tomaszewski. 1984. Selection of artificial insemination sires to maximize profits. *J. Dairy Sci.,* 67:2407.

Wilton, J. W., L. D. Van Vleck, R. W. Everett, R. S. Guthrie, and S. J. Roberts. 1972. Genetic and environmental aspects of udder infections. *J. Dairy Sci.,* 55:183.

16

Crossbreeding

Crossbreeding has long been popular with swine breeders and in the 1960s and 1970s became popular with beef breeders. In general, dairy breeders have not been very excited about crossbreeding, although a number of crossbreeding experiments have been conducted.

Crossbreeding is generally thought of as a cross between breeds; often more than two are eventually involved. Linecrossing is nearly the same thing, but the crosses are between unrelated inbred lines of the same breed. The possible advantages are similar for both crossbreeding and linecrossing. Crossing of inbred lines, however, has not been practical with farm animals because of the expense of maintaining the lines with the usual reduction in fertility as inbreeding increases.

16.1 REASONS FOR CROSSBREEDING

Heterosis

Heterosis is the measure of the effects of crossbreeding and can be defined as the difference of crossbred animals from the average of their purebred parents. Heterosis may be positive (the crossbreds exceed the parent purebred average) or negative (the purebred parents average more than their crossbred progeny).

Heterosis on a percentage basis is defined as follows:

$$\text{heterosis} = 100\left[\frac{(AB + BA)/2 - (A + B)/2}{(A + B)/2}\right]$$

where A and B are the averages for parent breeds A and B, and AB is the average for the cross resulting from sire breed A mated dam breed B, BA is the average for the reciprocal cross, and the difference as a fraction of the parental average is multiplied by 100 to form a percentage. Estimates of heterosis from crosses between Ayrshire, Brown Swiss, Holstein, and Jersey breeds as summarized by R. E. McDowell are listed in Table 16.1. A primary reason for crossbreeding is

TABLE 16.1 WEIGHTED AVERAGE HETEROSIS AND RANGES FOR TWO-BREED CROSSES AMONG AYRSHIRE, BROWN SWISS, HOLSTEIN, AND JERSEY

Trait	Number of herds	Number of breed groups	Weighted average heterosis (%)	Range among groups (%)
Production traits				
Milk yield	4	14	5.1	−9.0 to 18.9
FCM yield	4	14	6.6	−5.0 to 17.5
Fat yield	4	14	5.5	−2.0 to 16.6
SNF yield	3	11	5.2	−17.0 to 18.1
Protein yield	2	8	5.7	−1.0 to 15.6
SCM yield	2	8	5.0	−5.1 to 17.3
Milk composition				
Fat (%)	4	14	−0.3	−2.5 to 3.0
SNF (%)	3	11	−1.1	−5.0 to 0.4
Protein (%)	2	8	−1.5	−3.4 to 1.0
Lactation characteristics				
Age first calving	4	14	0.2	−3.0 to 6.1
Lactation length	4	14	1.0	−1.0 to 6.0
Persistency milk yield	3	11	2.3	−4.3 to 8.0
First day of milk yield	1	6	5.8	0.9 to 13.9
Peak daily yield	1	6	6.8	4.1 to 8.4
300th day yield	1	6	17.8	13.9 to 20.0
Day of peak yield	1	6	1.4	−5.4 to 12.9
Size first lactation	1	6	−6.9	6.8 to 8.4
Weight gain first lactation	1	6	−3.0	−8.0 to 7.0
Milking rate				
Time start to strip	1	6	−3.9	−9.8 to 2.3
kg/min to strip	1	6	1.4	−1.7 to 6.6
kg/min for total time	1	6	0.6	−1.3 to 3.8
Feed efficiency				
Mcal milk/Mcal intake	1	6	1.2	−2.0 to 4.0
kg FCM/Mcal intake	1	6	1.5	−1.0 to 6.0
kg SCM/Mcal intake	1	6	1.3	−1.0 to 3.0
Body characteristics				
First lactation				
Condition score (fatness)	1	6	−7.7	−7.4 to 7.9
Dairy character	1	6	−5.2	−15.8 to 27.2
Body shape	1	6	9.0	2.2 to 20.1
Udder characteristics	1	6	6.6	−3.3 to 9.8

TABLE 16.1 *(continued)*

Trait	Number of herds	Number of breed groups	Weighted average heterosis (%)	Range among groups (%)
Teat placement	1	6	−3.7	−4.6 to 8.3
Correlation milk yield and body dimensions	1	17	8.0	−4.1 to 18.7
Viability				
Gestation length	2	9	0.9	−1.7 to 2.4
Calf survival 0–3 months	4	14	4.6	−0.1 to 11.2
% survival to first calving	4	14	6.2	−12.3 to 27.6
Calves born dead	4	14	3.7	3.2 to 9.6
Lactation termination				
Health	4	14	−1.0	−11.4 to 17.8
Sterility	4	14	−1.2	−8.7 to 9.4
Death	4	14	0.1	−7.6 to 4.7
Health disorders				
Birth to calving	1	6	10.8	−32.0 to 25.3
First lactation	1	6	−2.3	−33.0 to 41.0
Breeding efficiency				
Days open	4	14	1.7	−21.1 to 33.0
Calving interval	4	14	2.7	−12.1 to 5.7
Days parturition to first estrus	4	14	3.5	−8.1 to 7.0
Days first breeding to conception	4	14	2.2	−9.0 to 4.1
Services per conception	4	14	0.8	−3.1 to 4.9
Proportion pregnant				
<90 days	4	14	5.0	−14.0 to 12.0
<120 days	4	14	4.2	−13.0 to 7.0
<145 days	4	14	2.3	−13.0 to 6.0
<200 days	4	14	2.1	−16.0 to 10.0
Overall breeding efficiency	4	14	3.2	−6.4 to 3.6
Culled for sterility	4	14	2.3	−8.7 to 11.6
Reproductive disorders	4	14	−1.4	−6.1 to 8.9
Calving difficulty	4	14	1.3	−17.4 to 5.6
Growth rate				
Weight				
Birth	1	9	5.7	3.0 to 10.0
20–60 days	1	9	3.4	0.0 to 6.5
80–160 days	1	9	3.2	1.5 to 6.2
6 months	1	9	3.0	−0.6 to 6.6
12 months	1	9	4.6	2.5 to 7.7
18 months	1	9	4.9	3.4 to 5.8
30 months	1	9	3.5	2.4 to 4.7
Skeletal dimensions				
6 months	1	9	0.9	−0.8 to 3.2
12 months	1	9	0.9	−0.9 to 2.3
18 months	1	9	1.4	−0.2 to 2.7
30 months	1	9	1.1	−0.8 to 2.8

Source: R. E. McDowell, 1982, *South. Coop. Series Serv. Bull.* 259, Louisiana Agricultural Experiment Station, Baton Rouge, LA, Table 36.

to obtain positive heterosis. If sufficient numbers of the best pure breed are available for breeding, then whether heterosis exists is not very important; the important question is whether crossbred animals are superior to the best pure breed. For one trait this question can be answered by comparing many crossbred and pure breed animals, but for total economic merit the answer is more complex. Even if crossbreds do not exceed the best parent breed for any trait, total economic merit, considering all economic traits of the crossbreds, may still be greater than that of either parent breed.

Upgrading

The other primary reason for crossbreeding is to change the characteristics of a population quickly. Such rapid change is useful in upgrading native cattle or changing traits that do not improve from selection within a breed, as is the situation when no genetic differences exist within a breed but genetic differences do exist between breeds.

Breed Formation

A third reason to cross breeds is to form the basis of a new breed—a synthetic breed. The foundation breeds are chosen to bring desirable traits to the new breed. After the initial crossbreeding, selection continues among the new population of crossbred animals.

16.2 THEORY OF HETEROSIS

The theories of why crossbreeding might result in heterosis generally fall into three classes.

Dominance

Crossbreeding is simply an extreme example of outcrossing, which is the opposite of inbreeding. Inbreeding increases the proportion of homozygous gene pairs, and, if the genes are recessive with deleterious effects, inbreeding will increase the frequency of such individual unfavorable effects. So-called pure breeds, although they are not homozygous at all loci, are necessarily somewhat inbred to differentiate themselves from other breeds. Crossbreeding breaks up such homozygosity and would lead to heterosis. The dominance theory of heterosis relies on reducing the proportion of loci with homozygous recessive gene pairs.

Overdominance

A similar theory is the overdominance theory, which is based on the value of the heterozygote exceeding the value of the best homozygous gene pair. Crossbreeding ensures a maximum number of heterozygous loci and would result in superior animals if overdominance is important. Because overdominance is not common for simply inherited traits, this theory is unlikely to hold for traits affected by genes at many loci.

Epistasis

The interaction or epistatic theory requires favorable effects resulting from new combinations of genes from different loci. For example, a gene at one locus together with a gene at another locus would give more response than would be expected. Such epistatic response could be favorable or unfavorable.

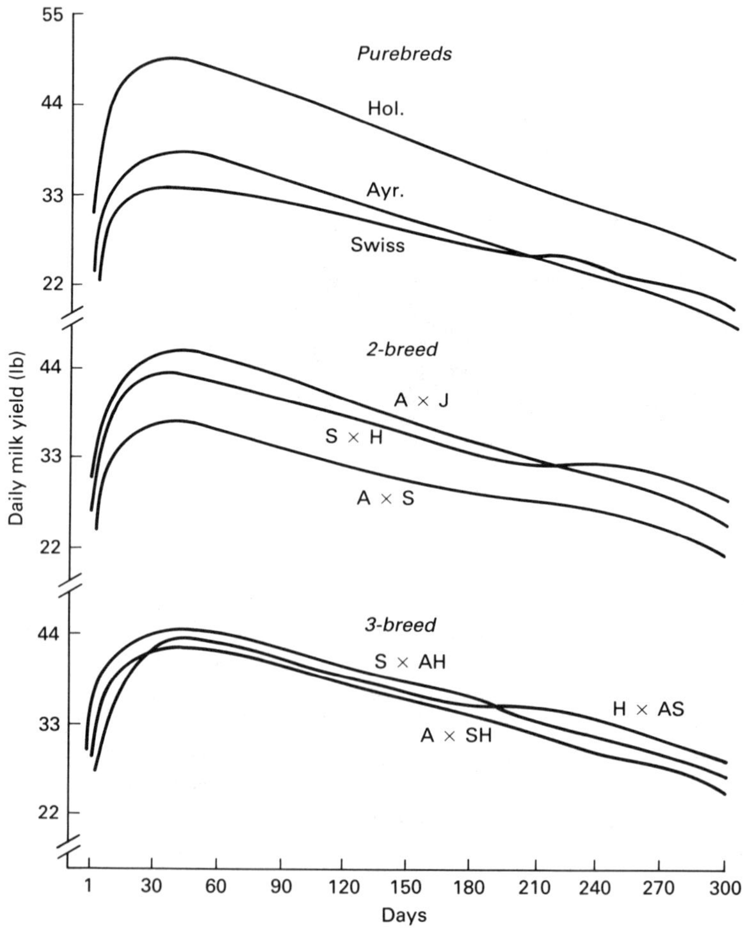

Figure 16.1 Lactation curves for Ayrshire, Brown Swiss, Holstein, and three-breed crosses. (Beltsville): Source: R. E. McDowell, 1982, *South. Coop. Ser. Bull.*, 259. Louisiana Agricultural Experiment Station, Baton Rouge, LA.

16.3 EXPERIMENTAL RESULTS

Pearson and McDowell (1968) have reviewed published material on the crossbreeding of dairy cattle in temperate regions. Because of the high cost of such experiments, very few crossbreeding projects have been conducted with dairy cattle in contrast to the large number that have been conducted with swine. Generally, three-breed crosses have not been as good as two-breed crosses, although crossbred dams might be expected to have more maternal ability than purebred dams if maternal traits are influenced by heterosis.

In most studies crossbreds have not exceeded the best parent breed for productive traits and particularly not for milk yield when one parent breed was Holstein (Figure 16.1). Heterosis for viability and fertility traits, however, suggests possible advantages for crossbreds since the parent breeds generally have similar averages. For growth traits, as for most traits, heterosis is greater earlier in life than later.

Results of a study by McDowell and McDaniel (1968) of 137 two- and three-breed crosses of Ayrshire, Holstein, and Brown Swiss conducted in Beltsville, Maryland, show that total merit, based on economic values of the various traits of crossbreds as compared with pure breeds, may be quite different from results for single traits. These researchers studied 15 pricing combinations for feed, milk yield, and milk composition and assigned health costs for veterinary visits and treatment. Table 16.2 shows average net returns from all combinations of feed costs. The results vary only slightly from the comparison of crossbreds with Holsteins using the income-over-feed cost situation most favorable to Holsteins.

Crosses with half of their inheritance from Holsteins compared favorably with Holsteins, although none of the other crosses had net returns that were even close to the net value of Holsteins. The problem with this experiment is simply the small number of animals involved. The results do show that all traits should

TABLE 16.2 DOLLAR RETURNS FOR FIRST LACTATIONS, COMPUTED AS A DEVIATION FROM HOLSTEINS

Breed group	Income over feed cost	Health cost per lactation	Animal losses per lactation	Dry cow costs[a]	Net returns[b]
Ayrshire (A)	−62	6	6	3	−47
Brown swiss (S)	−73	−8	−27	−4	−112
A × H	2	8	−11	12	11
S × H	3	2	−2	6	9
A × S	−69	−11	−16	3	−93
A × (S × H)	4	−25	−23	5	−39
S × (A × H)	14	−38	−95	16	−103
H × (A × S)	7	6	−11	8	10

[a] Differences in average days open × $0.50/day.
[b] Returns per lactating cow over Holsteins, using average differences in income over feed cost.
Source: R. E. McDowell, and B. T. McDaniel, 1968, *J. Dairy Sci.*, 51:1649.

be considered in determining the most profitable breed of cow for a particular set of conditions.

A summary of 20 years of crosses between Guernsey and Holstein breeds at the Illinois Agricultural Experiment Station included a total of 2015 births. Fewer crossbred calves (6.2%) than purebreds (9.0%) were dead at or within 24 hours of birth. Of the females, a much larger fraction of crossbreds lived to calving (92%) than purebreds—71% of the Guernseys and 85% of the Holsteins. Heterosis was approximately 7% for productive traits, although the crossbreds' average did not exceed the average of Holsteins for any of the yield traits (milk, fat, nonfat solids, and protein).

Performance of the purebred and crossbred groups was evaluated by establishing monetary values in terms of economic values of differences in livability, growth, and production. After assigning prices and costs to milk production, veal calves, and feed costs, the values in Table 16.3 were estimated for 100-cow herds of Guernseys, crossbreds, and Holsteins based on experimental results for livability, growth, and production. The analyses show a sizable amount of heterosis for monetary returns for crossbreds; 22% for $\frac{1}{2}$ crossbreds, and 16% for $\frac{3}{8}$ and $\frac{5}{8}$ backcrosses, over the purebred averages. In this experiment the crossbreds did not exceed the Holsteins. However, Touchberry (1970) extrapolated the heterosis values to Guernsey and Holstein DHIA averages in Illinois to estimate a slight advantage for the crossbreds over Holsteins. Touchberry also concluded that crossing grade Guernseys, Jerseys, Ayrshires, and Brown Swiss with Holsteins would be profitable but that crossbreeding grade Holsteins would be advantageous only if proved sire programs were as intensive in the other breeds as in Holsteins. Given the relative numbers of animals of the dairy breeds, no other breed can be as selective in proving young bulls as the Holstein breed.

Velasco (1971) summarized the results from 6870 reproductive records from the USDA crossbreeding project at four locations: Beltsville, Maryland; Jeanerett, Louisiana; Reidsville, Georgia; and Clemson, South Carolina. The breeds used in crossing were Holstein, Brown Swiss, and Jersey. Although this was the largest study ever made of this type, crossbreeding did not appear to result in significant improvement in reproductive ability, although there was a tendency for the crossbreds to have shorter calving intervals than purebreds.

TABLE 16.3 DOLLAR COSTS AND RETURNS FOR 100-COW HERDS, BASED ON THE ILLINOIS CROSSBREEDING PROJECT

Factor	Guernseys	Crossbreds $\frac{1}{2}$	Crossbreds $\frac{3}{8}$ and $\frac{5}{8}$	Holsteins
Calves died	−2,351	−790	−1,026	−1,601
Calves sold before calving	1,595	2,264	2,220	2,161
Cows sold and died	2,213	6,190	6,357	5,439
Heifer maintenance	−5,796	−8,885	−8,747	−8,639
Feed costs for cows	−16,754	−21,134	−20,842	−23,252
Value of milk	40,050	56,620	55,587	64,913
Herd total	18,957	35,265	33,547	39,020

Source: R. W. Touchberry, 1970, *Proc 19th Annu. Natl. Breed. Roundtable.*

A breeding project (Southern Regional Research Project S-49) to utilize the adaptability of Zebu breeds to tropical conditions in crosses with European breeds showed no advantage for crossbreds. The conclusion was that in the southern United States, selection within European breeds was recommended over crossbreeding to either the Red Sindhi or the Brahman breeds. Although $\frac{1}{2}$ and $\frac{3}{4}$ Zebu groups were more resistant to parasites and heat, they had lower performance than Holstein, Jersey, and Brown Swiss breeds. Groups with more than one-half European blood produced less than European breeds but had approximately the same adaptability to semitropical conditions.

Artificial insemination provides a method of reducing the cost of crossbreeding since semen from purebred bulls can be obtained without each breeder having the cost of maintaining all breeds—assuming, of course, that crossbred cows and not purebred cows are maintained in the breeding herd. Maintaining crossbred cows implies some form of rotational crossbreeding with an attempt to maintain an optimal fraction of inheritance from the various breeds. For example, the results of the Beltsville and Illinois experiments suggest the advisability of maintaining a majority of inheritance from Holsteins. The unanswered question is whether crossbreeding can, over several generations, provide more income than selection within a breed—in this example, within the Holstein breed. McDowell

TABLE 16.4 PREDICTED PERFORMANCE (MATURE EQUIVALENT BASIS), ESTIMATED GROSS INCOME, AND EXPECTED YEARLY PRODUCTION OF FCM OF TWO-BREED CROSSES BASED ON PARENTAL BREED AVERAGES IN USDA SIRE SUMMARY LIST, 1975

Breed of sire	Breed of dam	Milk (lb)	Fat (lb)	FCM (lb)	Fat (%)	Income[a]	FCM/yr[b] (lb)
				Crosses			
Ayr	Hol	14,640	535	13,970	3.77	$1912	13,070
Hol	Ayr	12,610	477	12,210	3.79	1651	11,430
Jer	Hol	12,740	530	13,160	4.27	1689	12,310
Hol	Jer	12,280	539	12,460	4.30	1629	11,660
Swiss	Hol	13,850	519	13,460	3.77	1811	12,590
Hol	Swiss	13,490	515	13,430	3.80	1766	12,550
Ayr	Swiss	11,660	460	11,570	3.93	1531	10,830
Swiss	Ayr	11,470	466	11,600	3.92	1507	10,860
Swiss	Jer	10,890	506	11,970	4.41	1451	11,200
				Purebreds			
Ayr	Ayr	10,520	416	10,450	3.95	1382	9,530
Swiss	Swiss	11,840	477	11,900	4.03	1560	10,850
Hol	Hol	14,100	513	13,330	3.63	1839	12,160
Jer	Jer	8,780	444	10,180	5.05	1195	9,290

[a] Milk price $13.00/100 lb with differential for fat test of $0.166 per 0.1% from base 3.50%.
[b] FCM/year = calving interval/365 × lactation FCM (calving interval 400 days for purebreds and 390 days for crosses).
Source: R. E. McDowell, 1982, *South. Coop. Serv. Bull.* 259, Louisiana Agricultural Experiment Station, Baton Rouge, LA, Table 39.

(1982) in summarizing crossbreeding results has projected production and income from various breed crosses under average U.S. conditions, as shown in Table 16.4. Only one cross, Ayrshire by Holstein, was projected to exceed the Holstein in net income.

16.4 SYSTEMS OF CROSSBREEDING

The two main classes of crossbreeding are those that require maintenance of the pure breeds and those that are rotational, which means that crossbred (after the first cross) cows are mated to purebred bulls.

Pure Breed Crosses

Two-way crosses. In this system, two breeds are mated together and the offspring are used for production but not for breeding. Such a system is unlikely to work with dairy cattle because of the difficulty in obtaining enough replacements to maintain sufficient numbers of both the pure breeds and the crossbreds.

Three-way crosses. Two-way cross cows are mated to bulls of a third breed to utilize any heterosis (hybrid vigor) for maternal ability of the crossbred dams. Unless crossbred females are allowed to produce replacements, not enough replacements will be available to maintain the breeding program.

Double two-way crosses. Two-way cross cows are mated to two-way cross bulls with four breeds represented. Although corn and poultry breeders have successfully used this technique with inbred lines, there seems to be no reason to use this system in dairy breeding, except possibly to have four breeds represented as foundation animals for developing a synthetic breed.

Rotational Crosses

Crisscrosses. Crisscrossing is similar to backcrossing except that the two parent breeds are used alternatively in each generation. Such crosses maintain heterosis between two breeds while allowing use of crossbred dams to produce replacements. After several generations the fraction of genes from the last-used male breed will tend toward 0.67 and 0.33 from the other. The fractions for several generations are shown in Table 16.5.

Three-way rotation. Bulls of three breeds are used in successive generations on the crossbred dams of the previous generation. The fraction of genes from the last used male breed will approach 0.57; from the next-to-last-used male, 0.29; and from the third breed, 0.14. The breed with the greatest genetic contribution to the crossbreds changes in each generation. Adding a fourth breed to

TABLE 16.5 FRACTION OF INHERITANCE RESULTING FROM CRISSCROSS MATING

Generation	Cross	Fraction from A	Fraction from B
1	$A \times B$	0.50	0.50
2	$A \times (AB)$	0.75	0.25
3	$B \times (AAB)$	0.375	0.625
4	$A \times (BAAB)$	0.6875	0.3125
5	$B \times (ABAAB)$	0.34375	0.65625
⋮			
Limit	$A \times$ crossbred	0.66667	0.33333
	$B \times$ crossbred	0.33333	0.66667

the rotation changes the fractions only slightly to 0.53 for the last breed of sire. Thus a three-breed rotation is nearly optimal for maintaining heterozygosity.

Indeterminate rotation. Continued use of bulls of new breeds with crossbred cows will maintain heterozygosity at a maximum, with the last breed of sire contributing 0.50 of genes; the next-to-last contributing 0.25; and so on. Obviously, dairy breeders would soon run out of suitable breeds to use as a source of sires.

Grading-up

Grading-up is a method used to change a population quickly from one breed type to another. Bulls of the breed of the newly desired type are continually backcrossed to cows from the previous generation. Table 16.6 shows how rapidly the genes in the population come to resemble those of the breed of the male parents. After four or five generations there is little difference between the genetic material of the graded-up cattle and the pure breed. Selection should accompany the grading-up process, but how much selection is optimal is unknown. In many situations the environment frequently must be changed so that the introduced breed and the graded-up animals can express their genetic potential. Often the original cattle would have been selected for productivity in an environment that would not allow survival of the introduced breed.

TABLE 16.6 CHANGE IN FRACTION OF INHERITANCE OBTAINED BY GRADING-UP (CONTINUAL BACKCROSSING)

Generation	Grading-up cross	Fraction from A	Fraction from B
1	$A \times B$	0.50	0.50
2	$A \times (A \times B)$	0.75	0.25
3	$A \times (AAB)$	0.875	0.125
4	$A \times (AAAB)$	0.9375	0.0625
5	$A \times (AAAAB)$	0.96875	0.03125
⋮			
n	$A \times n$th cross	$1 \times (\frac{1}{2})^n$	$(\frac{1}{2})^n$

16.5 WHICH BREEDS TO USE IN CROSSES

The basis for any crossbreeding or selection program is knowledge of what is wanted. In general, high milk yield is the major goal, but if other traits are important, some compromise must be achieved to select for several traits. In choosing breeds for crossbreeding, a knowledge of the potential of each breed for economically important traits is essential. Then the performance of various crosses can be predicted from the average of parent breeds plus any estimated increase resulting from heterotic effects. McDowell has ranked the major dairy breeds in the United States for several economically important traits. His results, shown in Table 16.7, can be used to decide which breeds when crossed are most likely to give a desired result. If, for example, milk yield is most important but percentage of protein content is to be improved, the table suggests a cross between Holsteins and Brown Swiss. Both breeds rank high in milk yield, and although the Holstein ranks last in protein content, the Brown Swiss ranks second. Thus the Brown Swiss–Holstein cross should improve protein content without a substantial reduction in total milk yield.

TABLE 16.7 GENERAL RANKING OF U.S. DAIRY BREEDS FOR ECONOMICALLY IMPORTANT TRAITS

Yield			Percent								
Milk	Fat-corrected milk	Fat	Fat	Nonfat solids	Protein	Feed efficiency	Income over feed	Days open	Vet. cost	Viability	Mature weight
Hol	Hol	Hol	Jer	Jer	Guer	Hol	Hol	Ayr	Hol	Ayr	BS
BS	BS	BS	Guer	Guer	BS	Jer	Ayr	Jer	Jer	Jer	Hol
Ayr	Ayr	Guer	BS	BS	Jer	Ayr	Jer	Hol	Ayr	Hol	Ayr
Guer	Guer	Jer	Ayr	Ayr	Ayr	Guer	BS	Guer	Guer	Guer	Guer
Jer	Jer	Ayr	Hol	Hol	Hol	BS	Guer	BS	BS	BS	Jer

Source: R. E. McDowell, 1972, personal communication.

On the other hand, if a shorter calving interval and increased viability are important in addition to milk yield, crossing Holsteins with Ayrshires would be expected to improve livability and reproductive efficiency of the Holsteins. Such a cross would not, however, be very efficient in improving fat, nonfat solids, or protein content of the milk.

The use of such a table for tropical countries with special environmental problems would require measurement of such physiological indicators as heat resistance and parasite resistance as well as potential for production in those environments.

16.6 RESULTS OBTAINED IN TROPICAL COUNTRIES

In the tropics even improved environment may not allow maximum production of a pure breed. For such improved environments the optimal inheritance from the pure breed may be close to $\frac{1}{2}$ or $\frac{5}{8}$. Amble and Jain (1967) found that performance of half-breds (Ayrshire or Holstein) was most effective as compared to other fractions of European breeds in crosses with native tropical cattle, such as Sahiwal in India. Many tropical cattle require that the calf either stimulate milk ejection or be present at milking. Apparently, a single cross to an European breed overcomes this inherited behavior.

The potential of crossbreeding in tropical countries where selection of cattle has been primarily for survival or for meat or draft under severe conditions has been demonstrated by Moulick's (1972) study of Deshi cattle of India, in which body weight averages only 450 lb at maturity, and of Jersey–Deshi crosses. Some average performances observed at the Haringhata breeding station are shown in Table 16.8.

TABLE 16.8 COMPARISON OF DESHI AND DESHI × JERSEY COWS IN THE FIRST LACTATION

Trait	Deshi	Crossbred
Lactation milk (pounds)	824	2858
Lactation length (days)	276	317
Age at calving (months)	47	32
Days open	187	91
Calving interval (days)	452	368

Source: S. K. Moulick, R. E. McDowell, L. D. Van Vleck, and H. Guha, 1972, *J. Dairy Sci.*, 55:1148.

The response from crossbreeding is rather dramatic, especially when compared with the potential for selection within the Deshi breed. Moulick's evidence suggests that genetic variability, possible selection intensity, and generation intervals allow for a 1% genetic increase in milk yield per year. If the average lactation yield were 824 lb in pure Deshi, the increase would be only 8 lb per year. On this basis, an increase to the level of the crossbreds would require more than 240 years.

A practical problem is whether to try to maintain a Jersey share of inheritance of approximately 50% for that environment by crisscrossing and backcrossing or to mate the crossbreds among themselves with selection for optimal economic production for that environment. Another possibility would be to introduce other European and native breeds to the cross before such inter se mating and selection.

REFERENCES AND FURTHER READINGS

Amble, V. N., and J. P. Jain. 1967. Comparative performance of different grades of crossbred cows on military farms of India. *J. Dairy Sci.*, 50:1695.

Bereskin, B., and R. W. Touthberry. 1966. Crossbreeding dairy cattle: III. First lactation production. *J. Dairy Sci.*, 49:791.

BRANTON, C., R. E. MCDOWELL, and M. A. BROWN. 1966. Zebu–European crossbreeding as a basis of dairy cattle improvement in the USA. *South. Coop. Ser. Bull.* 114. Louisiana Agricultural Experiment Station, Baton Rouge, LA.

DONALD, H. R., and W. S. RUSSELL. 1968. Some aspects of fertility in purebred and crossbred dairy cattle. *Anim. Prod.*, 10:465.

FOHRMAN, M. H., R. E. MCDOWELL, C. A. MATTHEWS, and R. A. HILDER. 1954. A crossbreeding experiment with dairy cattle. *USDA Tech. Bull.* 1074.

JOHANSSON, I., and J. RENDEL. 1968. *Genetics and Animal Breeding*. W. H. Freeman and Company, Publishers, San Francisco.

MCDANIEL, B. T., and R. E. MCDOWELL. 1968. Economic aspects of crossbreeding. *J. Dairy Sci.*, 51:969.

MCDOWELL, R. E., and B. T. MCDANIEL. 1968. Interbreed matings in dairy cattle: III. Economic aspects. *J. Dairy Sci.*, 51:1649.

MOULICK, S. K., R..E. MCDOWELL, L. D. VAN VLECK, and H. GUHA. 1972. Potential of Deshi cattle of India for dairy production. *J. Dairy Sci.*, 55:1148.

PEARSON, L., and R. E. MCDOWELL. 1968. Crossbreeding of dairy cattle in temperate zones: A review of recent studies. *Anim. Breed. Abstr.*, 36:1.

TOUCHBERRY, R. W. 1970. A comparison of the general merits of purebred and crossbred dairy cattle resulting from twenty years (four generations) of crossbreeding. *Proc. 19th Annu. Natl. Breed. Rountable*. Poultry Breeders of America, Kansas City, MO.

VELASCO, J. A. 1971. Comparison of the reproductive efficiency of purebred and crossbred dairy cattle. Ph.D. thesis. Cornell University, Ithaca, NY.

17

Basic Principles of Dairy Cattle Nutrition

The principal function of the dairy cow in the food chain is to convert feeds into milk that can be consumed by humans. A major advantage of the dairy cow and other ruminants is their ability to convert feeds not usable by humans, such as hays, silages, pastures, and industrial by-products, into human foods. This ability results from breakdown of feed by microorganisms in the ruminant stomach and utilization of the resultant end products by ruminants for their physiological functions. The small and large intestines also play a role in the conversion of feeds. All changes in feed that occur in the digestive tract in preparation for absorption and use by the body are referred to as digestion.

17.1 DIGESTION IN THE RUMINANT STOMACH

The major prehension organ is the tongue, which draws feed into the mouth; the lips are not effectively used. Feed consumed by cattle is briefly chewed and swallowed with some saliva. The swallowed material moves through the esophagus to the rumen and reticulum, two of the four compartments in the ruminant stomach, shown in Figure 17.1. The other two parts of the ruminant stomach are the omasum and abomasum (Figure 17.2). The rumen is also called the paunch and the reticulum is referred to as the honeycomb because of the nature of its

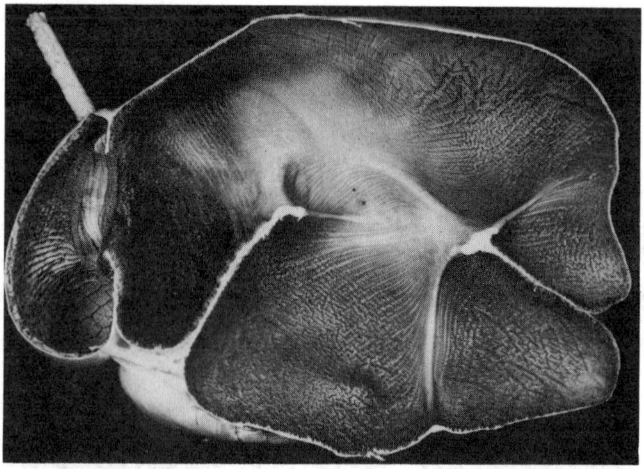

Figure 17.1 Internal arrangement of the rumen and reticulum. Ret., reticulum; D, dorsal sac of rumen; V, ventral sac of rumen; P, posterior sacs of rumen; E, esophagus; EG, esophageal groove. From N. J. Benevenga et al., 1969, *J. Dairy Sci.*, 52:1294.

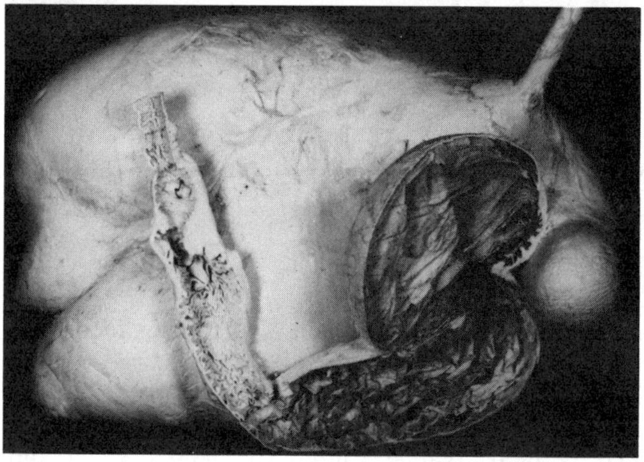

Figure 17.2 Right side of the ruminant stomach, showing a cross section of the omasum (O) with its many plies; the abomasum (Ab.); the beginning of the small intestine (SI); and the exterior of the reticulum (Ret.), rumen (R), and esophagus (E). From N. J. Benevenga et al., 1969, *J. Dairy Sci.*, 52:1294.

lining. The omasum is referred to as many-plies because it has many leaves or plies in its structure. The abomasum is the true stomach because it secretes gastric juice containing hydrochloric acid and enzymes.

Swallowed feed and saliva are added to the liquid mass in the rumen and reticulum. The rumen and reticulum are large in the mature animal and hold 151 to 227 liters (40 to 60 gal) of liquid. Regular contractions (about one per minute) of the rumen and reticulum result in the mixing of the swallowed material with the rumen contents. After eating, the cow rechews her feed by a process of regurgitation and rechewing (called rumination) and frequently is said to be "chewing her cud." In this process, a core of feed and liquid is passed from the rumen and reticulum through the esophagus to the mouth, where it is rechewed for approximately 1 minute and reswallowed. A cow spends approximately 8 to 10 hours a day in rumination. The amount of time varies with animals and rations.

Cows secrete large volumes of saliva. In one study (Stallcup, 1971), cows grazing grass secreted 178 liters (47.0 gal) per day, 148 liters (39 gal) when eating hay, and 108 liters (28.5 gal) when fed silage. Saliva has a low surface tension, which helps to prevent froth formation in the rumen, promotes rapid wetting of material, and results in a more rapid attack on chewed materials by rumen microorganisms.

Feed stays in the rumen and the reticulum for several hours to several days. During this time large numbers of microorganisms play important roles in the digestion of the feed. The microorganisms convert a large part of the available carbohydrates in the feed into volatile fatty acids (VFA), methane, and carbon dioxide. Major VFA are acetic acid (50 to 70%), propionic acid (20 to 30%), and butyric acid (10 to 15%). Small amounts of valeric acid and other short-chain acids are also produced. These short-chain VFA are absorbed by small fingerlike projections called papilli, metabolized by the rumen wall, and the products absorbed by the blood. Acetic acid is used for energy and oxidized by muscle tissue. Propionic acid is converted into blood glucose by the liver. Glucose is used by the animal for energy and as a precursor for milk lactose. Butyric acid is converted into β-hydroxybutyric acid by the rumen epithelium or liver and used for the formation of milk fat by the mammary gland. Carbon dioxide and methane are the major rumen gases given off during digestion. Gases are eliminated through the process of belching (eructation) and absorption into the blood. If gases accumulate in the rumen because of entrapment, bloat results. If the bloat becomes severe, death results. The rumen acts as a storage place for feeds until they are broken down. Coarser particles are broken down into finer particles in the rumen and by rumination so that digestive juices in the abomasum and intestines can act on them.

Microorganisms in the rumen play a significant role in carbohydrate digestion and in the synthesis of proteins. These microorganisms utilize nitrogen from nonprotein and preformed protein sources to synthesize amino acids for their own use. As the organisms move to the intestines, they are digested and the amino acids are utilized by the animal for its nutritional needs. In this way ruminant animals can obtain part of their protein needs from such nonprotein feed sources as urea and ammonia. Microbes in the rumen also are involved in the manufacture of vitamins, especially the B vitamins. Microorganisms synthesize vitamins for their own needs. When they are digested in the small intestine, the liberated B vitamins can be used by the ruminant animal. Vitamin K is also synthesized in the rumen.

Material not broken down by the microorganisms in the rumen passes to the third compartment of the stomach, the omasum. This compartment consists of many plies of tissue, comparable to leaves in a book. The omasum holds approximately 18 liters (20 qt) of liquid. A major function of the omasum is to absorb water from the digested material. From the omasum, the material goes to the abomasum, the true stomach. The abomasum secretes the gastric juice containing hydrochloric acid, which decreases the pH (acidity) of the abomasal contents to approximately 2.5. The abomasum also secretes two enzymes, pepsin and rennin. An enzyme is an organic catalyst that helps to bring about changes in other organic compounds without being changed or broken down itself. Because pepsin requires an acid medium for action, the low pH of the abomasal contents allows pepsin to break proteins down into simpler compounds, mainly peptones and peptides. Rennin clots milk and is important for the young calf to digest milk properly.

17.2 DIGESTION IN THE INTESTINES

Material that goes from the stomach to the intestines is known as chyme. The small intestine of the cow is approximately 40 m (130 ft) in length and holds approximately 37 liters (10 gal) of material. The walls of the intestine are lined with villi, small fingerlike projections that increase the absorptive surface of the intestinal walls. The upper part of the small intestine is concerned primarily with secretion of enzymes; the lower part is concerned primarily with absorption of nutrients from the ingested material.

Chyme undergoes important changes in the small intestine that prepare feed ingredients for absorption from the small intestine. Digestion of the chyme is aided by pancreatic juice, bile, and intestinal juice. Pancreatic juice is secreted by the pancreas, a slender gland lying just below the stomach. Pancreatic juice is alkaline and neutralizes the chyme coming from the abomasum. The acid chyme acts on the intestinal lining to cause it to release the hormone secretin. Secretin is absorbed by the blood and is carried to the pancreas, where it causes the secretion of pancreatic juice into the small intestine.

Pancreatic juice contains three proteolytic enzymes. One of these is trypsinogen, which when activated by the enzyme enterokinase from the intestinal wall, forms trypsin. The optimal pH for trypsin is 8; this indicates the necessity of an alkaline pancreatic juice to neutralize the acid chyme from the stomach. Chymotrypsinogen, a second enzyme, is converted into chymotrypsin by the action of trypsin. Chymotrypsin converts proteins and partially hydrolyzed proteins into amino acids and short chains of amino acids known as peptides. Chymotrypsin also coagulates milk. Carboxypeptidase, a third proteolytic enzyme, works on peptides and splits off amino acids.

Lipase is another enzyme secreted by the pancreas. The amount of fat digestion occurring in the ruminant stomach is relatively small. Most fat digestion occurs in the intestines. Lipase separates fats into fatty acids and glycerol, allowing them to pass into the walls of the small intestine. Released fatty acids are insoluble in the intestinal contents. However, when fatty acids are combined with bile salts, water-soluble diffusible compounds (micelles) are formed. Bile contains no digestive enzymes and functions primarily to emulsify fats, breaking them up into small globules and increasing the surface area so that lipase can split fats more easily.

Bile is a thin, greenish liquid secreted by the liver and stored in the gallbladder. Bile is released from the gallbladder in a manner similar to the process of pancreatic juice secretion. Bile is alkaline in reaction and helps to neutralize the acid chyme coming from the stomach. It also furnishes salts that combine with fatty acids to form micelles so that these acids can be absorbed more easily by the animal's body.

A third enzyme in pancreatic juice, amylase, converts starches into maltose. Saliva of the ruminant animal contains little or no ptyalin. In nonruminant animals, ptyalin changes starch into maltose.

As chyme moves down the intestinal tract, it comes into contact with a number of enzymes secreted by the intestinal wall. These are enterokinase, which activates trypsinogen; peptidases, which convert peptides into amino acids; maltase, which converts maltose into glucose; lactase, which splits lactose into glucose and galactose; lipase, which splits fats into fatty acids and glycerol; and amylase, which converts starches into maltose. Intestinal juice contains enzymes that break down nucleic acids and their associated proteins.

Amino acids, peptides and peptones, simple sugars, vitamins, and minerals are absorbed by the intestinal wall, move into the bloodstream, and are transported by the portal blood system to the liver. From the liver, they go to all parts of the body, where nutrients are used for the physiological functions of the animal.

Most fatty acids and glycerol are absorbed in a manner different from other compounds. Fatty acids and glycerol move to epithelial cells of the villi of the small intestine, where fatty acids are esterified with glycerol to form neutral fat. The neutral fat droplets (micelles) move from the cells to the tissue spaces and then to the lymph capillaries of the villi. Lymph then moves through the lymphatic ducts to the thoracic duct, where it enters the venous blood system anterior to the heart. The fine absorbed fat micelles in the lymph give it a milky appearance called chyle.

Material leaving the small intestine travels to the large intestine, where it remains a relatively long time. The large intestine contains no digestive juices; however, digesta reaching the large intestine contain undigested nutrients. The digestive processes that are of benefit to the animal have been nearly completed in the small intestine. Bacterial fermentation of the undigested food takes place in the cecum and large intestine, which causes the offensive odor of the feces. The large intestine and the cecum, a blind pouch located at the junction between the small and large intestines, absorb water and bacterial products from the digestive material. The large intestine also acts as a reservoir for waste materials that constitute the feces.

17.3 NUTRIENT UTILIZATION

Definition of "Nutrient"

A nutrient is defined as any food constituent or group of food constituents of the same composition that aids in the support of life. Nutrients are classified as proteins, carbohydrates, fats, minerals, and vitamins. A "digestible nutrient" is the portion of the nutrient that is digested and absorbed by the body.

Nutrients are classified as essential or nonessential. An essential nutrient cannot be synthesized by the body in sufficient amounts and must be supplied by the animal's diet. Nutrients are sometimes synthesized in the body, but the rate of synthesis is inadequate to meet daily requirements. Nonessential nutrients are compounds needed by the animal for physiological functions, but can be synthesized by body tissues or microflora of the digestive tract in sufficient quantities for the animal's physiological functions.

A nutrient is essential or nonessential depending on the species of animal as well as the animal's stage of growth. For example, microflora of the rumen in the mature ruminant synthesize sufficient quantities of B vitamins to supply the nutritional needs of the animal. In comparison, the young calf, in which the rumen microflora have not been established, must rely on its diet for B vitamins. Simple-stomached animals are completely dependent on the diet for essential amino acids and B vitamins.

Use of Nutrients by the Ruminant Animal

A number of chemical compounds are available to the animal body as a result of digestive processes in the rumen, reticulum, omasum, abomasum, and small intestine. Acetic, propionic, and β-hydroxybutyric acids are absorbed from the rumen. Glucose that escapes rumen fermentation, amino acids, vitamins, fatty acids, and glycerol are absorbed from the small intestine. Vitamin D can be obtained from feed as well as from irradiation of the animal by sunlight to convert provitamin D into vitamin D in the skin. Other nutrients are also essential for the proper physiological function of the animal's body; however, they are absorbed in smaller amounts than the major nutrients mentioned.

Nutrients absorbed from the rumen and small intestine, except for fats picked up by the lymphatic system, travel to the liver, where some of the compounds are changed or modified. From the liver, nutrients travel to the body for various physiological functions.

Energy, protein, vitamins, and minerals are required to maintain the normal body processes without accomplishing any productive work. The amount of these nutrients required to maintain the body is referred to as the maintenance requirement. Energy is required for physiological functions such as respiration, muscular contraction, heart beat, body heat, digesting feed, and body movement. Protein is required daily to replace cells that are broken down. Minerals are required by new cells and to replenish minerals from the bones. If the animal has insufficient energy, protein, or other nutrients for maintenance; body tissue is broken down, resulting in a loss of body weight.

The main nutritional needs of the lactating dairy cow are maintenance of her normal physiological functions, production of milk, development of a calf, and growth in young cows. The young calf requires nutrients for maintenance and growth; the bred yearling heifer must support maintenance, growth, and reproduction.

Determination of Nutrient Requirements

Observing deficiencies. Information is required on nutrient needs and nutrient contents of feeds to formulate rations that will meet nutrient requirements of animals for various physiological functions. A number of methods have been used to determine requirements of animals and the nutritional value of feeds. A common method is to observe naturally occurring deficiencies and add compounds to the ration to cure the deficiency. Mineral and vitamin needs in the animal's diet have been determined by this method. The amount of nutrients required in the diet must be determined in a feeding trial.

Feeding trials. A feeding trial is a controlled experiment in which the amount and composition of feed fed to an animal are measured and the physiological response of the animal is determined in terms of maintenance, milk production, growth, and reproduction. Two feeds can be compared, with one feed serving as a control and the second feed having an added or a deleted nutrient. Response to the two rations can be used as a measure of the quantity of that component needed to meet the animal's requirements.

Digestion trials. A digestion trial is used to determine the digestibility of chemical components of a feed. The simplest digestion trial involves feeding animals a certain feed or feeds for a fixed period so that the animal becomes accustomed to the feeds. As part of the actual digestion trial, the amount of feed fed daily and chemical composition of that feed are determined. Feces are weighed and analyzed daily. Digestibility of the feed component is determined by subtracting the amount of the component in the feces from that fed to the animal. Digestibility is expressed as the difference between the amount of the nutrient fed and that in the feces divided by the amount in the feed. An example of a digestion trial follows. A steer was fed 10 kg (22 lb) of a grass hay containing 12% protein per day. Thus 1.2 kg (2.6 lb) of protein were consumed per day. If the animal voided 25 kg (55 lb) of feces per day containing 1.4% protein, a total of 0.35 kg (0.77 lb) of protein was voided. The difference between the amount fed (1.2 kg) and the amount in the feces (0.35 kg) is the amount that is assumed to be absorbed by the body (0.85 kg). Dividing 0.85 kg by 1.2 kg and multiplying by 100 gives the digestion coefficient as a percentage, which, for this hay, is 70.8%. Similar determinations and calculations can be made for the other feed components.

Balance trials. Nutritional balance trials are similar to digestion trials except that nutrients voided in all excreta are measured. To determine nitrogen balance, nitrogen voided in urine as well as nitrogen voided in feces must be measured. For the lactating cow, nitrogen secreted in milk must also be determined. If the animal is in negative nitrogen balance (the amount voided in feces, urine, and milk is greater than that ingested), an additional amount of protein is fed to balance intake with requirements. Requirements for nitrogen can be determined in this way. A problem with mineral balance trials is changes in body composition occur. These changes (especially calcium and phosphorus deposition in the skeleton) cannot be measured by external measurements.

Rumen digestion techniques. Researchers measure the effect of microbial degradation and digestion on feedstuffs. Through the use of feed and microbial markers, these changes can be measured. Rumen volume and fluid turnover rates can be measured with isotopes, inert earth elements, and dilution techniques. Cannules (externalized openings) inserted in the rumen and the small intestine allow sampling of digesta in the rumen and after it leaves the abomasum. A dacron bag (5 × 15 cm) with small pore sizes (20 to 40 μm) containing feed is placed in the rumen through a fistula. Rumen fluid and bacteria can pass through the bag and attack the feed. The bag can be removed at various times (2, 4, 8, 12, and 24 hours) to recover feeds not broken down. Chemical analysis conducted on the residue in the bag can determine protein degradability, amino acid composition of undegraded protein, fiber digestibility, and organic matter solubility.

Slaughter trials. In balance and feeding trials concerned with changes in growth and body composition, slaughter trials accompany feeding trials. Changes in skeletal, muscle, and fat composition that result from changes in diet can be measured. These trials are applied to growing and fattening animals.

Fasting catabolism. Maintenance requirements are determined by fasting catabolism studies, which are performed in the absence of feed. Nutrients required to support respiration, circulation, maintenance of muscular tonus, production of internal secretions, and other activities required of the fasting animal come from the breakdown of body tissue. Fasting catabolism is measured by the amount and composition of waste products eliminated through various paths of excretion. Energy used by the fasting animal for these processes is dissipated as heat and can be measured in a respiration calorimeter or by indirect calorimetry. Heat liberated during fasting is called the basal metabolic rate of the animal and is related to body surface area. Actual maintenance requirements are higher than fasting catabolism values because animals require additional energy, protein, and other nutrients for body movement.

Feeding standards. Information on nutritive requirements gathered from various sources is the basis for formulation of feeding standards. Feeding standards are tables showing the amount of each nutrient required by different species to perform physiological functions of maintenance, growth, fattening, reproduction, and lactation. These requirements, along with feed composition and digestibility, provide the essential information needed to formulate rations to meet the animal's needs.

17.4 VOLUNTARY FEED INTAKE

The feed intake of a ruminant animal is controlled by physiological mechanisms that attempt to maintain a constant internal environment. The mechanisms governing feed intake are interrelated with mechanisms that regulate blood glucose, plasma amino acids, body temperature, and body composition. Energy balance is regulated by feed composition and metabolic changes that cause the animal to consume feed and to stop eating.

The hypothalamus of the brain contains centers that initiate consumption of feed and stop animals from eating. Increased neural activity in the feeding center located in the lateral hypothalamus initiates consumption of feed. Neural activity in the ventromedial hypothalamus, the satiety center, stops feed consumption.

Several metabolic and physical events control the feed intake of ruminants. Gut fill (distension) and concentration of volatile fatty acids resulting from rumen fermentation are two factors. Volatile fatty acid concentration of rumen fluid increases during and after a meal. Increased concentration of acetic acid is detected by the rumen wall and increased production of propionic acid is detected by the liver. This information is relayed to the hypothalamus, which integrates it with other metabolic information to determine when feed consumption will be started and stopped. Information on body stores that describe energy balance homeostasis is also integrated by the hypothalamus to control feed intake. Distension of the rumen and reticulum also limits feed intake and is related to the proportion and quality of forage in the diet. Blood glucose and heat generated during feeding have little effect on the regulation of feed intake in ruminants.

The two types of controls that affect intake are physical limitations (rumen distension or fill) and metabolic results associated with absorption of nutrients from a meal (Figure 17.3). In distension, feed intake is related to rumen fill, especially with forage diets. Feed intake increases with increased digestibility of feeds. When the nutritive value of the feed is further increased by adding concentrate, dry matter intake decreases, but energy intake remains constant. In the chemostatic-thermostatic range, rumen fill no longer limits feed intake, but some chemostatic or thermostatic mechanisms control feed intake.

Dry matter intake becomes a limitation in high-producing dairy cows. Suggested dry matter intakes are listed in Table 21.3. These values should be reduced in early lactation (first 4 weeks postpartum) because intake builds gradually. Several factors that can increase dry matter intake include increasing feeding frequency, feeding smaller amounts of concentrate per meal, changing the feed se-

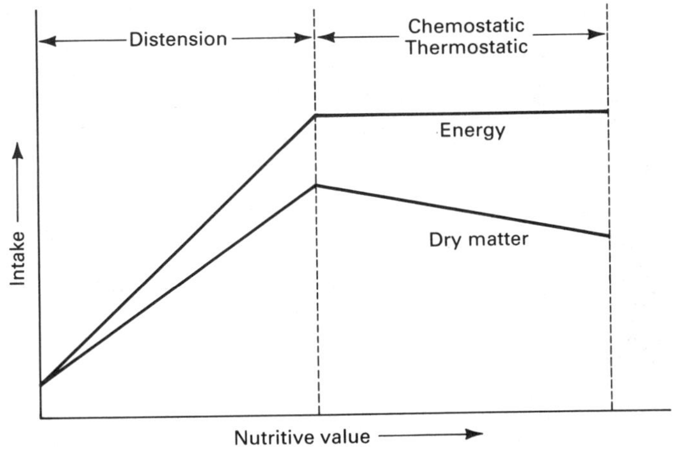

Figure 17.3 Relationship between nutritive value of rations and feed dry matter and energy intake. From B. R. Baumgardt, 1970, *Proc. Cornell Nutr. Conf.*

quence with forage fed first, blending concentrate and forage (total mixed rations), maintaining feed availability, increasing feed palatability, and keeping rations below 50% moisture. New York researchers (Chase, 1979) suggest a decrease of 0.1 kg of dry matter per 45 kg of body weight as ration moisture increases 10 percentage points over 45 to 50 dry matter. Depression in total dry matter intake due to moisture is related to feed pH, smaller particle size, less saliva production, and fermentation products associated with wet feeds.

REFERENCES AND FURTHER READINGS

BAUMGARDT, B. R. 1970. Voluntary feed intake by ruminants: model and practical applications. *Proc. Cornell Nutr. Conf.*, p. 85.

CHASE, L. E. 1979. Effect of high moisture feeds on feed intake and milk production in dairy cattle. *Proc. Cornell Nutr. Conf.*, p. 52.

NATIONAL RESEARCH COUNCIL COMMITTEE. 1988. *Nutrient Requirements of Dairy Cattle*, 6th ed. National Academy of Sciences, Washington, DC.

STALLCUP, O. T. 1971. Why we feed dairy cattle as we do. *Hoard's Dairyman*, 116:65.

18

Nutritive Requirements of Dairy Cattle

Many nutrients are required for proper physiological functions. The two most important nutrients are energy and protein. Minerals and vitamins are required in small amounts. Fiber is required for optimal rumen function and milk-fat production. Water is also required, but is not considered a nutrient.

18.1 ENERGY

The most important nutrient in the formulation of rations for dairy animals is energy. Many measurements of energy are used. Each measure should have requirements for various physiological processes and energy values of feeds needed to meet the requirements.

Total Digestible Nutrients

A common measure used in the United States is the total digestible nutrients (TDN) standard, which is expressed in both percentages and weight. F. B. Morrison (1960) summarized many feeding trials and made the data available in his book, *Feeds and Feeding*. The TDN measure has the advantage of being one of the easier energy values to determine.

The TDN content of a feed is defined as the sum of digestible fat multiplied by 2.25, digestible protein, digestible fiber, and digestible nitrogen-free extract (NFE). Fat has 2.25 times more energy per unit of weight than that of protein or carbohydrate. The TDN value of a feed is usually expressed as a percentage. Information on the TDN content of feeds comes from digestion trials and chemical analysis of the feed and feces. Forages range between 45 and 70% TDN, while concentrates range between 65 and 90% TDN on a dry matter basis.

The total digestible nutrient measure of energy does not give the amount of energy available for physiological processes because it does not consider energy losses in combustible gases or heat increment from the feed. The TDN system does not accurately evaluate the energy in feeds since a kilogram of TDN in forage provides less energy for physiological processes than does a kilogram of TDN from concentrate. Therefore, other measures of energy have been used, such as digestible energy, metabolizable energy, and net energy. The relationships between these measures are shown in Figure 18.1.

Digestible Energy

Gross energy of a feed is the total amount of heat liberated when feed is completely combusted. This value is usually obtained by combustion of the feed in a bomb calorimeter. The digestible energy content of a feed is obtained from a digestion trial similar to that conducted to obtain TDN values of a feed (see Section 17.3). Digestible energy is determined by subtracting energy in feces from gross energy intake. Because this measure is similar to the TDN value of a feed, it has comparable advantages and disadvantages.

Metabolizable Energy

Metabolizable energy of a feed is determined by subtracting energy lost in the urine and combustible gases (primarily methane) from digestible energy. Urine

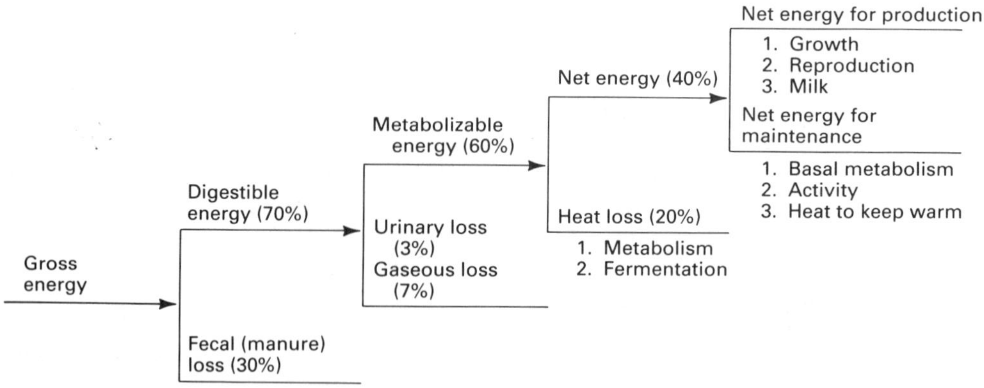

Figure 18.1 Energy losses and uses of feed consumed by a lactation cow. From J. G. Linn, D. E. Otterby, R. D. Appleman, and M. F. Hutjens, 1980, *Ill. Ext. Circ.* M-1183.

losses are greater when animals are fed forages than when they are fed concentrates. The amount of combustible gases eliminated by the animal is measured by a respiration calorimeter or chamber. A respiration calorimeter is used to measure the amount of gaseous material inhaled and exhaled by the animal.

Net Energy

Net energy is calculated by subtracting the heat increment from metabolizable energy. Heat increment is the amount of heat dissipated to make the nutrients in the feed available to the animal. Net energy represents the real value of the feed for productive purposes, such as growth, reproduction, maintenance, and milk and fat production.

The net energy value of feeds is expressed in megacalories (Mcal) per kilogram of dry matter. A megacalorie is the amount of heat required to raise the temperature of 1000 kilograms of water 1°Celsius (°C). Because obtaining net energy values of various feeds is expensive, Morrison estimated the net energy value of feeds from studies of the relative value of feeds for productive purposes. He established an estimated net energy standard (ENE) as well as a TDN standard for feeding livestock. The net energy standard provides an improved measure of the value of a feed for productive purposes and also evaluates more accurately feeding values of forages and concentrates. The energy content of feed in the ENE system is *not* equal to the net energy system; the ENE values are lower and expressed as therms of energy.

Factors That Determine Energy Requirements

The energy requirements of a lactating cow depend on her maintenance requirement, amount of milk produced, energy content of milk as indicated by its fat content, reproductive requirements, and growth rate if she is not mature. Recommended nutrient allowances are based on minimum requirements for a physiological function as determined from feeding and digestion trials plus an added amount to protect against individual animal differences and variation in feeds. A minimum requirement is the amount of a nutrient needed in the ration that is adequate in all other nutrients to support a body function for a species of animal at an optimal rate.

The TDN and net energy values of various forages, concentrates, protein supplements, and by-product feeds are given in Tables 19.2 and 19.3. The net energy value of a feed depends on its use for maintenance, growth, or milk production. Net energy has been expressed in two ways. The procedure of Lofgreen and Garrett (1968) expresses the requirements for the physiological functions, such as net energy for maintenance or net energy for weight gain. The system of Moe and Flatt (1969) expresses net energy requirements for the various functions to that required for milk production. One energy value is expressed for each feed given the lactating cow whether it is used for maintenance, pregnancy, or milk production (Tables 19.2 and 19.3). This value can be used to calculate the amount of feed to meet requirements for maintenance, milk production, and reproduction. These tables also include net energy values for gain for growing animals.

Maintenance requirements. Maintenance values in feeding standards have been found from feeding trials which determine the amount of energy and protein required to maintain a constant weight. When body weight change is the sole criterion for measuring the maintenance requirement, the requirement is underestimated if a shift in body composition occurs. Animals can lose body tissue and replenish the weight by addition of water or body fill.

Other maintenance requirements are determined by fasting catabolism studies. The maintenance requirement depends on the energy required by the fasting resting animal and the animal's physical activity. The maintenance requirement of a grazing animal is higher than that of a barn-fed animal. Table 18.1 lists maintenance requirements that provide sufficient energy for cows housed in individual stalls. If an animal is on pasture, allowances for maintenance should be increased by 25 to 100%.

Production requirements. Energy requirements for milk yield with different fat percentages are given in Table 18.1. The requirements in Table 18.1 are National Research Council (NRC) standards.

Reproduction requirements. Energy and protein requirements during pregnancy increase as gestation advances. The amount of nutrients depends on size of fetus, volume of fluids, size of membranes, and size of uterus. Most feeding standards do not recommend additional nutrients until the last 2 to 3 months of pregnancy. Recommended allowances for maintenance and during the last 2 months of pregnancy are shown in Table 18.1. Thin cows should receive energy in addition to maintenance and reproduction requirements to replace their body stores (see Table 18.1).

Growth requirements. The amount of nutrients required daily for the maintenance of growing heifers is separated from the amount required for growth because of differences in utilization of energy for the two processes (Table 18.2). First- and second-lactation cows require additional energy for growth in addition to their maintenance and reproduction requirements. The NRC recommendations are to add 20% to the maintenance allowance for first-lactation cows and 10% for second-lactation cows.

Fiber and Energy Relationships

Fiber in itself is not considered a nutrient, but is necessary in a ration. The chemical composition of fiber varies because it is made up of several components. Fiber controls feed intake, stimulates rumination, and maintains optimal rumen pH and digestion. If the fiber is lignified, it is less digestible, slows passage through the digestive tract, and decreases dry matter intake. An optimum amount of fiber is needed to avoid energy shortages or excesses. The effectiveness of fiber for stimulating rumination in the diet is destroyed if the particle size is reduced by grinding, pelleting, or excessive chopping of forages.

Three methods are available to measure fiber content of feeds. Crude fiber is the oldest method and is the residue of a feed after successive boiling with

TABLE 18.1 DAILY NUTRIENT REQUIREMENTS OF LACTATING DAIRY CATTLE

Body weight (kg)	Crude protein (kg)	Energy NEL (Mcal)	Energy TDN (kg)	Calcium (g)	Phosphorus (g)
		Maintenance requirements[a]			
400	0.39	7.16	3.13	16	11
500	0.49	8.46	3.70	20	14
600	0.59	9.70	4.24	24	17
700	0.68	10.89	4.76	28	20
800	0.77	12.03	5.26	32	23
		Requirements for maintenance and pregnancy (last 2 months of gestation)			
400	0.69	9.30	4.08	26	16
500	0.80	11.00	4.82	33	20
600	0.94	12.61	5.53	39	24
700	1.07	14.16	6.20	46	28
800	1.21	15.64	6.85	52	32
Requirements for milk production (nutrients per kilogram of milk for different fat percentages)					
Milk fat (%)					
3.0	0.072	0.64	0.280	2.73	1.68
3.5	0.079	0.69	0.301	2.97	1.83
4.0	0.084	0.74	0.322	3.21	1.98
4.5	0.091	0.78	0.343	3.45	2.13
5.0	0.099	0.83	0.364	3.69	2.28
5.5	0.104	0.88	0.385	3.93	2.43
		Body weight change during lactation (nutrients per kilogram of weight change)			
Weight loss	−0.32	−4.92	−2.17	—	—
Weight gain	+0.50	+5.12	+2.26	—	—

[a] To allow for growth of young cows, increase the maintenance allowances for all nutrients by 20% during the first lactation and by 10% during the second lactation.
Source: Adapted from National Research Council Committee, 1988, *Nutrient Requirements of Dairy Cattle*, 6th ed., National Academy of Sciences, Washington, DC.

dilute acid and alkali treatments. Crude fiber is not an accurate measure of total fiber or cell wall content because crude fiber fractions contain cellulose and variable amounts of hemicellulose, lignin, silica, and pectin. Many forage testing labs no longer analyze for crude fiber; but it is required on feed tags of purchased feeds. In NRC nomenclature, feeds that contain 18% or more crude fiber are classified as forages or roughages.

Acid detergent fiber (ADF) consists of cellulose, lignin, lignified nitrogen compounds (heat-damaged protein), and insoluble ash. Like crude fiber, ADF does not represent the total fiber content in feed. ADF can be used to predict the digestibility and energy content of feedstuffs.

Neutral detergent fiber (NDF), which gives the best estimate of total fiber,

TABLE 18.2 DAILY NUTRIENT REQUIREMENTS OF GROWING HEIFERS (S = SMALL BREED, L = LARGE BREED)

Body weight (kg)	Daily gain (kg)	Feed DM intake (kg)	Crude protein (kg)	NE_m (Mcal)	NE_g (Mcal)	TDN (kg)	Calcium (g)	Phosphorus (g)
100 (S)	0.4	2.4	0.39	2.72	0.91	1.67	15	8
100 (S)	0.6	2.9	0.46	2.72	1.40	1.98	17	9
100 (L)	0.6	2.6	0.42	2.72	1.22	1.84	17	9
100 (L)	0.8	3.0	0.48	2.72	1.66	2.11	18	10
200 (S)	0.4	4.2	0.58	4.57	1.26	2.76	19	13
200 (S)	0.6	5.0	0.72	4.57	1.95	3.23	20	14
200 (L)	0.6	4.4	0.70	4.57	1.65	2.95	20	14
200 (L)	0.8	5.0	0.80	4.57	2.25	3.34	22	15
300 (S)	0.4	6.3	0.76	6.20	1.56	3.87	22	16
300 (S)	0.6	7.4	0.89	6.20	2.43	4.52	23	17
300 (L)	0.6	6.3	0.75	6.20	2.02	4.01	23	17
300 (L)	0.8	7.1	0.88	6.20	2.77	4.52	25	19
400 (S)	0.4	9.0	1.08	7.69	1.84	5.12	24	18
400 (S)	0.6	10.5	1.26	7.69	2.87	6.00	25	19
400 (L)	0.6	8.4	1.01	7.69	2.37	5.12	25	19
400 (L)	0.8	9.5	1.14	7.69	3.24	5.77	26	21
500 (L)	0.6	10.9	1.31	9.09	2.69	6.34	28	20
500 (L)	0.8	12.3	1.48	9.09	3.68	7.16	29	21

Source: Adapted from National Research Council Committee, 1988. *Nutrient Requirements of Dairy Cattle*, 6th ed., National Academy of Sciences, Washington, DC.

consists of ADF plus hemicellulose. Commonly called cell walls, NDF is highly correlated with feed intake, rumination, and total chewing times, and provides the best measurement of fiber for formulating dairy rations.

The amount of fiber varies with plant species, stage of growth, and growing conditions. In general, the higher the fiber content of a feed, the lower the energy value. Recommended fiber levels are listed in Table 18.3.

18.2 PROTEIN

Protein is required in the animal's diet to supply amino acids for its physiological functions. The biological value of a protein is its ability to supply amino acids in the relative amounts needed by the animal for a specific body function.

Role of Microbial Digestion

The ruminant animal does not depend solely on its diet for protein because microorganisms in the rumen convert nitrogen from feed and nonprotein nitrogen (NPN) sources into amino acids. When microorganisms are digested in the small intestine, the released amino acids can be utilized by the animal. The protein requirement of the mature ruminant depends on the amount of nitrogen supplied,

TABLE 18.3 RECOMMENDED NUTRIENT CONTENT OF DIETS FOR DAIRY CATTLE

	Lactating cow diets: (kg 4% FCM/day)			Dry pregnant cows	Calf starter	Growing heifers and bulls: age (months)			Maximum tolerance levels
	10	30	50			3–6	6–12	>12	
Energy									
NEL (Mcal/kg)	1.42	1.62	1.72	1.25					
NEM (Mcal/kg)					1.90	1.63	1.47	1.31	
NEG (Mcal/kg)					1.20	1.03	0.88	0.74	
TDN (% of DM)	63	71	75	56	80	70	65	60	
Protein equivalent									
Crude protein (%)	12.8	16.2	17.7	12.0	18.0	16.0	14.0	12.0	
Undegraded (%)	4.4	5.7	6.2			11.1	6.9	4.2	
Degraded (%)	7.8	9.6	10.4			4.2	6.0	6.9	
Fiber content (minimum)									
Crude fiber (%)	17	17	15	22		13	15	15	
Acid detergent fiber (%)	21	21	19	27		16	19	19	
Neutral detergent fiber (%)	28	28	25	35		23	25	25	
Ether extract (minimum)	3	3	3	3	3	3	3	3	
Minerals									
Calcium (%)	0.55	0.65	0.75	0.37	0.60	0.50	0.42	0.32	2.00
Phosphorus (%)	0.35	0.38	0.44	0.26	0.40	0.32	0.30	0.25	1.00
Magnesium (%)	0.20	0.20	0.25	0.16	0.10	0.16	0.16	0.16	0.50
Potassium (%)	0.90	0.90	1.00	0.65	0.65	0.65	0.65	0.65	3.00
Sodium (%)	0.18	0.18	0.18	0.10	0.10	0.10	0.10	0.10	2.00
Chloride (%)	0.25	0.25	0.25	0.20	0.20	0.20	0.20	0.20	3.00
Sulfur (%)	0.20	0.20	0.20	0.16	0.20	0.16	0.16	0.16	0.40
Iron (ppm)	50	50	50	50	50	50	50	50	1000
Cobalt (ppm)	0.10	0.10	0.10	0.10	0.10	0.10	0.10	0.10	10.00
Copper (ppm)	10	10	10	10	10	10	10	10	100
Manganese (ppm)	40	40	40	40	40	40	40	40	1000
Zinc (ppm)	40	40	40	40	40	40	40	40	500
Iodine (ppm)	0.60	0.60	0.60	0.25	0.25	0.25	0.25	0.25	50.00
Selenium (ppm)	0.30	0.30	0.30	0.30	0.30	0.30	0.30	0.30	2.00
Vitamins									
A (IU/kg)	3200	3200	3200	3200	2200	2200	2200	2200	66000
D (IU/kg)	1000	1000	1000	1200	300	300	300	300	5000
E (IU/kg)	15	15	15	15	25	25	25	25	2000

Source: Adapted from National Research Council Committee, 1988, *Nutrient Requirements of Dairy Cattle*, 6th ed., National Academy of Sciences, Washington, DC.

which can be utilized as protein (Figure 18.2). Approximately 60% of the nitrogen in the dairy cow diet is broken down to ammonia by microbial digestion. The rumen microbes must convert the ammonia to microbial protein for the animal to benefit. Fermentable energy must be available for the microorganisms to grow and synthesize the necessary amino acids. If rumen ammonia is excessively high, ammonia is absorbed into the blood and recycled or excreted as urea. All feed protein sources are not degraded in the rumen to the same extent (see Table 19.3). The optimal ration will meet the nitrogen requirement for maximum rumen microbial protein synthesis, avoid loss of excessive ammonia from the rumen, and provide undegraded (also called bypass or escape) protein (typically 40% of dietary protein). Protein synthesis by rumen microbes depends on feed intake, organic matter digestibility, feed type, protein level, and feeding method. About 1.6 kg (3.5 lb) of microbial protein can be synthesized per day in a high-producing cow. The remainder of the protein must be derived from undegraded protein sources. Young fast-growing heifers and high-producing cows may require additional undegraded protein sources beyond their normal diet to meet their total protein requirements. Alfalfa hay, brewers' grain, distillers' grain, corn gluten meal, or heat-treated soybeans are examples of feeds that could improve performance where microbial protein synthesis and normal undegraded protein sources may be limiting.

Nonprotein Nitrogen Sources

NPN (nonprotein nitrogen) is any compound that contains nitrogen not in the form of amino acids. Most NPN compounds consist of carbon, nitrogen, oxygen, and hydrogen (sometimes phosphorus). Common NPN compounds fed to dairy cattle are listed.

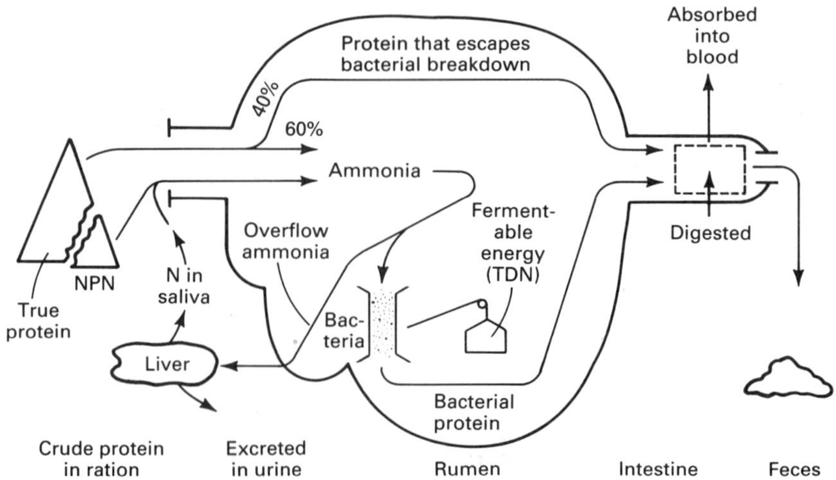

Figure 18.2 Schematic flow and breakdown of nitrogen through the rumen. From L. D. Satter and R. E. Roffler, 1975, *J. Dairy Sci.*, 58:1219.

- Urea is a manufactured product that is white and bitter tasting. Feed-grade urea contains 45% nitrogen and has a protein equivalence of 281% (45% nitrogen times 6.25).
- Biuret is formed by heating urea. Although it is an acceptable nitrogen source, biuret cannot be fed to milking cows because it appears in milk.
- Ammoniated products are feeds that have been treated with ammonia. Ammoniated wheat straw, citrus pulp, and corn silage are examples. Ammoniated feeds are improved nutritionally since the ammonia may increase fiber digestibility, nitrogen content, and feed intake.
- Monoammonium phosphate is an excellent phosphorus mineral source (11% nitrogen and 26% phosphorus).
- Liquid supplements are commercially manufactured feeds consisting of molasses, urea or ammonia, minerals, and vitamins. These products vary from 20 to 60% protein (on a nitrogen equivalent basis). Liquid supplements are palatable, which leads to varied intake when offered free-choice from a lick wheel. Force feeding (controlled intake) of liquid supplements mixed with forage or concentrate is the recommended feeding approach.

NPN is used in ruminant rations since it is usually two to four times cheaper than feed protein sources per unit of nitrogen. To be economical and beneficial, rumen microbes must convert NPN to a microbial source of protein. Guidelines for feeding NPN include (1) limiting urea to 1% of the grain mixture or 0.2 kg (0.45 lb) of urea per cow per day, (2) feeding NPN-containing feeds frequently, (3) using NPN with a high-energy feed (such as corn silage or shelled corn), and (4) adding NPN to diets that are less than 13% crude protein to avoid high rumen ammonia. Commercial protein supplements that contain NPN must list the crude protein equivalence on the feed tag (Figure 18.3). In the example feed tag, supplement XYZ contains 36% crude protein, of which 11.2% is supplied by NPN. The level of urea can be calculated by dividing 11.2 (percent protein from NPN) by 281, which is 4%. If a dairy cow consumes 2 kg (4.4 lb) of supplement XYZ, she would consume 0.08 kg (0.18 lb) of urea per day.

Factors That Determine Protein Requirements

In calculating nutrient requirements for protein, both body losses of protein and fecal losses during digestion of the feed must be considered. Crude protein content of feeds is determined by multiplying nitrogen content by 6.25 because protein

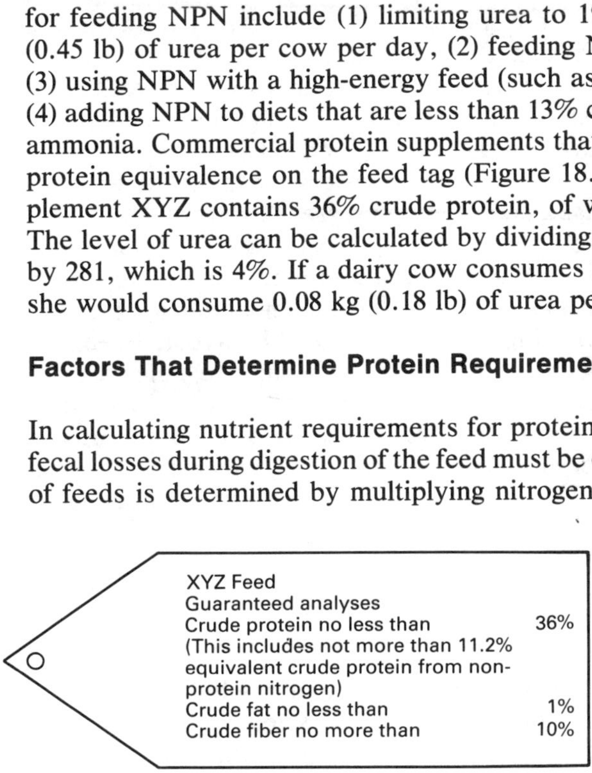

Figure 18.3 Sample feed tag.

contains approximately 16% nitrogen. The apparent digestible protein of feed is that absorbed by the animal's body and is obtained by subtracting fecal nitrogen loss times 6.25 from the amount of crude protein consumed.

Fecal nitrogen is composed mainly of metabolic fecal losses, which are proteins broken down by the body or bacteria and excreted in the feces. For example, walls lining the intestinal tract are being constantly replaced and protein from the discarded cells is lost in the feces. The amount of nitrogen contributed by the body to the feces is related to the amount of feed consumed and body size. In addition to metabolic fecal loss, endogenous urinary loss also occurs.

Protein requirements of lactating cows given in Table 18.1 are expressed in terms of crude protein. Digestible protein has been used in ration formulation, but has been deemphasized because nondietary fecal nitrogen from endogenous and microbial origin underestimates true digestibility, and chemical analysis for digestible protein is not possible. NPN is also difficult to evaluate as an effective digestible protein source. A measure of metabolizable protein or absorbable amino acids from the intestine seems to be most desirable. Until these systems are available, crude protein is the most desirable system for ration formulation.

Maintenance requirements for protein are calculated from feeding trials and nitrogen balance data. Feeding trials are inaccurate in estimating protein requirements of milking cows because of difficulty in partitioning protein requirements for body maintenance and milk production. In addition, some body protein can be utilized for the production of milk (Table 18.1).

Protein requirements for maintenance are determined by the amount of nitrogen in urine, which indicates protein that was broken down to support vital function. Protein requirements also depend on growth of hooves, hair, horns, and skin. Protein requirements are also directly related to milk yield and protein content of the milk, as related to fat content (Table 18.1).

Protein is one of the most expensive ingredients in the dairy ration. Adequate protein is required because a deficiency depresses the rate of growth and decreases milk production. Gestation can be interrupted by a severe protein deficiency. The values given in Table 18.1 are guides to the amount required for the cow's physiological functions and serve as general recommendations for the formulation of dairy cow rations. Overfeeding of protein should be avoided because protein is expensive and additional energy is required to deaminate amino acids and excrete nitrogen in the form of urea. The carbon skeleton of the amino acids can be utilized for energy. Excessive nitrogen intake can increase blood and uterine levels of ammonia, urea, and other nitrogenous compounds, which can affect reproductive performance through delayed conception, embryonic death, or increased services per conception.

18.3 FATS AND OILS

The essential fatty acids (linoleic, linolenic, and arachidonic acids) and choline are essential in the diet of monogastric animals, but these substances have not been proven to be essential in the diet of ruminants. Microorganisms in the rumen of the mature animal can synthesize these fatty acids from carbohydrates and other fats. From a practical standpoint, providing essential fatty acids to dairy animals requires no special attention because normal feeds contain sufficient amounts.

Research has focused on the benefits of fats in the diets of ruminant animals, especially since the solvent process of oil extraction has reduced the fat content of such feeds as soybean meal in comparison with that obtained when the oil is removed by expeller or hydrolic processes. Soybean meal obtained by solvent extraction has less than 1% oil; the expeller process leaves between 4 and 5% oil.

Ohio research (Palmquist, 1985) has demonstrated that adding fat or oil (0.45 to 0.7 kg per cow per day) can increase milk production 5 to 7%. Total fat and oil content should not exceed 6 to 7% of the total ration dry matter. Because fat and oil are 2.25 times higher than carbohydrates in energy content, energy can be increased without increasing starch and decreasing fiber. Sources of added fat and oil include calcium salts of free fatty acids (e.g., Megalac), oil seeds (soybean, cottonseed, sunflower, peanut, and rape), animal fats (tallow and lard), and spray-dried vegetable or animal products (vary from 20 to 80% fat or oil). Animal fat is superior to vegetable oil since it contains more saturated fatty acids, which do not depress fiber digestibility as much as does unsaturated oil. Higher calcium, magnesium, fiber, and nondegraded protein levels are recommended when fat and oil are added. The nutrient composition of oilseeds and fat sources is listed in Table 19.3. Two to three kilograms (4.4 to 6.6 lb) of soybeans or whole cottonseed or 1 to 1.5 kg (2.2 to 3.3 lb) of sunflower seeds or rapeseed are recommended. Only high-producing dairy cows in negative energy balance will respond economically to added fat or oil with more milk, a higher fat test, improved reproduction, or less metabolic stress.

18.4 MINERALS

Mineral elements are needed for structural needs, constituents of proteins and lipids that make muscle and tissues, enzyme systems, osmotic relationships, and muscle and nerve transmission. Large amounts of minerals are secreted in milk. Macrominerals, needed in relatively large amounts (grams per day), include calcium, phosphorus, sodium, chlorine, magnesium, sulfur, and potassium. Microminerals, required in smaller amounts (milligrams per day), include iodine, iron, copper, cobalt, manganese, zinc, and selenium. A third group of minor elements considered essential for normal function comprises chromium, vanadium, tin, nickel, bromine, fluorine, and molybdenum, but these are not required as dietary supplements. Excessive concentrations of most minerals can be detrimental, toxic, or increase the requirement for other minerals. Lead, cadmium, and mercury are nonessential minerals that can be toxic. Recommended levels of minerals are listed in Table 18.3.

A mineral in a diet has limited value unless it is available to be digested, absorbed, and transported to the site of function. Biological availability is the amount or percent of a mineral used to support some physiological process and is influenced by age, sex, nutritional status, mineral form and level, hormonal and homeostatic control, environment, and diet relationships. Inorganic minerals occur in the form of a salt, such as calcium carbonate or copper chloride. A chelated mineral is formed in a complex or ring-type organic compound, such as zinc methionine. If the chelate is water soluble, it is referred to as a sequestering agent. Common chelating agents are amino acids and polysaccharides. The availability of the chelated mineral depends on the stability of the compound. Although the bioavailability of chelated metals is usually higher, response has been variable and the cost of these mineral forms are higher.

Macrominerals

Calcium. Calcium is a white, alkaline metal widely distributed in nature. Approximately 99% of calcium in the animal body is found in the bones and teeth. The calcium content of blood is homeostatically controlled within a narrow range by the parathyroid hormone, calcitonin, and metabolically active vitamin D. When intake is inadequate, cattle are able to draw calcium from the skeleton for a short period of time. Dietary calcium absorption decreases as the dairy animal increases in age. In older animals, true absorption ranges from 22 to 55%, with an average of 40%.

Bone formation is the most important function of calcium. The other 1% of body calcium is involved in blood coagulation, muscle contraction, myocardial function, normal neuromuscular irritability, and enzyme activation. Sources of calcium include legumes, limestone, and dicalcium phosphate.

A calcium-deficient diet in young calves prevents normal bone growth and retards growth. Bones deficient in calcium fracture spontaneously. In mature cows calcium deficiency depletes bone calcium reserves, leading to easily fractured bones and reduced milk yield. No reduction in calcium in milk occurs with calcium-deficient diets. Milk fever is caused by a disturbance in calcium metabolism (see Section 22.3). Cattle can tolerate large excesses of calcium because absorption decreases. Excessive calcium has an antagonistic effect on other minerals, including manganese and zinc.

Magnesium. Approximately 60% of total body magnesium is located in bone, which is slowly mobilized. Dietary magnesium sources vary from 70% availability (in young milk-fed calves) to 17% (in pastured dairy cows). Concentrate sources of magnesium are more available than are forage sources. Magnesium is needed for all phosphate transfer reactions (ATP), activation of enzyme reactions, and fat and protein metabolism.

Signs of deficiency include loss of appetite, reduced gain, hyperexcitability, soft-tissue calcification, and grass tetany. Grass tetany (also called grass staggers, winter tetany, and wheat pasture poisoning) can be a major problem in cattle fed lush, rapidly growing forage highly fertilized with nitrogen or potassium. The affected animals become susceptible to convulsions (tetany) and neuromuscular abnormalities because magnesium is tied up. Older cows are more susceptible to grass tetany due to decreased bone mobilization of magnesium. Magnesium toxicity is not known to be a practical problem.

Phosphorus. Phosphorus is widely distributed in nature, but never in the free state. In addition to its major role with calcium in the formation of bones and teeth, phosphorus is located in every cell. Phosphorus is involved in muscle, energy, protein, and fat metabolism; is a component of nucleic acids; and is involved in coenzyme systems. Phosphorus availability declines from 90% in calves to 50% in older animals. Phytate (plant) phosphorus is utilized by ruminants as readily as are inorganic sources. Except for dry cow rations, a high ratio of calcium to phosphorus can be fed as long as phosphorus requirements are met. Protein supplements, by-product feeds, dicalcium phosphate, and monoammonium and monosodium phosphate are good phosphorus sources.

Symptoms of phosphorus deficiency include fragile bones, a decline in feed intake, depraved appetite (chewing on wood and bones), stiff joints, anestrus, low conception rates, and reduced feed efficiency. High levels of phosphorus can cause bone resorption and calcium imbalance.

Potassium. Potassium is highly variable and widely distributed in feed sources. Forages are high in potassium if properly fertilized. Weather damage can reduce potassium in forage. Potassium chloride is a common mineral source. The main functions are osmotic equilibrium, acid-base balance, enzyme reactions, and maintenance of heart and kidney tissue.

Deficiency symptoms include a decrease in feed intake, reduced weight gains, lower milk production, pica (licking and chewing on objects), and loss of hair gloss. Potassium toxicity is not likely under normal situations. Young, lush-growing forages may be high in potassium, which can interfere with magnesium metabolism, leading to grass tetany. Under heat stress, higher potassium amounts (1 to 1.5% in the total ration dry matter) can increase feed intake and milk production.

Sodium and chlorine (salt). Sodium and chlorine have been found to be essential minerals for dairy cattle. Although each mineral has a specific requirement, most research has involved salt, the most common source of both minerals. Buffers and other minerals are sodium sources. Sodium is the main ion of extracellular fluid in maintaining osmotic pressure, body fluid balance, and tissue hydration. Heart action and nerve impulse depend on sodium. Chlorine maintains fluid balance and is used to produce hydrochloric acid in the stomach.

Signs of sodium and chlorine deficiency include an intense craving for salt and pica, loss of appetite, unthrifty appearance, weight loss, and decreases in milk production. Excessive salt reduces milk yield, increases water intake, causes udder edema in dry cows, and leads to weight loss. Cattle can tolerate a higher level of salt in the ration if water is available. Cows will consume more loose salt than block salt.

Sulfur. Sulfur is an essential component of proteins, vitamins, and enzymes. Rumen microbes can incorporate inorganic sulfur into sulfur-containing compounds. Sulfate minerals are effective sulfur sources; elemental sulfur is less effective. A nitrogen-sulfur ratio of 10 to 12:1 is needed for microbial protein synthesis. Inadequate sulfur reduces feed intake, lowers digestibility, slows gains, and depresses milk production. Excessive sulfur can also decrease feed intake and interfere with the metabolism of other minerals. Dairy cattle can tolerate higher levels of sulfur found in feed than when added as inorganic mineral forms.

Microminerals

Cobalt. Cobalt is essential for ruminants because it is incorporated into vitamin B_{12} by rumen microbes. Symptoms of cobalt and vitamin B_{12} deficiencies are the same (loss of appetite, body weight loss, reduced milk production, muscle incoordination, rough hair coat, stumbling gait, and anemia). Vitamin B_{12} is needed to metabolize propionate. Cobalt sulfate and cobalt chloride are common sources. Under practical conditions, cobalt toxicosis is not a problem.

Copper. Copper is needed for hemoglobin formation and numerous copper-dependent enzymes. Most symptoms of copper deficiency are nonspecific, including reduced growth and milk production, severe diarrhea, rough hair coat, changes in hair coat color and texture, stiff joints ("pacing gait" in older cattle), and reduced reproduction. Copper toxicity can occur when excessive amounts of supplemental copper or contaminated feeds are fed, resulting in a hemolytic crisis. Sources of copper include copper sulfate, copper carbonate, and copper nitrate. Molybdenum and copper are antagonistic. Major symptoms of excessive molybdenum are similar to copper deficiencies with scouring (called "teart pasture" or "peat scours"). The maximum amount of molybdenum is 6 ppm in the total ration dry matter.

Iodine. The main requirement for iodine is for the synthesis of hormones by the thyroid gland. Iodine deficiencies occur in the Great Lakes and northwestern regions of the United States, which are known as the "goiter belt." Deficient cattle show an enlargement of the thyroid glands (goiter), especially in newborn calves, which may be dead or weak. Long-term deficiencies can cause decreased milk yields and hypothyroidism. Total iodine concentrations of less than 40 mg per liter of serum or 10 to 20 μg per liter of milk are indicators of an iodine deficiency. Corn silage, soybean meal, kale, and rape contain goitrogenic substances. Iodized salt and EDDI (ethylenediamine dihydroiodide) are sources of supplemental iodine. EDDI has been used to prevent and treat foot rot and soft-tissue lumpy jaw. Excessive iodine can cause toxicity symptoms, including excessive tearing from the eyes and coughing. Approximately 10% of consumed iodine is excreted in milk, which can be a potential problem (high-iodine-containing milk).

Iron. Iron is involved in cellular respiration and oxygen transport. Hemoglobulin, myoglobulin, and enzyme systems contain iron. Anemia and low hemoglobin are early signs of iron deficiencies. Other deficiency symptoms include reduced weight gain, listlessness, and reduced appetite. When calves are fed only milk, iron-deficiency anemia develops, causing a pale color to the meat (called fancy veal). Parasitic disease can cause anemia and iron deficiency. Cattle can retain iron when a deficiency occurs. Excessive iron intake is not a common problem, although high iron levels in water can produce detrimental effects. Ferric oxide, ferric chloride, and ferrous carbonate are poor sources of iron; ferrous sulfate, ferrous chloride, and ferric ammonium citrate are available sources.

Manganese. Manganese is needed for reproduction and birth of normal calves. Reduced growth, skeletal abnormalities, and deformed calves at birth are deficiency symptoms. Diets low in manganese may also cause silent heats and lower conception rates. Cattle can tolerate high dietary manganese without negative effects.

Selenium. Early research with selenium centered on toxicity leading to sudden death or signs of distress (labored breathing, ataxia, and diarrhea). Symptoms of chronic selenium toxicity (alkali disease) include lameness, emaciation, sore feet, deformed hooves, and loss of hair from the tail. However, selenium-deficient areas in the United States are more common than toxic areas. Selenium is an integral component of glutathione peroxidase. Deficiency results in white muscle disease (nutritional muscular dystrophy) in young calves. Cardiac and skeletal muscles degenerate with necrosis, leading to heart failure and paralysis of the hind legs. Supplemental selenium can also reduce the incidence of retained placenta, improve calf liveability, improve reproductive performance, and decrease the severity of mastitis when selenium levels are marginal. Blood selenium can be measured to evaluate the selenium status in cattle. Calcium in the diet (over 1% or under 0.5%) decreases selenium absorption. Vitamin E can reduce the amount of selenium needed (see Section 18.5).

Zinc. Zinc is required for normal growth; it activates some enzymes and is a component of a metalloenzymes. Deficiency symptoms include skin dermatitis (parakeratosis) on the legs, neck, and head; lesions; and failure of wounds to heal. Contamination of feed and water with zinc can occur when they are stored in galvanized containers, particularly under acidic conditions. This exposure can be fatal, but usually causes feed refusal and gastrointestinal distress. Other potentially toxic sources of zinc include pesticides, fungicides, and industrial pollution. Most zinc salts are biologically available sources of zinc. Zinc methionine (e.g., ZinPro) is an organic complexed zinc and methionine compound that hardens hooves and promotes wound healing.

18.5 VITAMINS

Dairy cattle require the same vitamins as other animals. The B vitamins and vitamin K are synthesized by microorganisms in the rumen. Vitamin C is synthesized by body tissue. The fat-soluble vitamins (A, D, and E) are normally added to the diet (Table 18.3).

Vitamin A

Vitamin A deficiency in cattle results in degeneration of epithelial cells of mucosa of the respiratory tract, salivary glands, mouth, tear glands, eyes, intestinal tract, kidneys, vagina, and urethra. Degeneration of these cells makes the areas more susceptible to infection. A severe vitamin A deficiency also causes diarrhea, loss of appetite, and emaciation.

Vitamin A is supplied naturally in feeds in the form of its provitamin, carotene, which is found in relatively high levels in green feeds such as hay crop silage and hay. One milligram of carotene is equal to 400 IU of vitamin A. The minimum carotene requirement for maintenance is 19 mg per 45 kg (100 lb) of live weight. Increasing the carotene content of the feed increases the carotene and vitamin A content of milk. Vitamin A deficiency occurs if poor-quality forage or low levels of forage are fed to dairy cows, or in calves fed limited amounts of colostrum or whole milk.

Vitamin D

Vitamin D is required for maintenance, reproduction, lactation, growth, and proper bone formation. Vitamin D plays a role in absorption of calcium from the gastrointestinal tract (see milk fever, Section 22.3). A deficiency of vitamin D (rickets) produces symptoms similar to those produced by calcium or phosphorus deficiency. Vitamin D deficiency causes a weakening of the bones, swollen and stiff joints, and a humped back. In advanced states of the disease, stiffness of gait, dragging of the hind feet, irritability, tetany, weakness, anorexia, and retardation of growth in young animals occur.

Animals receive vitamin D from two sources. Vitamin D_2 is formed by the ultraviolet irradiation of the plant sterol, ergasterol. Vitamin D_3 is formed from animal sterol on exposure to ultraviolet light. Both forms have equal biological value.

Vitamin E

Vitamin E is a group of closely related fat-soluble compounds called tocopherols. One milligram of α-tocopheral is equal to 1 IU of vitamin E. Green forages, well-cured hay, dehydrated alfalfa meal, and wheat germ oil are good sources. Synthetically produced concentrates are available from feed and drug industries. The metabolic role of selenium is linked to vitamin E. Deficiency symptoms include white muscle disease, retained placenta, higher incidence of mastitis, and decreased reproductive performance. Feeding vitamin E (400 to 1000 mg per cow per day) can reduce oxidized flavor in milk.

18.6 WATER

Even though water is not considered to be nutrient, it is the most critical element in sustaining life. Water is required for several basic processes of life, including maintaining body fluid balance, eliminating waste materials, maintaining proper ion balance, aiding in heat loss and temperature control, and in digesting, absorbing, and metabolizing nutrients. During pregnancy, water provides a fluid environment for the developing fetus. Milk contains 87% water. Water also provides a fluid base for transporting nutrients to the mammary gland. Illinois researchers (Woodford et al., 1984) developed an equation to estimate the daily requirement for water:

$$\begin{aligned}\text{water intake (kg/day)} = {} & 15.98 + (1.58 \times \text{kg DM}) + (0.90 \times \text{kg milk}) \\ & + (0.05 \times \text{grams of sodium}) \\ & + (1.20 \times \text{minimum daily temperature in degrees Celsius})\end{aligned}$$

A lactating cow consuming 18.2 kg (40 lb) of dry matter, producing 29.5 kg lb (65 lb) of milk, consuming 85.1 g of sodium, and exposed to an average minimum temperature of 20°C would consume 99.5 kg (219 lb) of water. The amount of water needed by cattle for growth, maintenance, pregnancy, and milk production will vary (Table 18.4).

Water intake also depends on the availability of water and the temperature of the water. Cows having water available at all times drink more per day and

TABLE 18.4 GUIDELINES FOR DAIRY WATER INTAKE (LITERS/DAY) FOR DAIRY CATTLE

		Air temperature [C° (F°)]		
Weight [kg (lb)]	Milk [kg (lb)]	4° (40°)	15° (60°)	27° (80°)
Heifers				
91 (200)	—	8	9	12
182 (400)	—	12	17	23
364 (800)	—	24	30	40
545 (1200)	—	33	41	55
Dry cows				
636 (1400)	—	37	45	61
728 (1600)	—	39	48	68
Lactating cows				
636 (1400)	9 (20)	45	55	68
	27 (60)	83	91	94
	36 (80)	102	121	147
	45 (100)	121	141	173

Source: J. G. Linn, D. E. Otterby, R. D. Appleman, and M. F. Hutjens, 1980, *Ill. Ext. Circ.* M-1183.

give more milk than when they have access to water once or twice daily. High-producing cows show a more marked increase in milk production than do low-producing cows as a result of more frequent watering. On cold days cows will consume considerably more water if it is warmed slightly than if the temperature of the water is just above freezing. Hard water (as compared to soft water) has not been shown to affect the performance of lactating cows.

Water quality is important for optimal performance (Jimenez, 1981). Nitrate-nitrogen over 100 ppm is unsafe, especially for young animals (200 ppm nitrate-nitrogen should not be used). Most animals can tolerate a total dissolved solid concentration (hardness) of 10,000 mg per liter. Salt above 2500 ppm affects milk yield and water palatability. Bacteria (such as coliform) should be less than 1000 colonies per 100 ml. The pH, a measure of acidity or alkalinity, should range from 6.5 to 8 (7 is neutral). Calcium and magnesium in water are not detrimental to animal health, but contribute minerals to the diet. Mature dairy cattle can develop tolerance to high sulfates in water (2000 to 2500 ppm) after a period of time. Disinfection of water with chlorine up to 220 ppm poses no serious problems; however, 50 to 100 ppm is more palatable. Iron is troublesome since it stains floors, fixtures, and clothing and reacts with phosphorus to form insoluble salts. Water sources above 3 ppm iron should be treated to remove iron. Pesticide screening tests are also recommended from time to time.

REFERENCES AND FURTHER READINGS

JIMENEZ, A. A. 1981. Keep that water flowing. *Dairy Herd Manage.*, 19, No. 1, p. 64.

JORGENSEN, N. A. 1978. Mineral and vitamin sources and evaluation. Applied Dairy Nutrition. *11th Ann. Conf. Proc. AABP*, p. 17.

LINN, J. G., D. E. OTTERBY, R. D. APPLEMAN, and M. F. HUTJENS. 1980. Feeding the dairy herd. *Ill. Ext. Circ.* M-1183.

LOFGREEN, G. P., and W. N. GARRETT. 1968. A system for expressing net energy requirements and feed values for growing and finishing beef cattle. *J. Anim. Sci.*, 27:793.

MOE P. W., and W. P. FLATT. 1969. Net energy value of feedstuffs for lactation. *J. Dairy Sci.*, 52:928.

MORRISON, F. B. 1960. *Feeds and Feeding*. Morrison Publishing Company, Ithaca, NY.

NATIONAL RESEARCH COUNCIL COMMITTEE. 1980. *Mineral Tolerance of Domestic Animals*. National Academy of Sciences, Washington, DC.

NATIONAL RESEARCH COUNCIL COMMITTEE. 1988. *Nutrient Requirements of Dairy Cattle*, 6th ed. National Academy of Sciences, Washington, DC.

PALMQUIST, D. L. 1985. Using fat as an energy source in lactating cows. *Energy Fats Proc. NFIA*.

WOODFORD, S. T., M. R. MURPHY, and C. L. DAVIS. 1984. The importance of water for lactating cows. *Ill. Dairy Rep.*, p. 25.

19

Nutritional Value of Feeds

Most feeds are sold on the basis of their chemical composition, from which estimates of their nutritional values can be made. Much research has been devoted to estimating the nutritional value of feeds for maintenance, growth, fattening, or milk production. The various types of nutritional analysis that are used and feed types are discussed in this chapter.

19.1 METHODS OF ANALYSIS

Proximate Analysis

Proximate analysis is a system of chemical analyses used to predict the total digestible nutrients (TDN) content of feed. In these analyses, crude fiber is determined by treating feed with a weak acid followed by an alkali solution. Protein content is measured by multiplying the nitrogen content by 6.25. Mineral content is determined by burning the sample and weighing the ash residue; lipids are extracted with ether. The four components—crude fiber, ether extract, protein, and ash—are subtracted from the total dry matter of the feed and the remainder is called nitrogen-free extract (NFE).

To obtain the TDN value of feed, the digestibilities of crude fiber, NFE, ether extract, and protein must be determined. Digestibility values are obtained from digestion trials with animals. For ration formulation, digestibility figures from many research trials are averaged and are available in tables found in Morrison's Feeds and Feeding (1960) and from the National Research Council (NRC). Digestibilities of concentrates found in the literature are accurate because there is little variation in the digestibility of these components. Large variations in the digestibility of forage components occur, especially in fiber and NFE fractions.

The TDN measurement and proximate analysis have limitations. Treatment of a feed with a weak acid and an alkali dissolves some of the plant lignin, a compound that the animal cannot digest. The dissolved portion of fiber also includes most of the hemicellulose and some cellulose, which becomes part of the NFE portion. In some analyses, the fiber portion of the feed has a higher digestibility than the NFE portion, which represents the most digestible carbohydrate components of feed. The TDN method does not adequately characterize the fibrous fraction of the forage into digestible and indigestible fractions.

Lignin, which is indigestible, is present in higher concentrations in forages than in concentrates. Because of the inability of proximate analysis to recover lignin, TDN values of forages overestimate their productive values for milk production in comparison with concentrates.

Detergent Analysis

Because the TDN method does not effectively separate less digestible nutrients from more digestible nutrients, attention has been devoted to identifying the chemical components and physical structure of plants that determine the nutritive value of feeds. This work has been pioneered by P. J. Van Soest at the USDA laboratory in Beltsville, Maryland, and at Cornell University and by D. R. Mertens at the USDA Forage Laboratory in Madison, Wisconsin.

The detergent fiber method divides the carbohydrate fraction of the plant into cell contents and cell wall constituents (Table 19.1). The feed is first treated with a neutral detergent solution at pH 7. The fraction dissolved in the neutral detergent solution contains cell contents, which include lipids, sugars, organic acids, water-soluble mineral, pectin, starch, nonprotein nitrogen, and soluble protein. These components are almost completely digested by the animal. The residue after neutral detergent treatment is composed of cell walls and is called neutral detergent fiber (NDF). Neutral detergent fiber contains the structural portion of the cell. This portion determines the volume that the feed will occupy in the digestive tract and limits the amount of feed consumed.

The feed sample is treated with an acid detergent solution, which dissolves the hemicelluloses and some fiber-bound protein. These fractions are partly digestible, depending on the degree of lignification. The insoluble material is called acid detergent fiber (ADF) and is composed of heat-damaged protein, cellulose, lignin, cutin, and silica. Silica and lignin depress the digestibility of plants. Lignification increases as the plant matures and depresses digestibility. Cell wall silica appears to function by adding strength to the plant in a manner similar to that of lignin. Determination of lignin and silica contents of forages has helped explain

TABLE 19.1 DIVISION OF ORGANIC MATTER BY DETERGENT SYSTEMS

Fraction	Detergent fiber	Component	Digestibility
Cell contents		Lipids	98
		Sugars	
		Starches	
		Pectin	
		Soluble protein	
		Nonprotein nitrogen	
		Soluble minerals	
Cell wall constituents	NDF[a]	Hemicellulose	60
	NDF[a], ADF[b]	Cellulose	30
	NDF[a], ADF[b]	Lignin	0
	NDF[a], ADF[b]	Heat-damaged protein	0
	NDF[a], ADF[b]	Insoluble mineral	0

[a] Neutral detergent fiber
[b] Acid detergent fiber

differences between temperate-zone and tropical forages. Temperate-zone forages are higher in digestibility, higher in soluble carbohydrates, and generally higher in crude protein than are tropical forages, which are higher in cell wall materials.

Analyses of chemical composition of plants, especially those using neutral and acid detergent fiber methods, have been used to estimate the digestibility and net energy values of feeds. These values are given in Tables 19.2 and 19.3. Since the digestibility of the cellular contents is approximately 98%, most attention is directed to the cell wall of the forage plant. The digestibility of the cell wall depends on the degree of lignification and amount of heat damage, which results in the formation of indigestible protein bound to carbohydrate.

NIR Analysis

NIR (near-infrared reflectance) analysis is a nondestructive technique that uses light wavelengths. The reflected light rays from a prepared sample are used to estimate nutrient content. When a beam of light is directed on the sample, chemical bonds between elements in a compound vibrate and some of the light is absorbed. The reflected light is directed to a sensor which "reads" the wavelengths of light and sends the information to a computer. A computer loaded with a series of equations determines the nutrient values for the feed. Speed is the main advantage, as a feed analysis can be done in 10 to 15 minutes. It does not require a laboratory or chemicals. However, the equipment is expensive to purchase and requires a skilled technician to operate.

In Vitro Methods

In vitro methods have been used to determine the digestibility of dry matter of forage samples. A small amount of ground forage is incubated with an inoculum

TABLE 19.2 NUTRIENT CONTENT OF FORAGES FED TO DAIRY CATTLE (expressed on a 100% dry matter basis)

Feed	Stage of growth	TDN (%)	Net energy (Mcal/kg)			ADF (%)	NDF (%)	Crude protein (%)	Ether extract (%)	Calcium (%)	Phosphorus (%)
			Lact.	Gain	Main.						
Alfalfa	Bud stage	66	1.50	0.92	1.51	28	38	23.0	4.0	1.54	0.29
	Mid bloom	58	1.30	0.68	1.24	35	46	17.0	2.6	1.41	0.24
	Full bloom	55	1.23	0.58	1.14	37	50	15.0	2.0	1.25	0.22
Barley	Dough	56	1.25	0.62	1.18	34	50	8.7	2.1	0.23	0.26
Bermuda grass	30-day growth	50	1.11	0.42	0.97	38	76	12.0	2.1	0.32	0.20
Bromegrass	Late veg.	68	1.55	0.97	1.57	35	65	16.0	2.6	0.32	0.37
	Late bloom	59	1.33	0.70	1.27	37	68	10.0	2.3	0.30	0.35
Clover, red	Full bloom	64	1.45	0.85	1.44	41	56	14.6	2.9	1.53	0.25
Corn silage	Well-eared	64	1.45	0.85	1.44	24	51	8.1	3.1	0.30	0.33
Corn stover	No ears or husk	59	1.33	0.70	1.27	34	67	6.6	1.3	0.38	0.31
Fescue	Full bloom	58	1.30	0.68	1.25	39	67	12.1	5.3	0.51	0.37
Oats	Late veg.	62	1.40	0.80	1.38	38	58	12.8	2.5	0.34	0.25
	Dough	57	1.28	0.64	1.21	35	48	10.0	4.1	0.24	0.22
	Straw	50	1.11	0.42	0.97	47	70	4.4	2.2	0.24	0.06
Orchard grass	Early bloom	65	1.47	0.88	1.47	34	61	15.0	2.8	0.58	0.35
Sorghum silage	All analysis	60	1.35	0.74	1.31	28	53	7.5	3.0	0.35	0.21
Sudangrass	Early veg.	70	1.60	1.03	1.63	29	55	16.8	3.9	0.43	0.41
	Full bloom	56	1.25	0.62	1.18	42	68	8.0	1.8	0.55	0.30
Soybean	Dough stage	61	1.38	0.77	1.35	35	47	16.8	4.1	1.36	0.30
Timothy	Early bloom	66	1.50	0.92	1.51	32	61	15.0	2.9	0.53	0.25
	Full bloom	58	1.30	0.68	1.25	38	68	8.1	3.1	0.43	0.20
Wheat	Early veg.	57	1.28	0.64	1.21	39	60	11.9	2.5	0.27	0.27

Source: Adapted from *United States-Canadian Tables of Feed Composition*, 3rd. ed., 1982, National Academy of Sciences, Washington, DC.

TABLE 19.3 NUTRIENT CONTENT OF GRAINS, PROTEIN SUPPLEMENTS, AND BY-PRODUCT FEEDS OF DAIRY CATTLE (expressed on a 100% dry matter basis)

	TDN (%)	Net energy (Mcal/kg)			ADF (%)	NDF (%)	Crude protein (%)	Undegraded protein (% total)	Ether extract (%)	Calcium (%)	Phosphorus (%)	Upper limit (% conc mix)	Morrison energy	Constant protein
		Lact.	Gain	Main.										
Home-grown grains														
Barley	84	1.94	1.40	2.06	7	19	13.5	30	2.1	0.05	0.38	80	0.908	0.093
Corn, shelled	87	2.03	1.48	2.16	3	9	10.0	50	4.3	0.03	0.29	80	Base	Base
Corn, ear	83	1.91	1.37	2.03	10	28	9.0	50	3.7	0.07	0.27	100	0.918	−0.018
Sorghum	86	1.99	1.45	2.12	5	23	12.4	60	3.1	0.04	0.33	80	0.916	0.056
Oats	77	1.77	1.22	1.86	16	32	13.3	30	5.4	0.07	0.38	50	0.806	0.095
Wheat	88	2.04	1.50	2.18	4	13	16.0	25	2.0	0.04	0.42	50	0.875	0.125
Protein supplements														
Cottonseed meal	76	1.74	1.18	1.88	19	26	45.6	40	1.3	0.21	1.16	40	0.128	0.773
Linseed meal	78	1.79	1.29	1.88	19	25	38.3	40	1.5	0.43	0.89	40	0.201	0.699
Soybean meal (solvent)	84	1.86	1.40	2.06	10	14	49.9	35	1.5	0.29	0.68	40	Base	Base
Soybean meal (mech)	87	1.96	1.43	2.10	10	14	47.7	50	5.3	0.29	0.68	40	0.153	0.854
Soybean meal (dehulled)	87	2.01	1.48	2.16	6	10	55.1	25	1.0	0.29	0.70	40	−0.116	1.123
Soybeans	91	2.11	1.57	2.27	10	14	42.8	25	18.8	0.28	0.66	20	0.352	0.746
By-product feeds														
Beet pulp	74	1.69	1.14	1.76	33	54	9.7	35	0.6	0.69	0.10	50	0.931	−0.051
Brewers' grain (dry)	66	1.50	0.91	1.51	24	46	29.4	60	7.2	0.33	0.55	50	0.374	0.464
Corncobs	50	1.11	0.42	0.97	35	89	3.2	50	0.7	0.12	0.04	25	NA	NA
Corn distillers' grain	86	1.99	1.45	2.12	18	44	23.0	50	9.8	0.15	0.71	40	0.710	0.350
Corn gluten feed, dry	83	1.91	1.37	2.03	16	38	25.6	35	2.4	0.36	0.82	50	0.456	0.434
Corn gluten meal	89	2.06	1.52	2.20	5	14	67.2	65	2.4	0.08	0.54	25	−0.322	1.287
Cottonseed, whole	96	2.23	1.69	2.41	29	39	23.9	35	23.1	0.16	0.75	15	0.656	0.303
Hominy feed	87	2.01	1.48	2.16	13	55	11.5	NA	7.7	0.05	0.57	80	1.043	0.012
Malt sprouts	71	1.62	1.06	1.67	18	47	28.1	55	1.4	0.23	0.75	20	0.908	0.093
Molasses, beet	75	1.64	1.08	1.69	0	0	8.5	0	0.2	0.17	0.03	5	NA	NA
Molasses, cane	70	1.60	1.03	1.63	0	0	10.3	0	0.9	1.00	0.11	5	1.058	−0.169
Wheat bran	70	1.60	1.03	1.63	15	51	17.1	NA	4.4	0.13	1.38	35	0.619	0.218
Wheat midds	69	1.57	1.00	1.60	10	37	18.4	50	4.9	0.13	0.99	25	0.743	0.222
Whey	81	1.87	1.32	1.97	0	0	14.2	0	0.7	0.92	0.82	5	0.839	0.140

Source: Adapted from *United States-Canadian Tables of Feed Composition*, 3rd. ed., 1982, National Academy of Sciences, Washington, DC.; F. B., Morrison, 1960, *Feeds and Feeding*, Morrison Publishing Company, Ithaca, N.Y.; and R. L., Preston, 1986, *Feedstuff*, 58:18.

collected from rumen ingesta and a buffered "artificial" saliva, and fermentation is allowed to continue for 48 hours in a carbon dioxide atmosphere. At the end of fermentation, the disappearance of dry matter or cell walls is determined and correlated to some other index of forage value, such as digestible dry matter, digestible energy, or TDN. The *in vitro* method is preferable to digestibility trials because it allows a larger number of samples to be evaluated at less cost.

Net Energy

The best measure of energy of a feed is its net energy value, because a unit of net energy from concentrate is equal to a unit from forage. The determination of net energy values by calorimetry is expensive, and few values for feeds have been obtained. Indirect methods of estimating net energy have been used by researchers.

19.2 FORAGES

The main portion of the dairy feeding program is forage to meet energy and protein needs. Forages keep total feed costs to a minimum. Because high-producing cows cannot meet energy needs, high-quality forages must be utilized. Low-quality forages result in lower milk production or the need for concentrates, which increase cost of production. The choice of forage depends on yield per acre, growing conditions, production costs, and dairy cattle needs. Forage nutrient values are listed in Table 19.2.

Legumes (alfalfa and clover are examples) are excellent sources of protein, minerals, and vitamins if harvested and stored correctly. Cell wall contents (NDF) of legumes are lower than those of grass, while lignin levels can be higher. Grasses are higher in cell wall content and silica, which depress plant digestibility. Legumes generally have a higher protein content than do grasses. Energy levels of legumes and grasses are similar if harvested at similar stages of maturity. Legumes produce more dry matter per acre than do grasses; have more seedling vigor, which enhances rapid stand establishment; and dominate legume–grass mixtures during the seeding and first year following establishment. Grasses reduce bloat problems, can tolerate wet conditions, reduce soil erosion due to their fibrous root system, speed curing time, spread the risk in establishment, and dominate older stands.

Corn silage contains nearly 40 to 50% grain on a dry matter basis and is an excellent source of energy. If this feed is properly made, cows will eat large amounts of it. Corn silage requires balanced supplementation of protein and mineral for high milk production.

Hybrid corns that yield the largest amounts of grain are generally best for silage. To attain maximum yield, corn should be harvested for silage at physiological maturity; kernels are fully dented and cells at the base of the kernel (when dissected) are black. Dry matter content should be approximately 35% (the ear is 55 to 60% dry matter when the whole plant is 32 to 38% dry matter). Immature corn silage (lower than 30 to 32% dry matter) yields less total dry matter per acre, and seepage losses will occur after ensiling. If corn becomes too dry before ensiling, field losses are greater and the feed may not ensile as well (poor compaction, molding, and lower palatability).

Sorghums can be used for silage in areas adapted to long growing seasons. Forage sorghum equals corn in yield, but grain sorghums usually do not yield as well as corn except during droughty conditions. The energy and intake potential of sorghums are lower than those of corn silage.

Oats, barley, wheat, and rye (small grains) can be used for forage, although yield per acre is usually less than corn or legume silage. They should be harvested in boot to early milk stage for high protein content. The total dry matter yield may be 50% lower in the boot stage than when harvested at the soft dough stage. Peas or beans can be planted with small grain forage to increase the protein content.

Straws are low in energy, protein, minerals, and vitamins and should be used only when additional fiber is needed. If adequately supplemented, some straw can be fed to dry cows and older heifers.

Corn stover (residue after harvesting corn grain) can be used in heifer and dry cow rations if properly supplemented. It is low in protein and minerals and is considered a forage extender.

19.3 CONCENTRATES

The concentrate mixture, composed of grains, protein supplements, by-products, minerals, vitamins, and additives, is an important component of dairy cattle rations. Grains, protein supplements, and by-products are higher in energy and lower in fiber than is forage (Table 19.3). In most feeding programs, cows are fed forages *ad libitum*. Concentrate is fed to make up the difference between nutrients required by the animal and those supplied by the forage. Most concentrates have TDN values between 65 and 90%, which is higher than that of forage.

Grains

Shelled corn is a common grain source and one of the highest-energy grains available. Where corn can be grown successfully, it is generally the most economical source of energy. Because of its high starch content, frequent feeding is needed to optimize digestibility and minimize digestive disturbances. Shelled corn must be processed (breaking the seed coat) for dairy cattle over 1 year of age to avoid passage of whole seeds. Ear corn (corn and cob meal) is another form relished by cows. Although it is lower in energy, the added fiber from the cob can maintain the fat test and keep cows on feed. Approximately 20% of ear corn consists of cob. Ear corn must be ground to avoid separation of the grain and cob particles. The cob is processed to a pea-sized particle. An alternative method of harvesting corn as dry grain is to ensile it as high-moisture corn. The corn can be picked, shelled, and the whole kernels stored in an oxygen-limiting storage unit. The ideal moisture content of the kernel is approximately 25 to 28%, but ranges from 24 to 30%. Shelled corn can also be cracked or ground before it is stored in conventional storage. The entire ear can be stored by first running it through a roller or hammer mill. The ideal moisture content is 28%, with a range of 26 to 35%. Harvesting high-moisture corn with over 30% moisture results in more degradable carbohydrate and protein in the rumen, which can decrease feed utilization. High-moisture ear corn harvested with a snapper head (called snaplage) is inferior to picked or combine-harvested ear corn because it is higher in fiber (husks, leaves, and stalks), which can separate in storage, causing mold and variation in feed quality.

Use of high-moisture corn has several advantages. Harvesting of this type of corn can begin 2 to 3 weeks earlier than that of bin or cribbed corn. Total losses of high-moisture corn are 5 to 15% lower because field and harvesting losses are less. Handling costs for high-moisture corn are reduced because it is ready for feeding. High-moisture corn can also be used effectively with mechanized feeding. The feeding value of high-moisture corn is superior to dry corn expressed on a dry matter basis. Adjustments for moisture levels must be made.

High-moisture corn can be stored in oxygen-free or conventional silos, but dry matter losses will be slightly higher in conventional storage. The conventional silo must be in good condition and the doors must fit tightly so that air is excluded. Rapid filling and even distribution of the corn are essential. Uniform filling of high-moisture ear corn is necessary to prevent separation of the kernels from the cob. After the silo has been filled, it should be sealed with a plastic cap to prevent surface spoilage. At least 3.8 to 5.1 cm ($1\frac{1}{2}$ to 2 in.) of high-moisture corn must be removed from the silo daily (7 to 10 cm per day during the summer) to prevent surface spoilage in conventional silos. Fifty or more cows are required to consume adequate quantities of high-moisture corn to prevent surface spoilage even with a small-diameter silo.

High-moisture corn can be treated with chemical preservatives such as propionic acid to permit storage in an open bin. The amount of acid (usually 1% propionic acid) needed depends on the moisture content of the grain. Propionic acid–treated corn has no palatability or handling problems, and this can be an economical method for storing early harvested corn.

Milo or grain sorghum is a high-energy grain raised in the southern and central areas of the United States. Processing milo is important because the seed is round and hard. The seed coat must be broken, but powdery grain with fine particles should be avoided. Reconstituting milo (adding water and storing) can improve digestibility and palatability.

Oats contain 85% of the energy value of shelled corn. The starch content of oats is lower (44%) than that of shelled corn (75%). The bulky nature limits the amount in the concentrate mixture. Adding oats to high-energy concentrate mixtures can improve texture, palatability, and digestion. Hull-less oats (grouts) are higher in feed value.

Barley is a popular crop in western and northern areas of the United States, especially if corn does not grow well. It is slightly lower in energy than corn and the starch is rapidly converted in the rumen to VFA. Barley can be substituted for corn and is higher in protein. Rolling barley is superior to fine grinding to maintain digestibility and intake. High-moisture barley (25 to 30% moisture) can be fed successfully and is stored in oxygen-limiting units.

Wheat is not commonly used in dairy grain mixtures because the price is usually too high. Wheat is an acceptable ingredient when limited to 50% of the concentrate mixture. The wheat seed should be processed to break the seed coat while not powdering the grain, which would cause it to form a pastelike consistency when consumed. High-moisture wheat can be utilized in dairy rations. Rye and buckwheat can be used, but they are less palatable.

Protein Supplements

Soybean meal is the most common and usually most economical protein supplement in the United States. Most commercial supplements contain substantial amounts of soybean meal. Three types of soybean meal are available: soybean meal—solvent extracted (called 44% since it contains 44% protein on an as-is basis), oil is removed with an organic solvent; dehulled soybean meal (called 48% because of its higher protein level on an as-is basis), oil is extracted with a solvent; and soybean meal-expeller processed (contains 44% protein and 5% oil on an as-is basis), oil is removed by pressing.

Unprocessed soybeans can be included in the grain mixture when economically favorable. They are slightly lower in protein than meal and contain 18% oil. The maximum amount of soybeans should not exceed 20% of the concentrate mixture (1 to 2 kg of raw beans or 2 to 3 kg of heat-treated soybeans). Cattle should be adjusted to beans gradually to avoid diarrhea and off-feed problems. Raw beans contain the enzyme urease, which releases ammonia when mixed with urea; urea and raw beans should not be mixed together. Roasting, extruding, or heat processing of soybeans decrease protein degradation in the rumen, increase palatability, destroy urease and trypsin inhibitor, and decrease rancidity problems. The economics of heat treatment must be considered.

Linseed meal is a product of the flax industry, a good protein source, extracted by solvents, and usually higher in price per unit of protein compared to soybean meal. Expeller-processed linseed meal (old processed) is used in fitting cattle for shows or sales because it adds shine to hair coats. Cottonseed meal, sunflower meal, and cannola meal can be used if prices are favorable.

Urea is a simple compound containing nitrogen, carbon, and oxygen and is broken down in the rumen to form ammonia and carbon dioxide. Ammonia is utilized by rumen microorganisms to synthesize microbial protein, which is then broken down in the intestinal tract and the amino acids are utilized by the animal. Feed-grade urea contains 45% nitrogen and furnishes approximately 2.8 times as much nitrogen as a comparable amount of plant protein. One percent urea will increase the crude protein content of a feed 2.8%. Because urea has no energy value, additional energy must be fed to enable the microorganisms to synthesize the amino acids. A convenient rule of thumb is that 50 kg of a protein feed, such as soybean meal, can be replaced by 7 kg of urea and 43 kg of a high-energy feed, such as shelled corn. The major reasons for using urea in dairy cattle rations are to reduce the cost of the protein supplement and to add a source of rapidly degradable nitrogen.

Rapid conversion of urea to ammonia and carbon dioxide presents problems because a high level of ammonia is toxic to the animal. Ammonia toxicity is characterized by uneasiness, muscle tremors, and increased salivation occurring 20 to 30 minutes after feeding. The progressive symptoms cause uncoordination (especially in the front legs), slobbering, heavy breathing, failure to stand, tetany, and death within 1 to 2 hours. The antidote for ammonia poisoning is a dilute solution of acid (a 5% solution of acetic acid is effective if given orally before tetany occurs). Under emergency farm conditions, a drench of 4 liters of household vinegar is the most available antidote. Under most conditions, an animal should not receive more than 0.2 kg (0.45 lb) of urea per day. If concentrates containing urea are fed twice daily, two high ammonia peaks are produced in the rumen. If microorganisms cannot utilize all of the released ammonia, some is absorbed by the blood and either recycled through saliva or eliminated from the body. If more feedings occur, smaller ammonia peaks occur throughout the day and microorganisms can make better use of the released ammonia. Cattle require 2 to 3 weeks to become accustomed to urea-containing feed. The adaptation should be gradual; a general recommendation is to feed one-third the level for 1 week, followed by two-thirds for 1 week, followed by concentrates containing the desired level of urea.

By-Products

By-product feeds are residual feedstuffs remaining after components such as starch, sugar, or oil have been extracted. Some nutritionists call these feeds coproducts because the nutritional value can be superior than the initial feed. Excellent buys may be available for dairy producers in certain areas. The main considerations are cost (relative to locally available grains and forages), nutrient quality, and optimal level (Table 19.3).

Several corn by-products are available. Dry corn gluten feed (DCGF) is produced during the wet milling of corn to produce starch for various uses. Approximately 6 kg (13 lb) of DCGF are produced from one bushel of corn. DCGF is a medium-protein feed and high in digestible fiber, consisting of two-thirds corn bran and one-third steep water. Wet corn gluten feed (WCGF) is the wet form of DCGF, containing 55 to 65% moisture. WCGF is higher in energy than DCGF on a dry matter basis because digestibility decreases during drying. WCGF can be ensiled with forage or grain. The level of WCGF mixed with other feed and ensiled will depend on type of cattle fed and feeds used. If WCGF is stored in a pile, it should be fed within 7 to 10 days, as mold formation will occur reducing palatibility. Corn distillers' grain (CDG) is an excellent protein supplement which is low in rumen degradability. Oil content varies from 8 to 12%, which makes distillers' grain equal to shelled corn in energy. Both wet and dry CDG with and without solubles are available. A dark-to-black CDG may indicate that heat damage occurred during the drying process, with decreased feed value and palatability. Corn gluten meal is a high-protein supplement (varies from 40 to 60% protein). The meal is low in protein degradability in the rumen, making it a bypass (low-rumen-degraded) protein source. Hominy, a corn product produced by the partial extraction of starch, is high in oil (8 to 10%) and can be substituted for shelled corn. Corn by-product feeds are low in lysine (an essential amino acid), which should be considered when used as a low-degradable protein source.

Wheat by-products include bran, midds (middlings), and shorts. Bran consists of the outer portion of the wheat seed, is high in phosphorus, and may serve as a laxative. Shorts consist of portions of the outer coat and endosperm. Middlings include wheat shorts and germ portions of the embryo. Wheat by-products are good sources of protein, fiber, and phosphorus.

Barley by-products are produced by the brewing industry. If producers are located near a brewery, wet brewers' grain (WBG) may be obtained at a reasonable cost. WBG contains 70 to 85% moisture depending on whether the grain was pressed or not prior to shipping. Dried brewers' grain is a good source of low-degraded protein. WBG should be fed within 7 days if piled outside during the summer due to mold formation. WBG can be stored in large plastic bags to avoid losses and molding. Malt sprouts, a barley by-product, are bitter and are usually mixed with other feeds. The protein in malt sprouts is low in rumen degradability.

Whole cottonseed is an excellent ration ingredient, with moderate protein, high oil (over 23%), and digestible fiber. No processing of the seed is needed prior to feeding and cows will eat it topdressed. The linted form (containing a small cotton fuzz wrap) is desirable, but makes mechanical handling difficult. The recommended amount is $2\frac{1}{2}$ to $3\frac{1}{2}$ kg (5.5 to 7.7 lb) per day. The oil can cause rumen digestive problems when fed at higher levels. Cottonseed hulls are a forage extender and a marginal source of nutrients.

Whey (dried or liquid forms) is available in certain areas at economical prices. Dried whey can be incorporated at 5 to 10% of the concentrate mix. Dried whey can also be added to silage prior to ensiling at a rate of 9 to 45 kg (20 to 100 lb) per wet ton of forage as a preservative and nutrient source. Liquid whey can be offered to cattle on a free-choice basis along with free-choice water. Because liquid whey is over 94% water, large amounts must be consumed to obtain substantial amounts of dry matter intake. Whey contains large quantities of lactose (milk sugar) and a small amount of protein and minerals. Liquid whey should not be stored longer than 24 to 36 hours because it will become acidic, lowering consumption and mottling teeth. Flies can be a problem if strict sanitation is not practiced.

Cane and beet molasses supply energy and are used to enhance acceptability of the ration. The amount should be limited to 5 to 7% of the concentrate mix (10% in pelleted feeds) to maintain flow characteristics in feeding equipment and to avoid undesirable rumen effects.

Beet pulp is a popular by-product feed that can be fed dried or wet (85 to 90% moisture). Molasses can be mixed with the beet pulp. Dried citrus and orange pulp are similar in feed value to beet pulp. Citrus pulps should be introduced gradually to accustom cows to the taste.

Almond hulls vary in quality due to contamination with dirt, shells, and foreign material. The soft almond hull is worth 80% of the energy value of barley grain.

Pomace is the refuse in the production of grape, apple, and tomato juice. It consists of seeds, stems, skins, and some fruit. The quality varies depending on type of product and processing methods. Pomace should be limited to 15 to 20% of the ration.

Rice and sunflower hulls have no practical value as livestock feed. Soybean hulls (also called flakes) are an acceptable feed source, containing 71% TDN and 12% crude protein.

Additives

Some ingredients are added to the ration that do not contain nutrients, but function to improve performance in other ways. Several additives used in dairy rations are discussed below.

Antibiotics are not essential nutrients for animals. Antibiotic feeding to young dairy calves can increase appetite and improve feed efficiency (see Chapter 26). Antibiotic feeding to cows has been suggested as a way of increasing milk production; however, research results have not been consistent. The risks of drug residues in milk and the development of resistant bacteria are concerns. Feeding of penicillin to dairy cows may be helpful in preventing bloat of cows on pasture (see Chapter 22).

Bentonite is a clay mineral that swells 5 to 20 times in the rumen. It can adsorp minerals (keep minerals available in the rumen), adds bulk in the ration, slows the rate of feed passage, and is a binder in pelleted fed. Feeding 0.5 kg (1.3 lb) can correct milk fat depression due to heavy concentrate feeding. Cows may consume significant amounts of bentonite when offered free choice, which can reduce *ad libitum* soil consumption.

Isoacids (e.g., Isoplus) are a combination of short-chain and branched-chain rumen fatty acids which may be deficient in some rumen fermentations. Supplementing 85 g (3 ounces) can increase milk yield by 2.7 kg (6 lb). Field response has been variable. Management for optimal response includes feeding isoacids 2 weeks before calving, removing isoacids from the diet 225 days postpartum, and supplementing high-producing older cows (over 7300 kg of milk). The response in first-lactation cows is economically marginal.

Magnesium oxide is a source of magnesium (54% by weight) and can correct milk fat depression through a buffer effect in the rumen and mammary gland uptake of milk fat precursors. The recommended level is 0.1 kg (0.2 lb) per cow per day in combination with sodium bicarbonate (2 to 3 parts bicarbonate to 1 part magnesium oxide). Magnesium oxide is unpalatable; cows will not consume it free choice and concentrate intake may decline with levels over 1% in the mixture.

Methionine hydroxy analog (e.g., MHA) is a form of the essential amino acid methionine. Fat test may increase when 0.25 to 0.3% analog or 30 g is added to the concentrate mix. Fat test response occurred in herds fed high-concentrate diets (60%) in early lactation. The addition of methionine hydroxy analog to farm rations is not recommended.

Monensin (e.g., Rumensin) is an antibiotic that alters rumen fermentation and increases propionic acid production. Dairy heifers fed 200 mg of monensin gained more body weight per day, had improved feed efficiency, and calved earlier. The product is cleared by the FDA for feeding to replacement heifers starting at 180 kg (400 lb) and up to calving (clears the digestive tract in 48 to 72 hours). No negative effects on milk production, reproduction, or calving occurred in monensin-supplemented heifers. Monensin is not legal in feed for dry or lactating dairy cows. Lasalocid (e.g., Bovatec) was cleared as a growth promoter for dairy heifers in 1987. Replacement heifers can be supplemented with 60 to 200 mg of lasalocid per head per day with no weight restriction. Monensin and lasalocid are similar in function and cost.

Niacin is a B-complex vitamin. Adding 6 g per day from 2 weeks prepartum to 10 to 12 weeks postpartum reduces ketosis. The function of niacin is to reduce fat mobilization, stimulate appetite, and increase synthesis of protein in the rumen. For ketotic cows, 12 g of niacin are recommended.

Propionic acid is a volatile fatty acid that also functions as a feed preservative which acidifies feed and inhibits the growth of mold. Adding $\frac{1}{2}$ to $1\frac{1}{2}$% propionic acid to baled hay and high-moisture corn is an effective preservative. The level of acid added will depend on the feed moisture level and length of feed storage. Economics and methods of application are management considerations.

Propylene glycol is a product converted by the liver to glucose. If 0.4 to 0.5 kg (0.9 to 1.1 lb) is administered prior to symptoms of clinical ketosis, propylene glycol is effective in preventing primary ketosis by maintaining blood glucose. The product is unpalatable and may require drenching twice a day.

Sodium bicarbonate and sodium sesquicarbonate (e.g., S-carb) are rumen buffers which neutralize acid in the rumen and maintain a desirable pH (6.2 to 6.5). Maintaining a favorable pH increases rumen acetate and butyrate, which maintains milk-fat test, avoids acidosis and going off-feed, and stimulates dry matter intake. The recommended level is 0.2 to 0.3 kg (0.4 to 0.5 lb) per cow per day. Buffers should be introduced gradually to avoid feed refusal. Some cattle may consume buffers when offered free choice, but force feeding is recommended.

Yeast is a source of vitamin D and B-complex vitamins and may improve the fermentation pattern in the rumen (research results have been variable). The amount is 0.1 kg to prevent rumen acidosis or 0.45 kg to treat rumen acidosis per cow day. Yeast will not persist in the rumen because of the acidity, must be fed daily, and is not routinely recommended.

Zinc methionine (e.g., Zinpro) is a chelated zinc–protein complex not degraded in the rumen. Preliminary research indicates that it produces harder hooves and promotes healing of wounds. The suggested level of zinc methionine is 4.5 g per cow per day of the 4% zinc product.

REFERENCES AND FURTHER READINGS

CHANDLER, P. T. 1978. Roughage and fiber. *Large Dairy Herd Management*. University of Florida Press, Gainesville, FL, p. 354.

HUTJENS, M. F. 1985. Feed additives. *Ill. Dairy Guide* 15.

HUTJENS, M. F. 1986. By-product feeds. *Ill. Dairy Guide* 14.

MILLER, D. A. 1984. *Forage Crops*. McGraw-Hill Book Company, New York.

NATIONAL RESEARCH COUNCIL COMMITTEE. 1988. *Nutrient Requirements of Dairy Cattle*, 6th ed. National Academy of Sciences, Washington, DC.

20

Production of High-Quality Forage

Even though dairy cows are being fed higher levels of concentrate, forage still plays an important role in supplying energy and protein. Nutrient cost is considerably less with forage than with concentrate. The backbone of the dairy feeding program is supplying high-quality forage.

Because energy requirements of high-producing cows are great and there is a limited physical capacity of the digestive tract for feed, it is critical to provide the best-quality feeds available. If poor-quality forages are fed, either milk production is reduced or additional concentrate must be fed to maintain milk production.

Inclusion of forage in dairy cow rations is essential. A minimum amount of forage is required to maintain a normal fat test in milk. In addition, a ration high in concentrate and low in forage can cause digestive upsets.

20.1 LEGUME–GRASS FORAGES

Legumes, especially alfalfa, are widely used as forage for dairy cattle. Direct comparisons among species of forages are difficult because comparisons are affected by other variables. The energy content of forages cut at a comparable stage of maturity is similar. Grass is higher in fiber, which reduces forage intake. In legume–grass mixtures, grass may mature earlier than legumes, which can lower forage quality. The protein content of legumes is higher (Table 19.2). Legumes also play an important role in improving soil fertility because of their ability to fix nitrogen.

Stage of Maturity

The most important factor that determines the quality of forages is the growth stage of the plant at time of cutting. The best measure of the physiological growth stage is the stage of maturity (Table 20.1). As the plant matures, cell wall and lignin contents increase, making the plant less digestible. This general relationship between maturity and digestibility applies to perennial forages with the exception of a few late-maturity varieties.

Four factors affect fiber formation and digestibility of forage plants: temperature, age of plant, light intensity, and nitrogen fertilization. Increased temperature and advanced age increase lignin formation and decrease digestibility of the plant. Increased light intensity and nitrogen fertilization increase digestibility and decrease cell wall content. Decreased digestibility of first-growth forage in the northern part of the United States is caused by increased temperatures and the advanced age of plants. Light intensity and nitrogen fertilization play minor roles in comparison with temperature and age of plant.

The digestible dry matter (DDM) of second and later cuttings range from 57 to 64% regardless of growth stage. The reason for the slower decrease in digestibility with the advancing age of later cuttings in comparison with first-growth forages is that the temperature remains approximately the same. The digestibility of plants grown in late summer and early fall may increase because the effect of decreasing temperatures on the lignification process may be greater than the effect of age.

Many feeding trials have evaluated nutritive value of forages harvested at different stages of maturity (Table 20.2). Wisconsin researchers studied various qualities of forage fed with four different levels of concentrate. Cows fed prebud alfalfa produced more milk compared to low-quality forage at all levels of concentrates fed. High-concentrate feeding did not compensate for mature low-qual-

TABLE 20.1 RELATIONSHIP OF FORAGE QUALITY AND MILK YIELD TO STAGE OF ALFALFA MATURITY[a]

	Stand crop					Crop as-fed				
Stage of maturity	DE (%)	CP (%)	NDF (%)	DMI (%)	Milk (kg)	DE (%)	CP (%)	NDF (%)	DMI (kg)	Milk (kg)
Midbud	71	24	37	20.4	30.8	64	21	<40	19.1	23.9
First bloom	67	20	41	17.0	21.7	61	18	44	15.9	16.1
Midbloom	64	18	48	15.3	16.4	57	16	51	13.4	9.7
Full bloom	62	16	53	12.6	10.2	54	<14	56	11.5	5.3

Source: J. R. Kawas, et al., 1983, *J. Dairy Sci.*, Supp. 1, 66:181.

[a] DE, digestible energy; NDF, neutral detergent fiber; CP, crude protein; DMI = dry matter intake.

TABLE 20.2 IMPACT OF ALFALFA FORAGE QUALITY AND CONCENTRATE INTAKE ON MILK YIELD

	Alfalfa stage		Daily difference	
Concentrate (% ration)	Prebloom kg 4% FCM	Full bloom kg 4% FCM	kg/day 4% FCM	kg/day delay[a] 4% FCM
20	36.2	23.7	12.5	0.57
37	37.9	25.2	12.7	0.58
54	39.6	29.4	10.2	0.46
71	39.1	31.6	7.5	0.34

Source: J. R. Kawas, et al., 1983, *J. Dairy Sci.*, Supp 1, 66:181

[a] 22 days between prebloom and full bloom.

ity forage. A 22-day lag occurred between prebloom and full-bloom stages, which resulted in 12 kg less milk per day, a decrease of 0.6 kg per day of delay in harvest.

Leaf Content

Total dry matter yield per acre increases with the maturity of the plant. Leaf dry matter increases at a constant rate through the early bloom stage, while stem dry matter increases at a constant rate throughout all growing stages. This results in an increase in dry matter yield, but a decrease in leaf/stem ratio. Alfalfa leaf protein varies from 30% (prebud) to 23% (full bloom), while stem protein content drops from 18% (prebud) to 9% (full bloom). Energy follows a similar pattern. Retention of leaves during harvest is also critical.

Field and Harvesting Losses

Harvesting methods affect the amount of forage consumed and the efficiency of conversion into milk (Table 20.1). Dry matter losses and mechanization of forage system are important in determining how a crop should be harvested. This topic is discussed in more detail in the next section. Dry matter losses resulting from various harvesting methods will be discussed briefly.

Harvesting, storage, and feeding losses of various systems are shown in Table 20.3. These losses are averages and variation will be found among farms and harvesting methods. Harvesting losses increase as the plant remains in the field after cutting. Barn-dried hay, whose moisture content is 25 to 35% before baling, has lower field and harvesting losses. Losses for field-cured hay are much higher because the hay remains in the field longer and is more susceptible to rain damage. Dry matter losses for field-cured hay range from 26 to 54%, depending on amount of rain damage, method of harvest, and amount of leaf loss during harvesting.

Hay crushers and crimpers (hay conditioners) are used to decrease field-curing time. The crusher consists of two rollers that break or crush the entire stem of the plant. Crushing reduces field-drying time by increasing surface area

TABLE 20.3 EXPECTED DRY MATTER LOSSES IN LEGUME–GRASS FORAGE HARVESTING, STORING, AND FEEDING

	Dry matter losses			
Method	Harvest (%)	Storage (%)	Feeding (%)	Total (%)
Hay, conventional bale				
Rained on	32.6	4.0	5.2	41.8
Average	25.0	3.8	5.2	34.0
No rain	17.4	3.6	5.2	26.2
Barn dried	13.4	1.8	5.2	20.4
Large packages				
Field cured	25.0	14.2	15.3	54.5
Acid treated	15.0	10.7	5.5	31.2
Hay crop silage moisture (%)				
70+	2.0	21.2	11.0	34.2
60–69	5.0	10.1	11.0	26.1
Under 60	11.5	8.2	11.0	30.7

Source: F. J. Benson, 1980, *Univ. Minn. Silage Clinics Spec. Rep. 81*, p. 13

of the stem. Crimpers perform the same function as crushers except that two corregated rollers crimp the plant at intervals of approximately 2.5 cm (1 in.). Crushers and crimpers decrease field drying time by ½ to 1 day.

Chemical hay-drying agents are salt solutions sprayed on legumes to speed the rate of forage drying. Water is lost from the plant through stomata (pores), outer cell wall, and transfer from the stem to the leaves. The waxy cutin layer of the outer stem cell wall is broken down with chemical agents, allowing moisture to escape. The drying rate of the stem becomes about the same as the leaf-drying rate. Drying agents are applied as a spray to the stems at cutting. Potassium carbonate and sodium carbonate are drying agents that can be used with emulsifying agents and methyl esters. Under optimal conditions, drying time is reduced by 1 day with the drying agent. Hay quality is also improved with lower dry matter losses. Uniform application of the chemical is important. Economics, forage type, and drying conditions determine when drying agents can be used.

When a forage is harvested as silage, moisture content at storage is higher and forage does not need to remain in the field as long. Cutting forage and putting it directly in the silo (direct cutting) minimizes field and harvesting losses, but increases silo fermentation and seepage losses. Storage losses are lowered when moisture content is reduced at ensiling. As moisture content is reduced, field losses are increased, but fermentation and seepage losses are decreased. Silage, with moisture content as low as 50%, can be stored in conventional tower silos. However, oxygen-limiting tower (OLT) structures are recommended for silage with lower moisture content. Storing silages below 40% moisture is not recommended because of excessive heating, which leads to heat-damaged protein (browning reaction). Dry matter losses of silage stored in OLT units vary from 1 to 7% less than that of silage stored in conventional tower silos.

The decision to buy an OLT or conventional tower unit depends on savings in dry matter, type of forage to be stored, number of times the unit is filled, and type of feeding system. The OLT structure is suited for high-moisture corn and low-moisture hay crop silage. Because these units are unloaded from the bottom and filled from the top, they can be used for continuous feeding, even while being filled. The OLT units are considerably more expensive than conventional tower silos and have to be filled two times each year to justify their higher investment cost.

Harvesting Hay

Forage can be harvested as hay, silage, or pasture. Almost all hay is harvested as baled hay because loose hay requires too much labor to handle, and chopped hay is not as palatable as baled hay or as easy to handle. One major limitation of hay is the risk of weather damage. Because of the weather conditions existing in eastern and central parts of the United States, hay must be harvested during seasons when weather is adverse for hay making. During good drying conditions, hay has to be left in the field after being cut for approximately 2 days.

The moisture content of hay to be baled depends on the type of curing that will follow. If hay is baled without further curing, the moisture content must be less than 22% for conventional rectangular bales and 18% for large round bales to prevent spoilage and barn fires. If hay is barn dried by forced air, the moisture content can vary between 25 and 35%. Barn drying with forced air is effective on dry days when humidity is low. Hay can also be baled and barn dried with heat. The cost depends on the type of system used and number of tons dried per year. Solar heat is a possibility on some farms.

Propionic acid is another method to preserve wet hay. Adding $\frac{1}{2}$ to 2% propionic acid to hay containing 20 to 30% moisture prevents molding and heating. The acid is corrosive and should be washed off equipment after use. No health problems occur with animals fed acid-treated hay. Acid blends should contain over 60% propionic acid to inhibit mold formation. Other organic acids are available, but research results are variable.

Baling hay has several other disadvantages in addition to sizable field losses. Feeding baled hay is difficult to mechanize. Hay balers with bale throwers are used to throw bales into a wagon, but bales have to be unloaded by hand. Feeding bales out of storage is also a hand operation.

Baled hay has the advantage of being a practical way of shipping hay long distances. Hay wafers, pellets, and cubes are used for shipment in some sections of the country, but only a relatively small percentage of hay is shipped in this manner.

Hay Crop Silage

Many terms have been used to describe different types of silage systems. The term "hay crop silage" applies to silage made from perennial forages rather than from annual forages, such as oats and other small grains. The term "grass silage" is also used to describe hay crop silage, but this term is often confused with high-moisture, direct-cut, perennial silage. Other terms that have been used are "direct-cut silage," "wilted silage," and "haylage." Silages are more accurately described on the basis of their moisture content at harvesting. Two moisture ranges are used: high-moisture silages (direct cut) contain 60% or more moisture at ensiling; and low-moisture silages (wilted) contain less than 60% moisture at ensiling. Haylage refers to silage having a moisture content of 40 to 60%.

Silages with moisture contents as low as 50% can be satisfactorily stored in conventional silos, especially if silage having the lower moisture is stored in the bottom half of the silo. As the moisture content of the silage is reduced, more management skill is required to make successful silage. Some management recommendations for good ensiling practices are: (1) short chopping of the forage, (2) rapid filling of the silo, (3) storage in a silo in good condition, and (4) sealing with a securely placed and weighted plastic cap. Forage high in grass content and cut late in the season should be ensiled at a higher moisture content to ensure fermentation of the coarser material, which is difficult to pack. Immature forages contain more fermentable carbohydrates to produce acids to preserve the silage. An optimal fermentation process includes four phases. Phase 1 is the time period from ensiling through completion of aerobic fermentation. Oxygen is consumed and heat is generated. The longer silage remains in phase 1, the greater dry matter loss and potential for heat damage. Phase 2 occurs when acetic acid is produced, causing a decrease in pH from 6.0 to 4.2. Phase 3 results in lactic acid formation, a gradual decline in forage temperature, and a stable pH. Phase 4 represents a stable ensiled forage which is cool and preserved. If insufficient lactic acid is formed, butyric acid is produced in wet silage or mold formation can occur in dry silage.

Silage preservatives are designed to inhibit undesirable bacterial and mold activity and promote the type of fermentation desired. Fermentation stimulants include bacteria that direct the fermentation and rapidly reduce the pH to preserve the forage. Wet silages (over 60% moisture) may benefit from microbial inoculants if insufficient numbers of desirable bacteria are present to dominate undesirable organisms. Bacteria cultures need to be added at the rate of 100,000 colony-forming units per gram of silage. Enzyme additives break down the plant carbohydrates, making nutrients available for fermentation. Fermentation inhibitors consist of acids or products that inhibit bacteria or mold growth. Propionic acid, formic acid, blends of propionic acid and acetic acid, diacetate, and salts of weak acids are examples of inhibitors. These products are more effective in preventing molding or heating when added to dry silage than to wet silage.

Silage additives are products that improve the nutritional value of silage. Cereal grains (50 kg/ton), dried whey (10 kg/ton), or molasses (10 kg/ton) provide a source of fermentable carbohydrate. Adding carbohydrates improves fermentation and feed value of the forage.

Moisture of silage can be lowered below 45%, but field and harvesting losses are increased. Another limitation of low-moisture silage is the potential heating reaction. This condition, which results from heat damage, causes formation of undigestible protein and carbohydrate complex (also called artifact lignin) and reduces the nutritive value of silage. Two chemical tests are available to measure heat damage: acid detergent insoluble nitrogen (ADIN) and pepsin insoluble nitrogen. The unavailable nitrogen should be less than 10% of the total nitrogen in the forage; 5 to 7% is considered normal. Higher bound or unavailable protein can also be caused by grass and mature forage.

High-moisture silage is not recommended for several reasons. Dry matter intake from high-moisture silage is less than that from low-moisture silage and hay. High-moisture silage has higher fermentation and seepage losses, which result in dry matter losses. Fermentation usually produces objectionable odors. In addition, harvesting direct-cut or high-moisture material necessitates hauling large amounts of water with the forage. Protein in wet silage is also more soluble and converted to ammonia compared to drier forage.

Some variation in moisture content occurs at ensiling because harvesting all forage at one moisture content is impossible. For improved compaction, drier silages should be ensiled in the bottom part of the silo and wetter silages should be in the top part of the silo. Guidelines to estimate moisture in hay crop forages are listed in Table 20.4.

Silage can be stored in bunker silos, which have lower investment costs per ton and require less time to fill and feed out. Storage losses are higher in bunker silos than in tower silos; however, the extent of loss depends on the amount of silage stored in the silo. Becauses dry matter loss in bunker silos is potentially high, extra care must be taken to ensure tight packing of silage and adequate sealing of the silo. The silo should be filled rapidly and a heavy tractor should be driven back and forth over the silage to pack it. A plastic, airtight film should be placed over the top of the silage to reduce surface spoilage. The film should be weighted down, securely held in place at the edges of the silo, and protected from punctures.

The silage storing system to be used depends on the type of crops stored, number of times silos are refilled, dry matter losses in storage, size of operation, and comparative investment and annual costs. Hewitt (1986) made an economic

TABLE 20.4 ESTIMATING THE MOISTURE CONTENT OF HAY CROP FORAGE

Moisture level (%)	Condition
Over 70	Silage will have juice running when squeezed into a ball.
60–70	Silage will not juice, but will maintain a round or ball shape when squeezed.
50–60	Silage will slowly open up from a ball shape when squeezed.
40–50	Silage will not maintain a round or ball shape (falls apart).
30–40	Hay leaves begin to rustle and juice is difficult to extract by twisting.
20–30	Hay rustles and a bundle will shape with difficulty.
Less than 20	A bundle of hay will shape and leaves will shatter.

comparison of silage storage systems. Ownership cost and investment per ton of dry matter stored are summarized in Table 20.5. Annual costs vary by size and type of storage. Oxygen-limiting unit costs are highest but can be reduced if filled twice annually. Conventional tower silos combine high-quality forage with lower costs (compared to OLT units), but require better management. Horizontal concrete silos had the lowest cost per unit of dry matter. Because surface spoilage is a problem, 300 tons of silage dry matter or more is needed to make these units economical and efficient. High-moisture forage must be used to improve packing, exclude oxygen, and improve fermentation. Top management and large herd sizes are needed with horizontal silos.

Plastic bags are popular on farms where large silage units are not economical or when temporary silage storage is needed (Figure 20.1). The plastic bags hold 100 to 150 tons of wet silage. Dry matter losses and storage costs are similar to conventional silo storage. If bags tear, spoilage losses are high. Forage should be high in moisture content (over 60%) to ensure compaction and desirable fermentation. Since the diameter of the bag is small (2 to 3 m), daily feeding of fresh, palatable silage from the open end occurs. Cattle can self-feed directly from the bag.

Making hay silage in big round bales and sealing in a plastic bag (called balage) is another forage storage method. This approach allows producers to use existing hay equipment, especially when weather damage could occur. Bales are rolled up when moisture levels are from 50 to 70% and placed in plastic bags. Air should be expelled before tying. The seal must not be broken or the forage will mold and spoil. A high degree of management is needed since mold formation occurs if forage is too dry, spoilage develops if forage is too wet, wind-whipping can break the seal, and rodent damage can cause air exposure. Adding propionic

TABLE 20.5 ECONOMIC COMPARISONS FOR VARIOUS FORAGE STORAGE UNITS AT TWO CAPACITIES

Silo type and size	Capacity (tons DM)	Total investment	Ownership costs (per ton DM)
Metal oxygen-limiting			
25 × 80	200	$ 82,000	$67.65
25 × 90	325	113,800	57.77
Concrete oxygen-limiting			
20 × 72	170	62,000	60.18
30 × 100	510	120,000	38.82
Concrete stave			
20 × 70	155	30,250	32.20
20 × 80	425	52,500	20.38
Concrete bunker			
10 × 30 × 185	750	24,800	5.13
12 × 40 × 112	500	23,800	7.38
Bagger and bags			
5 bags	250	34,500	32.13
25 bags	750	38,000	11.91

Source: T. D. Hewitt, 1986, *Dairy Herd Manage.*, Vol. 24, No. 12, p. 29.

Figure 20.1 Storage of silage in a horizontal plastic bag. Photograph courtesy of the University of Illinois.

acid, bacterial additives, or anhydrous ammonia may improve fermentation and storage characteristics.

20.2 PASTURES AND GREEN CHOP SYSTEMS

Pastures continue to supply a substantial amount of energy and protein to dairy cattle. Pastures are important for small herds and some large herds when fields are near the barn or are too rough for mechanical harvesting. Pastures provide low-cost forage during the growing season. Pastures are economical when adequate land is available. There is a definite trend, however, toward feeding out of storage throughout the year.

No one summer feeding system is best for all farms. The proper type of summer grazing or feeding system depends on size of herd, amount of land available, amount of labor and management available, and location of fields in relationship to barns.

Grazing Systems

Three types of grazing systems are used. The most common method is continuous grazing. Cattle are turned out on pasture and the entire pasture is grazed at one time. Continuous grazing requires the least labor, but makes inefficient use of plants in the field because of trampling of existing plants and contamination of surrounding plants by manure droppings. More efficient use of the plants can be made by either rotational or strip grazing systems.

Rotational grazing is accomplished by dividing a pasture into several areas. Cows are allowed to graze one area for 1 to 2 weeks, after which they are moved to a second area. After grazing the second area, the cows are moved to a third and then to a fourth area. Plants in areas that have been grazed are allowed to regrow and are regrazed in the same rotation. This system makes more efficient use of plants because there is less trampling of forage and less manure contamination. Stages of growth of forage are easier to control. If excess forage is present in early spring, one or more of the areas can be harvested for hay or hay crop silage. Rotational grazing systems require additional fencing and water must be supplied to each area.

Strip grazing is a modification of rotational grazing. New pasture is given to the cattle daily by moving a temporary fence, such as an electric wire. This grazing system makes more efficient use of plants than rotational grazing. Strip grazing requires daily moving of fences, and water must be available to each strip or area that is being grazed.

Management of Pastures

Good management of cows and pastures is required to prevent the slump in milk production that often occurs during July and August in some parts of the United States. Most of the slump is caused by inadequate nutrient intake. An abundant source of clean water should be available close to the grazing areas at all times.

Cows should be afforded some protection from heat and solar radiation during summer months. Temporary shades or access to a wooded area should be provided. Flies and mosquitoes are annoying to cows and can cause a drop in milk production. Animals should be sprayed against insect pests.

Pasture also needs good management. A primary ingredient for successful pastures is the type of forage. Proper fertilization is also important. In addition, the pasture should be clipped periodically to encourage new growth and eliminate weeds. It is desirable to harrow the pastures after each grazing to spread manure droppings.

Several methods are available to supplement the nutrients supplied by pasture during hot summer months when the amount of permanent pasture is reduced. Hay and silage can be fed or some annual crops can be chopped daily (green chopped) and fed to supplement the pasture. During the later part of summer, early corn can be chopped and used as a supplemental feed.

Planting sudangrass or sorghum-sudan grass hybrids (SSG) is another way of supplying supplemental grazing. Planted in late spring, SSG produce large amounts of forage from late June to September and are drought-resistant. The SSG must reach a height of at least 46 cm (18 in.) before it is grazed. Immature SSG may contain prussic acid (hydrogen cyanide) which kills cattle. Plants that are 46 cm high or taller contain nontoxic levels of prussic acid. Growth of SSG occurring after frost must not be grazed by cattle because new growth has a high content of prussic acid. To spread maturity, three plantings at 10-day intervals are recommended.

Green Chopping

Many dairy managers have discontinued pasturing crops and have changed to either green chopping or stored feeding. Green chopping (zero grazing or soiling) is used if herd size is large, labor is plentiful, and land for pasture is scarce. Green chopping is direct harvesting of growing forage, once or twice each day, and feeding it directly to the cows. A high degree of crop management and skill is necessary to produce a continuous succession of crops that can be fed at their most nutritious stage of growth. Land is effectively used with this method if machinery is available.

Advantages of green chopping are: (1) it makes efficient use of the plant because all of the plant is harvested and fed to the cows; (2) less acreage is required per cow; (3) animals are near the barn where they can be closely watched; (4) the necessity of providing fencing and supplying water to each pasture is eliminated; (5) incidence of bloat is reduced; and (6) daily milk yield fluctuates less than with most pasture systems.

Green chopping also has several disadvantages. It requires daily or twice-daily trips to the field and entails high labor and machinery costs. Wet weather and machinery breakdowns create problems unless a pasture or stored forage is available. In addition, crops must be harvested at the correct stage of maturity to make most effective use of green chopping. This requires a high degree of management skill. Equipment costs are relatively high unless the same equipment is also used for harvesting silages. Large herds are required before the machinery can be afforded for green chopping. Sanitation can be a problem unless paved lots are present for feeding chopped material. Soil compaction occurs on certain soil types when green chopping is done on wet ground.

Stored Feeding

Most dairy farmers who have used green chopping change to stored feeding for most or all of the year. Stored feeding eliminates daily labor and equipment requirements of green chopping, as well as problems of wet weather and machinery breakdowns. Other advantages of stored feeding are: (1) less acreage per cow is required than any other feeding system; (2) all forage can be harvested near the optimal stage of growth; (3) storage facilities can be used throughout the year so that silos can be used for hay crop silage and corn silage; (4) milk production is uniform throughout the year because the summer slump in milk production is eliminated; (5) cows can be watched closely; (6) little fencing is required and

TABLE 20.6 COMPARISON OF SUMMER FEEDING SYSTEMS

	Strip grazing	Green feeding	Stored feeding
Milk yield/acre (kg)	2654	2784	3648
DM available/acre (kg)	2784	2478	3152
DM consumed/acre (kg)	1809	2418	2845
Loss of DM (%)	35	5	10
Cows per acre (no.)	1.3	1.4	1.8
Alfalfa reduction, 4 years (%)	46	15	15

Source: H. J. Larsen, M. F. Finner, R. F. Johannes, and J. M. Sund, 1964, *Wis. Exp. Stn. Bull.* 549.

water does not have to be supplied to each pasture; (7) fields can be far away from the barn; and (8) conditions are favorable for maintaining good forage stands.

Stored feeding has some disadvantages. Storage facilities have to be enlarged to handle forage fed during the summer. Manure removal is required. Large, well-drained pastures that could be used for exercise lots would help to alleviate this problem. Additional forage must be made during the forage harvesting season. This is particularly critical when first-growth material is harvested because forage must be harvested in a short period. This requirement is less critical when corn silage is used.

Comparisons of various summer feeding programs are summarized in Table 20.6. Stored feeding systems are superior because they produce more milk per acre, have less dry matter loss, and result in minimal damage to the alfalfa stand.

20.3 CORN SILAGE

Since 1960, marked increases in corn acreage grown for silage have occurred in corn-growing areas. In areas where corn is well adapted, corn harvested as silage results in an energy yield per acre that is unsurpassed by any other crop. In addition, corn can be harvested as grain, silage, or green chop material. Other advantages of corn silage in comparison with other forages are high digestibility, adaptability to harvesting and handling by mechanical systems, and ease with which high-quality silage can be consistently harvested and stored with minimal losses.

Stage of maturity at harvest of the corn plant determines dry matter yield per acre, field and storage losses, and acceptability of the silage to dairy cattle. Stage of maturity, however, has little effect on digestibility. Total yield per acre can be misleading because it does not accurately reflect dry matter and energy value of the crop. As the plant matures, dry matter continues to accumulate until physiological maturity is reached. As the plant matures, stalk and leaves make up a smaller percentage of the dry matter, but this decrease is compensated for by the increase in dry matter content of the ear. Approximately 25 days are required for silage to move from 19% dry matter to 30% dry matter. During this interval, each acre accumulates 2 tons of dry matter or approximately 73 kg (160 lb) per day.

The recommended dry matter content of corn silage is between 30 and 35%; the corn kernel is in the dent stage. There are several reasons for this recommended moisture content. Dry matter content is higher in the dent stage than in the early or late milk stage. Field and harvest losses are low. If the plant is left in the field longer, field losses increase. Below 32% dry matter, seepage loss occurs in the silo. Seepage loss also depends on the height of the storage unit (taller units require drier silage). Field and harvest losses of silages harvested between 30 and 35% dry matter are approximately 4%. Storage loss for conventional silos is 8% and 7% for the OLT units.

Silage intake by cows also depends on dry matter content. Above 35% dry matter, feed intake levels off and there is little advantage to increasing dry matter content of corn silage to 50%. The TDN content remains approximately the same at 68 to 70%. Digestibility of stalks and leaves decreases with maturity, but this decrease is compensated for by the increase in the proportion of grain in the plant.

When mature plants are used for silage, the rate of whole kernel passage through the digestive tract increases. Rolling corn forage (harvested above 35% dry matter) prior to feeding reduces whole kernel passage, but the added cost of equipment and power must be considered.

Research has been done on selection of plant varieties for corn silage. In general, any variety producing high grain yields is well suited for silage production.

All-Corn Silage Forage Ration

Corn silage has been used as the only forage for dairy cows for partial or complete lactations. When corn silage is adequately supplemented with proteins, vitamins, and minerals, milk production from corn silage is similar to rations based on hay or hay crop silage. In a Cornell trial, cows that received only corn silage as the forage for 3 years and liberal concentrate feeding left the herd at a much faster rate than did cows that received some hay in the ration. By the end of the third lactation, 9 of the 10 cows were lost from the corn silage group for various reasons. Dairy managers have fed corn silages for long periods, but cows usually receive some supplemental hay, are fed hay crop silage in addition to the corn silage, or receive some pasture or other forage during the dry period, which can alleviate difficulties encountered with an all-corn silage program. Physical form (fineness of chop), level of fiber, amount of concentrate, and frequency of feeding affect the success of all-corn silage forage diets. Dry cows and cows in late lactation on an *ad libitum* intake of corn silage receive more energy than they require and become fat.

Supplementation of Corn Silage

Corn silage is an excellent, high-energy feed, but has limitations. Because of the low protein content, cows that are fed corn silage require concentrate mixtures that contain high amounts of protein supplement to balance their protein needs.

Corn silage is also low in calcium and other minerals, such as potassium, sulfur, magnesium, and iodine. Carotene content of silage is dependent on stage of maturity of the plant at harvest. Carotene content is high when silage is harvested at the milk stage (141 mg of carotene per kilogram of dry matter). When corn silage is ensiled at the dent stage, carotene is reduced to 33 milligrams per kilogram. After frost, carotene content is 5 mg per kilogram of dry matter.

Additives for Corn Silage

Several methods can be used to increase nitrogen content of corn silage. Nitrogen fertilization increases yield of corn and nitrogen content of the silage. Excessive nitrogen fertilization, however, causes nitrate accumulation in the corn plant, which can cause nitrate poisoning of animals. However, other factors also influence nitrate content of the plant. Another hazard resulting from excess nitrate in the corn plant is accumulation of toxic gases in the silo. During fermentation, nitrate breaks down to produce nitrogen dioxide, which, at high levels, is toxic to people and animals. Extreme care must be taken when entering silos immediately after filling that contain corn silage with high nitrates.

A common method of increasing protein content of silage is to add a nonprotein (NPN) supplement such as urea or ammonia at ensiling or at the time of feeding. Thorough mixing of NPN with the silage is critical to prevent palatability problems. The major advantage of adding NPN to corn silage is that a relatively inexpensive source of nitrogen can be used to increase protein content of the silage. The addition of NPN to corn silage instead of to concentrate also minimizes the possibility of excess rumen ammonia because NPN intake is spread out during the day. If NPN-treated corn silage is the major source of forage for dairy cattle, the NPN content of the concentrate mixture should be reduced.

Corn silage can be supplemented with NPN at time of ensiling or when the silage comes from the silo at feeding time. Advantages of adding NPN at ensiling are: (1) adding NPN at ensiling ensures adequate mixing; (2) microbes in the silo utilize NPN as a nitrogen source sparing corn protein; and (3) NPN-treated silage does not heat up quickly and has a long bunk life when fed.

Some advantages of adding NPN to the silage as it comes from the silo are: (1) flexibility is retained in supplementing the corn silage for dairy cattle that do not need additional protein, (2) NPN is not lost through fermentation and seepage losses, (3) the possibility of NPN migration and concentration in part of the silo is avoided, and (4) money is not invested since NPN can be purchased as needed.

To use NPN effectively, chopped corn should contain between 30 and 40% dry matter. Seepage occurs if dry matter is less than 32% and leaches NPN out of the unit. If the percentage of dry matter is too high, added NPN lowers silage intakes due to ammonia odor and nitrogen losses occur through volatilization. The recommended level of additional NPN is 2 to 3 kg ($4\frac{1}{2}$ to $6\frac{1}{2}$ lb) of actual nitrogen per ton of corn silage at 33% dry matter. This level of NPN increases crude protein content of the silage approximately 3.0 to 4.3% on a dry matter basis, which decreases the amount of protein required in the concentrate mix.

Calcium in the form of limestone or calcium carbonate can be added to corn silage. Addition of calcium increases lactic and acetic acid and calcium content of the ensiled material. Results of feeding silage with added limestone to lactating dairy cows have not been consistent in terms of increased silage intake and milk production.

20.4 DEVELOPMENT OF A FORAGE PROGRAM

The type of forage program that should be established on a dairy farm depends on many factors. For example, the area of the United States and the crops grown in that area are important, as are herd size, labor supply, type of barn, and mechanization available. If a stall barn with overhead hay storage is present, baled hay will be one forage used. The type of machinery on the farm may dictate the forage program. If a baler and bale wagons are present and in good condition, the use of baled hay will be continued. The size of the herd is also important. Mechanization is not critical when herds are small, and baled hay can be effectively used. In larger herds, more mechanization is required. If good weather conditions prevail during hay harvest, less emphasis is placed on an all-silage harvesting program. Land available for forage also influences the type of harvesting program. If some feeds have to be purchased, more emphasis will be placed on dry hay because silage is difficult and expensive to transport.

For large herds where mechanization is required for efficient milk production and labor efficiency, an all-silage feeding program has many advantages. Silage is easier to mechanize than feeding hay. No hand labor is required to move silage from a bunker or tower silo to the feed bunk. Some manual labor is required to feed hay. Mechanization of silage feeding is best adapted to free stall barns or to feeding systems outside the housing area. Mechanization of silage feeding can be done in a stall barn, but it is expensive.

A second advantage of an all-silage system is that only one set of equipment is required. If a combination of corn silage and hay crop silage is used, one field chopper can be used for both crops if it has a hay silage head and a corn head. The forage wagons and silage blower can be used for both silages. If baled hay is used with corn silage, a baler and separate bale wagons must also be used.

If an all-silage system is established, there are several advantages with using a combination of corn silage and hay crop silage. A combination of the two forages facilitates balancing nutrient requirements of the rations. Hay crop silage, with high protein and calcium, balances the high energy and low protein and calcium content of corn silage. When corn silage and hay crop silage are varied by production groups, the protein content of the concentrate mix can be the same for all cows at various production levels.

A combination of hay crop silage and corn silage reduces the risk of total crop failure, and the work load is spread over two seasons instead of one. The majority of hay crop silage is harvested in late spring and summer; corn silage is harvested in late summer or early fall. Because legumes and grasses are usually grown in a crop rotation, there is a definite advantage to using hay crop silage with corn silage. An all-silage program also makes maximum use of storage structures because it is easier to feed out of storage and refill throughout the year.

20.5 FORAGE TESTING PROGRAMS

Forage testing laboratories have been set up around the United States in which samples of forage and other feeds are analyzed for chemical components depending on the request of the dairy manager. For an accurate forage analysis, a representative sample must be taken. Hay should be sampled as it is going into storage. Two to four bales from each load can be set aside and samples taken with a forage probe (a metal tube device) at the end of the day. A minimum of 15 to 20 bales should be sampled, mixed, and a composite sample taken. Silages should also be sampled during ensiling. Grab samples from each load can be placed in a large plastic bag. A composite subsample can be taken and frozen until all samples have been collected and are ready for analysis. If heat damage and fermentation losses occur, the fermented silage should be retested. A notebook is critical for matching forage test results with stored forages. Forage samples should be taken once a month or when major changes have occurred.

The major purpose of forage testing is to mix and feed concentrate more accurately. Forage testing gives an indication of the nutritive value of a feed. Energy prediction equations are listed in Table 20.7. Accurate estimation of forage consumption is still the biggest obstacle to feeding concentrate according to the needs of cattle. If the analysis includes a measure of cell wall content or neutral detergent fiber (NDF), it can be used to estimate voluntary feed intake.

TABLE 20.7 ENERGY PREDICTION EQUATIONS FOR FEEDS[a]

1. Legumes
   $NE_L(Mcal/kg) = [1.044 - (0.0123 \times \%ADF)] \times 2.2$
2. Mixed legume–grass and unknown forages
   $NE_L(Mcal/kg) = [1.044 - (0.0131 \times \%ADF)] \times 2.2$
3. Grasses
   $NE_L(Mcal/kg) = [1.085 - (0.015 \times \%ADF)] \times 2.2$
4. Corn silage (whole plant)
   $NE_L(Mcal/kg) = [1.044 \times (0.0132 \times \%ADF)] \times 2.2$
5. Sorghum and small-grain forage
   $NE_L(Mcal/kg) = [1.044 - (0.0123 \times \%ADF)] \times 2.2$
6. Complete rations
   $NE_L(Mcal/kg) = [0.866 - (0.0077 \times \%ADF)] \times 2.2$
7. Concentrate mixture
   $NE_L(Mcal/kg) = [(0.01598 \times \%TDN) - 0.36774] \times 2.2$

Source: R. S. Adams, 1975, Equations and conversions used by the Penn State Forage Testing Service, *Mimeo 1C:9:75*.

[a] NE_L, net energy—lactation; ADF, acid detergent fiber; TDN, total digestible nutrients (all values expressed on a 100% dry matter basis).

Forage analysis is helpful in formulating a ration. Analyses of protein and mineral contents of forages provide information on the amount of protein and minerals needed to be added to concentrate and other sources. A fiber analysis helps to maintain a minimum fiber content of the complete diet to produce normal milk fat test.

Forage analysis can also be helpful when protein and mineral contents of the ration might be lacking. The protein content of forages can vary markedly, and a chemical analysis would give the dairy manager information on the amount of protein required in the concentrate. Forage analysis is also helpful when heat-damaged silage is fed. Because the calcium content of corn silage and grasses is low and variable, its analysis can be used in determining the amount of calcium required in the concentrate mixture. Some dairy farmers overestimate the amount of legume in their hay or hay crop silage. If they are trying to maintain calcium and phosphorus intake, a forage analysis allows them to formulate the mineral supplementation program. If trace minerals are lacking in dairy rations, a trace mineral analysis of forage is valuable to determine if supplementation of these minerals is needed.

20.6 FORAGE INVENTORY

Forage inventory and access are important management factors. Feeding a balanced ration year round will maintain milk production and reduce feed costs. A feed inventory can allow correct decisions regarding current and future feeding programs. Forage access allows feeds to be fed as needed and be available as desired.

The yearly forage requirement (expressed on an as-fed or wet basis) with various forage combinations and milk yields are illustrated in Table 20.8. Forage

TABLE 20.8 YEARLY FORAGE REQUIREMENTS (TONS WET FEED/YEAR) OF A DAIRY COW (5900 KG OR 13,000 LB WITH A 3.5% MILK-FAT TEST)[a]

		Forage amount [kg/day (lb/day)]				
	Corn silage:	27.3 (60)	20.5 (45)	13.6 (30)	6.8 (15)	0 (0)
Milk yield [kg (lb)]	Alfalfa hay:	0 0	2.3 (5)	5.5 (12)	8.2 (18)	10.5 (23)
8,182	Corn silage	12.6	9.8	6.1	3.1	—
(18,000)	Alfalfa hay	—	1.1	3.5	3.6	4.8
6,818	Corn silage	12.9	10.1	6.2	3.2	—
(15,000)	Alfalfa hay	—	1.1	2.6	3.8	5.1
5,909	Corn silage	13.6	10.6	6.7	3.3	—
(13,000)	Alfalfa hay	—	1.2	2.7	4.0	5.3

Source: J. G. Linn and R. D. Appleman, 1980, *Univ. Minn. Ext. Folder* 545.

[a] Corn silage contains 35% dry matter and alfalfa hay contains 86% dry matter. All feed amounts expressed on an as-is or wet basis; include dry cow and feeding losses.

needs for young stock must be added to the totals listed in Table 20.8. Forage needs of replacement heifers can be estimated by adding 30% of the milking herd needs to the total. A second approach is to convert young stock to cow equivalents (two yearlings or four calves equal one cow unit).

After completing a feed inventory, the dairy manager can determine forage supply, sell any excess forage, purchase forages at the best-buy times of the year, substitute forage for concentrate, plan future cropping programs, and plan all feeding programs (milk cows, dry cows, heifers, and calves) based on forage supplies.

REFERENCES AND FURTHER READINGS

ADAMS, R. S. 1975. Equations and conversions used by the Penn State Forage Testing Service. *Mimeo 1C:9:75.*

BENSON, F. J. 1980. Economic comparison of silage structures. *Univ. Minn. Silage Clinics Spec. Rep.* 81, p. 13.

CROWLEY, J. W., N. A. JORGENSEN, and G. P. BARRINGTON. 1979. Corn silage for the dairy ration. *Wis. Coop. Ext. Pub.* A1178.

HEWITT, T. D. 1986. Silo economics. *Dairy Herd Manage.,* 24, No. 12, p. 29.

KAWAS, J. R., N. A. JORGENSEN, A. R. HARDIE, and J. L. DANELON. 1983. Changes in feed value of alfalfa with stage of maturity and concentrate level. *J. Dairy Sci.* Suppl. 1., 66, p. 181.

LARSEN, H. J., M. F. FINNER, R. F. JOHANNES, and J. M. SUND. 1964. Systems of summer feeding. *Wis. Exp. Stn. Bull.* 549.

LINN, J. G., and R. D. APPLEMAN. 1980. Feed inventory for dairy cattle. *Univ. Minn. Ext. Folder* 545.

MONTGOMERY, M. J. 1978. Use of forages in meeting nutrient requirements. *Large Dairy Herd Management.* University of Florida Press, Gainesville, FL, p. 321.

SNIFFEN, C. J., and L. E. CHASE. 1981. Forage testing. *Cornell Dairy Manage.* P225.00.

TRIMBERGER, G. W., ET AL. Effects of liberal concentrate feeding on health, reproductive efficiency, economy of milk production, and other related responses of the dairy cow. *N.Y. Food Life Bull.* 8.

21

Feeding Concentrate to Dairy Cows

Even though forage is the foundation of the dairy feeding program, concentrate provides additional energy, protein, minerals, and vitamins. A cow consuming an average amount and quality of forage can consume enough nutrients for maintenance requirements and for 10 to 20 kg (22 to 44 lb) of milk. The extra nutrients required for higher milk production must come from concentrate consisting of grains, by-product feeds, protein supplements, minerals, and vitamins.

21.1 FUNCTIONS OF CONCENTRATE

The major function of concentrate is to supply additional nutrients required for milk production above that obtained from forage. Higher energy intake can be achieved with concentrate because of its higher energy value per unit of dry matter and higher digestibility than forage. Concentrate is also less bulky, takes less time to pass through the digestive tract, and is more palatable than forage.

Another major function of concentrate feeding is to adjust the amount of protein fed to the dairy cow. The amount of protein required in the concentrate mixture depends on the quantity and quality of forage fed. The protein content of the concentrate mix must be higher when low-protein forage, such as corn silage, is fed rather than legume forage. General rules for protein content needed

TABLE 21.1 PROTEIN CONTENT (PERCENT) NEEDED IN CONCENTRATE MIXTURES FED WITH VARIOUS TYPES OF FORAGE

	Forage Consumption Rate		
Forage type	High	Average	Low
Nonlegume	18	16	14
Nonlegume and corn silage	20	18	16
Mixed grass–legume	16	14	12
Mixed grass–legume and corn silage	18	16	14
Legume	12	13	14
Legume and corn silage	14	15	16
Corn silage	22	20	18

in concentrate mixtures with various forage types and quantities are given in Table 21.1.

Another function of concentrate is to supply other nutrients, especially minerals and vitamins. Concentrate mixtures supply salt, phosphorus, calcium, other macro and micro minerals, additives, and vitamins required by the cow.

A concentrate mixture should (1) provide sufficient protein to supplement that furnished by the forage, (2) balance energy needs; (3) be palatable and of good quality, (4) provide sufficient minerals and vitamins, and (5) meet these requirements at the least cost. Some dairy managers feel that complex mixtures containing many ingredients are required for high-producing cows. Most research indicates that no difference in performance occurs when cows are fed simple mixtures containing four or five ingredients rather than mixtures containing many ingredients.

21.2 SUBSTITUTION OF CONCENTRATE FOR FORAGE

High-producing cows require large amounts of energy. Because grain production in the United States is high and forage making is becoming increasingly expensive, the gap between the cost per unit of energy in grain and forages has narrowed. Some dairy managers have expressed interest in substituting concentrate for forage in dairy cow rations.

Concentrate should not constitute more than 60% of the total dry matter intake. Above this level, factors other than physical capacity control consumption of feed. If the concentrate portion of the ration is increased, no further gain in digestible energy occurs.

With high-quality forage, total energy consumption tends to plateau when the concentrate portion of the ration approaches 40% of total intake. With poor-quality forage, this point is not reached until concentrate makes up 60 to 70% of the total ration.

Another factor that determines the upper limit of concentrate feeding is fat test. If the concentrate portion of the ration is too high, the fiber content of the total ration decreases, resulting in a lowered fat percentage. Cows should be fed at least 1.5% of their body weight daily as forage dry matter (DM) to maintain fat test.

A third consideration in maintaining adequate forage intake is prevention of digestive upsets. When forage intake is less than 1% of body weight, digestive upsets (reduced rumination, bloating, and constipation) occur. The incidence of displaced abomasum increases when high-concentrate, low-forage rations are fed.

21.3 BALANCING DAIRY RATIONS

Calculation of Nutrient Requirements (Step 1)

The first step in balancing dairy rations is to calculate the nutrient requirements. Requirements can be calculated for the average cow in the herd, the average cow in each group, or every cow. Even though many nutrients are required, calculations include dry matter, energy, crude protein, acid detergent fiber, calcium, and phosphorus.

In the example that follows, the assumptions are an average body weight of 600 kg (1320 lb) and daily production of 30 kg (66 lb) of milk containing 3.5% fat. Daily requirements, calculated from Table 18.1, are summarized in Table 21.2.

Estimating Dry Matter Intake (Step 2)

After calculating the nutrient requirements, the second step is to determine total dry matter intake that contains the needed nutrients. Dry matter intake is based on body weight and milk yield (Table 21.3). Kilograms of 4% FCM (fat-corrected milk) are calculated by multiplying kilograms of milk by 0.4 and adding kilograms of fat multiplied by 15. In the example, kilograms of 4% FCM is 30 kg times 0.4 (12 kg) plus 1.05 kg of fat (30 kg × 3.5%) times 15 (15.75 kg), which equals 27.75 kg of 4% FCM. Estimated dry matter intake would be 21.4 kg per day (Table 21.3).

1. 0·4 × 30 = 12·00
2. 0·35 × 30 = 1·05 + = 27·75.
3. 1·05 × 15 = 15·75

TABLE 21.2 NUTRIENT REQUIREMENTS FOR 600-KG (1320-LB) COW PRODUCING 30 KG (66 LB) OF MILK CONTAINING 3.5% MILK FAT

	Net energy (Mcal)	Crude protein (kg)	Calcium (g)	Phosphorus (g)
Maintenance (Table 18.1)	9.7	0.59	24	17
Milk production (Table 18.1)	20.7	2.37	89	55
Total	30.4	2.96	113	72

TABLE 21.3 DRY MATTER INTAKE GUIDELINES (PERCENT BODY WEIGHT)

Milk yield 4% FCM (kg)	Body weight (kg)		
	400	600	800
10	2.7	2.2	1.9
20	3.6	2.9	2.4
30	4.4	3.5	2.9
40	5.5	4.0	3.3
50	—	4.7	3.7
60	—	5.4	4.3

Source: Adapted from National Research Council Committee, 1988, *Nutrient Requirements of Dairy Cattle*, 6th ed., National Academy of Sciences, Washington, DC.

Determining Forage Consumption (Step 3)

The amount of nutrients supplied by forage is estimated, which introduces the largest error in calculating concentrate needs. Several methods can be used to determine accurate forage intake values.

1. Silage and hay amounts can be weighed. Waste must be weighed or estimated and subtracted. Hay waste varies from 5 to 15% depending on quality, feeding method, and amount offered. Silage waste is less, varying from 4 to 9%.
2. Feed bunks can be filled and several measured lengths weighed to estimate the total amount in the total bunk length.
3. Volumes (forks, baskets, or carts) can be weighed and multiplied by the number fed.
4. Disappearance rates out of silos (centimeters per day) and silo capacity tables can be used to estimate forage intake over a time period.
5. The amount of silage unloading can be caught for several 30-second periods from the silo, averaged, and multiplied by the time the silage unloader is running.

Forage dry matter intake varies from 1.5 to 3% of the cow's body weight, depending on forage quality, concentrate level, and feeding system. As a guideline, 2% of body weight can be used to estimate forage dry matter intake. In the example 12 kg (26.4 lb) of forage dry matter will be used. If the cow consumed 6 kg of alfalfa (midbloom) dry matter, 6 kg of corn silage dry matter could be consumed free choice. Nutrients from the forages are subtracted from nutrient requirements (Table 21.4).

Calculation of Concentrate Needs (Step 4)

After nutrients provided by forages are subtracted from requirements, the nutrient shortages must be provided by the concentrate mixture. The energy of the grain

TABLE 21.4 EXAMPLE CALCULATION OF A RATION FOR 600-KG COW PRODUCING 30 KG OF MILK CONTAINING 3.5% FAT FED ALFALFA (½ FORAGE DM), CORN SILAGE (½ FORAGE DM), SHELLED CORN (ENERGY SOURCE), SOYBEAN MEAL (PROTEIN SOURCE), AND DICALCIUM PHOSPHATE (MINERAL SOURCE)

	DM (kg)	Net energy (Mcal)	Crude protein (kg)	Calcium (g)	Phosphorus (g)	ADF (kg)
Requirements (steps 1 and 2)	21.0	30.4	2.96	113	72	4.0
Forages (step 3)						
Alfalfa	6	7.5	0.96	81	13	2.4
Corn silage	6	9.5	0.48	16	12	1.9
Forage nutrients	12	17	1.44	97	25	4.3
Shortage (requirements − forages)	9	13.4	1.52	16	47	(+0.3)
Concentrates						
Shelled corn (step 5)	4.6	9.3	0.46	1	14	0.1
Soybean meal (step 5)	2.2	4.1	1.09	6	15	0.2
Dicalcium phosphate (step 6)	0.10	0	0	21	18	0
Total (step 7)	18.9	30.4	2.99	125	72	4.6

mixture (Mcal/kg) is divided into the amount of energy needed. In the example, the energy content of shelled corn (2.03 Mcal/kg of dry matter) is divided into the energy shortage (13.4 Mcal) to find that 6.6 kg (14.5 lb) of shelled corn dry matter is needed to balance the energy needs (Table 21.4). Shelled corn also provides protein, fiber, and minerals, which are added to the forage nutrients. If all nutrient needs are met, no further calculation is needed.

Determining Protein Needs (Step 5)

If protein is deficient, a protein supplement replaces a portion of the grain mix. The amount of protein supplement needed is determined by dividing the amount of protein needed by the percent protein (on a DM basis) in the protein supplement. In the example, the amount of soybean meal (50% protein) is calculated by dividing the protein shortage (1.52 kg or 3.34 lb) by 0.50, which equals 3.0 kg (6.6 lb). Again, the fiber and mineral contribution of the soybean meal is added to forage and grain sources to check if all the nutrient requirements are met.

Simultaneous equations can be solved to determine the amount of grain (energy feed) and protein supplement needed to meet the energy and protein needs. In the example, shelled corn (2.03 Mcal/kg and 10% protein) and soybean meal (1.86 Mcal/kg and 50% protein) will be used. From Table 21.4 concentrate must supply 13.4 Mcal of energy and 1.52 kg of protein. If X equals kilograms of corn dry matter and Y equals kilograms of soybean meal dry matter, the following equations can be set up and solved.

For energy:

$$2.03X + 1.86Y = 13.4 \tag{1}$$

For protein:

$$0.10X + 0.50Y = 1.52 \tag{2}$$

Multiplying equation (2) by 3.72 (1.86 divided by 0.5 to calculate a constant) and subtracting it from equation (1) will cancel Y out of the calculations.

$$3.72 \text{ times } (0.10X + 0.50Y = 1.52)$$

The new equation (2) is subtracted from equation (1).

$$\begin{array}{rcrl} 2.03X + 1.86Y & = & 13.4 & \text{[Equation (1)]} \\ \underline{-(0.37X + 1.86Y} & \underline{=} & \underline{5.7)} & \text{[Equation (2)]} \\ 1.66X \qquad\qquad & = & 7.7 & \end{array}$$

so that

$$X = 4.6 \text{ kg of shelled corn}$$

The amount of soybean meal is determined by substituting the value of X (4.6) in equation (1).

$$(2.03) \times (4.6) + 1.86Y = 13.4$$

$$9.3 + 1.86Y = 13.4$$

$$1.86Y = 4.1$$

so that

$$Y = 2.2 \text{ kg of soybean meal}$$

Check Mineral Balance (Step 6)

The next step is to add a mineral supplement if phosphorus or calcium is deficient. Dicalcium phosphate or a commercial mineral mixture should be used to meet phosphorus needs first. The grams of mineral mix needed is calculated by dividing the amount needed by the amount of phosphorus in the mineral mixture. In the example, the shortage of 18 g of phosphorus is divided by 180 (grams of phosphorus per kilogram of dicalcium phosphate), which results in 0.10 kg (0.22 lb) of dicalcium phosphate. Then the calcium balance is checked. If a calcium deficiency exists, limestone is added, with the calculation of the amount similar to that for phosphorus.

Check Dry Matter and Fiber Levels (Step 7)

After calculating the amounts of feeds needed, the ration should meet the nutrient needs. If dry matter intake is not exceeded, the cow should be able to consume the feed. The calculations for the example indicate that the cow would be fed

18.9 kg per day. Step 2 indicated that the cow could consume 21 kg per day. Because of the potential to consume additional dry matter, the manager may want to substitute more forage for concentrate. A final check is to determine if fiber is adequate by dividing the amount of acid detergent fiber (4.6 kg) by the amount of total dry matter (18.9 kg) and multiply by 100. The percent acid detergent fiber (24.3%) is above the minimum level of 19%. The example ration is balanced.

21.4 RULES OF THUMB

Guidelines have been developed to estimate the amount and composition of concentrate for dairy cows. Thumb rules provide quick checks when a calculator or computer is not available or when troubleshooting a field problem. These rules do not follow a specific order or build on each other.

Thumb Rule 1. Concentrate amounts (expressed on a 90% DM basis) based on milk yield and breed are listed below.

	kg milk/kg grain
Holstein and Brown Swiss	
Less than 18.2 kg (40 lb) milk	4:1
18.2 to 31.8 kg (40 to 70 lb) milk	3:1
Over 31.8 kg (70 lb) milk	2.5:1
Jersey, Guernsey, Ayrshire, and Milking Shorthorn	
Less than 13.6 kg (30 lb) milk	3:1
13.6 to 27.3 kg (30 to 60 lb) milk	2.5:1
Over 27.3 kg (60 lb) milk	2:1

Amounts of concentrate (expressed on a 90% DM basis) also can be based on fat yield. Forage usually can support maintenance plus 0.45 kg (1 lb) of milk fat. If forage quality is above or below average, the constant (0.45 kg of milk fat) should be adjusted accordingly. The formula to calculate concentrate needs is as follows:

$$(\text{kg of milk fat} - 0.45) \times 15 = \text{kg of concentrate}$$

For example, a Jersey cow producing 20 kg of milk containing 5% milk fat produces 1 kg of fat. The formula would predict 8.3 kg [(1 − 0.45) × 15] of concentrate.

Thumb Rule 2. Protein percentage needed in the concentrate mixture can be estimated by subtracting the weighted average protein content of all forages (on a 100% DM basis) from 28. A forage program consisting of high-quality alfalfa forage (18% protein) would require a 10% protein concentrate mixture (28 − 18).

A ration of half corn silage (8% protein) and half grass forage (14% protein) would require a 17% protein concentrate mixture [28 − (8 + 14)/2]. Some nutritionists refer to this approach as the total 28 method.

Once the desired protein percentage in the concentrate is determined, the proportions of grain and protein supplement can be determined with the box or square method. The example (Figure 21.1) illustrates a 17% protein grain mixture (100% DM basis) using shelled corn (10% protein on a 100% DM basis) and soybean meal (50% protein on a 100% DM basis). The desirable protein percentage in the concentrate is entered in the middle of the box (step 1). Protein percentages of the grain and protein sources are entered in the upper left and lower left corners (step 2). Proportions of each component are calculated by subtracting diagonally (the smaller number from the larger number) and writing it in the appropriate corner (follow the arrows) (step 3). The feed proportions can be obtained by reading across or horizontally. In the example, a 17% protein grain mixture consists of 33 parts shelled corn (50 − 17) and 7 parts soybean meal (17 − 10) or 82.5% shelled corn [(33 ÷ 40) times 100] and 17.5% soybean meal [(7 ÷ 40) times 100] (step 5).

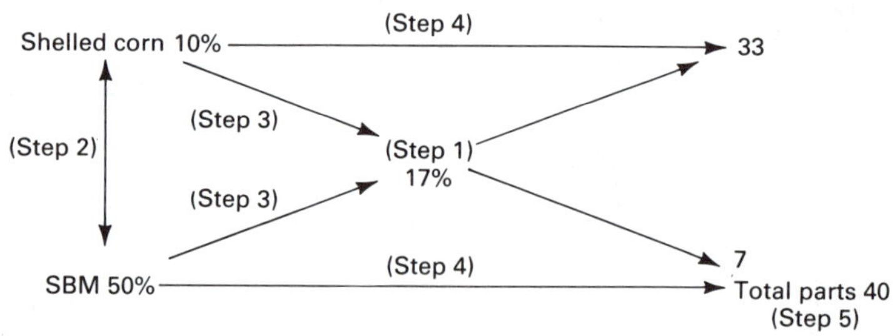

Figure 21.1 Illustration of the box method to determine the proportions of two feed ingredients to obtain the desired level of a nutrient.

Another way to determine the proportion of grain and protein supplement in a concentrate mixture is the algebraic method. If grain (energy feed) contains 10% crude protein and protein supplement contains 50% crude protein, the proportion of protein supplement is equal to X and the proportion of grain is equal to $1 - X$. An equation can be set up and solved for a concentrate mixture containing 17% crude protein.

$$50X + 10(1 - X) = 17$$

$$50X + 10 - 10X = 17$$

$$40X = 7$$

$$*X = 0.175 \text{ or } 17.5\% \text{ protein supplement}$$

The amount of grain can be determined by substituting 17.5% (0.175) for X in the formula $(1 - X)$ or $1 - 0.175$, which equals 0.825 or 82.5% shelled corn.

Thumb Rule 3. Topdressing of protein is needed when the protein in the concentrate mixture and forage does not meet the cow's needs. An additional 1 kg of protein supplement should be topdressed per 10 kg of milk when protein becomes limiting. The topdressing guide shifts to 1 kg of protein supplement per 5 kg of milk when maximum dry matter intake is reached. More protein supplement is added to meet the needs for higher milk yield and to replace the protein removed from substituting feeds.

Thumb Rule 4. Concentrate mixtures are formulated to supply 5 to 9 kg (12 to 20 lb) trace mineralized salt and 9 to 13.6 kg (20 to 30 lb) of dicalcium phosphate (or commercial calcium and phosphorus source) per ton. Additional limestone may be supplied when corn silage constitutes the only forage (see Section 20.3). Another guideline for salt is 28 g (1 ounce) for maintenance plus 28 g (1 ounce) for each additional 13.6 kg (30 lb) of milk per day. A vitamin supplement (2 to 4 kg per ton of concentrate) should provide 6000 IU of vitamin A, 1333 IU of vitamin D, and 25 IU of vitamin E per kilogram of concentrate.

21.5 CONCENTRATE FEEDING GUIDELINES

Concentrate feeding tables have been developed that account for milk production, milk fat content, age, forage consumption and quality, total dry matter intake, body weight change, and stage of lactation. Stage of lactation has two factors that influence concentrate recommendations. In the first 60 days after calving, milk production is increasing and peak dry matter intake has not been achieved. A cow should be challenged with extra nutrients to achieve higher milk yields. Dry matter intake is not at the maximum (10 to 18% below guides listed in Table 21.3) and nutrient concentration must be higher to achieve targeted nutrient needs.

Forage quality and intake have a large effect on concentrate needs. Table 21.5 illustrates the influence of three qualities of alfalfa and grass forages. Changes in the amount of concentrate and proportion of shelled corn (energy source) and soybean meal (protein source) occur. Economics of concentrate feeding must

TABLE 21.5 EFFECT OF FORAGE QUALITY ON CONCENTRATE FEEDING GUIDELINES (545-KG COW)

		Legume[a]		Legume–grass[b]		Grass[c]	
Milk yield [kg (lb)]	Fat test (%)	Shelled corn [kg (lb)]	Soybean meal [kg (lb)]	Shelled corn [Kg (lb)]	Soybean meal [kg (lb)]	Shelled corn [kg (lb)]	Soybean meal [kg (lb)]
13.6 (30)	3.9	3.1 (6.9)	0	4.0 (8.9)	0	4.5 (10.0)	0.5 (1.0)
22.7 (50)	3.7	6.8 (14.9)	0	7.0 (15.3)	0.9 (1.9)	7.2 (15.8)	1.6 (3.5)
31.8 (70)	3.5	9.7 (21.4)	0.6 (1.3)	9.5 (20.8)	2.0 (4.3)	9.6 (21.2)	2.7 (5.9)
40.9 (90)	3.3	12.1 (26.6)	1.6 (3.6)	11.9 (26.1)	3.0 (6.6)	12.0 (26.5)	3.7 (8.2)
50 (110)	3.1	14.3 (31.4)	2.6 (5.8)	14.0 (30.9)	4.0 (8.8)	14.3 (31.4)	4.7 (10.4)

[a] Legume forage contained 20% crude protein, 1.39 Mcal/kg, and 32% ADF.

[b] Legume–grass forage contained 15% crude protein, 1.30 Mcal/kg, and 35% ADF.

[c] Grass forage contained 12% crude protein, 1.21 Mcal/kg, and 40% ADF.

always be considered. Unlimited concentrate feeding to every cow is not recommended because only those cows that have the genetic ability respond to additional feed. A basic principle is to increase inputs (concentrates) to the point where the value of each additional output (milk yield) is equal to the value of the input.

Dairy Herd Improvement Feeding Guides

Dairy herd improvement (DHI) electronic data processing centers recommend the amount of concentrate to be fed to dairy cows. When DHI supervisors visit the farm, they estimate the amount of forage being fed to the entire herd, which is then averaged to a per-cow basis. Forage consumed is proportioned according to the body weight of cows on the assumption that cows eat forage according to their body weight. The amount of energy supplied by the forage is calculated according to individual body weights. Growth and maintenance requirements of each cow are subtracted from the energy supplied by the forage. The remaining energy is available for milk production. The amount of concentrate required for milk production above the amount supported by forage is calculated.

Other DHI centers compute the individual protein status of each cow (similar to energy) and recommend the protein needed in the concentrate or amount of protein supplement to be topdressed. Another option used at DHI computing centers is to estimate the amount of concentrate needed based on milk or fat yield with no forage considerations. Because DHI computing centers use different methods and factors, dairy farmers must carefully review how concentrate recommendations are determined in their state.

Linear Programming

Another method of formulating a concentrate mixture is by linear programming. The basic purpose of linear programming is to produce a concentrate mixture that meets specified minimum requirements at the least cost, and consequently these mixtures are called least-cost rations. Linear programming requires the use of an electronic computer because hand calculation using this procedure is laborious and time consuming.

The information concerning the proposed ration is programmed into the computer. Additional constraints may be placed on the ration, such as minimum fiber content, minimum and maximum number of feeds, and maximum amount of a particular feed that can be used. Other information on available feeds (cost, energy, protein, and fiber) is also programmed. If additional constraints on the ration, such as minerals and vitamins, are indicated; the contents of these in the feeds must also be given.

After the information has been programmed, the computer formulates a ration that meets the minimum specified requirements at least cost. Most computer programs also give other information, such as "price ranges" for each feed, or the range of prices between which each ingredient can be part of the ration without changing the least-cost formula. The computer can also produce "opportunity prices,'' which are the prices at which feeds not selected for the ration can be used in the ration and still produce the least-cost formula. The computer can also produce ''shadow prices,'' which are the cost of the last unit of each constraint used in the formulation of a ration. Shadow prices can be used to evaluate the relative cost of the various constraints to the total cost of the ration.

The least-cost formulation of rations had certain limitations. The major limitation is the problem of palatability of the ration. Sometimes rations are formulated that cows will not consume. Constraints on palatability of a ration are difficult to construct. Even if maximum amounts of a feed are used as a constraint, the combination of that feed with another feed may make the ration unpalatable. Another limitation is that data on the nutritive value of the various feeds vary among feeds and from year to year. Unless a chemical analysis is available, textbook figures have to be used. This limitation is less important in formulating concentrate rations than in formulating complete rations for the herd.

The formulation of dairy rations by computer is done by commercial computer organizations, feed companies and cooperatives, state agricultural experiment stations and extension services, and private nutrition consultants.

21.6 PHASE FEEDING

Phase feeding (Hutjens, 1986b) is an approach for designing feeding programs that are divided into five time periods based on milk yield, feed intake, body weight status, and gestation (Figure 21.2). Rations should be formulated for each phase

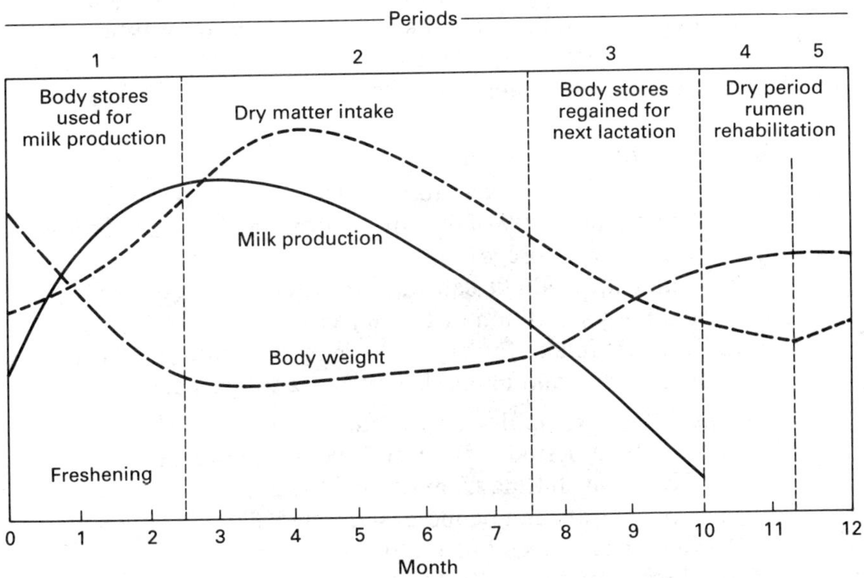

Figure 21.2 Nutrient and milk yield relationships in the lactation and gestation cycle. From M. F. Hutjens, 1986, *Ill. Dairy Rep.*, p. 2.

to optimize milk yields, minimize metabolic disorders, and control feed costs. Each phase is outlined below.

- *Phase 1* (peak milk occurs)
  a. The time period is from parturition to 80 days postpartum.
  b. Concentrate intake should increase 0.45 to 0.90 kg (1 to 2 lb) per day starting 3 to 5 days postpartum.
  c. Top-quality forage must be fed to maximize nutrient intake and digestability.
  d. Feeding supplemental fat or oil (0.45 to 0.90 kg per day) will increase energy intake.
  e. Adding a low-rumen-degraded protein supplement can increase protein for absorption in the small intestine and provide protein to utilize body fat mobilized as an energy source.
  f. Buffers can improve rumen pH and digestion.
- *Phase 2* (maximum dry matter intake reached)
  a. The time period is from 80 to 200 days postpartum.
  b. Nutrient intake should be calculated based on feed intake and nutrient concentration in feeds.
  c. Concentrate intake should not exceed 2.3 to 3.2 kg (5 to 7 lb) per meal.
  d. Body weight loss should stop and cows returned to a positive energy balance.
- *Phase 3* (late-lactation or tail-ender cow group)
  a. The time period is from 200 to 305 days postpartum (or drying off).
  b. Body condition lost in early lactation should be replaced.
  c. Economic rations can be fed, including more forage, less additives and concentrate, and NPN sources for protein.
  d. Concentrate changes should be based on 4% FCM yield, body condition status, and age.
- *Phase 4* (dry cow group)
  a. The time period is 45 to 60 days after drying off.
  b. A minimum of 1% of the cow's body weight should be long hay, preferably a grass-type forage.
  c. Concentrate intake can vary from zero to 3 kg depending on forage quality and body condition of the dry cow.
  d. Nutrient intake for dry cows must be controlled (Table 18.3).
  e. Excess salt and minerals can cause udder edema.
- *Phase 5* (close-up dry cow group)
  a. The time period is the last 2 weeks before calving.
  b. Cows are shifting from dry cow rations to early-lactation rations.
  c. Concentrate can be increased to 0.5% of the cow's body weight on a dry matter basis (lead feeding).
  d. Some long forage (0.5% of body weight) should be fed to avoid displaced abomasum.
  e. Six grams of niacin can minimize fatty liver syndrome.

f. Cows fed early in the evening (after 6 P.M.) will tend to calve between 5 A.M. and 7 P.M. when labor is more available to monitor and assist difficult births.

Body Condition Score

The amount of fat cover or body condition of a cow is one indication of her energy reserves. Changes in body condition reflect shifts in energy balance. Extremely fat cows are susceptible to fatty liver syndrome. Adequate body reserves must be available for mobilization to maintain high milk production. Overconditioned cows have more calving difficulties, increased metabolic disorders, and an impaired immune response. For optimum reproductive efficiency, cows must not be too fat or too thin. However, body condition has been difficult to evaluate; weight changes do not account for frame size and body conformation. Virginia researchers (Wildman and Jones, 1981) developed a scoring system of 1 (thin) to 5 (fat) which evaluates flesh covering over the ribs, hooks, pins, and tailhead regions. Virginia researchers reported a negative relationship between body condition of dairy cows and milk yield per unit of metabolic body weight. The more efficient milk producers deposit nutrients in milk, not in body weight. Dry cows scored 4 produce the most milk yield in the next lactation.

21.7 CONCENTRATE FEEDING SYSTEMS

A successful concentrate feeding system should balance the nutrient needs of each cow without overfeeding. The system selected will depend on herd size, labor availability, housing and milking systems, forage system, and type of concentrates used (Coppock, 1983).

Conventional Stanchion or Tie Stall Barn Feeding

Concentrate can be fed to individual cows, which allows for customized feeding. Several feeds (energy feed, protein supplement, minerals, and additives) can be fed individually to minimize feed costs and optimize milk response. Labor becomes a limiting factor, especially with multiple feedings each day. Cows should be restricted (by dividers or space) so that they cannot steal from other cows. Grouping cows in the barn by stage of lactation can minimize feed stealing. Mechanical feeding systems are available to feed concentrates with a cable device which drops a predetermined amount or with electronic feeders that travel on a track and identify and feed each cow individually.

Milking Parlor Feeding

When cows enter the parlor, each cow is fed concentrate in a feed bowl. The amount offered depends on the method (free choice, feeding a constant level to every cow, or feeding concentrate according to milk production). Feeding concentrate in a milking parlor represents a problem for high-producing cows. Generally, a cow does not spend sufficient time in the parlor to consume the amount of concentrate required for high production. Cows can consume approximately 0.23 to 0.45 kg (0.5 to 1 lb) of concentrate dry matter per minute. Pelleted concentrate is consumed faster (0.5 kg) than ground forms (0.23 kg per minute). In small herds, it may be possible to leave cows in the parlor for a few extra minutes to consume extra concentrate, but this is not feasible in large herds. Other disadvantages of feeding concentrate in the parlor include dust in the parlor, more manure droppings in the parlor, slow cow movement in and out of the parlor, and a tendency to feed cows the same amount of concentrate.

Concentrate Added to Forage

Adding some of the concentrate to silage at feeding time is a practice that has met with varying degrees of success. If the concentrate is topdressed on the silage, cows can selectively eat it and boss cows may eat more than their share. Mixing concentrate and forage will lessen the opportunity for cows to eat the concentrate selectively. Feeding each cow individually is not accomplished with this method.

Group Feeding

One possibility is to feed concentrate mixed with forage (as a complete ration or total mixed ration). Cows must be grouped to avoid overfeeding low-producing cows and underfeeding high-producing cows. Dry cows must be separated and fed a dry cow ration. Three groups of lactating cows allow for optimal and economic rations. Lead factors refer to the nutrient level (based on milk yield) to balance to maintain milk production for the high producers in the group (Table 21.6). For example, in a two-group system, the high group averaging 30 kg would be balanced for 35.1 kg (30 kg × 1.17, the lead factor from Table 21.6). The low group averaging 20 kg would be balanced for 24.6 kg (20 kg × 1.23 from Table 21.6). Group feeding strategies are listed. (1) Nutrient levels should not de-

TABLE 21.6 LEAD FACTORS FOR TOTAL MIXED RATION FORMULATION

Percent of cows in each group			Percent current milk for production group:		
High	Medium	Low	High	Medium	Low
100	0	0	1.32	—	—
70	0	30	1.22	—	1.21
50	0	50	1.17	—	1.23
30	0	70	1.14	—	1.25
33	33	33	1.14	1.10	1.21
25	25	50	1.13	1.07	1.23
25	50	25	1.13	1.14	1.21
50	25	25	1.18	1.08	1.21

Source: C. C. Stallings and M. L. McGilliard, 1984, *J. Dairy Sci.*, 67:902.

crease by more than 15% when shifting cows to a lower group. (2) The milk range within a group should be no more than 9 kg (20 lb) of milk to minimize nutrient differences when grouping cows. (3) Boosting concentrate by 2.3 kg (5 lb) for a week after switching cows can minimize milk drops due to ration changes. (4) Shifting cows frequently (once or twice a month) can minimize social interaction (fighting). (5) Large dairy operations may keep cows in the same group to avoid milk losses due to shifting cows. In this case, rations for the group are changed as milk yield declines. Cows can be grouped by age, milk yield, fat-corrected milk yield, body condition, stage of gestation, days in milk, reproductive status, and aggressiveness.

Some disadvantages with group feeding exist. More planning and rearranging of cows are required, resulting in additional expense for fencing, bunks, and waterers. Some free stall barns do not easily accommodate a herd that is split into groups. More than one concentrate mixture may be required and facilities are necessary to handle two or more mixtures. In addition, some cows may have difficulty when shifted from one group to another. Cows must find their place in the "butting order" as cows are moved from one lot to another, which may adversely affect feed consumption and milk production. Manure scraping also becomes difficult when cows are divided into groups because more fences and gates are present. Extra labor is required when regrouping is done. Aggressive eaters may also overeat and make inefficient use of feed. A crowding gate in the holding area is usually required because cows may be reluctant to enter the milking parlor if no concentrate is fed.

Complete Rations

Complete rations (CR) or total mixed rations (TMR) is a free-choice (*ad libitum*) feed system where forages, concentrates, minerals, vitamins, and additives are blended together on a weight basis. Each mouthful of feed contains the same nutrients, which are based on the requirements of a group of cows (Figure 21.3). Cows fed complete rations produce 4 to 7% more milk than cows fed each feed separately. Feed costs are reduced because nutrients are digested and utilized efficiently, less feed waste occurs, and cows produce more milk per unit of dry matter consumed. Less expensive feed ingredients can be used for lower production groups, while high producers can be fed diets higher in concentrates, additives, and high-quality forages. Less labor is required compared to feeding each feed ingredient separately, and cows with similar production have similar milking times so that milking time is reduced. Herd health can be improved since few digestive upsets occur in early lactation, and greater energy intake occurs due to higher dry matter intake. Because a more desirable rumen environment can be achieved, the fat test may increase. No selection of feeds occurs, which allows for accurate ration formulation. Complete rations mask the taste or odor of unpalatable components such as urea, minerals, buffers, and additives. When CR are fed free choice, cows eat three to five large meals (after each milking and each feeding) and several small meals, which reduce changes in the rumen environment.

Figure 21.3 Fence-line feeding of a total mixed ration. Photograph courtesy of the University of Illinois.

Disadvantages of complete rations include: long hay must be chopped before it can be blended, investment cost for a weighing and mixing unit is expensive, feeds must be analyzed on a regular basis, and ration rebalancing must be done by a competent individual.

Magnetic-Type Feeders

This system provides supplemental concentrate to cows wearing a magnet or chain. The cow wearing the chain or magnet enters the feeding stall, which turns on the feeder, dispensing 0.23 to 0.45 kg (0.5 to 1 lb) of concentrate per minute. Concentrate is dispensed as long as the cow remains in the stall, which depends on her appetite and other cow competition (pressure). Reduced forage intake and excessive concentrate intake can result in off-feed, acidosis, fat-test depression, and high feed costs. A high degree of management is needed for success with this device. Excellent forage quality is necessary to maintain forage intake. Adequate cow pressure (number of cows using the device) is needed to average 4.5 to 5.5 kg (10 to 12 lb) of concentrate per cow per day. A feedstall is recommended to avoid injury and fighting at the feeder box area. These units are inexpensive ($40 to $50 per cow using the feeder).

Calan Door Feeders

Another concentrate feeding device allows a cow with a key (identification unit) on her collar to open a door for access to feed. Concentrate, forage, or a complete ration can be fed behind the door. Although each cow could have a uniquely coded door, most commercial operations have one key for all cows getting extra feed. This system does not allow cows without a key to receive any leftover feed. The weakness of this system is free-choice access to the feed, which is usually a concentrate. Some dairy managers have successfully placed a high-nutrient complete ration behind the feeder door for high producers and a lower-nutrient complete ration in the bunk for the low-producing cows. This approach allows one group of cows to receive two different complete rations. This system is also inexpensive ($40 to $50 per cow).

Computer-Controlled Concentrate Feeders

Computer-controlled concentrate feeders (CCCF) are the newest device for dispensing concentrate to individual cows in a group situation. These units cost from $150 to $400 per cow depending on herd size and degree of sophistication. Each cow wears an identification device on her collar, in her ear, or under her skin that the feeding unit can "read" electronically. The concentrate is dispensed based on a predetermined amount and number of meals. This system regulates total concentrate consumption, controls meal size and frequency, and can print out the amount of concentrate consumed each day to identify cows that do not eat their programmed amounts.

A dual CCCF (the ability to feed two different concentrates to each cow) provides more flexibility and allows for individualized concentrate amounts and types compared to a single CCCF. The most common dual feeder pattern is an energy source (such as shelled corn) and a protein source (such as a commercial 40% protein supplement). Other alternatives include two different protein mixtures (14 and 18% protein mixtures), dry cow and milk cow mixtures, or concentrate and micronutrient pack combinations. Concentrate amounts should be adjusted daily for the first 3 weeks after calving and weekly for the next 8 weeks. Concentrate increases should be limited to 0.45 to 0.68 kg (1 to $1\frac{1}{2}$ lb) per day. Some CCCF systems automatically increase concentrate based on a predetermined lactation curve and concentrate feeding table. Feeding some concentrate with forage while using the CCCF as a top dressing unit is an alternative.

21.8 ECONOMIC CONSIDERATIONS

All dairy managers purchase some feed ingredients. Purchases can be grouped into several broad categories; (1) premixes (trace minerals and vitamins), (2) stress packs (sodium bicarbonate, magnesium oxide, limestone, bentonite, limestone, yeast, MHA or methionine, niacin, choline, or other combinations), (3) salt or trace mineralized salt, (4) a complete mineral supplement, (5) protein supplement,

(6) complete concentrate mixture, and (7) forages. Regardless of feeds purchased, several factors should be evaluated.

Cost-to-benefit ratios are critical. If a feed ingredient increases cost by $0.20 per cow per day, an increase in milk yield or fat test, a reduction in feed costs, or an improvement in health worth $0.20 or more is needed from every cow. If the response occurs only in cows during the first 120 days of lactation, the response in the fresh cows must also pay for the costs for nonresponding cows to be economical.

Several methods can be used to determine if a feed is competitively priced. One approach is to compare the purchased feed to ingredients locally available or used feeds. For example, costs can be compared for a mineral supplement and a mixture of dicalcium phosphate, limestone, trace-mineralized salt, and a vitamin premix. A second method is to calculate the cost per unit of the nutrient to be purchased. If a commercial 16% phosphorus supplement is priced at $26 per 45 kg (100 lb), the cost per unit of phosphorus is $3.59 per kilogram ($1.63 per pound) of phosphorus. A comparison can be made between commercial products and ingredients on a price per unit basis. A third method is to calculate a value based on the energy and protein content of a feed. Morrison feed constants (Morrison, 1959) can be used to determine the value of a feed based on an energy constant (using shelled corn as a price basis) and a protein constant (using soybean meal as a price basis). Several Morrison feed constants are listed in Table 19.3. For example, the value of dried brewers grain can be calculated by following these steps. (1) The energy value is calculated by multiplying the energy constant (0.374) by the price of 1 ton of shelled corn ($100), which equals $37. (2) The protein value is derived by multiplying the protein constant (0.464) times the price of a ton of soybean meal ($160), which is $74. (3) If dried brewers' grain can be purchased and delivered for less than $111 ($37 + $74), this feed is a good buy. Higher mineral contents, low rumen protein degradability, and oil as an energy source should also be considered when using this method.

Commercial feeds usually are slightly higher priced than the base feed ingredients. Several services provided by a feed company can make a commercial feed a "good investment": (1) forage testing is provided, (2) computer ration formulation is available, (3) nutritional consultation is locally available, (4) ingredients are accurately blended, (5) purchasing small amounts improve palatability and stability, (6) specific additives are available in recommended amounts, (7) delivery of feeds in bulk can be contracted, (8) feeds can be texturized to improve intake, and (9) herd problems can be be evaluated and corrected.

REFERENCES AND FURTHER READINGS

Coppock, C. W. 1983. Nutritional perspective on dairy cattle feeding systems. Dairy Housing II. *Proc. 2nd Natl. Dairy Housing Conf.*, p. 138.

Hutjens, M. F. 1986a. Get the most for your feed dollars. *Hoard's Dairyman,* 131:383.

HUTJENS, M. F. 1986b. Nutritional strategies with a phase feeding concept. *Ill. Dairy Rep.*, p. 2.

MORRISON, F. B. 1959. *Feeds and Feeding,* 22nd ed. Morrison Publishing Company, Ithaca, NY.

NATIONAL RESEARCH COUNCIL COMMITTEE. 1988. *Nutrient Requirements of Dairy Cattle,* 6th ed. National Academy of Sciences, Washington, DC.

SMITH, T. R., and D. E. PRITCHARD. 1983. An overview of recent developments in individual concentrate feeding equipment and management. Dairy Housing II. *Proc. 2nd Natl. Dairy Housing Conf.,* p. 150.

WILDMAN, E. E., and G. M. JONES. 1981. Are your dry cows too fat, too thin, or just right? *Hoard's Dairyman,* 126:1101.

22 Noninfectious Diseases

A number of diseases that affect the dairy cow appear to be related to nutrition and metabolism because no infectious pathogen has been associated with these diseases. Common noninfectious diseases found in dairy cows in the United States are ketosis, milk fever, bloat, displaced abomasum, fatty liver syndrome, and acidosis.

22.1 KETOSIS

Ketosis (acetonemia) affects high-producing dairy cows during the first 6 weeks after calving; the first 3 weeks after calving is the most critical period. Most ketosis occurs during the winter feeding period when cows are in the barn. The incidence of ketosis is greater in older cows than in first-calf heifers. Repeat cases of ketosis from one lactation to the next are common. Approximately 2 to 4% of cows in the United States are estimated to have ketosis each year. No breed or area differences for the disease have been observed.

Symptoms

Primary ketosis exists when no apparent pathological condition other than ketosis affects the dairy cow. Secondary ketosis is accompanied by other complicating ailments, such as metritis, mastitis, retained placenta, and hardware disease. Primary ketosis is a metabolic disorder, and consequently there is no elevation of body temperature. When an elevated body temperature accompanies a ketotic condition, complicating ailments are implicated. Approximately one-third of ketosis cases are of the secondary type.

The sequence of observable symptoms in ketosis includes loss of appetite first characterized by concentrate refusal, followed by silage refusal, and then by refusal of hay. The cow appears gaunt and sometimes develops a depraved appetite. Rumen inactivity and constipation also occur and the cow appears depressed and dull. Occasionally, the animal becomes highly excitable and is extremely difficult to handle. A distinctive acetone odor can be detected on the breath and in fresh milk. The loss of appetite and decreased feed intake result in decreased milk production, an increase in milk-fat test, and loss of body weight. Occasionally, a cow dies from ketosis.

Causes

A number of predisposing factors influence the incidence of the disease. The primary factors appear to be high glucose requirement for maintenance and lactose production and inadequate feed intake to meet this requirement. The exact mechanism that triggers the onset of ketosis is not known. Ketosis is characterized by metabolic changes centering around a decrease in blood glucose and an increase in blood ketone bodies. Some of the other common metabolic changes are also shown in Table 22.1.

Ketosis is named from the accumulation of ketone bodies in the blood and urine of cows. The primary ketone bodies are β-hydroxybutyric acid, acetoacetic acid, and acetone. Some ketone bodies are found in the body fluids of normal animals, but an abnormal elevation of ketone bodies is related to the disease. A normal elevation of ketone bodies in the fluids occurs when energy requirements exceed energy intake and body fat is mobilized. Fasting, underfed, and high-producing cows mobilize body fat and have increased ketone bodies in their tissue fluids.

In the normal ruminant animal, ketone bodies are produced by the rumen epithelium from rumen fatty acids, primarily butyrate. Silages containing high levels of butyric acid predispose the cow to ketosis. With clinical ketosis, most of the excess ketone bodies are produced from incomplete free fatty acid oxidation in the liver. During normal metabolism of fatty acids, these acids are broken down and completely oxidized to carbon dioxide and water, producing energy. During ketosis, a shift in fatty acid metabolism to a partial oxidation of the fatty acids to ketone bodies occurs. Ketone bodies accumulate and fat is deposited in the liver.

TABLE 22.1 BLOOD CHANGES IN KETOTIC COWS

Component	Normal	Ketosis
Blood	mg/100 ml	
Glucose	52	28
Ketones	3	41
Plasma		
Free fatty acids	3	33
Triglycerides	14	8
Free cholesterol	29	15
Cholesterol esters	226	150
Phospholipids	174	82
Liver	Percent by weight	
Glycogen	3	<0.8
Fat	3	>10

Source: Data from S. Yamdagni and L. H. Schultz, 1970, *J. Dairy Sci.*, 53:1046, and E. N. Bergman, 1971, *J. Dairy Sci.*, 54:936.

In addition to the change in metabolism of fatty acids by the liver, one of the major changes in ketosis is a marked reduction in blood glucose concentration and in liver glycogen content (Table 22.1). The reduction in blood glucose is responsible for increased free fatty acids in the blood. When blood glucose concentrations are normal, free fatty acid concentrations in the blood are low. As plasma glucose concentration decreases, an increase in plasma free fatty acids results because fatty acids are mobilized from adipose tissue. Hormones, including insulin, growth hormone, adrenocorticotrophic hormone (ACTH), adrenal steroids, and thyroxine, may be involved in free fatty acid mobilization and ketogenesis.

The exact reason for the shift from complete breakdown of fatty acids to partial oxidation to ketone bodies is not known. Availability of carbohydrates may play an important role and some researchers believe that a deficiency of oxaloacetate may be involved. The ruminant animal obtains its glucose from nonglucose sources, primarily propionic acid, lactic acid, some of the amino acids, and glycerol. Oxaloacetate is an intermediate in glucose formation in all of these precursors except glycerol. When a large drain of glucose for energy and lactose formation occurs within the animal's body, oxaloacetate might be preferentially used for glucose formation. This may leave a deficiency of oxaloacetate for acetyl coenzyme A oxidation, thus shifting the metabolism to formation of acetoacetyl coenzyme A and acetoacetate (a ketone body) and to partial oxidation of fatty acids.

Excessive body condition at calving is another predisposing factor. The excess condition depresses appetite, increases mobilization of body fat, increases fat accumulation in the liver, and increases the rate of ketone body production.

Inadequate energy intake after calving contributes to ketosis. If sufficient energy intake was possible for high-producing cows, the cow would minimize mobilization of body fat and would be capable of meeting glucose requirements from normal feed intake.

Other conditions that can cause ketosis are protein deficiencies and deficiencies of minerals and vitamins. A high intake of ketogenic materials could lead to the formation of ketone compounds, especially silages high in butyric acid.

Diagnosis

In addition to observable symptoms, a number of diagnostic tests are available. One common test is to determine ketone bodies in blood, urine, and milk. Normal levels of blood ketone bodies are less than 10 milligrams (mg) per 100 milliliters (ml); ketotic cows may have levels as high as 50 mg per 100 ml. The ketone bodies in milk are approximately one-half of those in blood and one-eighth of those in urine. Several ketone tests are on the market with differing degrees of sensitivity. Elevated ketone bodies in blood, urine, and milk do not necessarily indicate clinical ketosis; but high ketone levels, off-feed, and a drop in milk production provide strong evidence for ketosis. If the urine test is negative, ketosis can be ruled out as the primary disease. A milk test that becomes positive at 10 to 15 mg of ketone bodies per 100 ml of milk is an adequate warning.

In addition to an elevated level of ketone bodies, a decrease in blood glucose occurs during ketosis. Normal blood glucose level in the ruminant is 50 mg per 100 ml of blood. Values below 40 are considered subnormal and ketotic animals may have values as low as 25.

Treatment

Many treatments for ketosis are available, but their effectiveness is difficult to judge because many cows recover spontaneously. Most cows recover when milk production drops to a point where the glucose requirements are met by the glucose formed from feed. One common treatment, intravenous injection of 500 ml of a 50% glucose solution, increases blood glucose and is effective in mild cases of ketosis. Relapses are frequent and several treatments are usually needed in severe cases. Glucocorticoid hormonal injections are used either alone or in combination with intravenous glucose injections. Glucocorticoids are responsible for the production of blood glucose from amino acids. Glucocorticoid injections may upset the hormonal balance of the animal and utilize body protein. Supplementation with 12 g of niacin can be used to treat ketosis by feeding it for 1 week to ketotic cows.

Two glucogenic materials, propylene glycol or sodium propionate, can be administered orally. Feeding 0.2 to 0.5 kg per day is recommended, preferably divided into two administrations per day. The usual treatment period is 5 to 10 days. Because these compounds are unpalatable, the materials may have to be administered by drench. The limitations of this treatment are that considerable labor is required to drench cows and that a danger of producing inhalation pneumonia exists. Feeding sugar or molasses is ineffective because rumen microorganisms break them down to volatile fatty acids.

Prevention

Recommendations to reduce the incidence of ketosis by balancing feed intake with the nutritional requirements of high-producing cows during peak production are: (1) do not have cows excessively fat at calving; (2) increase concentrate feeding after calving to increase available energy and reduce mobilization of body fat; (3) do not make abrupt changes in the ration during the ketosis-susceptible period that may cause the cows to go off feed; (4) feed cows top-quality forage and avoid poor-quality silage having a high butyric acid content; (5) feed sufficient quantities of protein, minerals, and vitamins; (6) provide comfortable housing with adequate ventilation and a space for exercise; (7) feed propylene glycol or sodium propionate for 6 to 8 weeks after calving to cows that had previous cases of ketosis; (8) add 0.45 kg (1 lb) of supplemental fat or oil to the ration, and (9) feed 6 g of niacin per day from 2 weeks prepartum to 10 weeks postpartum.

22.2 BLOAT

Bloat is a noninfectious disease characterized by gaseous distension of the stomach, especially the rumen and reticulum, which is particularly noticeable on the left side of the animal (Figure 22.1). Death is common if the animal is not treated.

Cause

Bloat is caused by excessive accumulation of gas in the first two compartments of the ruminant stomach. Many different forms of bloat have been described, including legume, feedlot, toxic, pathological, and obstructive bloat. These types in order of severity are described as chronic, subacute, and acute. The two most common types of bloat, legume and feedlot, can have any of these three severities. However, feedlot bloat is usually not as severe as legume bloat. Chronic bloat results in mild distension of the rumen. Subacute bloat results from a specific dietary regime, such as consumption of succulent legumes or a high-grain diet. Considerably more distension results from subacute bloat than from chronic bloat. Acute bloat is a severe situation characterized by labored breathing, frequent urination, and defecation. Death commonly results if the animal is not treated.

Legume or pasture bloat is the most common form of bloat in dairy cattle. Feedlot bloat is usually limited to fattening beef cattle that are on high-concentrate, limited-forage rations.

The causes of bloat are complex, but the main causes appear to be linked to plants, animals, and the microbes in the rumen. Certain forages, usually alfalfa and ladino clover, produce bloat. Cutting these legumes as green chop can produce bloat; however, the disease occurs less frequently than when animals are allowed to graze the plants. Grass pastures such as bluegrass, brome grass, sudangrass, orchardgrass, and birdsfoot trefoil (a legume) do not cause bloat.

Animals differ in susceptibility to bloat. Susceptibility appears to be an inherited characteristic. Microbes in the rumen play a role since two types of bloat

Figure 22.1 The fistulated steer exhibits the distension (bloating) and explosive release of digesta when the rumen plug is removed. Photograph courtesy of *Hoard's Dairyman*.

have been observed in the rumen; free-gas bloat, in which the gas accumulates above the contents of the rumen, and frothy bloat, which is entrapped in the rumen contents. Frothy bloat is the more common type.

During the breakdown of feeds by rumen microorganisms, carbon dioxide and methane are released. During bloat, the animal is unable to remove excess gas by eructation (belching). Accumulation of gas causes pressure in the rumen and leads to acute symptoms.

Among the theories that have been advanced to explain the cause of bloat are (1) lack of coarse forages needed in the rumen to stimulate belching; (2) excessive consumption of dense feed, which lowers the end of the esophagus below the level of the rumen contents so that the animal has difficulty in releasing the gas; (3) presence of substances in feed or production of substances in the rumen that cause formation of a stable foam which interferes with belching; and (4) presence of substances in feed or production of substances in the rumen that interfere with nerve reflexes that accompany belching.

Among the chemical compounds believed to increase foam formation in rumen ingesta are cytoplasmic proteins. The primary cause of legume bloat is excessive foaming of rumen contents. However, the causative factor or chemical compound in bloat-provoking feeds has not been isolated. When bloat is induced by these feeds, the consistency of the rumen contents thickens and gas becomes entrapped, forming a stable foam that makes it impossible for the animal to eliminate the gas by belching. The gas builds up and causes the abdominal wall to swell.

Treatment

The most effective treatment of subacute legume bloat is the administration of antifoaming agents such as silicones, detergents, vegetable oils, animal fats, animal mucins, and liquid paraffins by stomach tube. A tube approximately 1.8 cm (0.75 in.) in diameter is passed through the mouth and into the rumen to release the free gas located above the rumen contents. Ingesta in the tube should be removed by periodic blowing into the tube. If the animal has a frothy bloat, 1 to $1\frac{1}{2}$ cups of vegetable oil can be administered through the stomach tube. If an animal is severely bloated and is lying down, the last resort may be a rumenotomy, in which the body and rumen walls are punctured on the left side slightly forward and 1 to 1.5 cm down from the hipbone. It is advisable to use a trocar and cannula to puncture the body and rumen walls and to leave the cannula in place so that the gas can escape. Liquid must not drain internally because it will cause peritonitis and death.

Prevention

Among preventive measures that are helpful to reduce the incidence of legume or pasture bloat are: (1) use pastures that contain a mixture of legumes and grasses (the incidence of bloat is less on these pastures), (2) withhold animals from pasture during the most likely bloat-producing stage (the rapid-growth stage) of plant growth, (3) provide dry feed prior to pasturing to reduce the amount of lush legume pasture consumed by the cows, and (4) use an antifoaming agent.

The best prevention method is feeding of a nonionic surfacant, poloxalene. Poloxalene fed at the rate of 10 g per 454 kg (1000 lb) of body weight per day, either as a topdressing to feed or in a molasses–salt block, prevents bloat in cattle grazing legume pastures. If poloxalene is fed in a molasses–salt block, no other source of salt should be available to ensure adequate poloxalene consumption from the salt block. One salt block should be available for every five animals. The animals should have access to these salt blocks before they go out on pasture. Poloxalene is effective even if it is used continuously for up to 80 days. No problem with poloxalene residues occurs in milk and the compound has no adverse effects on feed consumption or milk production. Poloxalene is also effective in preventing feedlot bloat in beef cattle.

22.3 MILK FEVER

Milk fever (parturient paresis) occurs at parturition or within 2 to 3 days after parturition. Milk fever is a disorder of high-producing animals and usually occurs in older cows. It is rare in first-calf heifers. The incidence is two to three times higher in Jerseys than in other breeds. Cows that have had previous cases of milk fever are much more likely to have milk fever at subsequent calvings than are cows with no previous history of milk fever.

Symptoms

Progressive symptoms of the disease are loss of appetite, static digestive tract, dull eyes, cold ears, a stilted gait, paralysis (in which the cow takes a characteristic position of lying on her sternum with her head tucked toward her flank), coma, and death (Figure 22.2). Death occurs in a high percentage of cases unless the animal is treated. This condition can lead to "downer" cow syndrome. Cows suffering from milk fever that lie in a cramped position can develop degenerative muscle changes that make them physically unable to rise. Ruptured ligaments and tendons and occasional bone fractures have been found, usually caused by the strain of the animal trying to get up. The problem is more severe when the animal is in a stanchion or when it tries to get up on a hard slippery surface with no bedding.

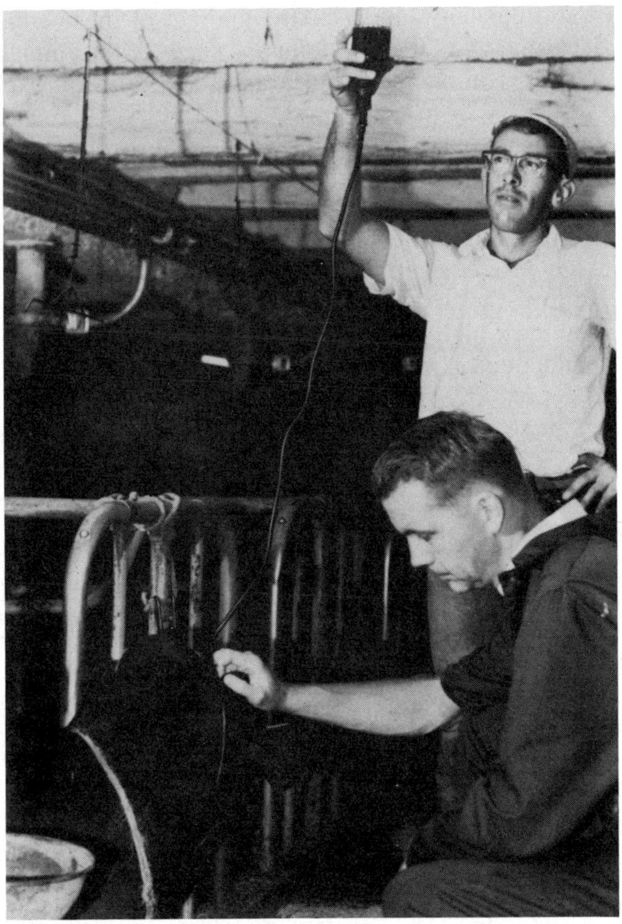

Figure 22.2 When a cow is down with milk fever, a veterinarian infuses a mineral solution intravenously to correct blood mineral imbalances and get the cow to stand. Photograph courtesy of *Hoard's Dairyman*.

Cause

The primary cause of milk fever is a sudden loss of calcium from the blood (Table 22.2). Normal serum calcium levels are between 8 and 12 mg per 100 ml of blood. At the time of calving and initiation of milk secretion, sizable amounts of calcium drain from the blood into the colostrum and the blood serum may drop to 7 mg per 100 ml. Milk fever occurs in animals that are unable to replenish the blood serum calcium at a sufficient rate to maintain adequate blood calcium.

Blood serum calcium is replenished by calcium absorbed from the intestinal tract and calcium mobilized from bone. Several factors that affect the ability of the animal to maintain normal calcium and phosphorus in the blood are the amount of calcium and phosphorus in the intestinal tract that is available for absorption, the adaptive ability of the cow to conserve calcium, the efficiency of the parathyroid glands to regulate blood calcium, the availability of bone calcium, and the amount of vitamin D in the feed. As the cow becomes older, the rate of calcium absorption from the intestinal tract decreases and calcium in the bone tends to be more tightly bound and less easily mobilized. Increasing calcium and phosphorus content in the ration usually increases the amount of these elements absorbed from the intestinal tract up to the physiological needs of the animal. Intakes above the animal's physiological needs result in an increased fecal excretion of calcium and phosphorus.

Vitamin D metabolites increase the rate of calcium absorption from the intestinal tract and the mobilization of calcium from bone. The parathyroid glands regulate mobilization of calcium and phosphorus from bone. Vitamin D and calcium intake have been shown to influence parathyroid activity. High vitamin D or high calcium intake may produce sufficient absorption from the gut to maintain normal serum calcium during the dry period without involvement of the parathyroid glands. A high calcium or high phosphorus intake in the diet during the dry period increases the incidence of milk fever. It has been postulated that high calcium intake suppresses parathyroid activity and increases secretion of calcitonin, which may cause bone resorption to be too slow to respond to increased

TABLE 22.2 CHANGES IN BLOOD CONSTITUENTS IN NORMAL AND VARIOUS STAGES OF MILK FEVER

	Blood serum (mg/100 ml)		
Stages	Calcium	Phosphorus	Magnesium
Normal nonparetic cow	8.0	3.9	2.8
Paretic cow			
Stage 1[a]	6.2	2.4	3.2
Stage 2[b]	5.5	1.8	3.1
Stage 3[c]	4.6	1.6	3.3

[a] Stage 1: lack of muscle control.

[b] Stage 2: unable to rise.

[c] Stage 3: comatose.

Source: J. W. Crowley and N. A. Jorgensen, 1978, *11th Annu. AABP Conf. Proc.*, p. 59.

calcium requirements at parturition. High serum calcium stimulates release of calcitonin from the thyroid gland, which acts to inhibit bone resorption. Low calcium intakes can deplete bone reserves which are not available at parturition. These interrelationships indicate that many factors prevent the susceptible cow from maintaining a normal blood serum calcium after calving.

Treatment

The most common treatment for milk fever is intravenous injection of calcium gluconate to replenish calcium in the blood as quickly as possible. The injection must be done quickly to avoid death of a cow that is in a coma. Recovery is spectacular; the cow is usually up within minutes after treatment unless she has complications caused by injury or a secondary disease or disorder.

Inflating the udder with air has been used with limited success. Air is pumped into the udder to stop secretion of milk and calcium drain from the blood. This procedure is effective if the inflation occurs before the cow goes into a coma. The disadvantage is that a high incidence of mastitis usually results because of introduction of bacteria into the udder by cannulation of the teat and pumping of air into the udder. The procedure requires 3 to 6 hours to be completely effective.

Prevention

Two methods of preventing milk fever have been quite successful. One method is controlling calcium and phosphorus in the diet of the dry cow. To reduce the incidence of milk fever, the dry cow ration should provide 60 to 80 g of calcium and 30 to 35 g of phosphorus per day for large-breed cows. Small-breed cows should receive 80% of these amounts.

When calcium and phosphorus intake is maintained in this range during the dry period, the cow can more easily replenish calcium that is drained from the blood for milk secretion at onset of lactation. High calcium intake occurs when alfalfa is the main forage during the dry period. Alfalfa has 11 g of calcium per kilogram (5 g per pound) of dry matter, which provides 100 to 150 g of calcium per day. To maintain an optimal calcium intake in the dry cow ration, alfalfa should be limited to 5 to 7 kg (11 to 15 lb) per day. Feeding this amount of alfalfa also requires a phosphorus source that contains no calcium, such as monosodium phosphate or monammonium phosphate. Commercial mineral sources should not be fed unless they have the correct calcium-to-phosphorus ratio. Corn silage has equal amounts of calcium and phosphorus. Feeding corn silage can produce a deficiency of calcium and phosphorus and should be supplemented with a calcium and phosphorus source. The substitution of a grass for alfalfa also lowers calcium intake.

Feeding cows a specific calcium-deficient diet (less than 11 g) 4 to 8 days prior to calving can reduce milk fever 50 to 74% in cows prone to the disorder. Accurate prediction of calving date is critical since extended feeding of calcium-deficient diets will deplete the bone reserves. The calcium-deficient diet stimulates the calcium hormonal system in the cow prior to calving.

Another preventive program that has been used is the feeding of large amounts of vitamin D prior to calving. Feeding massive daily doses (20 million units) of vitamin D for 3 to 7 days before calving affords 80% protection against milk fever in cows with previous histories of milk fever. These high doses of vitamin D should not be fed for more than 7 days or calcification of tissue may result. High doses of vitamin D also cause loss of bone strength resulting from depletion of minerals. The problem with this preventive program is timing the calving date so that vitamin D is fed for at least 3 days and no more than 7 days before calving. If massive doses of vitamin D must be stopped and the cow has not calved, the cow is more prone to develop milk fever.

Continuous feeding of vitamin D at between 300,000 and 550,000 units per day has been found to afford approximately 75% protection against milk fever in cows with previous histories of milk fever. When this amount of vitamin D is fed continually to cows with no previous histories of milk fever, a significant increase in the number of milk fever cases occurs. Feeding 30,000 units of vitamin D per day to dry cows are recommended.

Vitamin D must be metabolized to an active form, 1,25-dihydroxyvitamin D (1,25-D). Synthesis of the metabolite is increased by low serum calcium, high parathyroid hormone levels, and low plasma inorganic phosphorus. Conversion of vitamin D to 25-OH-D occurs in the liver; 25-OH-D is converted by the kidney to 1,25-D. The active form (1,25-D) causes bone mobilization and gut absorption of calcium. While the vitamin D metabolites show promise for milk fever prevention, they are not approved or available for use.

22.4 DISPLACED ABOMASUM

Displaced abomasum (DA) or twisted stomach is another nutritionally related disorder. A typical rate of occurrence is 5%. The disorder occurs from 1 week prepartum to 3 weeks postpartum in 90% of the diagnosed cases. A left-sided displacement or LDA occurs in 90% of the cases with the abomasum migrating from the low right side of the animal to the left side. A right displaced abomasum or RDA has a twist or torsion on the right side which restricts blood and nutrient flow and is more serious than an LDA.

Symptoms

Animals with DA go off-feed and refuse to eat normal quantities of feed. Ketosis is a common observation since feed intake is markedly reduced, rumen contractions slow, fecal material becomes pasty and dry, and the abomasum is dilated with fluid and gases. An LDA can be diagnosed by pinging (snapping the rumen wall with fingers, causing a hollow ringing sound which can be detected with a stethoscope). In some LDA cases, the abomasum may be seen high on the left side of the animal.

Causes

A combination of predisposing factors can lead to DA. Because a high incidence occurs near calving, expelling of the calf, fluids, and membranes allows room for the abomasum to move. The unborn calf also raises and compresses the rumen. Weak, flabby muscle support of the internal organs is another contributing cause. Heavy concentrate feeding, lack of bulk or long fiber, and abrupt ration changes can lead to gas formation and abnormal digestion. Any metabolic disorder or disease that causes the animal to go off feed increases the risk of DA.

Treatment

Treatment for DA consists of surgery; the abomasum is permanently sewn in place and scar tissue forms. This reattachment prevents the reoccurrence of the condition, but costs \$100 to \$150 per cow in lost milk production and veterinary expenses. Some animals are rolled on their back from right to left several times (casting) to allow the abomasum to correct itself. The success rate with casting is variable and the animal may redevelop a DA. If the animal is not valuable and has not received drugs, culling is another option.

Prevention

Dry cows should be shifted gradually from the dry cow diet to the milking ration. Long forage (hay) should be fed to maintain rumen fill and function. If lead feeding of concentrates occurs, the amount of concentrate should not exceed 1% of the cow's body weight. Prevention of all metabolic disorders and diseases is desirable to stimulate dry matter intake. Fat cows should be avoided because these cows have poor appetites and do not handle the stress of calving.

22.5 FATTY LIVER SYNDROME

Fatty cow disease is a term applied to cows that develop fatty livers. However, the disorder is not limited to fat cows.

Symptom and Causes

The syndrome occurs in high-yielding dairy cows when a depressed appetite causes an energy deficiency, leads to rapid loss of body weight after calving, and an accumulation of fat in the liver. Clinical signs include an increased incidence of metabolic, infectious, and reproductive disorders, which can cause death. The immune system is impaired with reduced white blood counts, which lowers the cow's resistance to disease and stress.

Diagnosis

One method is to measure liver fat content by taking a 0.2- to 1-g biopsy. Cows with over 20% fat in their livers are considered to have fatty livers. Measuring blood concentrations of free fatty acids, glucose, and liver enzymes is another method.

Treatment

Treatments include feeding 12 g of niacin to reduce fat mobilization and methionine (20 to 30 g) to mobilize fat out of the liver as lipoproteins, providing a broad spectrum antibiotic program for infection and disease control, and improving feed intake with top-quality feeds.

Prevention

Prevention is based on controlling energy intake. Dry cows should calve with optimal body condition (not too fat or thin). Dry matter intakes should be increased after calving to minimize excessive body weight loss. A balanced ration should maximize ration digestibility and nutrient availability. Low-rumen-degraded protein sources can be fed to metabolize body fat as an energy source and to avoid fat deposition in the liver. Adding 0.45 to 0.7 kg (1 to 1.5 lb) of supplemental fat or oil will provide extra energy. A preventive level of 6 g of added niacin can control body fat mobilization.

22.6 ACIDOSIS

Acidosis, milk-fat depression, off-feed, founder, and laminitis are terms applied to the general condition of low rumen pH.

Symptoms and Causes

Major rumen volatile fatty acid (VFA) changes occur as rumen propionate increases to 30 to 40%, rumen acetate decreases to 40 to 50%, and rumen pH drops below 6. Rumen acidosis (drop in pH) changes the rumen microbial population, slows rumen flow rates, and depresses rumen digestion, especially fibrous components. Hormonal shifts divert energy to adipose tissue, where fat deposition occurs, and away from mammary gland synthesis of milk fat. Lactic acid increases in the rumen, which can be absorbed by the bloodstream and can cause laminitis (inflammation near the top of the hoof), shift in blood pH, and death in severe cases. The degree of acidosis can be mild (several hours below a pH 6 with no visible symptoms) to a milk-fat depression (1 to 3 weeks of chronic acidosis) to death (acute acidosis). Cattle may exhibit diarrhea, off-feed, and sore feet.

Treatment

The rumen pH should be maintained above 6 by increasing saliva production and rumination with more long forage and fiber, lower soluble carbohydrate (less concentrate), and feeding or drenching with 0.45 kg (1 lb) of sodium bicarbonate.

Prevention

Optimal rumen fermentation should minimize rumen acidosis. Concentrate should be fed several times a day with a maximum of 2 to 3 kg (4.5 to 6.5 lb) per meal. A buffer, a compound that resists pH changes in a solution, can be added (0.5 to 0.75% sodium bicarbonate or its equivalent of total dry matter). Adequate acid detergent fiber (19%) and physical form (over 2.5 kg of forage dry matter with particles longer than 3.8 cm) should result in over 600 minutes of rumination and chewing time. Concentrate should not be finely ground, to avoid exposing starch and causing rapid rumen degradation. Forage (1 to 2 kg) fed 30 to 60 minutes prior to concentrate feeding stimulates saliva production and buffering in the rumen. Total mixed rations will avoid high concentrate intake and feed selectivity.

REFERENCES AND FURTHER READINGS

Bergman, E. N. 1971. Hyperketonemia-ketogenesis and ketone body metabolism. *J. Dairy Sci.*, 54:936.

Coppock, C. E. 1974. Displaced abomasum in dairy cattle: etiological factors. *J. Dairy Sci.*, 57:926.

Crowley, J. W., and N. A. Jorgensen. 1978. Role of feeding as related to metabolic disorders and reproduction. *Applied Dairy Nutrition.* In 11th Annu. AABP Conf. Proc., p. 59.

Hutjens, M. F. 1985. Metabolic disorders in dairy cattle. *Energy Fats NFIA Proc.*

Morrow, D. A. 1976. Fat cow syndrome. *J. Dairy Sci.*, 59:1625.

Reid, I. M. 1982. Subclinical fatty liver in dairy cows. *ARS Inst. Res. Anim. Dis.* Compton, Newbury, Berkshire, England.

Schultz, L. H. 1986. *Fresh Cow Problem—How to Control Them.* Hoard's Dairyman, Fort Atkinson, WI.

Yamdagni, S., and L. H. Schultz. 1970. Fatty acid composition of blood plasma lipids of normal and ketotic cows. *J. Dairy Sci.*, 53:1046.

23

Reproduction: Processes and Problems

High reproductive efficiency is a requirement for economical milk production on most dairy farms. Measures of reproductive efficiency include services per conception, conception rate on first service, days open, and calving interval.

A 12 to 13 month calving interval is desirable in most dairy herds. Milk production reaches a maximum between 3 and 6 weeks after calving and then gradually decreases. No practical ways have yet been found to maintain the peak production for long periods of time; therefore, a cow must have as many lactations with peaks as possible in order to attain high lifetime production. A cow cannot produce a calf more often then every 11.5 months since the gestation period of the cow is slightly longer than 9 months (278 to 283 days) and the uterus requires at least 45 to 50 days to return to normal before another pregnancy can take place. The ideal calving interval is usually stated as 12 months; in actual practice a 12- to 13-month calving interval is found in the better herds, whereas the calving interval for all herds on the DHI records program is closer to 13.5 months. A 12-month calving interval allows for a 10-month (305-day) lactation period and a 60-day dry period.

Another reason for a 12- to 13-month calving interval is to produce a maximum number of replacement animals for the herd. Approximately one-third of the milking cows are culled each year because of low production, mastitis, reproductive, and other problems. Since only 50% of the calves are females and since there will be some mortality among the calves and heifers before they reach milking age, a large percentage of the heifer calves are needed as replacements. Adding genetically superior heifers to the milking herd is the only practical and economical way of increasing the genetic potential of the herd. The average transmitting ability of cows entering the herd should be higher than that of the cows culled from the herd. Therefore, a long calving interval decreases the number of calves a cow produces in her lifetime, which decreases the opportunity in the selection of animals entering the milking herd.

The basic anatomy and functions of the male and female reproductive tracts are considered in this chapter. Practical recommendations for maintaining a high breeding efficiency and some of the common reproductive diseases will be discussed.

23.1 FEMALE REPRODUCTIVE TRACT

Anatomy and Functions

The female reproductive tract has three major functions: the production of sex cells (ova), providing a place for the fertilized ovum to develop, and production of steroid hormones. The reproductive organs of the cow with the exception of the vulva are located within the body cavity. The parts are located beneath the rectum; therefore, the organs can be located and manipulated by rectal palpation. The ovaries, uterus, and oviducts are suspended from a pelvic ligament, which allows them to move freely in the pelvic area and to accommodate easily a growing fetus.

The ovary is the primary organ of reproduction in the female. The paired ovaries are approximately $1\frac{1}{2}$ long and $\frac{3}{4}$ in. in diameter (Figure 23.1). The ovaries are suspended from the broad ligament near the end of the oviducts. The functions of the ovaries are to produce ova and some of the hormones involved in controlling the estrous cycle and pregnancy.

The ovary consists of an outer part, the cortex, and an inner part, the medulla, which contains blood vessels, nerves, and connective tissue. The cortex contains the germinal epithelium and produces the ova. The cow has the potential of producing thousands of ova during her lifetime; however, a cow will generally ovulate less than 100 since only one ovum is released at each cycle. By the injection of hormones, a cow can be induced to release more than one ovum at each estrous cycle. This is discussed under embryo transfer (see Section 24.2).

After puberty an ovum is generally produced every 21 days until the animal becomes pregnant. The ovum to be released is developed in the cortex of the ovary and is surrounded by a special layer of cells. This structure is known as a primary follicle. As the follicle matures, a cavity is formed within the center of the follicle and becomes filled with follicular fluid. This structure appears on the surface of the ovary and takes on the appearance of a large blister (known as the

Graafian follicle). This follicle can be palpated through the rectal wall. At maturity the ovum and follicular fluid are released from the ovary in the process called ovulation. After ovulation the walls of the follicle collapse and form the corpus luteum, commonly referred to as either the CL or yellow body. The yellow body gets its name from its deep yellow color in the cow due to the presence of β-carotene.

The released ovum is picked up by the funnel-like structure at the end of the oviduct known as the infundibulum (Figure 23.1). This structure partially surrounds the ovary in order to catch the released ovum, but there are instances when the released ovum is shed into the body cavity. The infundibulum has a fringed border, the fimbria, which helps to pick up the ovum from the ovary.

The oviducts, also known as the Fallopian tubes, are approximately 10 in. in length and $\frac{1}{4}$ in. in diameter. They are located between each infundibulum and each horn of the uterus. The released ovum caught by the infundibulum is moved to the enlarged upper end of the oviduct, the ampulla. The ampulla is the site of fertilization if sperm are present. Fertilization takes place about 12 hours after ovulation. The oviduct also transfers the fertilized ovum to the uterus, a process that takes 3 to 4 days.

The uterus consists of two parts, the body and horns. The uterus is suspended from the broad ligaments and is located in the posterior part of the body. The body of the nonpregnant uterus is about 2 in. in length and gives rise to the two horns. During most artificial insemination procedures, semen is placed in the body of the uterus. If semen is placed in the horn opposite the ovary from which the ovum was released, the chances of fertilization are very low.

The fertilized ovum develops and matures in one of the horns of the uterus. During pregnancy the uterine horn containing the fetus enlarges tremendously and drops down and forward into the body cavity. The endometrium (lining) of the uterus becomes very vascular after fertilization, in preparation for the implantation of the fertilized ovum. The uterus develops the maternal side of the placenta to protect and nourish the developing fetus. The caruncles (about 100) of the uterine endometrium interlock with the cotyledons of the fetal placenta and provide a passageway for the nutrients from the mother to the fetus and for the waste products to be removed from the fetus. Villi of the embryonic cotyledons fit into the crypts of the maternal caruncles to form placentomes, the functional units of exchange.

The developing embryo sets up its own placental membranes, which consist of the chorion (outer membrane), the amnion, the yolk sac, and the allantois. The amnion contains a cavity that surrounds the embryo and becomes filled with liquid that serves as a protective cushion to the fetus. The yolk sac supplies nutrients to the fetus during early development but functions for only a short period of time. The chorion contains a rich source of blood vessels and gives rise to the cotyledons. The allantois is an outpouching of the hind gut of the fetus and serves as a urinary receptacle for the fetus. A developing bovine fetus is shown in Figure 23.2.

The uterine wall contains several muscle layers, which aid the transport of semen from the body of the uterus to the oviducts. In addition, the contraction

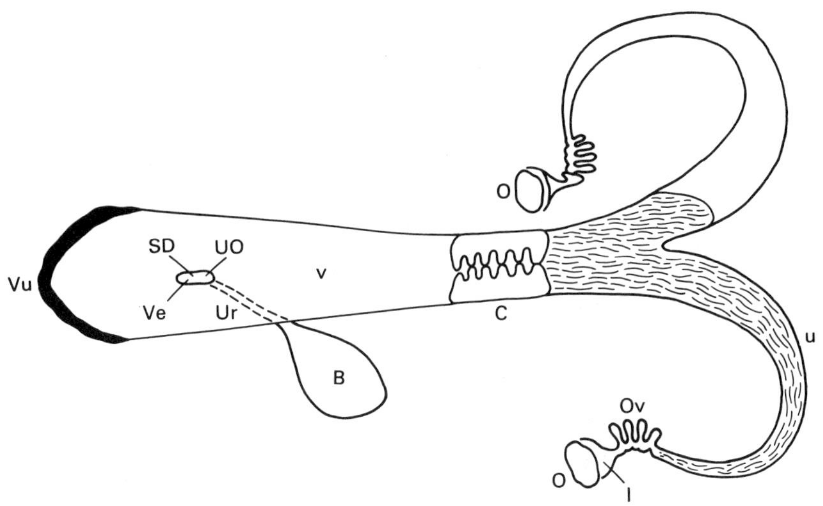

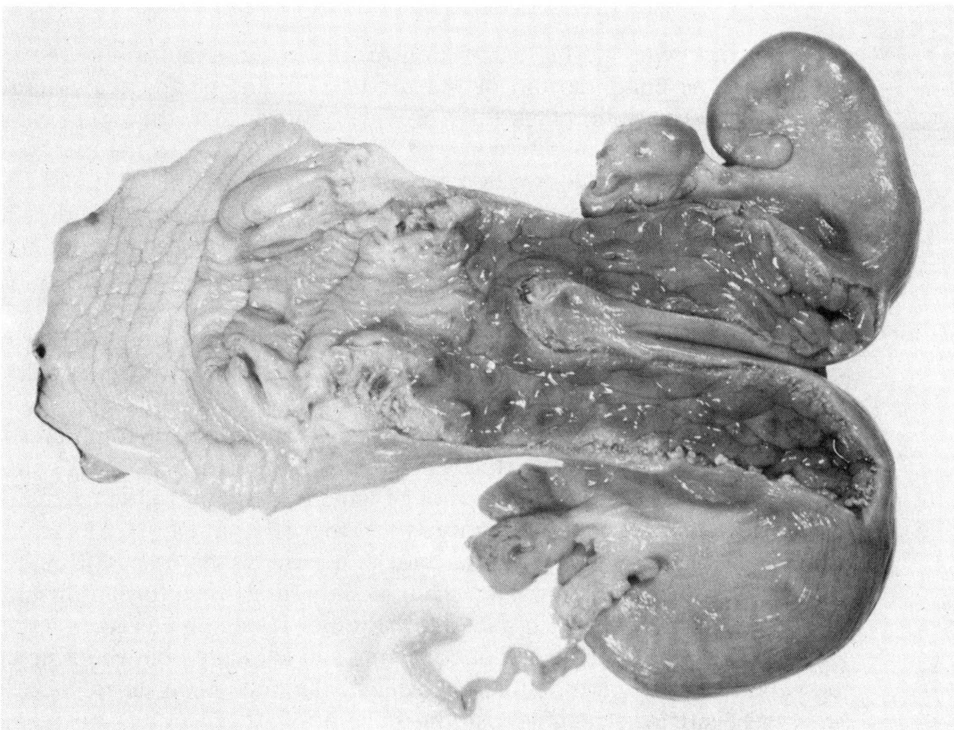

Figure 23.1 (a) Diagrammatic sketch and (b) photograph of the bovine female reproductive tract. Vu, vulva; Ve, vestible; SD, suburethral diverticulum; UO, urethral opening; Ur, urethra; B, bladder; V, vagina; C, cervix; O, ovary; U, uterus; Ov, oviduct; I, infundibulum. Photograph courtesy of Clifton Marshall, Select Sires, Plain City, OH.

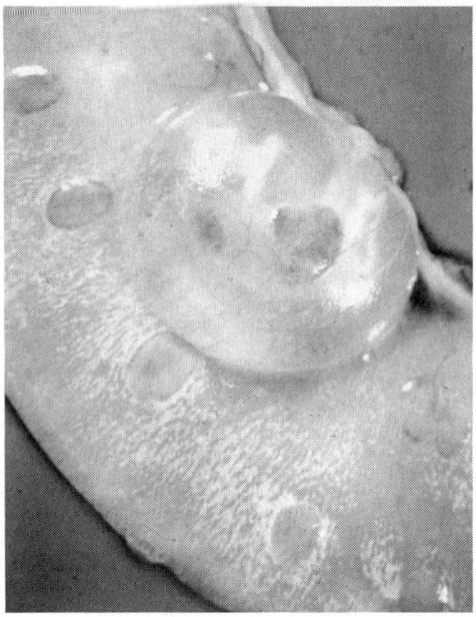

Figure 23.2 Photograph of a developing bovine fetus (F) and its placental membranes with the placentomes (P) for exchange of nutrients and waste products with the uterus. Photograph courtesy of Clifton Marshall, Select Sires, Plain City, OH.

of the uterine walls during parturition results in the expulsion of the fetus from the body. The endometrium of the uterus contains glands that secrete a fluid known as uterine milk. This fluid provides nutrients to the developing embryo before it develops its membranes and becomes attached to the uterine walls.

The cervix is a thick-walled structure about 4 to 5 in. long and 1 to 2 in. in diameter located between the uterine body and the vagina (Figure 23.1). The cervix contains ridges known as annular folds. These folds must be manipulated through the wall of the rectum during artificial insemination in order to pass the insemination rod into the uterus. The major function of the cervix is to restrict access to the uterus. The cervix serves as a place for semen deposition in artificial insemination when a possibility exists that the cow may be pregnant or if the estrous cycle length is other than 21 days. This is to avoid causing an abortion if the animal is pregnant. The cervix provides an environment conducive for sperm survival. The cervix also produces a secretion that is usually thick, but thins out at the time of estrus to facilitate the movement of sperm to the uterus. This fluid can be seen as part of the discharge from the vulva at estrus. During pregnancy the cervix forms a thick plug that blocks access to the uterus. Removal of this plug can result in abortion. The cervix and its fluids thus form a physical barrier and protect the uterus from bacteria and other foreign material.

The vagina, which is about 8 in. long and is located between the cervix and the vulva, has several functions. It receives the male penis during copulation and serves as a site for semen deposition. In addition, it serves as a birth canal during parturition. The epithelium of the vagina secretes fluids, which along with the fluids of the cervix, inhibit growth of bacteria and thus provides a line of defense against bacterial invasion of the uterus.

The vestibule starts at the urinary opening and contains the opening from the urinary bladder and an opening from a blind sac, known as the suburethral diverticulum (Figure 23.1). Care must be taken during artificial insemination that the rod is not forced into either the urethral opening or the suburethral diverticulum.

The vulva forms the external opening of the reproductive tract and consists of thickened folds of skin. The vulva is sensitive to changes in blood estrogen, which causes an increase in blood flow to the vulva and results in redness and swelling. These signs can be an aid in estrous detection.

Hormones

A number of hormones are directly involved in reproduction. These are produced by the pituitary, a small gland located at the base of the brain, the ovary and its structures, and the uterus. The hormones will be listed here, but they are described in more detail in the discussion of the reproductive processes.

The developing follicle produces estrogens, of which estradiol is the most important in the dairy cow. Estradiol stimulates the growth and motility of the reproductive tract, is responsible for secondary sex characteristics, such as the mammary glands in the cow, and produces the signs of estrus (commonly referred to as heat). After ovulation the remaining cells in the follicle form the corpus luteum, which produces the hormone progesterone. Progesterone is responsible for stopping the signs of estrus, preparing the lining of the uterus for the implantation of the fetus, and maintaining the lining of the uterus during pregnancy. If fertilization takes place, the corpus luteum and the production of progesterone persist throughout pregnancy; otherwise, the corpus luteum regresses and another follicle develops.

The anterior portion of the pituitary produces two hormones that play a direct role in reproduction. They are follicle-stimulating hormone (FSH) and luteinizing hormone (LH). FSH is responsible for the development of the follicle and for its production of estrogens. LH induces ovulation, causes development of the corpus luteum, and initiates progesterone production. The release of FSH and LH by the anterior pituitary is under the control of gonadotropin-releasing hormone (GnRH), a hormone produced by the hypothalamus. The hypothalamus is part of the brain and is located just above the pituitary gland.

Prostaglandins are produced by the uterus and cause the regression of the luteal cells of the corpus luteum at the end of the estrous cycle or pregnancy. Oxytocin, a hormone produced by the posterior portion of the pituitary, causes the contraction of the uterine walls to expel the calf at parturition. It also plays a role in the contraction of the myoepithelial cells during milk ejection (see Section 4.7). Relaxin is produced by the ovary near the end of gestation and is responsible for expansion of the uterus and relaxation of the cervix in preparation for parturition.

Estrous Cycle

The estrous cycle of the cow starts after puberty and occurs approximately every 21 days except for pregnancy. The cycle is divided into four parts: proestrus,

estrus, metestrus, and diestrus. Proestrus is the period between the regression of the corpus luteum of the previous cycle and estrus. It is the period of follicular development. When the corpus luteum starts to regress, an increasing amount of FSH is produced by the anterior pituitary, which in turn causes the development of the dominant follicle on the ovary and the production of estrogen (Figure 23.3). The blood then contains decreasing amounts of progesterone from the corpus luteum and increasing amounts of estrogen from the developing follicle. The continued production of estrogen by the follicle causes a surge in the release of LH and FSH by the anterior pituitary, which in turn stimulates maximum production of estrogen by the follicle. The high amount of estrogen in the blood produces the behavorial signs of estrus, such as the mounting of other cows, the willingness to stand while mounted by other cows, and general increase of activity. This is the period known as estrus and also described as the period of standing heat. The average length of estrus is about 17 hours. The surge in LH release during estrus causes ovulation, which takes place about 10 to 14 hours after the end of estrus.

Estrus is followed by a 3- to 4-day period referred to as metestrus. During this time the corpus luteum develops under the influence of LH and starts to produce increasing amounts of progesterone, which is responsible for preparing the walls of the uterus for pregnancy and inhibiting the signs of estrus (Figure 23.3). If the cow is not pregnant, the corpus luteum is induced to regress by the release of prostaglandins from the uterus. This occurs between days 16 and 18

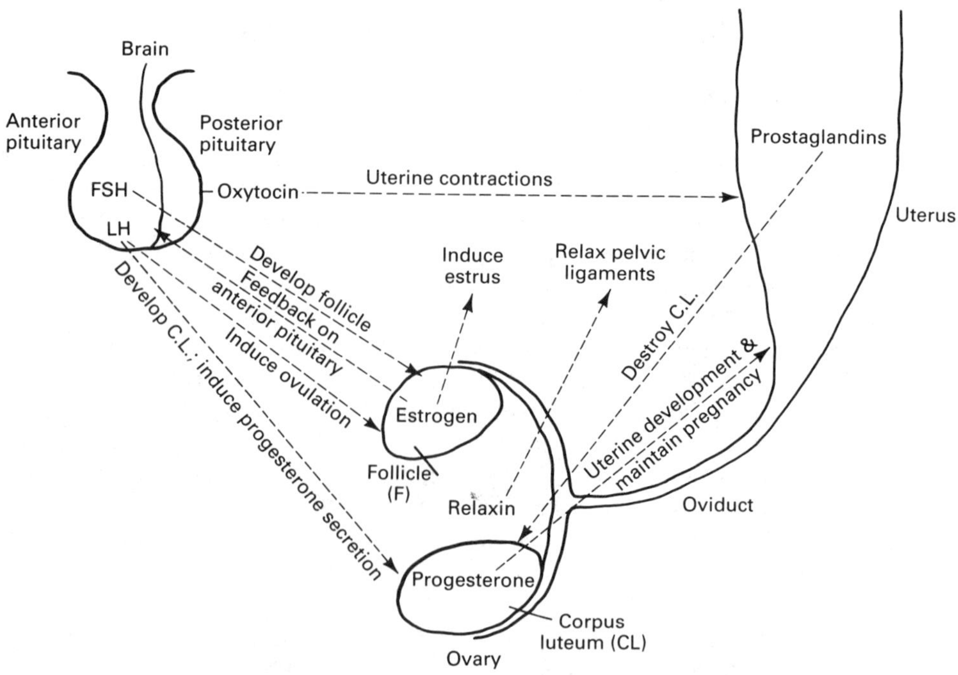

Figure 23.3 Relationship of the hormones of the ovary, pituitary gland, and uterus in the reproductive process.

after the start of estrus. The period between metestrus and the beginning of the regression of the CL is referred to as diestrus. The prostaglandins are transferred directly to the corpus luteum, where they interfere with progesterone production; thus the amount of progesterone in the blood decreases. This in turn allows FSH to stimulate the development of another follicle and the start of another proestrus. The cycle is repeated in a normal cow until she becomes pregnant.

Pregnancy

Dairy farmers who want their cows to become pregnant need to observe them carefully for the signs of estrus. Cows observed in estrus in the afternoon are usually bred the next morning, and those observed in estrus in the morning are bred in the afternoon. Breeding during the later part of estrus increases the conception rate. Sperm require only 10 to 15 minutes to travel from the vagina to the oviduct. Sperm undergo a series of biochemical reactions known as capacitation before they are able to fertilize an ovum. This requires about 6 hours. Sperm should be waiting in the ampulla of the oviduct at the time of ovulation so that fertilization can take place. The unfertilized ovum will live in the reproductive tract from 6 to 12 hours and the sperm will live in the female reproductive tract about 24 hours. If the ovum is not fertilized, it will be resorbed by the reproductive tract. If it is fertilized, it will move to the horn of the uterus, which takes about 3 days. Development of the fetal and uterine membranes then takes place, during which time the developing embryo must live on the uterine milk. Attachment of the fetus to the uterine wall begins about day 28 of gestation and is completed by day 45.

If the cow is pregnant, the blood progesterone remains high and the cyclic activity is blocked. It is believed that the embryo sends signals that maintain the integrity of the corpus luteum; thus there is continued production of progesterone. In the cow, progesterone production must continue for pregnancy to be maintained.

Even though fertilization takes place and the development of the fetal membranes proceeds normally, in some cases pregnancy is terminated and the developing fetus is either absorbed or expelled. If the termination is in early pregnancy, it is called early embryonic mortality, whereas it is called abortion in later pregnancy. The incidence of embryonic mortality has been reported to range between 5 and 28% for dairy cows and heifers, with 9 to 10% being a common figure. Some workers believe that many embryonic deaths are unavoidable because they may be due to genetic abnormalities and it is nature's way of eliminating unfit genotypes at a low biological cost.

The period from fertilization to the birth of the calf or parturition is the gestation period. It averages about 283 days for all breeds but some breed variation exists (Table 23.1). Brown Swiss have the longest gestation periods and Ayrshires have the shortest. The gestation period for cows carrying male fetuses is approximately 1.5 days longer than for cows carrying female fetuses. First-lactation cows have a slightly shorter gestation period than that of older cows.

Toward the end of pregnancy the fetus shifts position in the uterus and its

TABLE 23.1 AVERAGE GESTATION LENGTHS OF DAIRY CATTLE BREEDS

Breed	Days
Ayrshire	278
Brown Swiss	292
Guernsey	285
Holstein	280
Jersey	280

head lies between the front feet and points toward the cervix. During normal birth the front feet and head are delivered first. Relaxation of the pelvic ligaments occurs during gestation and a marked relaxation can be seen visually shortly before parturition by a dropping of the tailhead. Shortly before parturition the neck of the uterus and the vagina enlarge, the cervix starts to open, and a partial decalcification of the pelvic bones occurs in order to enlarge the birth canal. These events are under the control of relaxin (Figure 23.3). At parturition the fetus is pushed from the uterus through the cervix and vagina by the contraction of the muscles in the walls of the uterus and by some conscious straining of the abdominal walls by the cow. The contraction of the uterine walls is caused by oxytocin.

23.2 MALE REPRODUCTIVE TRACT

The primary organ of reproduction in the male is the testicle, of which the bull has two, as is the case in most animals. The testicles, or testes, are housed in the scrotum, an oval sac located outside the body cavity (Figure 23.4). The temperature of the scrotum is 1 to 4°C cooler than body temperature, which is a requirement for the normal development of spermatozoa. Sometimes one or both testes fail to move from the body cavity to the scrotum near the time of birth or shortly thereafter. This condition, cryptorchidism, causes the animal to be sterile if both testes fail to descend. Normal hormone production takes place in the cryptorchid bull.

The testes are made up of a large number of small seminiferous tubules, which are separated by interstitial tissue. The interstitial tissue gives support to the seminiferous tubules and also produces the male sex hormone, testosterone. Testosterone is responsible for producing the secondary sex characteristics of the bull and maintains the normal function of the accessory glands.

The seminiferous tubules are lined with germinal epithelium, the spermatogonia, which produce the spermatozoa. The tubules also contain Sertoli cells, which provide nutrients to the spermatozoa. The spermatozoa move toward the lumen of the tubules as they mature and then slowly move to the collecting tubes, the rete testis, and on to approximately 12 ducts, the vasa efferentia (ductus efferens). The spermatozoa then move through the head and body of the epididymis to the tail of the epididymis, where they mature and are stored until ejaculation. The epididymis is approximately 115 ft long in the bull and is lined with secretory

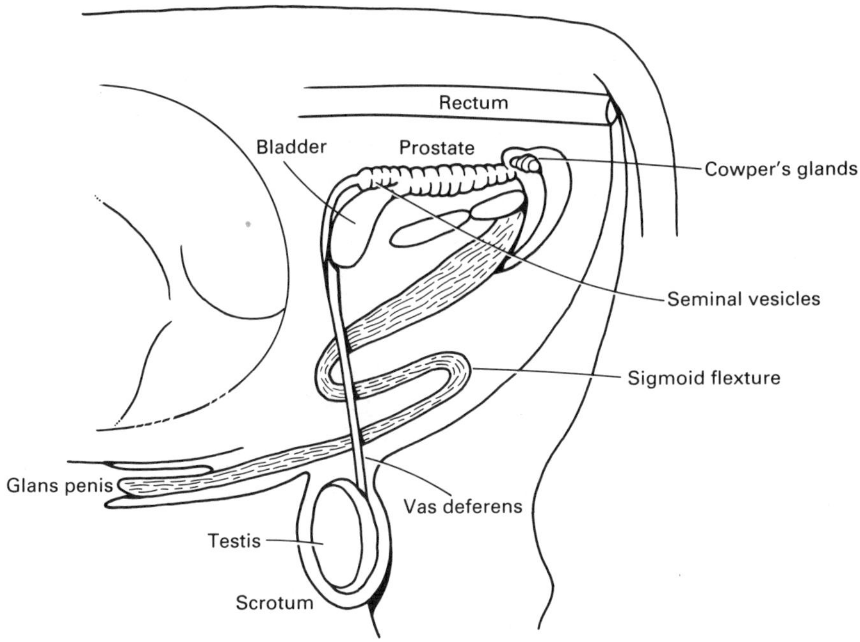

Figure 23.4 Reproductive tract of the bull. From *Principles of Dairy Science* by G. H. Schmidt and L. D. Van Vleck. Copyright © 1974 W. H. Freeman and Company. Used by permission.

cells. From here the mature sperm are carried through the vas deferens (ductus deferens) to the ampulla. One vas deferen is present for each testicle. The two vas deferens come together, enlarge, and become the ampulla. The sperm then pass through the urethra, which is a common passageway for sperm and urine. The urethra continues and passes through the penis to the outside of the body. The penis, when in a relaxed state, has a sigmoid flexure or S-shaped curve (Figure 23.4). During copulation the sigmoid flexure straightens and thus increases the length of penis for insertion into the vagina.

The bull has three sets of accessory glands, the seminal vesicles, the prostate, and bulbourethral glands (also called Cowper's glands). These glands produce a fluid that is high in carbohydrates and minerals. Thus the fluid provides nutrients for the spermatozoa and serves as a vehicle for their transport.

Semen is made up of the fluids from the accessory glands and the sperm (mature spermatozoa). The volume of the ejaculate varies from 1 to 15 ml, with 3 to 5 ml being an average range for each ejaculate from bulls collected twice with 30 to 60 minutes between ejaculations. The concentration will also vary considerably, but the usual range is 1.0 to 1.8 billion sperm per milliliter. Since only one sperm is required to fertilize an ovum, considerable dilution of the semen can be done. This has allowed extensive use of the best bulls through artificial insemination.

23.3 ATTAINING HIGH REPRODUCTIVE EFFICIENCY

Since a 12- to 13-month calving interval is the goal of most dairy farmers and is related to economical milk production, attention will be focused on the factors required to attain it. There are four major factors involved: reproductive disease-free herd, breeding policy of first breeding after calving, percentage of cows in estrus detected, and conception rate. Reproductive diseases are discussed in Section 23.4. The other three factors are considered below in some detail.

Policy of First Breeding after Calving

A minimum of 45 days is required after calving for the uterus to return to normal and attain a high conception rate. Some dairy farmers breed their cows as early as 40 days after calving. After 50 days the conception rate is not increased by waiting additional time before breeding a cow. Thus a breeding policy of not breeding a cow until she has been fresh 70 days automatically adds 20 days to the calving interval. The effect of this breeding policy is shown in Table 23.2. A routine reproductive tract examination is recommended for all cows 15 to 45 days after calving, to detect any infections or abnormalities.

Percentage of Cows in Estrus Detected

One of the major problems in herds with a poor reproductive record is the percentage of cows that are in estrus and are not detected. This presents no problem

TABLE 23.2 EFFECT OF DAYS TO FIRST SERVICE (DFS), HEAT DETECTION RATE (HDR), AND CONCEPTION RATE (CR) ON CALVING INTERVAL (CI), DAYS OPEN (DO), AVERAGE DAYS TO FIRST SERVICE (ADFS), AND SERVICES PER CONCEPTION (S/C)

DFS	HDR	CR	CI	DO	ADFS	S/C
50	50	45	13.2	121	81	2.16
		55	12.8	110	81	1.80
		65	12.5	101	81	1.54
	60	45	12.9	111	74	2.16
		55	12.5	101	74	1.80
		65	12.2	92	74	1.54
	70	45	12.6	103	70	2.16
		55	12.3	93	70	1.80
		65	12.0	85	70	1.54
70	50	45	13.8	141	101	2.16
		55	13.5	130	101	1.80
		65	13.2	121	101	1.54
	60	45	13.5	131	94	2.16
		55	13.2	121	94	1.80
		65	12.9	112	94	1.54
	70	45	13.2	123	90	2.16
		55	12.9	113	90	1.80
		65	12.7	105	90	1.54

to dairy farmers who have a herd bull that runs with the cows. For herds using artificial insemination estrous detection must have a high priority for economical reproduction and production. Increasing the estrous detection rate markedly reduces the calving interval and days open (Table 23.2).

Physical signs of estrus. The onset of estrus is brought about by the increasing amount of estrogen in the blood of the cow. The early signs of oncoming estrus are: increased activity, including nervousness and restlessness; attempting to mount other cows; nudging and licking other cows, including the sniffing of the genital areas; and swelling of the lips of the vulva. At estrus the cow will allow other cows to mount her without turning or trying to attack them. This is the period of standing heat. At the same time a clear watery mucus is usually discharged from the vulva. The signs of estrus vary among cows and may include a reduction in feed intake, a failure of milk ejection, and bawling or bellowing. After estrus the cow will usually have the hair on her tailhead roughened because of the mounting activity. Some cows will also show a bloody discharge on the vulva 2 to 3 days after estrus.

Estrous detection. The dairy farmer must be aware of the signs of estrus and observe the cows closely. A dairy farmer should set aside some time twice a day to observe cows for estrus. This should be at a time other then feeding or milking. Farmers should plan on at least $\frac{1}{2}$ hour each time. In addition, a heat expectancy chart is helpful in looking for those cows that had an estrous cycle 20 or more days previously. This calls attention to those cows that would be expected to be in estrus. Dairy farmers should be able to detect between 50 and 70% of cows that ovulate by visual observation.

A number of aids have been developed to help or replace human observations for estrus. In most cases, the aids have been helpful, but the use of the aids alone has not detected the percentage detected by visual observation alone. A common aid is a pressure-sensitive mount detector (Figure 23.5). This detector is glued to the rump of the cow expected to be in estrus and is activated by the pressure of the mounting of another cow. The color of the detector changes when activated. Some dairy farmers use tail chalk, crayon, or paint to accomplish the same result. The mounting by another cow will either smudge or rub off the material.

Recent interest has been developed in the use of a pedometer. One is strapped to a leg of a cow and then read each day when the cow enters the milking parlor or other place of observation. The distance walked is recorded manually or in some cases electronically. Cows in estrus show two to three times as much walking activity as do nonestrous cows. Two major problems with them are keeping them attached to the leg of the cow and keeping them clean enough for reading. Cost can also be a major detriment.

Detector animals are used on some farms. These consist of either surgically altered males or hormone-treated females. Males used are usually vasectomized or surgically altered so that the penis cannot enter the cow. Hormonally treated females, or in some cases steers, are also used to detect cows in estrus. These animals are usually treated with testosterone or testosterone and estrogen. Both

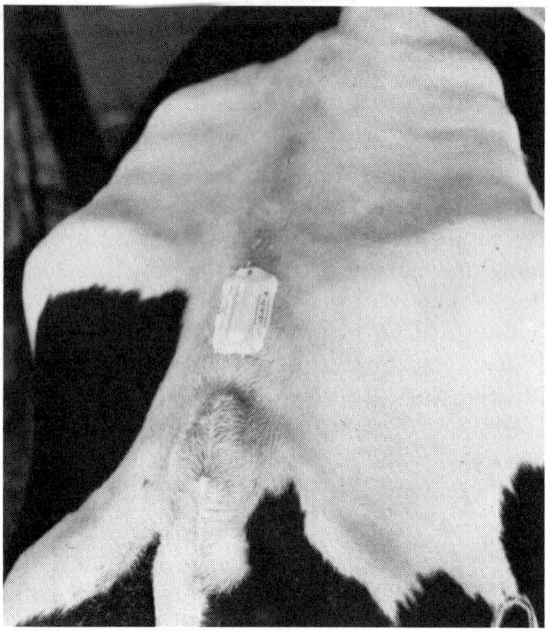

Figure 23.5 A heat mount detector to aid in the detection of estrus in dairy cows. The capsule releases dye and changes color if pressure is placed on it by the mounting of other cows. Courtesy of Kamar, Inc., Minnetonka, MN.

of these types of animals require some way to mark the cow that was mounted. Detection methods mentioned previously or use of a chinball marking device on the detector animal can be used.

One electronic aid that has been developed measures the electrical resistance of the vaginal mucus and has been found to be an effective way of detecting estrus. Experimental work is under way in training dogs to smell the genital tracts of cows and recognize the odors specific to cows in estrus. Also, efforts are under way to implant temperature-sensing devices in the bodies of cows and record the temperatures electronically. A rise in body temperature occurs at the time of ovulation.

Conception Rate

The conception rate on first service on many dairy farms is between 50 and 60%. A 55% conception rate results in about 1.8 services per conception (Table 23.2), which is the average for farms on a DHI records program. A number of factors influence the conception rate, including physiological factors of the cow, environment, and management practices by the dairy farmer. Physiological factors of the cow include milk yield, age, previous reproductive problems, and freedom from disease. Another factor is the quality of the semen. The fertility of bulls used in artificial insemination may vary as much as 20%. In most cases, however, the fertility of the semen of the bull in artificial insemination studs needs to be considered only with problem cows.

The most common environmental factor is high temperature at the time of breeding and even the 3 days before and after breeding. Consequently, the conception rate may drop as low as 10 to 20% in some herds in the South during July and August unless the stress from the heat is reduced. Extremely low winter temperatures may also decrease the conception rate. The sanitary conditions of the area in which the cow calves can also influence the conception rate. A clean environment with good hygiene practices reduces the incidence of retained placentas and uterine infections. Both of these conditions can lower conception rates.

The most critical factor from a management standpoint is the time of breeding in relation to ovulation. Here again accurate estrous detection is critical. Cows should be bred at least 6 hours before ovulation in order for sperm capacitation to take place. Since ovulation occurs 24 to 30 hours after the start of estrus, good conception rates are obtained if the cows are bred 10 to 20 hours after the start of estrus. The skill of the inseminator is most critical when using artificial insemination. The semen must be thawed and handled to minimize shock to the sperm. Also, the semen should be placed in the body of the uterus and not in one of the horns. Semen should be deposited in the cervix if there is a chance that the cow is already pregnant.

Pregnancy Diagnosis

A number of cows that are bred do not become pregnant and still are not detected in estrus. Some of the reasons include early embryonic mortality, silent estrus, anestrus (lack of estrus) and missed heats. Therefore, a number of leading dairy farmers routinely have a pregnancy diagnosis on all cows bred more than 40 to 45 days. An experienced veterinarian can determine pregnancy by feeling the uterine horns by rectal palpation on day 45 or later after breeding. Another aid to pregnancy diagnosis is a milk progesterone assay. The progesterone in the blood spills over into the milk and the amount in the milk is a good indication of blood amount. Cows that are not pregnant will have a very low amount of milk progesterone on days 21 to 23 after the last estrus. A cow with low milk progesterone on days 21 to 23 is almost certainly not pregnant, and about 85% of the cows with high progesterone are pregnant.

23.4 REPRODUCTIVE DISEASES

A number of reproductive diseases affect the cow and either cause abortion or impair the conception rate. A number of these also affect bulls.

Brucellosis

Brucellosis was a very costly disease in the early 1900s but has been greatly reduced because of government intervention. In 1934 the federal government started an eradication program in cooperation with state livestock sanitary officials. The program included blood testing of all dairy herds, vaccination of calves, and the payment of indemnities for discarded cattle. Brucellosis is also called Bang's disease and contagious abortion. The disease in cattle is caused by the bacterium *Brucella abortus*. When the organism is transmitted to humans either through infected milk or contact with an infected carcass in a slaughter plant, a disease known as undulant fever results. Its symptoms are very similar to the common flu; however, the resulting fever undulates up and down, thus the name for the disease.

Abortion after the fifth month of gestation is the most prominent symptom of the disease. If abortion does not occur, the calf is commonly weak at birth. Uterine infections and retained placenta are also common occurrences with the disease. Sterility occurs with severe cases of uterine infections. Reduction in milk production results from the alteration of normal lactation periods. Infected bulls show enlarged testes, loss of sexual desire, and infertility.

Brucellosis is a disease of sexually mature animals. Sexually inmature cattle are very resistant to the bacteria, but their susceptibility increases with sexual maturity and pregnancy. The most common form of transmission of the bacteria is the ingestion of the bacteria from the reproductive discharges from infected cows, be it the vaginal discharge or aborted fetus.

A number of serologic tests have been developed for the diagnosis of the disease; however, no one test is entirely reliable. A combination of two or more is highly effective. In addition to good sanitation, the most common method of prevention is vaccination of calves with Strain 19 vaccine. Dairy calves should be vaccinated at 3 to 6 months of age. Although the vaccine will not provide 100% protection in herds with a high degree of exposure, its value as a preventive measure is high.

A brucellosis ring test on bulk tank milk from herds is used by dairy plants as a screening test. It is so sensitive that it will detect the presence of one infected cow in a herd of 200. Most states also require that all herds be blood tested yearly and positive animals sold for slaughter. There is no known cure for brucellosis.

Vibriosis

Vibriosis is a venereal disease that is transmitted through natural mating or by the use of infected semen. It is caused by the organism *Vibrio fetus fenerealis*, which interferes with the blood circulation to the cotyledons of the fetal membranes. This causes early embryonic mortality and in some cases abortion after 2 or 3 months of gestation. The common symptoms are long, irregular estrous cycles with many repeat breedings. Bulls become infected but show no clinical signs of the disease. The organisms can be transmitted from one bull to another in artificial insemination studs if the same teaser animals are used for infected and noninfected bulls.

Isolation and identification of the bacteria are the most definitive diagnosis; however, they are difficult to isolate because they grow slowly. Detection of antibodies in the vaginal mucus is used to identify the disease in females. The best prevention method is using only artificial insemination with semen produced by disease-free bulls or with semen that has been treated with antibiotics. If animals are added to the herd, prepuberal animals offer the lowest risk, followed by pregnant heifers. Added cows should be isolated and tested before they enter the milking herd. Cows that are infected will usually develop an immunity against the disease and return to reproductive normalcy after 60 to 90 days of reproductive rest. Vaccination is possible for cows in problem herds.

Leptospirosis

Leptospirosis is caused by many serotypes of the bacterium *Leptospira*. These organisms can cause the disease in humans, domestic livestock, and wild animals. Therefore, exposure of animals to the organism is a continuous threat. The organisms tend to congregate in the kidneys and are then spread through infected urine and ingestion by noninfected animals.

Leptospirosis can vary in intensity from mild to severe. The incubation period is about 7 to 9 days and the symptoms include loss of appetite, loss in body weight, drop in milk production, elevated body temperature, bloody urine, abortions, and reduced breeding efficiency. The abortions usually occur about 1 to 3 weeks after the disappearance of the other symptoms and can be as high as 40% of cows in severely infected herds.

Isolation of the organism is the most definitive diagnosis; however, it is a difficult procedure since the organisms do not grow well on artificial culture media. A blood serum agglutination test is most commonly used for diagnosis. Since it is difficult to prevent animals from being exposed to the organisms, vaccination is usually practiced in most herds. Bacterins are available that contain the five most common serotypes affecting domestic animals. In general, annual vaccination is carried out, but a 6-month interval is recommended in problem herds. Only disease-free animals should be added to the herd. The treatment of leptospirosis is primarily to relieve the symptoms.

Listerosis

Listerosis is caused by the bacterium *Listeria monocytogenes*. It commonly invades the brain and the tissues covering the brain. It is not a common disease in livestock, but when it does occur it causes abortion in cattle, usually between the fourth and seventh months of gestation. The bacteria are spread by contaminated feed and water.

Bovine Viral Diarrhea

Bovine viral diarrhea (BVD) is caused by a virus and usually results in lesions of the digestive tract. The symptoms vary from mild to severe clinical. Severe cases are characterized by lesions in the digestive tract from the mouth to the anus, elevated body temperature, and profuse diarrhea. The disease is most common in animals from 8 to 18 months of age. Death may result in severe cases. The viruses are picked up by the mucous membranes of the noninfected animal from the contaminated environment.

The BVD virus can also cause abortion in pregnant animals, with the abortion occurring the first 3 to 4 months of gestation. A number of tests have been developed for diagnosis of the disease, but the virus is difficult to isolate from aborted fetus. Vaccination with a modified live virus vaccine is the best control measure and should be done in replacement animals from 8 to 12 months of age. Even though immunity may last for life, annual booster shots are recommended. Pregnant cattle should not be vaccinated since the vaccine may induce abortion. There is no known treatment for the disease.

Infectious Bovine Rhinotracheitis

Infectious bovine rhinotracheitis (IBR) is also caused by a viral agent and again is spread by noninfected animals picking up the virus through their mucous membranes. IBR is a disease of the respiratory tract and is widespread in the United States. The symptoms of the disease depend on the age of the animal, the stage of pregnancy, and the part of the body affected.

There are five different patterns or syndromes caused by the virus. The most common one is the upper respiratory infection, which is commonly referred to as "red nose." This is characterized by coughing, wheezing, fever, and nasal discharge. Diarrhea is the common symptom in the digestive version. Another form involves pustular vulvovaginitis, which is characterized by a swollen vulva and a pus discharge from the vulva. The infected bull develops pustules on the glans penis and the resulting pain usually prevents breeding.

Cows infected with the respiratory and vulvovaginitis types of syndromes often have the viruses invading the feti, which in turn results in abortion. The abortion occurs 3 weeks to 3 months after the onset of the infection. Retained placenta is a common result. Diagnosis is done either by isolation of the virus or use of the serum neutralization test. Vaccination with a modified live virus vaccine is an effective prevention measure. The vaccine was originally developed as an intramuscular injection, but is now commonly administered as a nasal spray. Vaccination is usually recommended for calves at 5 to 7 months of age, with booster vaccinations with the nasal spray in later life. The nasal spray is usually safe for use in pregnant animals.

Bluetongue

Bluetongue has become much more prominent in the last 30 years. It is a noninfectious viral disease of ruminants and seems to infect sheep to a much greater extent than cattle. Infected cattle are usually in a nonclinical state but shed the virus. When cattle become clinical, erosions of the mouth and feet develop with abortions and stillborn calves. Although the disease does not present a major health hazard, it does present economic problems of shipping animals and semen out of the United States. Vaccination with a modified live vaccine gives protection for 2 to 4 years.

23.5 OTHER REPRODUCTIVE PROBLEMS

Dairy cows experience a number of other reproductive problems. They include retained placenta, dystocia (difficult calving), irregular estrous cycles, anestrus, silent estrus, metritis, endometritis, and abortion. The first step in the control of these problems should be an excellent sanitation program. This is especially important in the calving area and in the handling of any equipment that may be inserted into the cow. Second, a complete disease control program should be developed in cooperation with a veterinarian. The farmer and veterinarian should work out a program for routine reproductive examinations as well as having an understanding of the handling of emergency situations. Third, the farmer should practice the sound management techniques for high reproductive efficiency outlined in Section 23.3. In addition to the above, the farmer should make sure that an adequate nutrition program is followed. Both deficiencies and excesses of certain nutrients can cause reproductive problems. Nutrients that have been found to be associated with poor reproductive performance are: deficiency of vitamins A, D, and E; deficiency of selenium, phosphorus, iodine, copper, manganese, and cobalt; inadequate protein; and excess intake of iodine and potassium.

REFERENCES AND FURTHER READINGS

BEARDEN, H. J., and J. W. FUQUAY. 1980. *Applied Animal Reproduction*. Reston Publishing Co., Inc., Reston, VA.

BRITT, J. H., N. M. COX, and J. S. STEVENSON. 1981. Advances in reproduction in dairy cattle. *J. Dairy Sci.*, 64:1378.

JORDAN, E. R., ed. Dairy integrated reproductive management. *W. Va. Coop. Ext. Serv.* IRM-16.

POPPENSIEK, G. C., and R. F. KAHRS. 1981. Twenty-five years of progress in understanding major infectious diseases of dairy cattle. *J. Dairy Sci.*, 64:1443.

SALISBURY, G. W., and N. L. VAN DEMARK. 1961. *Physiology of Reproduction and Artificial Insemination of Cattle*. W. H. Freeman and Company, Publishers, San Francisco.

SISSON, S., and J. D. GROSSMAN. 1953. *The Anatomy of Domestic Animals*, 4th ed., W. B. Saunders Company, Philadelphia.

24

Reproduction Management Tools

A number of reproductive physiology management tools have been developed to aid livestock producers. This is especially true for the dairy industry since many of the new developments were first applied to it. The development with the greatest application so far has been artificial insemination. Also receiving widespread use are embryo transfer and synchronization of estrous cycles. Most of the research work on sexing of fetuses and *in vitro* fertilization of ova has been done and the techniques are now being developed for application to the dairy industry. Sexing of semen is still in the research stage and may take a number of years before it can be applied to dairy cattle with satisfactory results.

24.1 ARTIFICIAL INSEMINATION

Introduction and History

Artificial insemination (AI) is probably the most valuable management tool for the dairy farmer. It is the process whereby semen is introduced into the female reproductive tract by artificial rather than by natural means. Its major contribution to the dairy industry is to make widespread use of sires with high genetic potential, which has been responsible, in part, for the steady increase in milk production by dairy cows.

Artificial insemination is a relatively old technique. The first recorded information on AI dates back to 1322, when an Arabian chief successfully obtained semen from a rival chief's stallion and artificially inseminated his prize mare. In 1780, an Italian physiologist, L. Spallanzani, successfully inseminated a bitch, which 62 days later produced a litter of three pups. Russian scientist E. I. Ivanoff began studies on artificial insemination in farm animals about 1900 and was the first to breed sheep and cattle successfully with it. Russians are credited with developing the practical aspects of AI and the discovery that semen could be preserved by cooling.

The artificial vagina was developed by G. Amantea, a professor at the University of Rome, for use with the dog. The artificial vagina is still an important component in the collection of semen from most farm animals. In 1937, veterinarians in Denmark developed the rectovaginal method of insemination. In 1936, the first cooperative AI association was formed in Denmark. E. J. Perry of Rutgers University, who visited the Danish cooperative and brought the idea back to New Jersey, was instrumental in organizing the first AI cooperative in the United States. During the first year, the New Jersey cooperative had 102 members and bred 1050 cows. Most states developed AI cooperatives during the next several years.

University of Wisconsin scientists P. H. Phillips and H. A. Lardy developed a yolk-phosphate diluter for semen which protected the sperm during cooling, provided nutrients for sperm metabolism, and prevented pH changes of the extender. Later, G. W. Salisbury and his co-workers at Cornell University substituted citrate for phosphate in the buffer, which allowed sperm to be seen under the microscope and observed for motility. The spread of reproductive diseases through the semen was prevented by adding antibiotics to the semen, and this was first demonstrated by John Almquist at the Pennsylvania State University. In the late 1940s British scientists perfected a method of freezing and storing semen at very low temperatures. The key to this process was the addition of glycerol to the semen and allowing it to equilibrate before freezing. Initially, dry ice and alcohol were used, but in 1957 the American Breeders Service starting using liquid nitrogen as the refrigerant. Semen was processed and stored in 1-ml glass vials until French workers developed the plastic straw for freezing, storing, and insemination of semen. Straws containing 1.0 ml were first developed and then replaced by 0.5 ml straws. This is the standard used by most AI organizations today.

Advantages and Disadvantages

Artificial insemination has a number of advantages. The most important is extending the use of good sires. Since many more cows can be bred with one ejaculate of bull's semen by AI than through natural means, greater use can be made of the superior bulls. This has also allowed for more accurate evaluation of bulls' transmitting abilities. A bull used twice a week in natural service can breed about

100 cows per year. Even with a high conception rate of 70%, only 70 offspring would result each year. With present-day collection techniques and semen-processing methods, a high-fertility bull can produce up to 40,000 breeding units of semen each year. With a 55% conception rate and if all semen were used, some 22,000 calves could be produced. A bull of superior genetic ability could then produce over 100,000 offspring in a 5-year collection period. AI studs hope to be able to collect good bulls for more than 5 years after they have been proven for transmitting ability, thus increasing the potential number of offspring even more.

A second major advantage of AI is the reduction or elimination of reproductive diseases transmitted through natural service, such as vibriosis and trichomoniasis. Bulls in AI studs are tested to make sure that they are free from these diseases. Also, antibiotics are added to the semen in case they contain infectious bacteria. Additional advantages of AI include the elimination of the danger of farmers keeping bulls, elimination of the cost of keeping a bull, young bulls can be proven more rapidly with AI than through natural service, and there is no risk of small cows and heifers being damaged when mounted by a large bull. In general, the record-keeping system is improved when AI is used. In some cases a bull that cannot be used naturally can still be collected artificially and thus his semen used.

The major disadvantage of AI is the added time and effort required to detect cows in estrus. It is the essential cornerstone of a successful AI program on farms and must be carried out diligently if adequate results are to be obtained. Detection of estrus appears to be a greater problem with yearling heifers than with milking cows. Another disadvantage listed by some farmers is the requirement for a trained person to do the inseminations. Either a dairy farmer must have attended an AI school to do his or her own insemination or a trained technician must be used. More and more dairy farmers are doing their own inseminations. This trend will probably continue in the future.

Semen Collection Techniques

The most common method of collecting semen from dairy bulls is the use of the artificial vagina. The artificial vagina (Figure 24.1) consists of a rigid rubber casing and an inner liner of thin rubber tubing. The length and diameter of the tubing varies and the one used is dependent on the size of the bull. Hot water is placed between the outer casing and inner lining to produce an internal temperature of about 113°F (45°C). A lubricating jelly is placed on the interior surface of the lining, which in addition to the temperature produces an environment similiar to the normal vagina and evokes ejaculation. One end of the artificial vagina is fitted with a rubber cone that leads to a graduated vial, in which the semen is collected. The artificial vagina is then placed in an insulated jacket (Figure 24.1b), which helps to maintain the temperature of the artificial vagina and protects the vial from breakage. The insulated jacket is usually covered by a layer of plastic to protect it from dirt and manure.

Before bulls are collected, they are sexually stimulated to ensure that they will mount and ejaculate semen into an artificial vagina and to ensure the largest

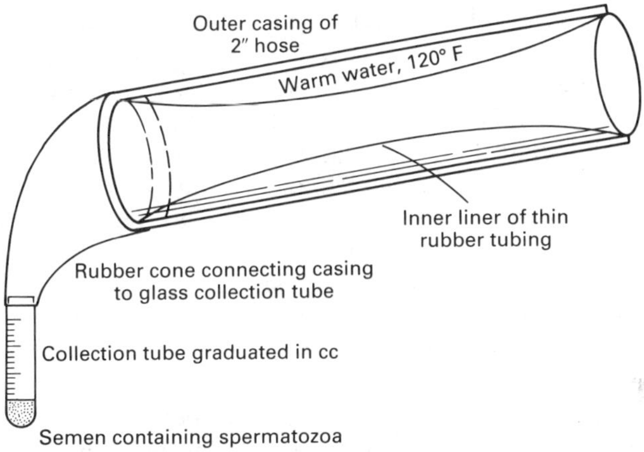

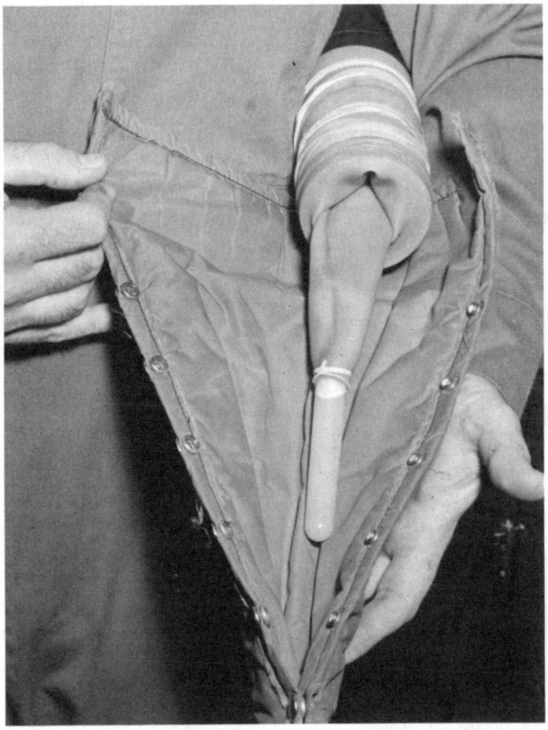

Figure 24.1 Semen is collected with an artificial vagina: (a) Drawing of an artificial vagina; (b) picture of an artificial vagina with a protective insulation bag to protect the semen from cold and light shock. Drawing from *Principles of Dairy Science* by G. H. Schmidt and L. D. Van Vleck. Copyright © 1974 W. H. Freeman and Company. Used by permission. Photograph courtesy of Clifton Marshall, Select Sires, Inc., Plain City, OH.

number of sperm in a high-quality semen. The stimulation is done by exposing the bull to a teaser animal and allowing the bull to mount without ejaculation. The teaser animal may be a cow, a steer, or another bull. Cows usually are not used because of the danger of actually breeding the cow and transmitting reproductive diseases. Since bulls vary in their preferences and time requirements, each bull must be handled individually to produce the maximum number of sperm. Usually, bulls are teased for ½ to 1 hour with intermittent exposure to a teaser animal for several minutes and then allowing a false mount. When collection takes place, the bull is allowed to mount an animal or a dummy animal and the handler guides the glans penis into the artificial vagina by placing the hand on the sheath of the bull (Figure 24.2). The semen is ejaculated in one upward thrust of the penis of the bull. The collector must be very careful to guide the penis into the artificial vagina and yet protect himself or herself against injury. After collection the semen must be protected from cold shock and light. Care must be taken to prevent contamination from urine, water, lubricating jelly, and bacteria.

In most cases the bull is collected twice on collection days with $\frac{1}{2}$ to 1 hour between collections. Bulls in full service will be collected 2 days per week with three collections per week done on highly fertile bulls that are in high demand. A normal ejaculate is between 5 and 7 ml in volume with a concentration of 1 to 1.8 billion sperm per milliliter. The motility (percent of sperm that move) will vary between 50 and 80%. Since the dilution is done on the basis of motile sperm, high motility is very important. After collection, the semen is taken to the laboratory for evaluation, dilution, and preservation.

Since some bulls will not mount another or dummy animal or ejaculate into an artificial vagina, two other methods of semen collection have been developed. These procedures are also used for bulls that may be incapacitated to mount or ejaculate because of advanced age or nonreproductive diseases. The two methods are electroejaculation and manual massage. The electroejaculation equipment consists of a bipolar electrode that is placed in the rectum of the bull above the accessory glands. The voltage is increased until ejaculation occurs. Massage is carried out by placing the hand in the rectum of the bull and massaging the prostate and ampulla until ejaculation occurs.

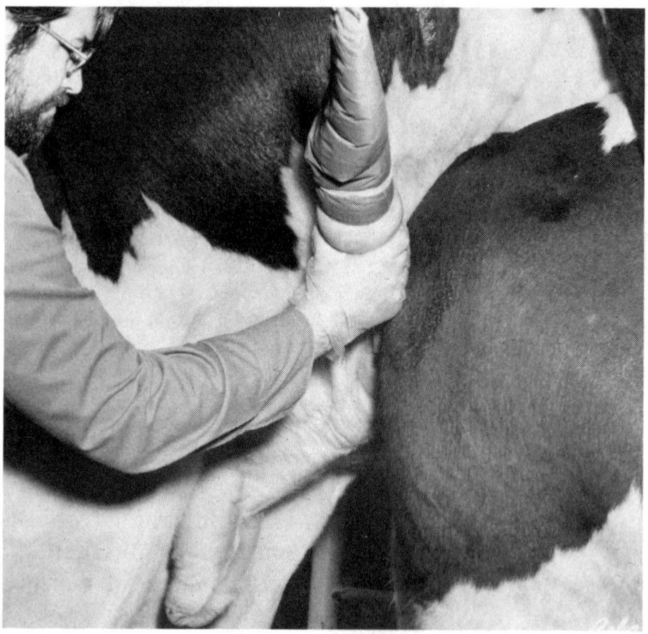

Figure 24.2 Semen collection procedure with the use of an artificial vagina. Photograph courtesy of Clifton Marshall, Select Sires. Inc., Plain City, OH.

Semen Evaluation

The best measure of the fertility of semen is the percentage of cows that conceive with it; however, this information requires months to obtain and laboratory technicians need to know immediately the desirable characteristic of the semen so that it can be processed. First, the volume of the ejaculate is noted and recorded. A drop in volume of ejaculate can be an indication of a problem with the bull, either due to health or inadequate preparation prior to collection. The gross appearance is also observed and a creamy white color and a swirling movement due to the motility of the sperm are desired.

In the laboratory a sample of semen is placed on a slide under a microscope and examined for progressive motility, which is the movement of sperm from one point to another in a straight line. This is expressed as percentage of motile sperm in multiples of 10. Circular, reverse, or rocking movement of sperm may be due to cold shock, water contamination, tail abnormality, or aged sperm. None of these conditions is desirable. Motility is an important measure since fertility is related to the number of motile sperm used in the insemination and motility also governs the dilution of the semen.

The concentration of sperm is obtained in one of three ways: the hemocytometer, photoelectric colorimeter, or the electronic particle counter. The latter is the most frequently used in AI studs. The hemocytometer was developed for counting blood cells and was adapted for sperm counting. This is a laborious procedure and not commonly used for routine determinations. With the photoelectric colorimeter, a small sample of semen is placed in the colorimeter and a beam of light passed through it. More concentrated samples have less light passing through; thus a measure of the concentration can be made. The electronic particle counter contains a capillary between two electrodes, through which one cell or particle passes at a time. The size of the capillary can be adjusted to the size of the sperm. The number of sperm in a sample is thus determined.

The morphology of the sperm is also examined to estimate the percentage of abnormal sperm. This may not be done on a routine basis, but periodically, to evaluate the performance of individual bulls. Fertility is usually not affected until the percentage of abnormal sperm reaches 20 to 25%. Since abnormal sperm are usually not motile, they are not counted as part of the motile sperm. For frozen semen, the integrity of the acrosome is also determined. The acrosome is the covering of the head of the sperm (Figure 24.3). Semen with a high percentage of sperm with intact acrosomes has a much higher fertility than that with damaged acrosomes. Acrosomes are damaged by freezing and thawing or other injury. Aging also increases the percentage of sperm with damaged acrosomes.

Semen Processing

Semen is commonly processed as frozen semen; however, some is still processed in liquid form. Semen to be stored in the liquid form is measured for its motile sperm concentration, and the amount of extender to be added is determined. The extender is usually an egg yolk–citrate combination, but whole egg and milk ex-

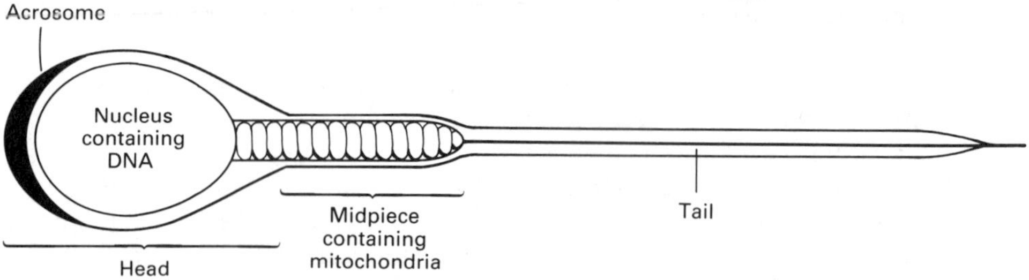

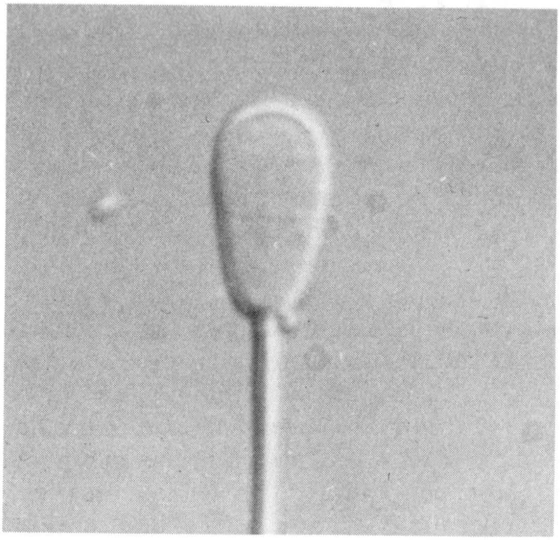

Figure 24.3 Sperm from a bull. (a) Drawing of a sperm with the component parts. Redrawn from H. J. Bearden and J. W. Fuquay, *Applied Animal Reproduction*, Reston Publishing Co., Inc., Reston, VA. (b) Photograph of the head and part of the tail of a sperm. Photograph courtesy of Clifton Marshall, Select Sires Inc., Plain City, OH.

tenders are as satisfactory. Antibiotics are added to inhibit or destroy bacteria that may be present. Liquid semen is diluted to contain 10 million motile sperm per milliliter. The semen is placed in 1-ml vials or straws and cooled to 5°C. The liquid semen can be stored and produces satisfactory fertility for 4 to 5 days.

The initial stages in the handling of frozen semen are similar to those of liquid semen; however, glycerol is added as a cryoprotective agent. The sperm are allowed to equilibrate with the glycerol at 5°C for periods of 4 to 18 hours, depending on the practices in a particular AI laboratory. Also, the sperm concentration will be two to three times as high as that with liquid semen. The diluted, equilibrated semen is then packaged in 0.5-ml plastic straws, although glass vials and plastic straws of other sizes may still be used. Plastic straws have the advantage over glass vials that they require less storage space per milliliter of semen stored. The plastic straws are then frozen by placing a single layer of straws on a tray and placing the tray above the liquid nitrogen of a large storage unit. The liquid-nitrogen vapors freeze the semen in about 2 minutes, at which time they can be placed on canes and lowered into the liquid nitrogen (Figure 24.4). Liquid nitrogen is the refrigerant of choice because frozen semen must be stored at temperatures below -75°C and liquid nitrogen has a temperature of -196°C.

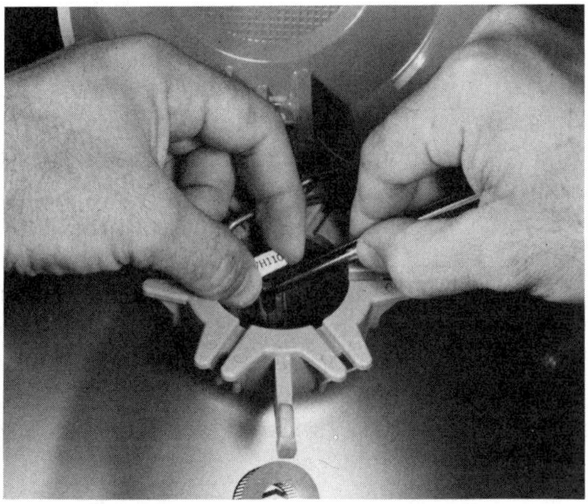

Figure 24.4 Removing a straw of frozen semen from a liquid-nitrogen tank. Photograph courtesy of Clifton Marshall, Select Sires, Inc., Plain City, OH.

Insemination of the Cow

The most common method of inseminating the cow is the rectovaginal procedure (Figure 24.5). In this procedure the gloved hand is inserted into the rectum and the cervix is picked up through the rectal wall. The insemination rod or "gun" is then inserted into the vagina and guided to the opening of the cervix. The insemination tool is then guided through the folds of the cervix by gently manipulating the cervix with the gloved hand. When the tip of the insemination tool reaches the body of the uterus, the semen is deposited and the insemination tool and hand removed.

Two types of inseminating tools are used. With liquid semen, a plastic dis-

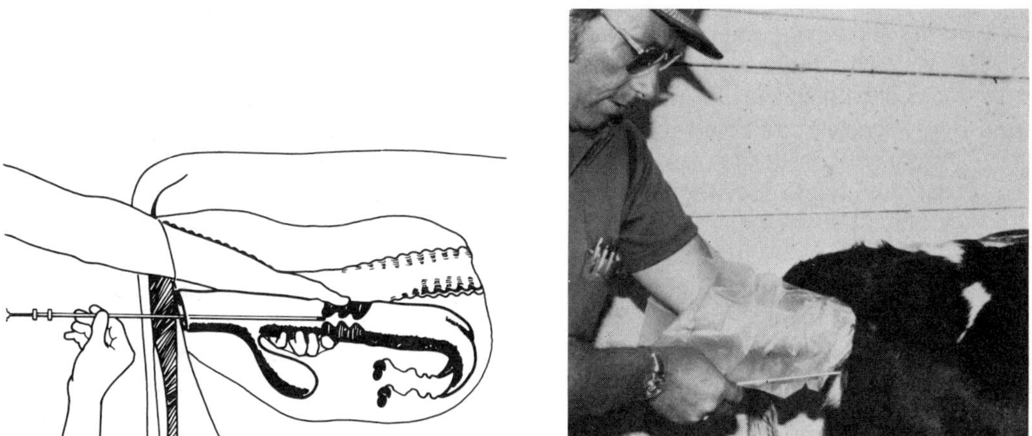

Figure 24.5 (a) Drawing and (b) photograph of the rectovaginal technique of inseminating the cow. Both courtesy of Clifton Marshall, Select Sires Inc., Plain City, OH.

posable insemination rod with an attached syringe is used. The semen is drawn from the vial into the plastic rod, whose tip is then inserted into the body of the uterus through the vagina and cervix and the syringe is used to force the semen from the rod. Semen stored in straws requires a special tool, commonly referred to as a "straw gun." It consists of a stainless steel rod that holds the straw. It also contains a plunger that forces the semen from the straw and a disposable plastic sheet to avoid contamination. The straw is inserted into the gun, the tip of the straw cut, and tip of the gun inserted into the uterus. In most breedings, the semen is deposited in the body of the uterus; however, most technicians deposit the semen in the cervix if there is a chance that the animal is pregnant. Depositing the semen in the vagina markedly reduces the conception rate, even though this is the place of semen deposition in normal service. In normal mating the number of sperm deposited there is many times greater than that used with artificial insemination.

24.2 EMBRYO TRANSFER

Although embryo transfer was first accomplished in the rabbit in 1890, it was not successful in cattle until 1951. Collection of bovine embryos had taken place in 1930, but the widespread application to the cattle industry did not take place until the 1970s. It is estimated that in 1984, 100,000 pregnancies resulted from embryo transfer. The first nonsurgical collection of embryos occurred in 1964, and the first *in vitro* fertilization of bovine ova occurred in 1983.

Embryo transfer is defined as the process by which an embryo is collected from one animal and transferred to another to complete the pregnancy process. Embryo transfer involves a number of procedures, which include: superovulation of the donor cow, collection of the embryos (commonly referred to as flushing), synchronization of the estrous cycles of the donor cow and the recipients, and transfer of the embryos to the recipients.

The donor cow from which the embryos will be collected must be watched closely for estrus. If only a single fertilized ovum is desired, the embryo of the cow can be collected if she is bred and conceives. In most cases a number of embryos are desired, and superovulation of the donor is practiced. This is done by the injection of the gonadotropin follicle-stimulating hormone (FSH) for 4 or 5 days between days 9 and 14 of the estrous cycle, with the day of estrus being 0. The FSH treatments are given twice a day to develop the multiple follicles. Prostaglandin injections are given 2 to 3 days after beginning the FSH treatment to the donor cow and the potential recipients to initiate synchronized estrous cycles in 2 to 3 days. Usually, 10 recipients should be available for embryo transfer if a high number of embryos are recovered. This means that about 14 recipient animals must be injected, since the prostaglandin injections will not be effective in cows or heifers that are not in the luteal phase of their cycle.

The donor cow is bred at standing estrus, usually with multiple inseminations at 12, 24, and 36 hours after the start of estrus. Sometimes semen from several bulls is used since the parentage of the offspring can be determined by blood typing. Six to 8 days later the embryos are collected by flushing them from the uterus. This is done by nonsurgical means. A catheter with an inflatable balloon is inserted into the donor's uterus and the embryos are flushed out with saline or a culture media (Figure 24.6). The embryos are collected and allowed to settle to the bottom of the container, which takes about $\frac{1}{2}$ hour. The fluid in the bottom of the container is collected, observed under a microscope, and the embryos removed. They are also examined and those appearing normal are transferred.

Two methods of transfer are still being used. The most successful to date is the surgical one, in which a small incision is made into the flank of the recipient and the horn of the uterus located. The embryo is transferred to the horn through a small puncture wound by a syringe or pipette. In the nonsurgical procedure the embryo is placed in the uterus through the cervix, similiar to artificial insemination. The embryo is placed in the horn of the uterus containing the ovary with the corpus luteum. To date, the pregnancy rate is somewhat higher with surgical transfer than with the nonsurgical procedure. With the surgical procedure, a pregnancy rate of 70% can be expected when good techniques are used. Embryos can be stored for short periods of time, such as 24 hours. They are also being frozen for use in a later transfer; however, 30 to 50% of the embryos are damaged by ice crystal formation. Embryos are now being split to produce two embryos having identical genetic makeup. This procedure requires some rather expensive equipment and a mastery of the manipulative techniques. It will probably be routine in 5 to 10 years.

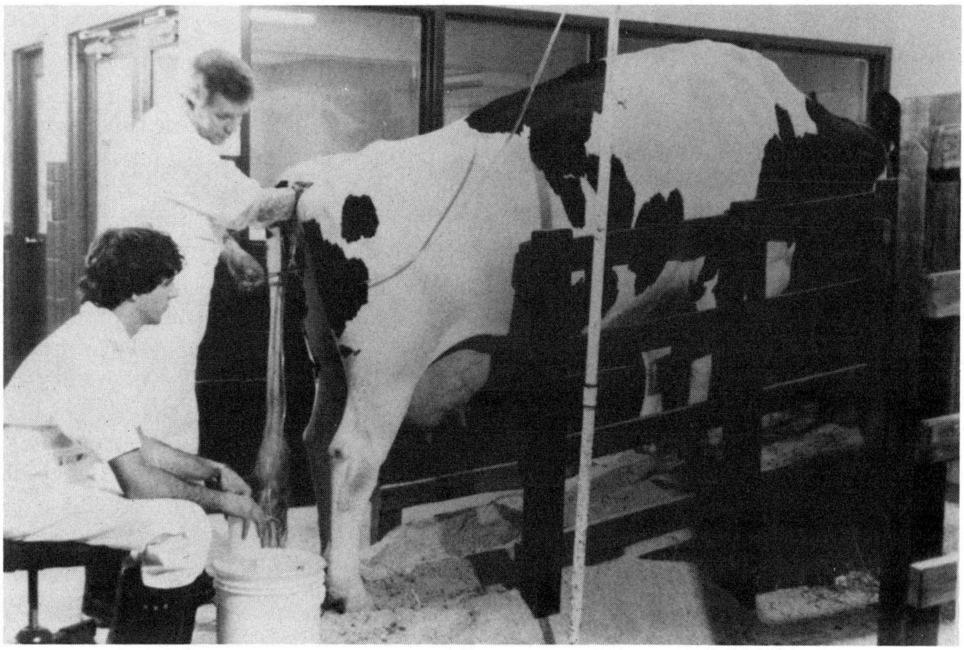

Figure 24.6 Nonsurgical method of collecting embryos from a dairy cow for use in embryo transfer. Photograph courtesy of Dr. Robert Baker, Select Embryos Inc. Plain City, OH.

The purpose and major advantage of embryo transfer is to increase the number of offspring of genetically superior females, similar to artificial insemination in males. Cows are now able to produce litters of calves instead of single births. A cow could produce up to 100 offspring during one year if she were superovulated at every third estrous cycle and averaged six pregnancies at each transfer. This is much higher than has been achieved in practice. So far the embryo transfer process has been profitable for registered purebred breeders who can sell the offspring for breeding stock at relatively high prices. It was estimated that an embryo transfer calf would have to sell for $2500 in 1978 to make the procedure profitable. The increased genetic improvement in milk production alone would support transfer fees of only $100.

Embryo transfer can be used in a number of other situations, such as testing bulls for genetic defects. A bull could be bred to a cow carrying the undesirable gene and six to eight of her offspring resulting from superovulation could be observed for the defect. The genetic defect, if any, would show up much faster than by breeding the bull to a random sample of cows. It is sometimes successfully used to recover ova from cows that are infertile because of disease, injury, or old age. In some cases it is used as a treatment of infertility. New genetic material can be brought into a country instead of importing animals. Similarly, it can be used to increase rapidly the genetic potential of a small herd. It also prevents the spread of some diseases, since the embryo does not spread them. It is also a way to store embryos for a long period of time, similar to semen.

24.3 ESTROUS SYNCHRONIZATION

Estrous detection of cows and heifers requires considerable time and effort. Also, on many dairy farms heifers are kept on pasture in the summer or at other barns and estrous detection is not done or is done in a haphazard manner. Consequently, a herd bull is used to breed these heifers. Synchronization of estrus (having a number of animals in estrus at the same time) has been a goal of research workers for a long time. If a group of animals could be brought into estrus during a 2- or 3-day period, the time required for estrous detection would be minimized and AI could be used more heavily, especially with yearling heifers. On the other hand, the synchronization process can cause some difficulties with dairy farmers. A group of heifers bred on the same day or within a 2-day period would also calve in a narrow period of time. Although it might be beneficial from the standpoint of checking on cows during calving, it would mean that a number of first-calf heifers would enter the milking string at about the same time. This might require more effort and test the patience of the dairy farmer.

Estrous cycles were first successfully synchronized by the injection or implantation of progesterone in cattle. This was done for a period of 14 to 20 days, at which time the hormone was removed. Cows would come into estrus 3 to 5 days later. This synchronization procedure was not adopted because of the long treatment time and the lowered conception rate in comparison to that when no hormones were used. Combining the progesterone injections with a corpus luteum–destroying agent (such as a conjugate of estradiol) reduced the time requirement for progesterone injections, maintained the synchronization success, and improved the conception rate; however, it was not approved for use in cattle.

Practical estrous cycle synchronization of cattle was made possible with the introduction of two synchronizing agents in the 1970s. One type is based on the natural hormone produced by the uterus, prostaglandin, and the other is a combination of a progesterone-like compound and estradiol. Prostaglandin $F_{2\alpha}$ ($PGF_{2\alpha}$) (e.g., Lutalyse) causes the regression of the corpus luteum (CL) and is produced at each estrous cycle. Cattle injected with $PGF_{2\alpha}$ come into estrus 60 to 72 hours after the injection (Figure 24.7). Because it causes regression of the

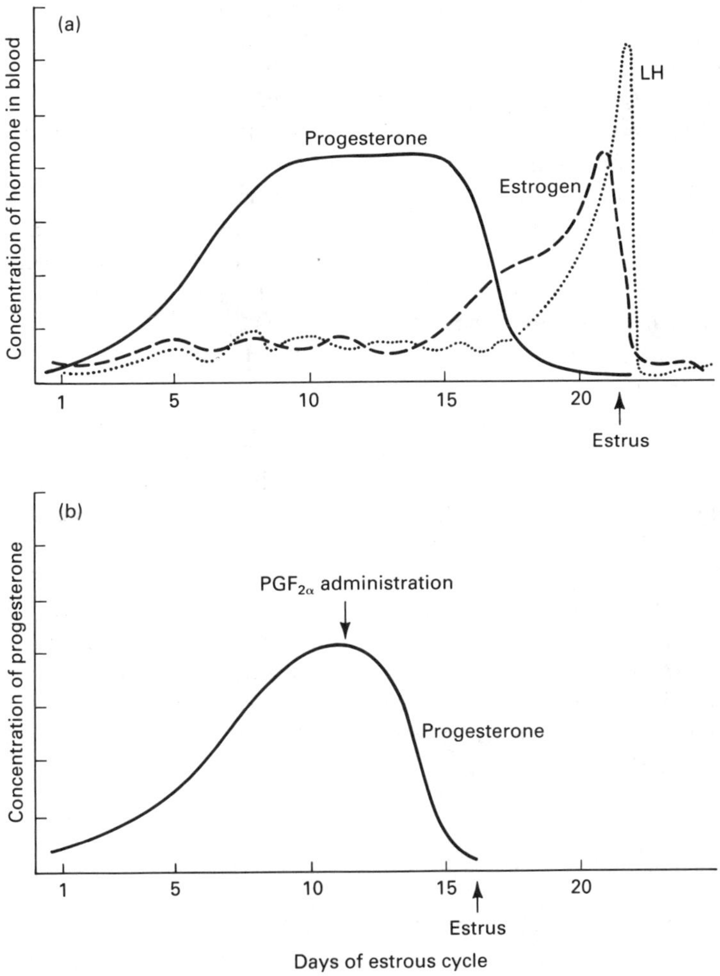

Figure 24.7 (a) Blood concentrations of hormones involved in reproduction during a normal estrous cycle; (b) the progesterone concentrations during a shortened estrous cycle caused by the administration of prostaglandin ($PGF_{2\alpha}$). From J. H. Britt, N. M. Cox, and J. H. Stevenson, 1981, *J. Dairy Sci.*, 64:1378.

CL, $PGF_{2\alpha}$ does not work during the first 4 days of the estrous cycle and is not used during the last 3 days of the cycle, since the CL is already regressing. Since it is effective for only days 5 through 18 of the estrous cycle, only 60 to 65% of a group of cows injected with it would be expected to respond with a synchronized estrus.

Several breeding schemes have been developed for use of $PGF_{2\alpha}$. In one scheme, a group of cows or heifers is injected with one injection of $PGF_{2\alpha}$ and observed for estrus 3 and 4 days later. Those responding to the treatment are bred. Some dairy farmers use fixed-time insemination by breeding all cows at 80 hours after the injection. The conception rate is lower than that resulting from breeding after estrous detection, but it eliminates all estrous detection. A second scheme involves rectal palpation of a group of cows and heifers and those with a palpable CL are injected with $PGF_{2\alpha}$. Fixed-time breeding or breeding after estrous detection is done on all injected animals. This reduces the amount and cost of the drug required. A variation of this scheme would be to observe a group of animals for 5 days, breeding those that came into estrus, and injecting the remaining animals.

A more common procedure is to inject a group of animals at one time. A second injection is given 11 or 12 days later. In this scheme all cows will be in the luteal phase of the estrous cycle at the second injection, and all should be in estrus within 72 hours after the second injection. The drug cost is twice that of the other schemes. The conception rates appear to be satisfactory and about equal to those when no drugs are used. Synthetic prostaglandin (e.g., Estrumate) is also on the market.

A second approach to current estrous synchronization is similar to the earlier attempts. A progesterone-like compound (e.g., Synchro-Mate B) is implanted in the ear between the skin and the cartilage. Several days before the implant an injection of the compound along with estradiol is given. The estradiol causes regression of the CL, probably by stimulating prostaglandin release from the uterus. The progesterone-like compound keeps the animals out of estrus until it is withdrawn. After 9 days, the implants are removed and the animals come into estrus 24 to 96 hours later, with peak activity occurring at 36 hours. Fixed-time breeding could take place 48 hours after the removal of the implant. The advantage of this scheme is that it works independently of the stage of the estrous cycle of the animals; however, it is approved only for heifers. About 80 to 90% of animals respond to the treatment.

24.4 OTHER TECHNOLOGIES

Of the remaining reproductive technologies, milk progesterone analysis as an aid in pregnancy diagnosis has been most widely used. Milk with low progesterone on days 21 to 23 after breeding is an almost perfect indication that the cow is not pregnant. If a cow has a high progesterone on the same days, the chances of pregnancy are about 85%. For the procedure to have widespread application, a cowside test is needed. Currently, a milk sample must be sent to a chemical laboratory for results. There is hope that the enzyme immunoassay procedure may be adapted as a farm test.

The procedure of sexing embryos was first demonstrated in bovine embryos in 1975. The application of this technique would have application to embryo transfer, especially if there were a great demand for one sex over the other. If only one sex of calves were desired, only those sex embryos would be transferred, which would reduce the cost of the recipients by one-half.

Two methods of embryo sexing are currently being used. One is associated with identifying the sex chromatin in the early morula stage of embryo development. The second involves the identification of the sex chromosome. The accuracy of the method is about 80%; however, the survival rate of sexed embryos is markedly reduced.

Much research effort has been expended in trying to separate sperm carrying the X and Y chromosomes in order for the resulting semen to produce either male or female offspring. Although great progress and periodic claims have been made in separating the sperm, no practical successful method is yet available.

In vitro fertilization has also received a lot of research attention since 1970. It has been successful with bovine sperm and ova and is being routinely used in human fertility clinics. *In vitro* fertilization will probably become very important when sexing of semen is successful, thus allowing the production of only male or female fertilized ova. It would also allow for the storage of frozen ova instead of frozen embryos. It may also develop into a laboratory procedure for measuring the fertilizing ability of bull sperm.

REFERENCES AND FURTHER READINGS

Bearden, J. H., and J. W. Fuquay. 1980. *Applied Animal Reproduction*. Reston Publishing Co., Inc., Reston, VA.

Britt, J. H., N. M. Cox, and J. S. Stevenson. 1981. Advances in reproduction in dairy cattle. *J. Dairy Sci.*, 64:1378.

Dailey, R. A. Synchronization of estrus in dairy heifers and cows. *W. Va. Coop. Ext. Serv*. IRM-8.

Kunkel, J. R. Embryo transfer. *W. Va. Coop. Ext. Serv*. IRM-26.

O'Connor, M. L. Artificial insemination technique. *W. Va. Coop. Ext. Serv*. IRM-12.

25

Herd Records

Every successful enterprise, agricultural or nonagricultural, needs a record-keeping system for effective management of the operation. Although management is defined in different ways, each definition contains three major components: decisions, resources, and goals or objectives. A management scheme is outlined in Figure 25.1 For effective management, a dairy farmer or any other business person must have a goal, hopefully a realistic one. The performance of the operation is then compared to the goal. Records are needed to do this. At this point a decision is made about the performance in comparison to the goal. If the performance is satisfactory, the management practices continue. If the performance is unsatisfactory, the reasons for the unsatisfactory performance must be defined and alternatives considered for improving the performance. Here the available resources must be considered as well as their cost. After the best alternative solution or solutions are chosen, they must be implemented. After the solution or solutions have been used for a period of time, the evaluation procedure must be repeated. The decisions can range in importance from questions such as: "Is my calving interval satisfactory?" to "Should I expand the dairy operation by increasing my herd size?"

Records form the backbone of the evaluation of the decision making steps in a management program. A record is nothing more than an accounting of events or data. When these data or events are summarized, they become information.

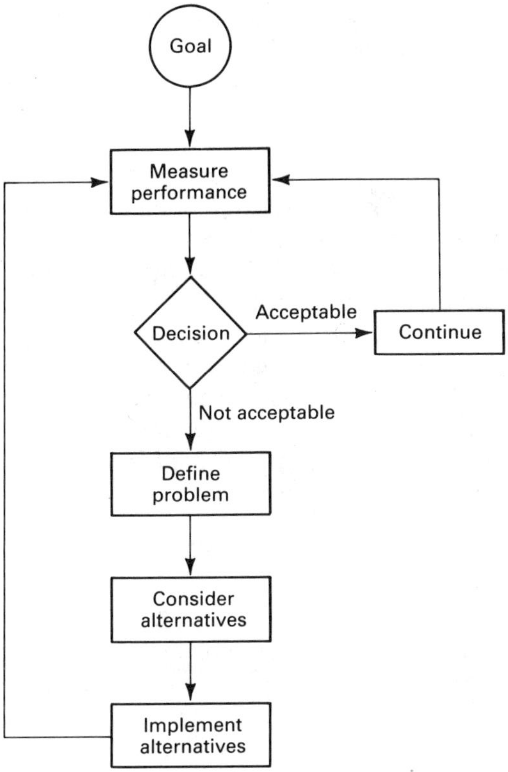

Figure 25.1 Decision-making scheme for use with dairy and other farm enterprises.

A list of data or events is usually not helpful in decision making, but information is. Hopefully, the records on dairy farms become information, which can be used for many different purposes. Some of these are: day-to-day management decisions, troubleshooting problems, measuring progress, enterprise evaluation, planning future actions, financial accounting and taxes, and genetic evaluation of animals. A good record-keeping system has the following characteristics: It is (1) current, (2) can easily be updated, (3) can easily be understood, and (4) can easily be summarized.

A dairy farmer needs two major types of records: those associated with the financial operation of the farm and those associated with the various enterprises, such as the milking herd. The required financial records are dependent on having accurate and complete information available to meet state and federal tax requirements. In addition, the farmer will probably want a farm accounting system that provides more information for decision making than that required for tax purposes. For example, enterprise evaluations, enterprise budgets, cash flow, depreciation, and fixed and variable costs are very helpful in total farm management.

In addition, the dairy farmer will probably want records for enterprises in addition to the milking herd and calves, such as the forage program, cash grain, and other livestock enterprises. Only the records associated directly with the dairy enterprise are discussed in this chapter; the reader is referred to one of the many current books in farm management for the other topics.

25.1 DAIRY RECORDS

Records on a dairy herd consist of two types: temporary and permanent. Temporary records are those recorded when the farmer is working with his or her animals and observes an event to be noted. It is recommended that the event be recorded in a small pocket notebook and later transferred to a permanent record. If a number of events occur before a permanent record is made, some of them may be forgotten. Some of these events are: cows in estrus, injuries, breedings, calvings, cows off-feed, and disease symptoms. These should be combined with the date and any other pertinent data. After the data have been transferred to the permanent record, the temporary data should be kept in a file in case the permanent record needs to be verified at a later date.

Identification

An important function of a dairy records program is identification of the animals in the herd. Daily management decisions concerning breeding, feeding, selection, calving, and culling depend upon accurate identification of animals. Proper identification is necessary for registration of animals with the purebred cattle associations and for keeping records of cows on production testing programs.

Identification of animals is also of two types: temporary and permanent. Permanent identification cannot be easily changed. Animals registered with the purebred cattle associations require permanent identification. Broken colored breeds, such as the Ayrshire, Guernsey, and Holsteins, require either a sketch or photograph of the color pattern of the animals. Mostly solid-colored breeds, such as the Brown Swiss and Jersey, require an ear tattoo. Branding is another form of permanent identification and consists of two types. Freeze branding is done by placing the branding irons in a refrigerant such as liquid nitrogen or dry ice and alcohol and then branding the animals. The freeze branding destroys the pigment cells in the hair follicles and thus produces white hair (Figure 25.2). For

Figure 25.2 Freeze brand on a dark-colored cow. From *Principles of Dairy Science*, by G. H. Schmidt and L. D. Van Vleck. Copyright © 1974 W. H. Freeman and Company. Used by permission.

white animals, the freeze branding iron is kept on the hair until the hair follicles are killed and the number or letters will appear by the absence of hair. Freeze branding has the advantage of being used on a daily basis to identify animals, whereas the ear tag is of little value for this and the farmer must remember the color pattern of his cows. With a herd of fewer than 50 cows, this may not be a problem, but it does present a problem in large herds. Hot iron branding is not used to any extent in dairy herds and is used primarily in beef herds to identify animals that belong to a farmer or rancher, not to differentiate among animals in the herd.

Most other types of identification are considered temporary, because they can be changed easily or are often lost. Among these are: ear tags, neck chains, ankle straps, brisket tags, and marking paint. Ear tags are of two types: small metal and large plastic ones. Small metal tags are placed in the ear of the heifer when she is vaccinated for brucellosis. They can also be used for identification; however, the animal must be caught before the tag can be read. Large plastic ones, about 2 by 3 in., have a shank that is inserted in the ear. These with large letters can be read from a distance to identify an animal in a barn lot or on pasture (Figure 25.3). This type of identification is receiving increasing use by dairy farmers. Another common method of identifying animals is a neck chain (or leather strap) with numbers. Neck chains with numbers are quite adequate for most conditions, except that they become lost and animals are difficult to identify in herringbone milking parlors because their necks are hidden by the bodies of other animals. Neck chains can become caught on fenceposts, fences, and bushes and at times may strangle the animal. Ankle straps are used primarily in milking parlors, but they are not very useful in identifying animals on pasture or in a drylot because they become caked with mud and manure. Brisket tags have limited use except for young stock and dry cows. They are not very useful in the herringbone milking parlor either. Marking paint is used primarily to mark a condition of an animal such as treatment for mastitis. It not generally used for cow identification.

Figure 25.3 Large plastic tag in the ear of an animal for identification purposes.

Permanent Record of Animal

Each animal in the herd should have a permanent record that is begun when the animal is born or when it is purchased. Information in the permanent record should include: animal's name and/or herd number, registration number if purebred, date of birth, purchase date, sire, dam, and vaccination dates. Added to this should be information that accumulates on the animal in her lifetime. These include: all health records, breeding and calving records, and feeding and production records. In addition to these becoming part of each animal's permanent record, some data will also be summarized and used on a day-to-day management basis.

Breeding and Calving Record

Complete breeding and calving records are required to breed cows to calve at 12- to 13-month intervals, to give cows adequate rest after each lactation, and to prepare the cow for the next lactation. A cow should have a complete breeding and calving record for each lactation and should include date of calving, sex and disposition of the calf, each estrus of the cow, date of service and sire used, pregnancy examinations, specific diagnosis and treatments for reproductive problems, expected calving date, and date the cow was dried off prior to the next calving. These data should become part of the cow's permanent record and the data also used to formulate management lists, such as cows due to calve, cows due to come into estrus, and cows due to be dried off. The prior records on the cow can be helpful in knowing what to expect at the next calving or lactation. It may also be of aid when making a culling decision on a cow.

Many types of breeding and calving records are available. A breeding wheel is used on many farms as a temporary file for information and as an aid in management decisions. The breeding wheel (Figure 25.4) consists of a revolving wheel with numbered pegs, which correspond to each cow in the herd. The wheel is marked off in time intervals and turned one notch each day. Each side of the peg has a different color and the cow's number. The colors usually correspond to (1) cows in estrus, (2) cows that calved but not yet bred, (3) cows bred and not diagnosed pregnant, and (4) cows that are pregnant. The peg is inserted into a hole corresponding to the current day, with the appropriate side toward the viewer. A dairy farmer can then look at the wheel and on any day see the cows that have calved and are not yet bred, cows that are expected to be in estrus, cows that were bred and not yet pregnant, and cows due to be dried off. As soon as the peg is moved or rotated, the data are lost unless they are recorded in a permanent manner.

Similar reproductive data can be stored in a permanent manner, and information for management decisions generated on a daily basis. One system is the Dairy Herd Improvement (DHI) program and the other is the use of an on-farm computer. In addition, a dairy farmer can use a notebook or filing system and generate the same information, but it becomes rather difficult in a large herd. With both the DHI and computer programs, management lists can be generated to aid the farmer in decisions and actions.

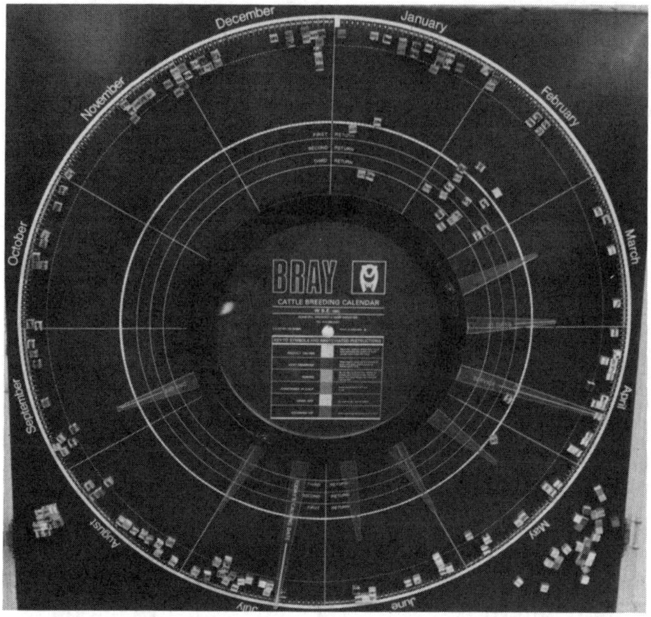

Figure 25.4 A breeding wheel aids in the management of the reproductive program on a dairy farm. The reproductive status of each cow in the herd can be seen at one time. Photograph courtesy of Donald Pritchard, The Ohio State University.

Health Records

A complete dairy records program should include a permanent record of the health of the animals. These data should include date of the problem, specific diagnosis, and treatment. Also included should be vaccinations and any other pertinent data that may be helpful in the future. Health records can be useful in tracking health problems in the herd, serve as an aid in a culling decision on an animal, and provide insights into potential future problems. The health records can be extremely helpful in tracking problems related to reproduction and mastitis, especially if all data, including estrous cycles and somatic cell counts, are included.

Production Records

Another important function of a record-keeping system is to record the production of individual cows in the herd. The information from these data can be used to allocate grain to each cow, evaluate the performance of cows on a daily or monthly basis, determine cows to culled, select replacement heifers, and help identify problems in the herd. Some dairy farmers keep daily milk production records, which is the ideal system because daily variation in milk yields can be followed and abnormalities detected more easily. The daily weighing of milk is a laborious procedure and it will probably not receive widespread adoption until milk weights can be obtained automatically and recorded on a computer. The equipment is now available to do this, but the technology to identify the animal automatically has not yet been completely developed.

A permanent record book containing individual pages on each cow is desirable. The minimum requirements of the record on each cow are: cow identification, sire, dam, calving dates, reproductive examinations, health problems, and milk production data, including yearly milk production and milk fat and protein percentages. Breakdown by monthly milk production provides additional valuable information.

Feeding Records

Some dairy farmers keep rather extensive feeding records. If cows are fed individually, the recording of these data will help to determine the profitability of each cow. In many cases, group feeding is practiced, and recording of the mixtures and amounts can be helpful for analyses of the dairy operation.

Other Records

Additional records need to be kept by dairy farmers. As already indicated, each business operation needs to keep the minimum records required for tax purposes. Purebred breeders need additional records in terms of registration papers, pedigree, type classification scores, and sales. Also, extensive records on incomes and expenses from individual farm enterprises can be helpful in the evaluation of the entire farm operation.

25.2 DAIRY HERD IMPROVEMENT PROGRAM

Most of the dairy records discussed previously are provided by participation in the DHI program. It is the most widely used record program in the United States and a form of it is also used in most western European countries as well as New Zealand, Australia, and Japan. In 1983, 30.7% of the milking cows in the United States were enrolled in one of the official plans and 12.6% were enrolled in one of the nonofficial plans of the DHI record program, making a total of 43.3%. In a survey of 660 Ohio dairy farmers, Schmidt and Smith (1986) found that in general farmers making greater use of the information on their DHI records had higher production per cow and above average herd performance records, such as calving interval, days to first service, and less mastitis.

The DHI program began as a cow testing association in Michigan in 1905. Cow testing associations originally developed as a group of approximately 26 dairy farmers who hired a milk tester. The tester visited each farm once a month, measured the milk yield on each cow for two milkings, and took a milk sample for fat determination with the Babcock test. The monthly milk and fat yields of cows were calculated from these daily yields by multiplying them by the number of days in the month.

The basic components of the cow testing association are still part of the program, with a supervisor visiting each farm once a month, weighing the milk, and taking a milk sample for fat, protein, and somatic cell contents. In addition, the supervisor collects all data that relate to the records, such as calving dates, service dates, dry dates, and other factors that influence the records. The milk is now analyzed at central testing laboratories (Figure 25.5) and the records processed with mainframe computers located at nine Dairy Records Processing Centers.

The DHI program is now run primarily by state associations under a set of uniform rules and regulations developed by the state and federal extension associations and the USDA, with considerable input from the Purebred Dairy Cattle Association (PDCA) and the National Association of Animal Breeders (NAAB). Most state DHI associations are members of the National Dairy Herd Improvement Association (NDHIA), which has become instrumental in policy and management decisions. It has given active leadership to the development of the DHI program and introduced a quality certification program for state organizations. It has also provided a series of seminars, talks, and meetings for its membership.

The policies of the DHI program are under the control of the National Cooperative Dairy Herd Improvement Policy Board, which consists of the following members: five dairy farmers from NDHIA, one PDCA representative, one NAAB representative, one representative from the USDA Agricultural Research Service, one representative from the USDA Extension Service, two state extension dairy specialists, and one state extension director. The federal, state, and local extension personnel have the responsibility for the education aspects of the program and the state and federal experiment stations the research aspects of the program.

Types of Records

The DHI program has three major records with a number of variations and minor records. The most common record is the standard DHI or DHIA record. The data

Figure 25.5 Machines that automatically determine (a) milk fat and protein composition and (b) somatic cell count of milk. Photographs courtesy of Ohio DHI Inc., Powell, OH.

are collected by a testing supervisor under the rules and regulations of the association. The records are processed at a Dairy Records Processing Center. The DHIR program is very similar to the DHI program in that it must meet the same basic minimum requirements plus several others required by the breed associations. These records are then accepted by the breed associations for their programs and advertisements. These two records are usually referred to as "official" records in that they are collected by an impartial supervisor and records calculated by a computer. Lactation records are corrected to account for the milk yield produced before the first and after the last visit by the supervisor. Variations of these records include the A.M.-P.M. programs, in which the supervisor visits each farm for one milking once a month on an alternating A.M., P.M. basis. Correction factors are used to account for the variations in milk yield and composition from A.M. and P.M. milkings. In most cases, the dairy farmer must have a meter attached to his milking unit that records the time of the start of milking.

A variation of the DHI record is the owner-sampler (OS) record. In this test, the dairy farmer weighs the milk of each cow and takes a sample for milk composition determination. The farmer also records all of the data concerning the records, such as calving, due, and dry dates. The data and milk samples are picked up by the DHI supervisor and the milk is analyzed and the data processed similar to the DHI and DHIR records. Because an impartial person did not record the milk yield and collect the milk sample, this record is usually referred to as an "unofficial record." These records have the same value for management decisions as the DHI and DHIR records if the dairy farmer accurately collects the data for the records.

A variety of other programs are offered by the various states. Among these are a weigh-a-day a month (WADAM) and milk-only records (MOR), in which milk yields only are collected and no attention is given to the composition of milk. Some state associations also have programs known as commercial programs, which can be a combination of any of the foregoing programs.

In addition to the basic information on the milk yield and composition of milk and the factors that influence them, the DHI program includes a number of different types of information. The types will again depend on the state association. Some of the common sources of information are feed and mastitis information. Most states conduct a somatic cell count on the milk sample on each individual cow. This information can be valuable in the mastitis control program. Most associations also allow for the recording of feed intake of individual cows, including the cost of the feeds consumed. From these data information such as income above feed costs for each cow and the recommended grain to be fed to the cow can be calculated.

Use of Records

The records generated by the DHI program have many uses. The daily use of the records for management decisions is discussed in the next paragraphs. The completed lactation record for each cow is used in several ways. The dairy farmer can use it to evaluate the merits of the cow in terms of culling, breeding, and a source of replacement animals. The dairy farmer can also use the record to advertise animals that are for sale. The completed record is sent to the USDA for use in the national sire-proving program to evaluate the transmitting ability of bulls, especially those located in bull studs. The records are used to identify the outstanding cows in the population. This information is extremely useful for persons involved in securing young sires for proving in bull studs.

The types of reports that the dairy farmer receives on a monthly basis will depend on the state association and the dairy records processing center handling the dairy farmer's records. Some of the most important include the individual cow record, the monthly herd summary, and the action lists. For the detailed components, each dairy farmer or student should check with his or her state association.

The individual cow record or page is produced when a cow completes her lactation record. It includes all of the pertinent data collected on the cow. If the cow is a second-lactation or later-lactation cow, information from the previous lactation records will be included. This record will be a complete record of the cow's production to date, but usually does not include health data and has limited reproduction data. This should become part of the permanent record on each cow.

The monthly herd summary is one of the most important sheets to help evaluate the present status of the dairy herd. In the survey by Schmidt and Smith (1986), Ohio dairy farmers indicated that receiving the monthly progress report was the most important reason for participating in the DHI testing program. It includes data on a herd basis and should be studied thoroughly each month. The production data for the herd are updated each month and a rolling herd average calculated for milk and fat production. In this calculation, the most recent month's production data are added to the total and that obtained 13 months ago subtracted. Thus it becomes a record for the last 12 months, consequently called a rolling herd average. Similar calculations are made on the other measures. Some of the other most important items of information that should be observed and evaluated monthly are: test day milk production on all cows, extrapolated 305-day mature equivalent milk production, average age of cows at first calving, average age of cows at last calving, average days dry, days in milk at first breeding, average calving interval, services per conception, days open, percent of cows with low somatic cell counts, and percentage of cows culled. Each of these measures should be compared with the goals of the dairy farmer and evaluated according to the management scheme shown in Figure 25.1.

Most computing centers also develop "action lists." Some common ones are: cows to breed, cows to calve, cows to dry off, potential culls, cows to check for pregnancy, and cows with high somatic cell counts. These can be used on a daily basis for working with the dairy cows. Most computing centers also provide a number of other reports, some of which are at the option of the dairy farmer. If the records are used by the dairy farmer, they can be a very valuable aid for decision making.

25.3 USE OF COMPUTERS IN DAIRY RECORDS

Some computing centers are now set up so that the dairy farmer can use a computer or terminal to be interactive by telephone with the mainframe computer at the center. In this way, the dairy farmer can update the data in the computer on a daily basis instead of waiting for the supervisor to enter the data on a monthly basis. In return, the dairy farmer can receive action sheets on a daily basis with up-to-date data.

Some dairy farmers have either substituted or are thinking about substituting their on-farm computer for the DHI program. To make this possible, an automatic milk recording system with automatic cow identification is almost a requirement. Otherwise, considerable labor is required to weigh and record the milk yield. Even so, there are no current ways to analyze milk for its components on an in-line system, which means that a milk sample would have to be taken and sent to a laboratory for analyses or milk composition data not obtained. In addition, many dairy farmers want the credibility of the records found with the DHI program. In the near future it does not appear that a records system confined to an individual farm is a very feasible alternative for the vast majority of dairy farms.

The computers can be very helpful in many other parts of the dairy management program (Figure 25.6). First, it can be useful in storing and calculating

Figure 25.6 An on-farm computer is being used with increasing frequency on dairy farms for the storage of data and for help in decision making. Photograph courtesy of Donald Pritchard, The Ohio State University.

the financial data on the farm. Also, a number of "software" programs have been developed to aid the farmer in evaluating the farm operation, including programs that are specifically designed for the dairy operation. In addition, a number of computer programs have been developed for future planning, commonly known as "what if" situations. Examples of these programs include herd expansion, buying more land, increasing production level of the cow, and borrowing of money. Many more will be developed in the next several years.

REFERENCES AND FURTHER READINGS

CADY, R. A. Records and record systems—the backbone of good management. *W. Va. Coop. Ext. Serv.* IRM-4 fact sheet.

SCHMIDT, G. H., and T. R. SMITH. 1986. Use of dairy herd improvement testing programs by dairy farmers. *J. Dairy Sci.*, 69:3156.

VOELKER, D. E. 1981. Dairy herd improvement associations. *J. Dairy Sci.*, 64:1269.

26

Raising Calves and Heifers

The production of vigorous, healthy heifers is essential to the economic success and future of every dairy herd. Adding genetically superior first-lactation cows is the only practical way of increasing the genetic base of the dairy herd. Because most dairy managers cull 25 to 35% of their herd annually, an adequate supply of replacement heifers is needed. Cows are culled or removed for the following reasons: 23% for low production (called voluntary cull); 13% for problems with reproduction; 8% for mastitis; 28% for injury, 5% by death, 8% for dairy; and 15% for poor disposition, ketosis, and other unknown reasons.

An alternative to raising calves and heifers is to buy replacement stock from other farms. Most managers prefer to raise their own replacements because (1) raising is usually less expensive than purchasing comparable heifers, (2) genetic merit is known, (3) chance of introduction of disease to the existing herd is less, (4) growth rate can be controlled (no fattening), (5) breeding age and season can be controlled, (6) mating sires can be chosen, and (7) disease control and vaccination programs can be instituted on time. The cost to raise a replacement heifer includes feed costs (42%), variable costs (12%), interest (4%), fixed costs (32%), and labor (10%) (Luening et al., 1984).

All replacement heifers should be bred to dairy bulls. If heifers are bred to beef bulls, replacement heifers from first-lactation cows will not be available. Calves from replacement heifers should be genetically superior to other calves because the dam and sire should have higher genetic values than the average of cows in the herd. To avoid calving problems, AI bulls with desirable calving-ease ratings should be selected for heifers freshening as 2-year-olds. Heifers should be large enough to avoid delivery problems. Breeding heifers to "jumper bulls" (genetically unevaluated bulls that breed all females) can lead to calving problems (large calves), spread reproductive disease among heifers, and result in calves with low genetic merit. Hormone injections or implants can be used to induce estrus and minimize time needed for estrus detection to facilitate AI. A veterinarion should be consulted if synchronization is used.

26.1 DIGESTIVE SYSTEM OF THE CALF

Even though the stomach of the new born calf has the same four compartments as a mature animal, the rumen and reticulum of the calf and mature animal differ considerably in degree of development (Figure 26.1). The abomasum makes up approximately 50% of the weight of the stomach of the newborn calf, whereas the rumen and reticulum make up 67% of the weight in 16-week-old calves. Relative proportions also apply to the volume of various compartments.

The stomach of the newborn calf is similar to that of the simple-stomached animal. Milk entering the stomach passes via the esophageal groove (reticular groove) to the reticulo-osmasal orifice. The esophageal groove conveys liquid or semiliquid materials from the cardia directly to the omasum and by-passes the rumen and reticulum. The esophageal groove closes by reflex action and serves as a conduit for milk in the young animal. The reflex action disappears as the animal matures so that ingested liquid and materials enter the rumen and retic-

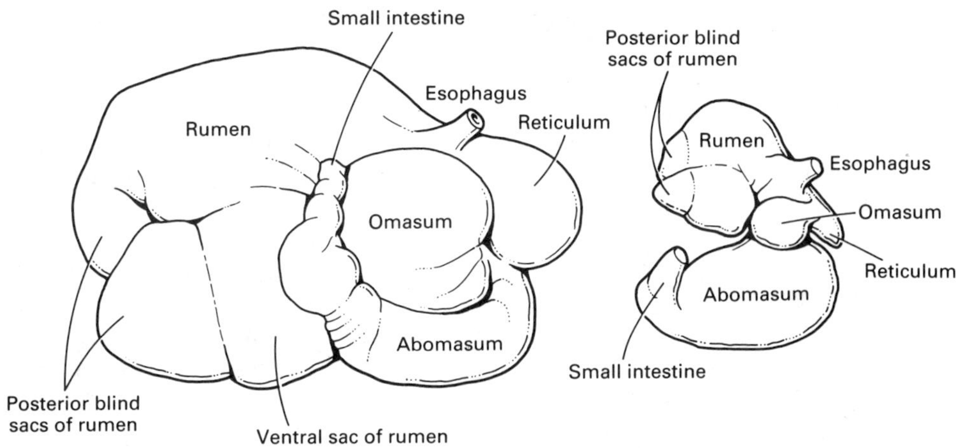

Figure 26.1 Relative sizes of the compartments of the stomach of the young calf (right) and the mature ruminant (left). Note the large size of the rumen in the mature animal and the relatively large size of the abomasum in the calf's stomach. Adapted from S. Sisson and J. D. Grossman, 1938, *The Anatomy of Domestic Animals*, W. B. Saunders, Company, Philadelphia.

ulum. If the esophageal groove remains open, the material falls into the rumen-reticulum area.

In the young calf, the rumen and reticulum are not developed to handle forage. Feed must contain B vitamins and high-quality protein with a proper balance of amino acids because these nutrients cannot be synthesized by the rumen microflora. The diet of the newborn calf must be milk solids because digestive enzymes to breakdown complex proteins and carbohydrates do not develop until 7 to 10 days after birth. Forage can be fed as the rumen and reticulum develop.

Several factors are responsible for development of the rumen and reticulum. End products of rumen fermentation are important stimuli for development of the papillae in the rumen and reticulum. Volatile fatty acids, in order of effectiveness in developing the papillae, are butyrate, propionate, and acetate. Glucose and salt solutions are ineffective in stimulating rumen development. Development of the rumen musculature depends on the presence of dry material and fiber in the rumen.

26.2 MANAGEMENT AT CALVING

Dry cows require adequate protein, energy, minerals, and vitamins (see Section 21.6). Nutrient deficiencies can reduce the effectiveness of the calf's immune system to combat disease. A balanced dry cow ration is essential for both the unborn calf and cow.

The dry cow should be placed in a maternity stall a week prior to expected calving. Stalls should be 16 to 25 square meters, well lighted, clean, and bedded with a nonslip surface. An overhead hook to lift downer cows is useful. One maternity stall per 25 cows is recommended. Maternity stalls should be disinfected between calvings. A clean, well-drained pasture area can be used successfully.

Approximately 5% of calves are born dead (stillborn). Many of these calves could be saved if assistance were available at calving. Cows should be monitored closely at calving to determine if help is needed. Small cows, large calves, abnormal presentations, and multiple births are situations that may require help. Cows should be checked every 2 hours after labor begins. Wisconsin researchers (Crowley et al., 1983) list the following stages of birth.

- *Stage 1: Restlessness.* A cow will get up and down frequently and is not comfortable. Visible labor begins when the water bag breaks. Slippery and viscous material found in the water bag is essential for normal delivery. For a mature cow, the calf will be born within 2 hours; heifers may require 4 or more hours.

- *Stage 2: Calf delivery.* The fetus should be expelled within ½ to 1 hour after it appears in the pelvic inlet, within 2 hours for heifers. Working with the cow as her muscles contract will minimize stress.

- *Stage 3: Expelling the placenta.* The placenta (after birth) is passed from 30 minutes to 8 hours after calving. Retained placenta occurs in 8 to 10% of uncomplicated births. Feeding vitamins A, D, and E, and selenium, phosphorus, and calcium minimize retained placenta.

The calf must begin breathing as soon as the umbilical cord is severed. Breathing can be initiated by removing mucus from the nostrils, rubbing to stimulate it, tickling the nostrils to stimulate breathing, or administering artificial respiration. The navel cord should be cut and tied about 5 cm from the body to prevent bacteria from entering the body. String disinfected in alcohol or special plastic clips can be used. Tying the navel too close to the body may cause a hernia to develop. The umbilical cord must be *thoroughly* disinfected with tincture of iodine solution to kill disease organisms. A cloth or paper can be used to dry the calf before moving the calf to a dry, draft-free area.

26.3 COLOSTRUM

The first milk produced by the cow after birth is colostrum, which is high in nutrients and antibodies (Table 26.1). A calf at birth lacks disease protection because antibodies (immunoglobulins) do not pass across the placenta from the dam to the circulatory system of the fetus. Antibodies provide the initial disease protection for the young calf until its system develops and produces antibodies. Two liters (2 quarts) of colostrum should be fed within 4 hours after birth (ideally within 30 minutes). Time is critical because the calf's digestive tract allows antibodies (large protein molecules) to pass directly into the blood without being broken down for only 24 hours. After 24 hours, antibodies cannot be absorbed intact by the small intestine (called closure). For optimal protection, a calf should absorb 60 g of immunoglobulins. A large breed calf should receive 2 kg (4.4 lb) of colostrum at each of the first two feedings (at birth and 12 hours). The efficiency

TABLE 26.1 COMPOSITION OF COLOSTRUM, TRANSITIONAL MILK, AND WHOLE MILK

	Milking number (cows milked twice daily)					
	1	2	3	4	5	11
Item	Colostrum	Transitional Milk				Whole milk
Total solids, (%)	23.9	17.9	14.1	13.9	13.6	12.9
Total protein (%)	14.0	8.4	5.1	4.2	4.1	3.5
Casein (%)	4.8	4.3	3.8	3.2	2.9	2.5
Immunoglobulins (%)	6.0	4.2	2.4	0.2	0.1	0.09
Fat (%)	6.7	5.4	4.9	4.4	4.3	4.0
Lactose (%)	2.7	3.9	4.4	4.6	4.7	4.9
Minerals (%)	1.11	0.95	0.87	0.82	0.81	0.74
Specific gravity	1.056	1.040	1.035	1.033	1.033	1.032

Source: J. W. Crowley, N. A. Jorgensen, and W. T. Howard, 1983, *Wis. Ext. Bull.* A1485.

of immunoglobulin absorption is maximum at birth (20 to 50%), drops to 10 to 20% by 6 hours after birth, to 5% at 12 hours, and is near zero at 24 hours. Calves should be force-fed a predetermined amount of colostrum; otherwise, the amount of colostrum consumed is not known, time of consumption after birth may vary, first-lactation cows may not allow the calf to nurse or may not let down their milk, the calf may be too weak to nurse, or the calf may suck the leg, flank, or other area of the cow and ingest disease organisms. An esophageal feeder should be used to force-feed colostrum to calves that will not consume colostrum after birth.

Colostrum quality varies greatly. Older cows (third and over lactation) produce colostrum higher in antibody concentration with a broader spectrum of disease protection than that of young cows. Color (yellow) and consistency (thick and creamy) are good indications of quality. A commercial colostrometer is available to measure the specific gravity, which is related to antibody concentration (Table 26.1). Bloody or mastitic colostrum should not be fed. A supply of high-quality colostrum from older cows should be saved and frozen in 2-liter (2-quart) milk cartons or plastic containers for calves from cows that die at birth, have mastitis or milk fever, do not let down their milk, or produce poor-quality colostrum.

Surplus colostrum should be fed since it is high in nutrients and the antibodies protect the lower digestive tract from disease. Fermenting colostrum is an economical method to handle extra colostrum, which can save $10 to $20 per calf (replaces whole milk or milk replacer). Twenty to 30 days of liquid feed can be provided from excess colostrum and nonsalable milk from the first eight milkings. Management guidelines are listed below.

1. Extra colostrum can be stored in clean plastic or plastic-lined containers (do not use galvanized metal containers).
2. Stirring the fermenting colostrum after every addition of fresh material avoids separation and layering.
3. The container should be covered to keep out contaminates.
4. Bloody milk, mastitic colostrum, and antibiotic-containing milk (antibiotics will stop the fermentation process) should be discarded.
5. The final liquid product (16 to 20% total solids) should be diluted (2 to 3 parts soured colostrum with 1 part warm water).
6. If calves refuse soured colostrum, adding sodium bicarbonate (0.5% by weight) prior to feeding will neutralize the acidic taste.
7. Adding propionic acid (1% by weight) to fresh colostrum in the summer minimizes abnormal fermentation and foul odors. The acid must be added slowly while stirring to avoid protein precipitation.
8. Acidified colostrum can be added to previously acidified colostrum.
9. Soured colostrum should be used within 3 to 4 weeks.
10. Fermented colostrum can be combined; do not mix fresh colostrum with soured colostrum because a second fermentation will begin.

26.4 RAISING CALVES FROM BIRTH TO WEANING

Liquid Diets

After the calf has received colostrum for three days, several liquid programs are available (Table 26.2).

Soured colostrum is a high-quality feed for the young calf. If bull calves are sold 3 to 5 days after birth, adequate amounts of soured colostrum may be available for replacement heifers from birth to weaning. Calves should be gradually shifted from soured colostrum when changing to whole milk or milk replacer.

Whole milk is a highly digestible feed for calves. Calves should be fed 8 to 10% of their body weight. Weaning at 8 weeks of age will have used 250 kg (560 lb) of milk plus the initial colostrum. High-solids milk (Jersey and Guernsey milk) can be diluted with 1 part warm water to 3.5 parts milk to avoid feeding problems. Excessive milk intake will limit dry feed intake and can cause digestive disorders.

Mastitic milk or antibiotic-containing milk (called waste milk) can be fed to calves. Abnormal milk or first milk after mastitis treatment (which is high in drug residue) should not be fed. Calves fed mastitic milk should not be allowed to suck another heifer or themselves because mastitic organisms could enter the undeveloped udder through the immature teat. Calves fed antibiotic-containing milk cannot be sold for slaughter until all antibiotic residues have cleared the calf's system.

Milk replacer is an economic alternative (half the cost of feeding marketable whole milk). Calves are typically fed 0.45 kg (1 lb) of powder per day diluted with 3.1 kg (7 lb) of water. The mixture should contain 12 to 15% solids. Quality of the milk replacer is critical (Table 26.3). Young calves do not have the digestive enzymes at birth to utilize nonmilk ingredients. The feed tag will list nutrients and ingredients.

Milk replacer should contain 22% protein (all-milk sources) or 24% protein (some plant protein). Protein quality has a significant impact on growth and health. When calves are over 2 weeks of age, some plant protein can be included as digestive enzymes are being produced to break down complex proteins.

TABLE 26.2 AVERAGE COMPOSITION (PERCENT) OF COLOSTRUM, WHOLE MILK, AND RECONSTITUTED MILK REPLACER

Component	First milk colostrum[a]	Pooled excess colostrum[a,b]	Whole milk[a]	Reconstituted milk replacer[c]
Dry matter	28.3	16.0	12.1	12.0
Fat	6.0	5.5	3.5	2.3
Protein	18.8	5.5	3.3	2.7

Source: J. G. Linn, D. E. Otterby, R. D. Appleman, and M. F. Hutjens, 1980, *Ill. Ext. Circ.* M-1183.

[a] Based on samples from Holstein cows.

[b] Average composition from first six milkings.

[c] One part milk replacer (24% protein, 20% fat) diluted with 7 parts water.

TABLE 26.3 QUALITY OF PROTEINS AND FATS IN MILK REPLACERS

Best	Acceptable[a]	Inferior
	Protein sources	
Skim milk	Specially manufactured soy flour	Unprocessed soy flour
Buttermilk	Soy concentrate	Meat solubles
Whole whey	Hydrolyzed fish protein	Fish flour
Delactosed whey		Distiller solubles
Casein[b]		Brewers' yeast
Milk albumin		Oat flour
Whey protein concentrate		Wheat flour
	Fat sources	
Lard	Hydrogenated vegetable oils	Liquid vegetable oils
Tallow		
Stabilized greases		

Source: J. W. Crowley, N. A. Jorgensen, and W. T. Howard, *Wis. Ext. Bull.* A1485.

[a] These specially processed products are acceptable when used with one or more sources from the "best" column.

[b] It is desirable that at least half of the protein be derived from casein.

The fat content of milk replacer can range from 3 to 25%. Ten percent fat provides optimal performance. Higher fat (15 to 20%) can improve performance when calves are exposed to cold weather, disease conditions, and stress. Fat also can lower the incidence of scours by minimizing irritation to the small intestine. Animal fats (lard, tallow, or grease) are superior fat sources, while unsaturated vegetable oils are less satisfactory (Table 26.3).

High crude fiber (over 1%) indicates that nonmilk ingredients were utilized. Young calves cannot digest these complex carbohydrates until they are 2 to 3 weeks of age. Milk replacers with less than 0.5% fiber should be fed for the initial 2 to 3 weeks before shifting to a higher fiber replacer to lower feed costs. Additional sources of minerals and vitamins (A, D, E, and B complex) should be included in the milk replacer. Warm water improves the mixing of milk replacer.

Acidified milk replacer has been used successfully in Europe. Organic acids and preservatives are added to a specially manufactured milk replacer. Calves are allowed unlimited access to the liquid with a common nipple. Because the milk replacer is acidified and fed cold, restricted intake at each feeding and frequent nursings occur through the day. The product remains wholesome for 3 days at room temperature. Benefits include lower gastric pH, improved digestibility, and less overloading of the digestive system. Disadvantages include greater consumption of milk replacer and a higher feed cost.

Antibiotics

Feeding low levels of antibiotics to dairy calves is helpful because growth rate is stimulated, feed efficiency is improved, physical appearance is enhanced, and the incidence of scours and colds is reduced. The beneficial effect of feeding antibiotics occurs during the first 8 weeks of life when the young calf is most susceptible to intestinal disturbances. Antibiotics improve growth rate and feed efficiency until it is 4 months of age, little advantage to feeding antibiotics occurs after that time. Most commercial milk replacers and calf starters contain antibiotics. Recommended withdrawal times must be followed to avoid drug residues in calves that are sold for slaughter.

Prestarter

Commercially produced prestarter is a dry pelleted feed, similar to milk replacer, made from dried whey, skim milk, sodium caseinate, fat products, and additives (minerals, vitamins, and antibiotics). Prestarter encourages calves to eat dry feed by day 4 and initiates rumen development. Prestarter intake should be limited to a maximum of 0.23 kg per day (supplement with calf starter when the calf will eat more).

Water

Opinions vary as to whether young calves need free-choice water before weaning and if it causes scours. Water does not increase the incidence of scours, but calves with scours consume more water (Kertz, 1984). Increased water intake may be an early sign of disease. After colostrum feeding was stopped (day 4 after birth), calves initially consumed more water. Calves offered water free choice consumed more calf starter, gained more weight per day, and had fewer scour days per calf. Water must be kept fresh and clean, especially during hot weather.

Calf Starter

Calves should be encouraged to consume calf starter or prestarter beginning at 4 to 6 days of age. High-quality, palatable ingredients should be used in calf starter mixtures (Table 26.4). Crude fiber can be increased to 10% in a complete starter if no forage is fed during the first 8 weeks. Calves can be weaned when they consume 0.45 to 0.91 kg (1 to 2 lb) of starter per day. Intake of starter should be limited to 1.8 to 2.3 kg (4 to 5 lb) per day.

Calves do not consume significant amounts of hay or forage during the first 8 weeks of life after birth. Hay can be fed as soon as the calves begin eating it, usually at 1 to 2 weeks of age. A medium-quality, palatable hay is preferable. Haylage can be fed to calves if it is kept palatable, fresh, and free of mold. The calf should be encouraged to consume forage because it will become the main ingredient in the ration of the calf after 4 months of age. If the calf consumes forage at the expense of starter, the amount of forage should be reduced to encourage consumption of starter. Corn silage and pasture should not be fed to animals until 4 to 6 months of age.

TABLE 26.4 RECOMMENDED COMPOSITION OF CALF STARTER ON A DRY MATTER BASIS

Nutrient	Minimum level
Protein	16–18%
TDN	75%
Fat	2%
Crude fiber	5%
Calcium	0.7%
Phosphorus	0.5%
Vitamin A	1136 IU/kg
Vitamin D	114 IU/kg
Vitamin E	1.1 IU/kg

Source: Adapted from J. W. Crowley, N. A. Jorgensen, and W. T. Howard, *Wis. Ext. Bull.* A1485.

Housing

A number of housing systems are available for dairy calves and each has its own management requirements. Individual calf stalls in an insulated building represent one of the more conventional types of housing. The stalls are 1.2 by 1.8 m (4 by 6 ft) and have solid partitions between the pens. The advantages of these stalls are that calves cannot suck each other after milk feeding, the spread of the disease organisms is minimized, and more individual attention can be given to the calves. However, labor requirements are high. Calves should be kept free from drafts, dampness, and any sudden changes in temperature to prevent pneumonia. Air should move from young to older calves to avoid disease exposure to young calves. Four air exchanges per hour should occur in the winter, with 20 to 30 air exchanges per hour in the summer to keep air fresh and low in humidity.

Use of the elevated stall has become common on farms. The stall has a chain to hold the calf's head and a grate so that urine and feces fall on the floor. These stalls can be easily moved and taken outside when not in use. Stalls are used primarily during the milk feeding period. Calves in elevated stalls should be protected from drafts, sudden changes in temperature, and dampness. One major advantage of elevated stalls is that they can be thoroughly cleaned after they have been used; two sets of stalls can be used alternately. An elevated stall should be 51 cm (20 in.) wide and 122 cm (48 in.) long. The bottom should be at least 30 cm (12 in.) above the floor. An expanded metal screen floor is satisfactory. The stall should be set on concrete to facilitate cleaning.

Housing calves outside is another possibility. During the summer, calves require relatively little protection; however, during the winter, they must be protected against wind, rain, and snow. Calves are housed in small individual hutches that are enclosed on three sides (Figure 26.2. Hutches have been used successfully as far north as Minnesota. The system minimizes the incidence of scours and

Figure 26.2 Calves can be raised individually in hutches in the cold temperatures in the northern areas of the United States. Photograph courtesy of the University of Illinois.

pneumonia. Growth rates of calves are comparable to those of calves raised in heated barns. During cold weather, milk or milk replacer intake should be increased 10% per 5°C (10°F) drop below 0°C (32°F). A third feeding at noon should be added when liquid intake is increased 25%. Hutches should be moved from one area to another to avoid disease contamination.

Management

Sanitation and general cleanliness are important to prevent digestive upsets and diarrhea. Pens and stalls should be thoroughly cleaned and disinfected after one calf is removed and before another calf enters. Feed boxes and hay mangers should be cleaned daily and milk utensils should be kept as clean as possible.

Dehorning of calves, usually done when the calf is 2 to 4 weeks of age, reduces injury calves can inflict on other animals with their horns. Several methods of dehorning are available, but the electric dehorner (Figure 26.3), a hot iron that fits over the horn button and destroys it, is the most efficient method. Another method is to apply caustic potash. The hair over the button is clipped and the horn button is rubbed with the caustic potash stick until it begins to bleed slightly. Care must be taken to prevent the caustic material from running into the calf's eye. The clipped area can be encircled with Vaseline or grease to avoid eye damage. Horns of older heifers can be sawed off, but this method is less desirable.

Figure 26.3 An electric calf dehorning iron is a safe and humane way to remove the horn. A copper-colored ring should be burned to kill the horn and prevent bleeding. Photograph courtesy of the University of Illinois.

Calf Diseases

Death losses during the first 2 to 3 weeks of life can be high. Losses on some farms are as low as 1%, but may be as high as 20 to 25%. Some diseases do not cause death, but leave the calves in a weakened and unthrifty condition, making them more susceptible to other diseases.

A number of procedures are important to reduce the incidence of calf diseases. The dam must be fed properly to produce a calf that is healthy at birth. The calf should receive colostrum, a source of antibodies, as soon as possible after birth.

Milk should be fed according to a definite schedule and the amount per day should not exceed 10% of the calf's weight. The temperature of milk and the feeding routine should be uniform from one day to the next. The addition of antibiotics to the milk or milk replacer helps combat disease. Sanitation of the calf's quarters and utensils used for milk is important. If a calf becomes sick, it should be separated from the rest of the calves to prevent transmission of infectious organisms. Three diseases commonly affecting young calves are acute calf septicemia, common calf scours, and pneumonia. A program for prevention and treatment of calf diseases should be formulated in cooperation with the local veterinarian.

Acute calf septicemia. This highly virulent disease affects the calf within 3 to 5 days after birth. One characteristic of the disease is diarrhea, in which the feces has a thin consistency, has a foul-smelling odor, and is grayish white in color. The color of the feces is the reason this disease is called white scours. The diarrhea produces a weakened condition and severe dehydration. Acute calf septicemia is an infectious disease and death is common. The disease is caused by a virus and is followed by an *Escherichi coli* infection of the intestinal tract.

To prevent acute calf septicemia, cows should calve in clean, dry, draft-free maternity pens. The calf should receive colostrum immediately after birth and the navel should be disinfected to eliminate it as a path of entry into the animal body. Feeding of antibiotics to the calf, vaccination of the dam prior to freshening, and vaccination of newborn calves with bacterins, serums, or antitoxins reduce this disease problem. The infected calf should be treated by a veterinarian. Because of the dehydration, the fluid balance of the calf's body should be returned to normal by infusion of balanced electrolytic solutions. Commercial preparations are available or a homemade mixture of 3 teaspoons of sodium bicarbonate, 4 teaspoons of table salt, and $\frac{1}{2}$ cup of corn syrup made up to 4 liters with water can be used. Another popular homemade electrolytic solution consists of one package of fruit pectin, 1 teaspoon of salt, 2 teaspoons of sodium bicarbonate, a 298-g ($10\frac{1}{2}$-oz) can of beef consomme, and water to make 2 liters of total fluid. Electrolytes can be fed at the rate of 1 to 2 liters per feeding with three to four feedings per day for 1 to 2 days depending on the calf's size. Signs of dehydration include inelastic skin and dry mouth (at 6% weight loss), sunken eyes (at 8% weight loss), cold legs and mouth (at 10% weight loss), shock (at 12% weight loss), and death (over 12% weight loss). Combinations of antibiotics and sulfa drugs are administered along with electrolytes. Administration of an antibacterial serum along with a blood transfusion from the dam is also helpful. The calf should be isolated immediately to avoid exposing other calves. A second cow should not be allowed to calve in the same maternity stall until the pen has been thoroughly cleaned and disinfected.

Common scours. Common scours do not usually result in death, but decrease vitality and growth rate. Common scours can occur at any age, but are prevalent during the milk feeding period. Feces are normal in color, but usually have a thin consistency. A calf with scours appears listless, eyes are dull, ears droop, and temperature and respiration rate are elevated.

A cause of common calf scours is overfeeding, but the disease can also be caused by irregularities in feeding milk (changes in composition, temperature, or the amount fed). Dirty utensils and damp quarters also cause scours. When a calf shows symptoms of scours, the amount of milk solids should be reduced by one-half at the next feeding and liquid balance maintained with electrolytes. The amount of milk fed should be gradually increased when the calf has improved. Antibiotics and sulfa drugs decrease the incidence of scours.

Pneumonia. Pneumonia, the most common cause of death in calves, usually occurs 3 to 8 weeks after birth and is often preceded by scours or other symptoms such as coughing, increased respiration rate, elevated body temperature, dullness, rough hair coat, loss of appetite, weakened body condition, and foul-smelling nasal discharge.

Pneumonia is caused by a virus. Cold, damp, poorly ventilated quarters may predispose the calf to pneumonia. Sudden changes in temperature that cause chilling or shipping calves from one farm to another may also result in the disease.

Prevention of pneumonia includes providing a dry environment for the calf during the milk feeding period. Treatment of pneumonia consists of removing predisposing factors and moving the calf to warm, dry quarters. Sulfa drugs and antibiotics may be helpful under veterinarian guidance. The calf should be isolated so that the spread of pneumonia to other calves is minimized.

26.5 RAISING CALVES FROM WEANING TO ONE YEAR OF AGE

Care of calves from weaning to 1 year of age is one of the easier management jobs. Shortly after weaning calves are placed in groups of 8 to 10 and housed in box stalls. Super hutches [3.7 m (12 ft) by 5.8 m (19 ft) semiportable units built on skids] can also be used. Calves should be watched to ensure that they do not suck each other because continued sucking may lead to mastitis or improper udder development. Clean and dry quarters are important. The feeding program for calves up to 4 months of age should include a calf starter containing 16% or more crude protein. Between 1 and 2.5 kg (2 and 5.5 lb) of starter should be fed per day, depending on forage quality and intake, rate of weight gain, and body condition of the calves. High-quality hay and hay silage can be fed. Water should be available at all times.

At 4 months of age the calves can be shifted to a concentrate containing 14 to 16% protein. In most cases, 2 kg (4.4 lb) of concentrate per day produces adequate growth and body condition. When calves are 10 months old, concentrate

TABLE 26.5 PREVENTATIVE VACCINATION SCHEDULE

Age	Disease	Condition for use
Birth–2 months	IBR, BVD, and PI-3	Intranasal vaccine or commercial serum if pneumonia is a problem
	Reo-corona virus, *E coli* scours vaccine	When epidemic conditions occur
	Johne's	In infected herds if available
2–6 months	Brucellosis	All replacement heifers before 4 months of age
4–6 months	IBR and PI-3	Intranasal vaccine
	Pasteurella and *Haemophilus somus* pneumonia	Parenteral vaccines
	Blackleg	In problem areas
6–8 months	IBR and BVD	Parenteral vaccine
13 months (2 months before breeding)	Leptospirosis	Five serotypes
	IBR, BVD, and PI-3	Second vaccination
	Vibriosis	If natural service is used

Source: Adapted from Hoard's Dairyman, 1982, *Calf Care and Raising Young Stock*, Hoard's Dairyman, Fort Atkinson, WI.

TABLE 26.6 HEALTH PROBLEMS OF CALVES

Problem	Cause	Treatment
Johne's disease (paratuberculosis)	Bacteria settles in the small intestine spread by contaminated colostrum, feed, water. Symptoms (weight loss, intermittent diarrhea, and milk losses) appear at 2 to 4 years of age.	Vaccination of baby calves (if available in your state) and isolation of calves are preventive methods. Affected cattle are culled.
White muscle disease (stiff calf disease)	Selenium and vitamin E deficiencies cause stiffness, weakness, and a white appearance to skeletal, cardiac, and diaphragm muscles.	Selenium and vitamin E should be added to the diet.
Goiter (big neck)	Shortage of iodine in the diet of the dry cow or heifer results in a large swelling in the calf's neck.	Supplemental iodine to the dry cow should minimize the problem.
Contracted tendons (knuckling over or under)	Toes of the calf are pulled back so the calf walks with its hoof knuckled under or on its knees. Causes are unknown.	Calf may correct it by itself. A veterinarian may apply a supporting cast. In severe cases, partial cutting of the superficial digital flexor tendons may be needed. Vitamins A, D, and E, selenium, magnesium, and manganese levels should be checked.
Pinkeye	Irritation of the eyes caused by foreign objects and several bacteria and viruses. Symptoms include swelling, excess tear flow, clouding of the cornea, and blindness.	Antibiotic treatment prescribed by a veterinarian and covering the eye to restrict light are warranted. Fly control will reduce the spread between animals. Vaccine is also available.
Parasites (grubs, lice, worms, and coccidiosis)	Presence of internal and external parasites due to exposure, contact, or consumption from contaminated feeds or areas cause the problem. Poor growth, rough hair coats, and respiratory problems occur.	Infected animals should be treated with approved products along with improved sanitation and isolation of infected animals.
Ringworm	Skin disease caused by fungus with gray crusty patches, especially in the winter in confinement housing.	Commercial preparations are available to apply externally or administer intravenously and orally. Increasing vitamin A levels and sanitizing buildings to kill the fungus are beneficial.

TABLE 26.6 *(continued)*

Problem	Cause	Treatment
Colic	Calves develop abdominal pain which lessens within a few hours due to overeating or drinking cold water.	Calf feeding and management procedure should be changed.
Warts	Raised skin lesions caused by a virus.	Small warts can be surgically removed with a scissors by cutting off close to the base. A vaccine may clear the warts quickly.
Navel hernia	The abdominal wall at the navel does not close properly and may be related to a navel infection. Part of the intestine protrudes resulting in inflammation with strangulation.	A veterinarian should examine to determine treatment or surgery.
Infected navel	Exposure of the navel at birth to disease organisms that cause symptoms including fever, listlessness, and swelling around the navel.	The maternity area must be kept as clean as possible. Navels should be dipped in iodine at birth and shortly after.

can be discontinued if high-quality forage is fed. Calves should receive salt as well as calcium and phosphorus supplements, trace minerals, and vitamins A, D, and E. Phosphorus supplementation is important after concentrate is discontinued.

A vaccination program is essential to minimize disease losses (Table 26.5). Dairy managers should work with their veterinarian to plan a program to meet specific condition and needs. Several other health-related problems are outlined in Table 26.6.

26.6 RAISING HEIFERS FROM ONE TO TWO YEARS OF AGE

Raising heifers older than 1 year is simple and economical. Nutrient requirements are listed in Table 18.2. The rumen of the animal has developed so that energy and protein can be obtained from hay, silage, or pasture. If grass hay or corn silage is fed, additional protein must be provided. Mineral supplementation is also required. Animals receiving legume forage will have sufficient calcium intake, but phosphorus must be supplied. If corn silage is used, both calcium and phosphorus should be given. Trace mineralized salt should be provided free choice. Concentrate should be fed to heifers 2 to 3 months before the expected calving date to provide nutrients for additional fetal growth and to maintain heifers in optimal body condition at calving. The heifers should have daily access to fresh water.

The objective of the heifer-raising program is to have the animals large enough to breed by or before 15 months of age. To produce large heifers calving at 24 months of age, the minimum desired weights at breeding are: Holstein and Brown Swiss, 330 kg (725 lb); Ayrshire, 273 kg (600 lb); Jerseys, 227 kg (500 lb); and Guernseys, 250 kg (550 lb).

Heavy feeding produces heifers that are fat at calving, produce less milk, and lack development of secretory tissue. Animals fed below recommended requirements from birth to calving do not produce as much milk as the animals that are fed normally. Heifers that have been underfed prior to calving have more calving difficulties because they are smaller. Underfeeding also delays the onset of puberty. If moderate-to-high nutrient intake is maintained, heifers should be bred earlier than 15 months of age. Excessive feeding of yearling heifers and breeding later than 15 months of age are not economical practices. Growth curves for large breed heifers (weight and height) are illustrated in Figure 26.4. Large breed heifers should gain 0.7 to 0.8 kg (1.5 to 1.8 lb) per day; small breed heifers should gain 0.6 to 0.7 kg (1.3 to 1.5 lb).

Housing facilities for heifers need not be elaborate and vary with climate. Minimum space requirements are listed in Table 26.7. Cleanliness reduces diseases and controls flies. Manure packs should be dry and removed prior to cold weather and in the spring. Freestalls should be scraped twice a week as a minimum, and more frequently in cold weather to prevent freezing and if they become sloppy. Four heifer groups are needed after weaning for optimal feeding programs. One water space per 40 animals should be provided. Waterers should be heated or insulated during cold weather to avoid freezing.

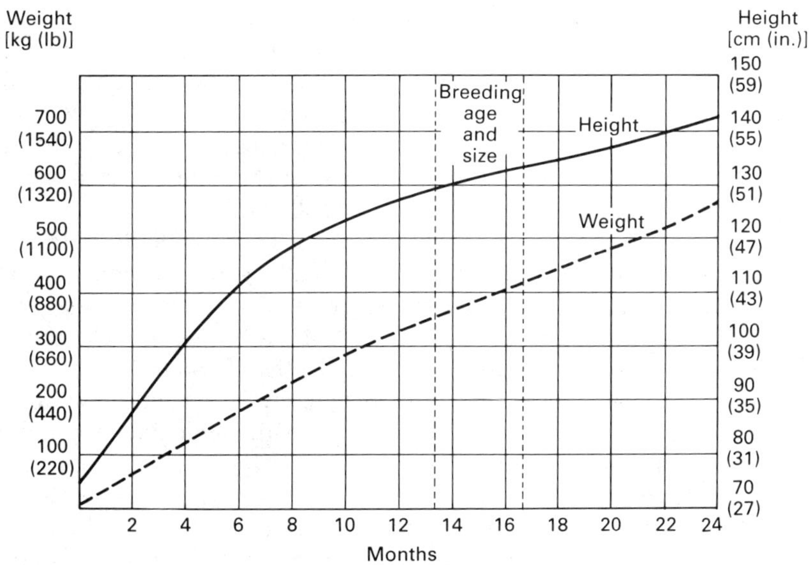

Figure 26.4 Holstein and Brown Swiss heifer growth curves for height and weight at various ages. Adapted from *Hoard's Dairyman*, 1982.

TABLE 26.7 MINIMUM HOUSING SPACE

Type of facility	Age (months) 3–4	5–8	9–12	13–15	16–25
			Square feet/animal		
Resting area and	20	25	28	32	40
paved outside lot	30	35	40	45	50–75
Solid floor (total confinement)	20	25	30	40	60
Slatted floor (total confinement)	11	12	13	17	25
			Dimensions		
Free stalls (width × length)	2′ × 4′6″	2′6″ × 5′	3′ × 5′6″	3′6″ × 6′6″	3′6″ × 6′6″
Minimum manure alley width	6′	8–10′	8–10′	8–10′	8–10′

Source: J. W. Crowley, N. A. Jorgensen, and W. T. Howard, 1983, *Wis. Ext. Bull.* A1485.

26.7 VEAL CALF PRODUCTION

Most dairy farmers sell bull calves and cull heifer calves shortly after birth. Some of these "bob" veal calves are slaughtered for meat. Larger calves are raised and fed on milk or milk replacer for 3 to 4 months to produce fancy veal. Fancy veal calves are marketed at 136 to 182 kg (300 to 400 lb). The meat must be light colored, indicating that the calves have not been fed hay or grain. Other calves are raised as feeder calves and finished as beef animals.

Because few dairy farmers raise their own veal calves, this discussion of veal calves will be concerned with the feeding and management of calves that are purchased from auction sales. The market demands a well-fleshed calf that has a blocky conformation and a light covering of fat. The meat must be light in color for the best price. The live weight range from 136 to 182 kg (300 to 400 lb). To achieve this weight, calves should gain more than 1.1 kg (2.5 lb) per day from birth to time of slaughter. Toward the end of the feeding period, calves gain up to 1.5 kg (3.3 lb) per day. A 136-kg (300-lb) calf should be ready for market between 10 and 11 weeks of age. Calves do not reach 182 kg (400 lb) until approximately 15 to 17 weeks of age. A successful veal calf operation requires that the losses from culling, diseases, and death be less than 4% of the calves that are purchased. Only calves weighing more than 40 kg (88 lb) at birth can make the desired rates of gain. Veal calves are limited primarily to the Holstein breed.

Because veal calves may come from many farms and travel through auction sales, they are exposed to a wide variety of pathogenic bacteria and viruses. The veal calf must receive colostrum shortly after birth. Calves that do not receive colostrum have high culling and death rates. The navel should also be dry when the calf is purchased. A strong prophylactic treatment, worked out in cooperation with a veterinarian, is recommended to suppress any diseases that calves may have picked up. Treatments contain broad-spectrum antibiotics and doses of vitamins A, B complex, D, and E. Serums and vaccines have limited success in suppressing infections after the calf arrives on the farm. Rectal temperatures should be taken daily for the first 2 to 3 weeks. Any calf showing a temperature of more than 39°C (103°F) should receive treatment. Electronic thermometers are available to reduce chore time. When severe diarrhea develops, the amount of milk powder being fed should be reduced, but the amount of electrolytes should be increased to prevent dehydration.

The correct feeding program is critical if economical and rapid growth is attained. A milk replacer containing at least 15 to 20% fat is desirable for optimal gains. The milk replacer should also contain 20% protein as well as vitamins A, D, and E. Clean equipment, close observation, and immediate treatment of calves with high temperature or other abnormal condition are essential for a successful veal calf operation.

Several types of housing are possible. Calves are commonly placed in slatted, elevated stalls. The internal environment of the calf barns in terms of temperature, humidity, and rate of air change is critical. The temperature should be between 15.5 and 21°C (60 to 70°F) and the relative humidity should not exceed 50%. Proper ventilation is essential to maintain fresh air with a low ammonia content.

For maximum disease control and to reduce risk, fewer than 50 calves should be housed per barn. Some modern barns house up to 300 calves, but require excellent sanitation, ventilation, and management. A group of calves should be brought into a barn, fed for the required period, and sold as a group. The barn is thoroughly cleaned and disinfected and should remain idle for a few days before another group of calves is bought to minimize the spread of disease between groups.

26.8 DAIRY BEEF

Feeding surplus dairy calves to a marketable weight of 455 to 545 kg (1000 to 1200 lb) may be more profitable than marketing them as veal calves, feeders, or finishing steers. In areas where corn silage is the most economical feed source, it can be the only forage fed to dairy steers. Adding concentrate increases rate of gain and improves the market grade. Dairy steers gain about 0.8 kg (1.7 lb) daily on forage and protein supplement, and 1.1 kg (2.5 lb) daily when concentrate is added at the rate of 1% of body weight. When alfalfa or grass forage is available, similar results can be obtained when it replaces part of the corn silage.

Steers finished on forage require a longer feeding period. This practice increases labor, housing costs, and interest on borrowed capital. On small dairy farms, the number of animals is limited and dairy steers are frequently raised with replacement heifers on an all-forage ration. The use of an electronic concentrate feeder can provide supplementary concentrate to dairy beef to increase rate of gain without overfattening replacement heifers in the same pen. If dairy beef animals are maintained in the same pen with replacement heifers, a longer feeding period will be required to finish them properly to obtain a higher grade. Growth promoters approved for finishing beef should be used when they are separated from the replacement heifers.

26.9 CONTRACT RAISING OF HEIFERS

Increased specialization in the dairy industry has resulted in alternative ways of raising heifers. Some dairy managers sell their young calves and buy bred heifers. Others find that contract raising of heifers has definite advantages over either buying replacements or raising their own heifers. The advantages of contract raising of heifers to another farm are utilizing labor, capital, and management for the milking herd; keeping a larger number of milking cows than would be possible if herd replacements were raised on the farm; and developing a definite breeding program rather than buying replacement animals.

The dairy farmer who raises heifers on contract can become more specialized in heifer raising, has a smaller capital investment compared to milking cows, and works well as a part-time enterprise. Heifer-raising specialists can usually raise heifers more cheaply than can dairy farmers. A contract with cooperating dairy managers also provides the heifer-raising specialist with a source of well-bred animals.

Some disadvantages are encountered in a contract heifer-raising program. When many animals from several farms are concentrated in one group, the possibility of spread of disease increases. Some heifer-raising specialists cannot raise heifers as well as the dairy farmer who owns the heifers. A suitable contract covering all contingencies must be drawn up or difficulties may arise in the interpretation of the responsibilities of the two partners. The heifer-raising specialist may have uncertainty about the duration of the agreement and may have difficulty in finding sufficient heifers to make adequate use of facilities.

Two usual types of contracts are drawn up between the dairy farmer and the heifer raiser. In one type, the heifer specialist agrees to raise the heifer for the dairy farmer for a specified period and at a specified monetary rate, either in terms of dollars per month or cents per unit of weight gain. The dairy farmer keeps title to the animals. A second type of contract includes an option-to-purchase clause. The heifer is sold to the heifer specialist, but the dairy manager has the first option to buy the animal prior to freshening. The success of contract raising of heifers depends primarily on the ability of the grower to raise the heifers satisfactorily and the nature of the contract between the owner and the grower.

REFERENCES AND FURTHER READINGS

Crowley, J. W., N. A. Jorgensen, and W. T. Howard. 1983. Raising dairy replacements. *Wis. Ext. Bull.* A1485.

Hoard's Dairyman. 1982. Calf care and raising young stock. Hoard's Dairyman, Fort Atkinson, WI.

HUTJENS, M. F. 1985. Nutritional management of calves. *Mod. Vet. Pract.*, p. 451.

KERTZ, A. F. 1984. Calves need water before they are weaned. *Hoard's Daryman*, 129:826.

LINN, J. G., D. E. OTTERBY, R. D. APPLEMAN, and M. F. HUTJENS. 1980. Feeding the dairy herd. *Ill. Ext. Circ.* M-1183.

LUENING, R. A., R. M. KLEMME, and W. T. HOWARD. 1984. Wisconsin Farm Enterprise Budgets—Dairy Cows and Replacements. *Wis. Ext. Bull.* A2731.

NATIONAL RESEARCH COUNCIL COMMITTEE. 1988. *Nutrient Requirements of Dairy Cattle*, 6th ed. *National Academy of Sciences*, Washington, DC.

27

Dairy Cattle Housing

Housing is one of the most expensive capital investments in the dairy cattle operation. In the northern United States where considerable protection of cattle against winter climatic conditions is required, the investment in housing represents approximately one-third of the total investment, including land, cattle, and equipment. The type of housing and milking arrangement also influences the number of cows cared for per person.

Regardless of the type of housing used, each system should (1) provide comfort to cows so that they can produce up to their maximum genetic ability; (2) minimize such injuries as stepped-on teats, swollen hocks, stiffness, and lameness; (3) minimize exposure of cows to disease and stress; (4) have a feeding system that enables cows to obtain sufficient quantities of feed to meet their nutrient requirements; (5) provide for production of good-quality milk; (6) provide a comfortable environment for people caring for the cows; (7) result in high labor efficiency in terms of cows handled per person and milk produced per person; and (8) be economically feasible.

27.1 MAJOR COMPONENTS OF A HOUSING SYSTEM

The four major components of the housing system are (1) the milking center, (2) protection from the weather, (3) the feeding center, and (4) facilities for manure collection, removal, and disposal. These will be discussed in more detail.

Milking Center

Milking cows occupies a higher percentage of the dairy farmer's time than any other management practice. When a milking parlor is used in conjunction with a free-stall barn, up to 70% of the farmer's time is spent in milking. When cows are milked in a stanchion barn, 50 to 60% of the total time is spent in milking cows. Efficient milking procedures are required for high production per cow and for production of quality milk. The milking operation should also be efficient so that the dairy farmer milks the largest number of cows per hour that is compatible with high production per cow.

Various milking arrangements are used to milk cows and transport milk to the bulk tank. Details of these were provided in Chapter 7.

Weather Protection

The type of weather protection required by dairy cows varies with the region of the United States. In the north, emphasis must be placed on protecting cows from harsh weather conditions of snow, sleet, hail, strong winds, and extremely cold temperatures. A roofed shelter provides protection from most elements; however, side walls with the roof are required for protection from the cold. Cows are usually not adversely affected by the cold until the temperature drops to 5°F. Below this temperature a drop in milk production occurs. Above this temperature, no drop in production occurs, but cows consume more feed to maintain body temperature.

In the southern part of the United States protection from winter weather is relatively unimportant, but cows must be protected from the heat, solar radiation, and wind in the summer months. The cost of this housing is less than that in the northern part of the country. Shades that protect cows from solar radiation and wind breaks that provide some protection from strong winds are usually sufficient; however, sprinkler systems to cool cows are gaining in acceptance. Air conditioning for cows is very expensive and usually not cost-effective.

Feeding System

Feeding ranks next to milking in percentage of time required to care for cows. All winter feeds and all grain during the summer are fed in the barn in most parts of the country. Since more emphasis is being placed on feeding cows in the barn throughout the year, use of an efficient feeding system is becoming increasingly important. Because feeding can be mechanized, the housing should be carefully planned to utilize as much mechanization as is economically feasible. See Section 21.7 for details.

Manure Disposal

Disposal of manure is a major problem for many livestock producers, especially in heavily populated areas. In addition to the problem for the livestock producer, manure disposal presents an environmental problem because improper disposal can result in pollution of air, land, streams and lakes, and underground water supplies. Consequently, more societal constraints are being placed on manure disposal.

A dairy cow produces between 7 and 8% of her body weight as manure each day. A 454-kg (1000-lb) cow produces 1.32 cubic feet of manure each day. Manure provides some fertilizing components for the soil. The daily manure from a 454-kg dairy cow contains 0.19 kg of nitrogen, 0.08 kg of phosphorus, and 0.15 kg of potassium. Since manure must be removed from barns and lots, a disposal system should be employed that recovers as much of the fertilizer value of the manure as possible. In areas with low rainfall the manure can be dried or composted. The dried manure can be used as feed or bedding. Even though the nitrogen loss from dried or composted manure is very high, it is in high demand as garden or greenhouse fertilizer and mulch. However, only a very small portion of the total manure produced can be used in this manner.

Some of the manure systems currently used will be discussed; however, the best system for a farm depends on a number of factors, including (1) size of the dairy herd, (2) land available for disposal, (3) population density near the dairy, (4) climate of the area, and (5) type of housing arrangement. The various possibilities of handling, collection, storage, transport, and utilization of manure from different types of housing systems are shown in Figure 27.1.

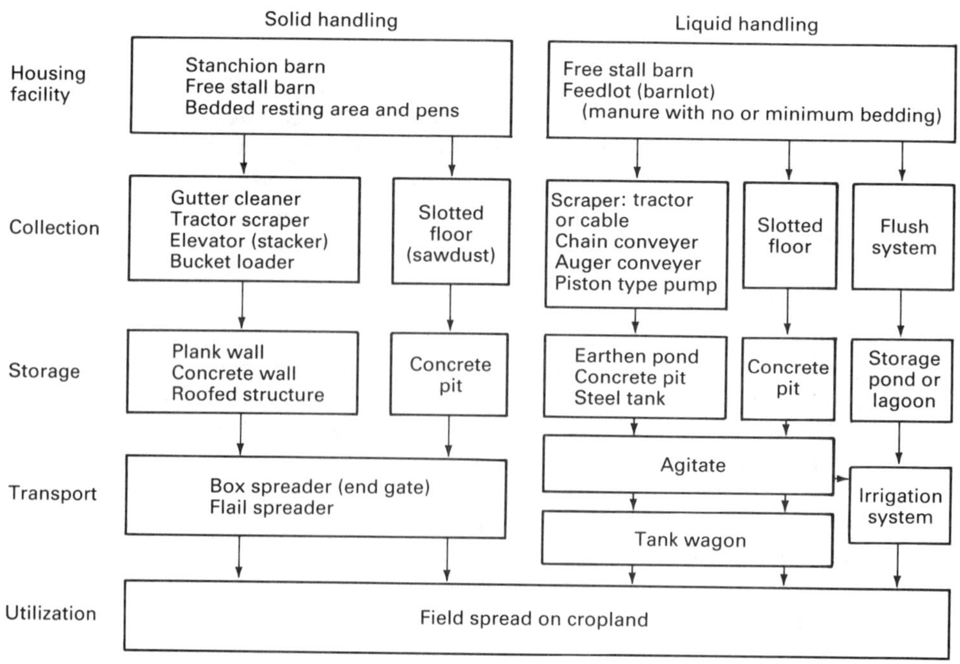

Figure 27.1 Alternative methods of hauling, collection, storage, transport, and utilization of manure from a dairy system. From R. K. White, et al., 1975, *Ohio Coop. Ext. Bull.* 604.

Daily hauling. Daily hauling of manure has been the most popular method for manure disposal for the past several decades. It is the most common type of manure handling in stanchion barns in which bedding is used. When sufficient bedding is used, the consistency of the manure and bedding allows use of a conventional manure spreader for hauling and disposal (Figure 27.2). The more liquid type of manure found in free-stall housing where less bedding is added must be handled in a box or flail type of spreader, which holds liquids as well as solids. The more liquid type of manure is also found in stall barns, where minimum bedding is used.

Daily hauling has several advantages. Labor is distributed throughout the year; however, this becomes a disadvantage during planting and harvesting seasons. Unsightly manure piles are not present and fly-breeding areas and odor problems are minimized. Daily hauling is the lowest-cost method of manure disposal because the investment is confined to a barn cleaner, a manure spreader, and a tractor, which is usually used for many different purposes.

Daily spreading, however, requires that manure be spread under weather conditions that are adverse to humans and land. On cold winter days, daily hauling is very undesirable for the farmer. Runoff from frozen or snow-covered ground causes loss of fertilizer nutrients as well as pollution of streams. Spreading during wet weather can damage the condition of the soil. It is usually recommended to have a short-time storage facility to avoid spreading manure during the most adverse conditions.

Storage of solid manure. Solid manure can also be stored for a period of time (Figure 27.1). This is one of the oldest methods of manure storage and

Figure 27.2 The daily hauling of manure is still one of the most common methods of spreading manure. Photograph courtesy of New Holland, Inc., New Holland, PA.

handling. The manure stack was in an area close to the barn and the manure was spread once or twice a year. This was very common when wheelbarrows were used to haul manure out of the barn—before the advent of mechanical gutter cleaners. Stacking manure eliminates the daily spreading, especially during bad weather and poor field conditions. Stacking manure has several disadvantages. Most of the liquid runs off, causing much of the fertilizer value to be lost as well as causing stream pollution. The manure pile is unsightly and may cause problems with neighbors, odors, and fly-breeding areas.

In some newer housing systems stacking is used because the investment in facilities and equipment is usually lower than that of a liquid manure system. Approximately 2.5 cubic feet of storage space should be planned per cow per day. The storage area is usually surrounded by a plank or concrete wall and covered with a roof. This reduces runoff and loss of fertilizer nutrients. Some systems combine stacking with a liquid system in which the runoff is collected and handled as liquid manure.

Another form of solid storage of manure is the bedded pack, where bedding is added to the manure pack periodically and the manure is spread on the fields once or twice a year. This procedure saves a great deal of labor as well as most of the manure nutrients; however, with rising bedding costs and shortage of straw in many areas it is used sparingly for dairy cows. A number of heifer barns still use the bedded pack.

Liquid manure. Handling manure in liquid form gained considerable acceptance in the United States and Europe in the 1960s. Mixed feces and urine from cows contain approximately 85 to 87% moisture. Slurries of 85% moisture can be pumped with present-day pumps. Water used for washing and sanitizing pipelines and washing milking equipment can be drained directly into the liquid manure storage tank, thus eliminating the cost of a separate disposal system for milking parlor wastes. Human waste, however, must be handled by a septic system.

Liquid manure requires a storage tank or holding area, a pump, and a manure distribution system, which is usually a tank spreader pulled by a farm tractor (Figure 27.3). A 150- to 180-day storage period is recommended for farms located in the northern part of the country. About 2 cubic feet of storage space for each cow each day are recommended for Holstein cows. This provides for the excreta and the milk house wastes. Daily pumping of liquid through an irrigation system is a possibility in the southern part of the United States, but it is not possible in areas where the ground freezes. Daily or periodic pumping of the storage tank greatly reduces the required storage capacity.

Liquid manure can be held in several different types of storage. They are: (1) storage tank underneath the barn, (2) outside below-ground storage tank, (3) above ground storage tank (silo), and (4) earthen storage basin. These storage systems present odor problems, especially during agitation and spreading. Mechanical agitation for aeration of manure ponds or lagoons to prevent odors is no longer economically feasible.

Figure 27.3 A liquid manure spreader is required to haul manure from a liquid manure system. Photograph courtesy of Donald Pritchard, The Ohio State University.

Underneath barn storage of manure is usually combined with slotted floors to eliminate the cost and labor of scraping the alleys because the manure passes through the slots to the holding area. Manure does not build up on the floors and the cows remain relatively clean. Underground manure tanks should be adequately ventilated because harmful gases can be released. This is especially true during agitation and pumping.

Outside storage areas should be built at least 100 ft from any water supply and on soil conditions that prevent uneven settling of the structure. Provision must be made for access to the area for pumping and protection from human and animal drownings. In addition, the bottom and sides should be watertight to avoid underground water pollution. Manure from dairy cows usually produces a good sealant for an earthen pond. Undiluted manure from dairy cows will usually form a crust of floating solids in the storage facility, which will greatly reduce the odors normally associated with manure ponds. Runoff from the surrounding barnlot should be minimized because of the cost of the storage structure. The choice of the type of outside holding area will depend on the availability and topography of adjacent land, the cost of the various structures, and the desires of the farmer.

A modification of the liquid manure storage system is a liquid–solids separation procedure. The liquid is separated from the solids by either settling in a holding area; screening, squeezing, or filtering by mechanical means; or evaporation. The solids part can be spread on the fields directly and the liquid stored as liquid manure.

The movement of the manure from the barn to the liquid storage area can be done in several different ways. They include: (1) dropping through slotted floors directly into the holding area, (2) tractor scraper, (3) mechanical scraper, (4) barn cleaner, (5) pumping from a hopper or tank into the storage structure, and (6) flushing with water. In most cases the manure is handled as it comes from the cow with a minimum amount of bedding. Most methods of moving manure can handle some bedding, but excessive amounts create mechanical problems for many systems. A piston-type pump with an underground pipe is becoming a common way to transport manure from the barn to the storage facility. Most of these pumps handle relatively dry materials.

Flushing alleys with water provides another method of moving manure to the storage area (Figure 27.4). To reduce the amount of water in storage and to be spread on the field, some systems use clarified waste water from the manure as flush water. Flushing has the advantages of requiring relatively little labor and of being easily automated. Because of the large amount of water in this system, a sprinkler irrigation system is usually used for land application of the manure. However, because of freezing conditions during the winter, it may not work in all parts of the United States.

Scraping of manure can be either by tractor or automatic systems. When done by tractor, it should be done once or twice daily. The manure can either be scraped directly into the holding area or to a holding tank and then pumped to the storage facility. Scraping becomes a problem on cold days when the manure becomes frozen. Either frequent scraping with automatic scrapers or installation of heating units in the floor can overcome the problem. The latter might not be cost-effective for most operations. The automatic alley scrapers can be designed to scrape either in one or both directions. They can also be programmed to scrape as often as desired. Cows soon learn to avoid the scrapers when they are in use.

The last two steps in manure disposal from liquid storage systems are agitation and spreading. Again adequate ventilation is required to avoid gases toxic for animals and humans. The manure in the storage area must be adequately agitated to get the solids into a slurry that will be removed by the pump. If not,

Figure 27.4 Flushing of manure in alleys is a way of moving the manure in some housing arrangements. Here it is shown with the feeding alley in an open-lot housing arrangement.

solids will accumulate in the storage area and reduce its capacity. Considerable power is required to accomplish this with heavy-duty agitators. In many cases the pump is used for agitation and the manure is recycled in the storage area until an adequate slurry is attained. Some agitators also have chopper attachments to break up the solids in the manure. The manure can be removed by either a liquid tank wagon or through an irrigation system. The moisture content of the manure for irrigation systems must be quite high and no bedding can be added. The tank wagons are filled by either centrifugal or vacuum pumps—in some cases gravity filling is possible. The manure is then spread on the fields, but in areas of high population density or where required by local law, the manure can be knifed into the soil to avoid odor problems.

Other Components of Dairy Barns

In addition to the four components listed above, all dairy housing units need some additional areas. Most milking parlor operations require a hospital and breeding area, usually located next to the milking parlor so that cows can be diverted into the area at the end of milking. If used, the area should include stanchions to hold cows until they have been bred or examined and treated by the veterinarian. Some housing systems contain stanchions in the feeding area to avoid the use of a hospital or breeding area (Figure 27.5). The stanchions can be locked at one time

Figure 27.5 Lock stanchions in the feeding area of a free-stall barn are used to catch cows for breeding and veterinary attention. It eliminates the necessity of a hospital area in the housing system. Photograph courtesy of Donald Pritchard, The Ohio State University.

and individual cows released. Thus at feeding time, all cows can be locked up and those not needing attention released. This avoids a lot of the confusion during the milking process.

Most dairy farmers also have calving stalls for their cows. This is a requirement in the colder climates during the winter months. One calving stall for 20 cows is recommended. During spring, summer, and fall an outdoor clean calving area is preferable from a disease-prevention standpoint. Although some dairy farmers have their cows calve in the free-stall barn, it is not recommended because of the lack of cleanliness in the area and the possible damage to the newborn calf.

Facilities for raising young calves are also needed. In some cases they may be near or part of the hospital facilities. The requirements for these facilities were discussed in Chapter 26.

27.2 TYPES OF HOUSING ARRANGEMENTS

Many types of housing arrangements are used for dairy cows in the United States. The best type of housing depends on the size of the herd, the degree of labor efficiency desired, the climate of the area, and the personal goals and preferences of the dairy farmer. The basic arrangements of stall barns, loose housing as either bedded pack or free-stall barns, and open-lot housing will be discussed in terms of feeding, milking, and cow comfort. Manure disposal for each system was covered previously.

Stall and Stanchion Barns

Stall barns are the most common form of housing dairy cows in the northern part of the United States, where considerable protection must be provided against adverse weather conditions. This is probably the most economical housing arrangement for the smaller herds (fewer than 50 cows). Cows in a typical stall barn are shown in Figure 27.6. Stall barns are housing arrangements where cows have individual stalls and are tied up with a neck chain or strap. Stanchion barns are similar except that cows are restrained with a metal yoke.

The major advantage of this type of barn is the individual attention that can be given to each cow. Controlled feeding is possible, sickness of the cow is easier to observe, and cows are more easily identified. Because cows are in the same stalls, they can be arranged in an orderly manner for milking and exhibition. It is easier to keep cows in neat appearance and for that reason it is preferred by many purebred breeders.

In addition, cows are more readily available and more easily restrained for breeding and veterinary work. Because cows are restrained, boss cows present no problems to the more timid cows, especially during feeding.

Milking is the most time-consuming chore in a stall barn and is usually the most difficult one in terms of effort expended. Most cows are milked in their stalls and the dairy farmer has to stoop at least four times while milking each cow.

Figure 27.6 Arrangement of cows in a stanchion barn. Cows can be shown to better advantage than in other types of housing arrangements.

Unless an around-the-barn pipeline is used, the milk also has to be lifted from underneath the cow to a dumping station or carried to the milk house.

Several pieces of equipment can be used to reduce the effort required to milk cows in a stall barn. Dumping stations can be used to transport milk from the stall barn area to the bulk tank. Dumping stations contain a receiving vessel that is attached to a plastic hose that runs to the bulk tank. The milk is moved through the hose either by a vacuum from the tank or by a small milk pump at the dumping station.

An around-the-barn pipeline can be used to transport the milk from the cow to the bulk tank. The effort to carry the milk is saved, but the stooping of the dairy farmer to milk each cow is still required. The initial cost of the pipeline is rather high because a portion of line must be installed for each cow. Also, there is little flexibility for increasing herd size. Many older barns do not lend themselves to installing pipelines because the ceilings are too low to accommodate the proper slope in the pipelines. To avoid agitation of the milk, the pipelines should not contain risers. Also, the farmer cannot see the amount of milk produced by each cow at milking.

Another possibility to reduce the effort to milk cows in a stall barn is to add a milking parlor to the existing barn or build one in conjunction with a new stall barn. This is usually not economically feasible for the smaller herds (fewer than 60 cows). Adding a milking parlor reduces the effort required to milk cows and increases the number of cows milked per milker per hour. During the winter months, the milking operation usually requires two people, one to do the milking and the other to move cows to and from the parlor. If the cows are on pasture during the summer months and held in a holding area prior to milking, one person can do the milking. However, provisions must be made to feed grain to them.

Considerable planning should be done before adding a milking parlor to an existing stall barn. Traffic patterns within the barn must be carefully worked out and the holding area must be located where cows can enter and leave the parlor with minimum difficulty. In addition, planning should allow for possible expansion of the herd.

Stall barns have the advantage that feed can be fed according to the cow's needs and the cow has no competition for feed from other cows. Stall barns, however, do not lend themselves well to mechanized feeding, although some systems are available. Individual feeds are usually fed by hand, which requires a large amount of labor and effort. Hay is usually handled as individual bales and fed to each cow. Silage carts are used and the silage forked to each cow. Concentrates are weighed or measured and fed by hand.

Mechanized feeding in stall barns is possible but rather expensive. Silage augers have been used to distribute silage to individual cows; however, the low ceilings in many dairy barns make the system rather difficult to install. Motorized feed carts also reduce the physical effort required to feed cows. The labor required for feeding of silage can be reduced if cows are group fed outside the barn, where mechanized feeding can be used to more advantage.

Stall barns give cows and humans maximum comfort during winter. Most dairy barns are maintained at 50°F during the winter and the comfort range for dairy cows is between 40 and 70°F. Ventilation is required to maintain the desired temperature and to avoid water condensation. Insulation of walls and ceiling is required to avoid excessive lowering of temperatures. Inadequate ventilation is a major problem on most dairy farms and produces dampness, which is uncomfortable for cows and people.

During the summer months the barn is not needed for protection of the animals against cold weather. In general, the barn then serves only as a milking area, a place for cows to receive concentrates, a breeding and hospital area, and a possible place to minimize heat stress.

Construction of new stall barns should provide adequate stall length and width to minimize injuries and maximize cow comfort. Most stepped-on teats, swollen hocks, and stiffness and lameness result from inadequate length or width of stalls. Details on recommended dimensions, use of stanchions or neck chains, and the sizes of feed and litter alleys can be obtained from bulletins of state agricultural colleges, agricultural extension personnel, and commercial building contractors.

Loose Housing with Bedded Pack

Loose housing gained considerable attention in the late 1940s and early 1950s as milking parlors increased in popularity. This allowed for mechanized feeding and reduced labor to milk cows. Because of the shortage of bedding in many areas, most of the loose housing barns have been converted into free-stall barns. The bedded pack loose housing system has four general areas: the milking center, the loafing area, the feeding area, and an outside paved exercise lot.

All loose housing systems require milking parlors, which reduce the labor required to milk cows and greatly increase the cows milked per person. The advantages and disadvantages of the various types of parlors were discussed in Chapter 7.

Feeding silage can be entirely mechanized in loose housing by using a conveyor that extends from a conventional tower silo or by using a side-unloading forage wagon to haul the material from a trench or tower silo. Hay is usually stacked in one area of the feeding barn. Either hay is fed to the cows daily or a movable bunk is located next to the hay so that the cows "eat" into the stack. Concentrate feeding is done in the milking parlor, which may present a problem of inadequate feed intake for high-producing cows.

Approximately 25 square feet of floor space and 2 linear feet of bunk space should be allowed for each cow in the feeding area. Part of the feeding area can be used as the holding area for cows entering the milking parlor, thus allowing double use of the space.

The bedded (manure) pack area is the resting area of the cows to which bedding is added frequently. The manure is allowed to accumulate during the winter and might reach a height of 5 to 6 ft if sufficient ceiling height is present. Approximately 75 square feet of bedded area should be provided for each cow. The warmth of the manure pack provides some protection from the cold. Doors away from the prevailing winds are usually left open so that the temperature is only a few degrees higher than that of the outside temperature.

Cows on the bedded pack usually have fewer teat and leg injuries than cows housed in conventional stall barns. One difficulty with this housing setup is catching a cow for breeding or other attention. A loose housing barn will probably require a hospital area next to the milking parlor. The other major disadvantage is the high bedding requirement.

Free-Stall Barns

Free-stall barns, which are basically a cross between a conventional stall barn and loose housing, were first constructed in the United States in the early 1960s. The free-stall barn consists of individual resting areas or stalls of approximately the same number as the number of cows in the herd. There is no feeding area in front of the cows as in a stall barn, but the feeding area is similiar to that of a loose housing barn. Since the cows are not tied up, they have the freedom of movement as in loose housing (Figure 27.7).

The components of a free-stall system—resting area, milking area, and feeding system—must be considered as a package. The major advantages of the free-stall barns are the labor efficiency in the milking operation, the mechanization of the feeding system, and the freedom of movement of the cows.

In planning the location of the milking parlor, considerable attention must be given to the holding area. Cows should have a straight entrance into the parlor to reduce the time required for them to enter. A crowd gate, preferably self-activated, helps to move the cows into the parlor. This is especially important if no grain is fed in the parlor. The exit doors from the parlor should be so located that the cows go directly to the feeding or resting areas.

Figure 27.7 Cows in a free-stall barn where the alleys are cleaned with a mechanical scraper.

Mechanization of the feeding system may take the form of a side-unloading wagon or the use of mechanical conveyors. The type of system chosen will depend on the type of silos, the layout of the system, and whether a complete feed is fed. For complete feeds, side-unloading mixer wagons are preferred. Mechanization of the feeding system is much easier with silages as the source of forage than with hay. Mechanization is required for high labor efficiency.

If cows are handled in groups, the feeding system must be arranged so that different feeds or combinations of feeds can be fed to the individual groups. Cows should have access to feed at all times. Considerable attention must be given to machinery traffic patterns for feeding the cows and for cow movement to and from the milking parlor in planning such a system.

One way to reduce the cost of free-stall housing is to have an outdoor feeding area. In severe weather cows have a tendency to stay in the resting area instead of going out to the feeding area. It is usually desirable to construct a roof over the feeding area, which in turn makes the cost of the area about the same as if it were part of the resting barn.

Cows appear to be more comfortable in free stalls, and less damage is inflicted to their feet, legs, and udders than in stall barns. The temperature in free-stall barns is 7 to 10°F above outside temperature; however, cows are not adversely affected by low temperatures until it drops below 5°F.

Free-stall barns usually contain some type of bedding, but the amount required is about half that of conventional barns. Most use either straw or sawdust; however, sand, recycled manure, and rubber mats with no bedding have been used successfully. Many materials have been used as surfaces for free stalls. Most common is concrete covered with a rubber mat. In most cases some bedding is added to the mat. Also used have been dirt, clay, wood planks, limestone, and sand as the base with added bedding. In general, cows prefer soft-surfaced stalls over hard stalls. Old car tires half buried in the dirt or sand in the stall and then covered with bedding have been used with success. This prevents cows from digging holes in the soft free-stall surfaces. Most cows adapt to free stalls and lie down in them but farmers report a few cows that refuse to use them and lie down in the manure alley.

In a New York study (LaDue, 1966), free-stall barns with herringbone milking parlors and high-silage feeding systems required 43 hours of chore labor per cow per year in contrast to 76 hours in a conventional stall barn. The winter labor requirements in the stall barn were only 20% higher than in the free-stall housing system; however, the summer labor requirements were 50% higher. Thus considerable expansion of the dairy herd can be done with free-stall housing systems using the same labor force.

Some dairy farmers have constructed warm free-stall barns in which the temperature is maintained at 50°F, similar to the stall barn. Insulation of the barn and forced-air ventilation are required for this type of barn. This is economically feasible in only a small part of the United States (northern Minnesota, northern North Dakota, and some higher elevations in the mountain areas), where the temperature is below zero a sufficient number of days to warrant the extra cost of insulation and ventilation. In some cases, the dairy farmer may feel that the extra cost is justified for the comfort of the workers.

Open-Lot Housing

Open-lot housing is used in areas of the United States where cattle need little or no protection from cold weather. An example is shown in Figure 27.8. The cost of this type of housing is much less than in areas where protection from the cold

Figure 27.8 Pie-shaped open-lot housing system that can be used in the warmer climates and costs little to provide shelter from the elements. Photograph courtesy of Babson Bros. Co., Oak Brook, IL.

must be provided. The feeding and milking areas are comparable to those required with free-stall housing; however, the feeding area may not be covered by a roof. Open-lot dairies usually have shade structures to protect animals from solar radiation and rain. A windbreak may also be constructed to protect animals against strong winds. Paved or dirt lots with adequate slope to provide sufficient drainage from rain are necessary; otherwise, the lots become mud holes. Unpaved dirt corrals require about 500 square feet per cow. Semipaved corrals with proper drainage and weekly manure hauling require about 350 square feet per cow. A 10-feet-wide paved strip should be located along the feed mangers.

Shade structures should be 12 to 20 ft high and oriented north and south to allow proper drying of the shaded earth. The width of the shades may vary between 14 and 20 ft. Corrugated metal roofs with top surfaces painted white and the underneath parts painted black are effective in reflecting solar radiation.

Manure from the milking parlor and holding area is handled in ways similar to those for a free-stall system. Corrals should be cleaned at least monthly and possibly more often. The area along the feed manger should be scraped every few weeks. The feed mangers should be so located that side-unloading wagons can be used for feeding.

27.3 PLANNING NEW OR REMODELED DAIRY BARNS

Most dairy farmers who are considering a change in housing for their dairy cows already have a barn for housing cows. The farmer usually considers a change in housing when (1) a barn fire has destroyed all or most of the existing barn (which means that the opportunity is different from the situation where the barn is still in a good state of repairs), (2) expansion of the herd is desired, or (3) more labor efficiency is desired. Building a new barn or remodeling an existing barn can seldom be justified on economic terms if the herd size remains the same. An increase in number of cows is required to pay for the cost of remodeling an old barn or building a new one.

Factors to Consider

The following questions should be considered before building or remodeling: (1) What is the value and state of repair of my existing buildings?, (2) Will the existing barn and milking facilities fit into an efficient operation?, (3) How many cows do I want to milk after remodeling and in the near future?, (4) What are the costs of the different types of housing systems?, (5) How much will it increase my yearly income and net worth?, (6) How much money can I borrow? (7) What is the length of time required to pay off the debt?, (8) Will I have sufficient cash flow to pay for the expansion and housing? (9) Are there sufficient resources such as land, forage, and labor available?, and (10) Do I have the managerial ability to handle the larger operation?

The age of the dairy farmer may affect the decision to build a new facility.

A middle-aged farmer may decide that the present herd size and facilities are adequate for his lifetime; however, if a son or daughter wants to join the dairy operation, expansion may be necessary. The farmer's goals and personal preferences should also be considered highly for some persons do not enjoy working in a cold environment. With any expansion, it should be planned for a 5- to 10-year time frame in an area where dairying will probably be carried out for the next 20 to 25 years.

Remodeling or New Construction

The problem of constructing new barns or remodeling existing barns faces most dairy farmers who want to expand their existing herds. A conventional stall barn usually has to be either extended to a larger stall barn or converted into free stalls. Use of the stall barn for milking cows housed in a free-stall barn may be used instead of a parlor but is frequently discontinued after a few years. The barn is used instead for dry cows and heifers. Part of the decision depends on the state of repairs of the existing barn and how efficient it would be as part of a new barn.

Sources of Information

A farmer can obtain information from various sources to aid in decisions regarding the type of housing desired. Other farms with different types of housing and milking systems should be visited to determine their relative advantages and disadvantages. The local county extension agent can be helpful in relating experiences of other farmers, supplying information from state agricultural colleges, and locating specific barn plans for the farmer. Private consultants, farm building contractors, and farm equipment representatives also provide valuable information. The farmer should look at all possible alternatives carefully to make sure that the best system is chosen.

Sanitation and Health Requirements

The production of milk is subject to regulations of one or more health departments, including city, state, and federal departments. A dairy farmer should have the approval, preferably in writing, of the health departments governing the farm before constructing the new system. This avoids future problems and misunderstandings.

Costs and Budgeting

After a plan has been selected, the dairy farmer should determine all of the costs and budget the expanded system for 5 or 10 years. Included in the budgeting process should be the net income, cash flow, and years required to pay off the loan. Only when realistic figures are used and the results appear attractive should construction begin.

REFERENCES AND FURTHER READINGS

BATES, D. W., ET AL. 1985. Dairy housing and equipment handbook (MWFS-7), 4th ed. Midwest Plans Service, Iowa State University, Ames IA.

IGONO, M. O., ET AL. 1985. Spray cooling effects on milk production, milk, and rectal temperatures of cows during a moderate summer season. *J. Dairy Sci.*, 68:979.

LADUE, E. L. 1966. Free-stall barn, herringbone-parlor, high-silage-feeding dairy chore systems. *Cornell A. E. Res.* 188.

MERRILL, W. G., and W. W. IRISH. 1986. There's no consensus on the best free stall surface. *Hoard's Dairyman*, 131:309.

SCHMIDT, G. H., and L. D. VAN VLECK. 1974. *Principles of Dairy Science*. W. H. Freeman and Company, Publishers, San Francisco.

SPEICHER, J., and L. H. BROWN. 1970. Warm free stall barns for dairy cows? *Hoard's Dairyman*, 115:269.

WHITE, R. K., ET AL. 1975. Ohio Livestock waste management guide. *Ohio Coop. Ext. Bull.* 604.

YECK, R. G. 1981. Managing dairy wastes. *J. Dairy Sci.*, 64:1358.

28

Milk Marketing

The marketing of milk is a very complex process because it involves a large number of organizations and agencies. The three stages in the marketing of dairy products are: assembly and transportation from farms to processing plants; processing, manufacturing, and packaging; and distribution of fluid milk and cream and manufactured products. Dairy products are channeled through or affected by milk producers, milk processors (handlers), manufacturing milk plants, distributors, retail stores, home delivery routes, schools, and restaurants, as well as local, state, and federal regulatory agencies. Local and state regulatory agencies and the U.S. Public Health Service are concerned primarily with establishing and maintaining sanitary requirements for the production of milk; the federal government is concerned primarily with the fair distribution of monies to producers and the pricing of milk.

Milk has many unique characteristics that make its marketing considerably different from that of most other agricultural products. Fluid milk is highly perishable and is produced every day of the year. It is bulky, which means that it has a low value in relation to its weight. Because milk is easily contaminated by foreign material and bacteria, it must be refrigerated from the time it is obtained from the cow to the time that it reaches the processing plant. Fluid milk must reach the market within several days after it is produced. Milk not used for fluid purposes reaches the consumer in many different products, each of which has a different elasticity of demand and price.

28.1 METHODS OF SELLING MILK

The milk producer has two methods of merchandising milk. One is processing it and selling it directly to the consumer, and the second is to sell it to a handler, who processes the milk either as fluid products or manufacturing products and sells them to the consumer. Only 1 to 2% of total sales of fluid milk is processed by the producer. The milk processor is faced with a number of problems. To ensure an adequate distribution of milk and milk products to the consumer, the processor must buy at least 15% more milk than is sold as fluid milk because of the fluctuating demand for milk during the week. Most milk plants do not operate on weekends; therefore, the plants need a large supply of milk on Thursday so that milk can be distributed to customers and stores on Friday for the weekend demands. Monday is another day of peak operation. Three-day weekends present an additional problem for the processor in terms of having adequate milk available for distribution to stores and customers. The milk processor must find alternative uses for the excess milk, usually in the form of manufactured milk products. Therefore, the processor must have some manufacturing capabilities in addition to the pasteurization and packaging of fluid milk.

In order for the processor to have a stable supply of milk, a commitment is usually made to a group of dairy farmers to buy all of their milk. The processor is then faced with the problem of seasonal variation in milk supply. The supply of milk is much greater in the spring and early summer than in the fall. In contrast to the supply the demand for milk and milk products is usually higher in the fall and winter than in the spring and summer. Again the processor must find alternative uses for the milk that is not used for fluid milk and cream.

The demand for milk and milk products is such that fluid milk receives a higher price than the milk used for the manufacture of such products as butter, cheese, and nonfat dry milk. The processor pays the producer a price for the milk that is based on the prices and percentages of milk used for fluid milk and manufactured milk products.

28.2 DEVELOPMENT OF MILK COOPERATIVES

Because the amount of milk produced by a dairy farmer does not greatly affect the total milk supply, an individual farmer has a very small voice in the decision-making processes that affect the dairy industry. Milk producers strengthened their bargaining power with the milk processors by forming cooperative associations in the early 1900s. The farm cooperatives were encouraged by the passage of the Clayton Antitrust Act in 1914 and the Capper-Volstead Act by the federal government in 1922. The latter act exempted agricultural cooperatives from normal antitrust contraints. Cooperatives also strengthened their position in the market during this time by helping to control the supply of milk produced and the quality of milk shipped by its members. Over three-fourths of the milk is marketed by cooperatives.

There are two general types of milk cooperatives: bargaining and operating or processing cooperatives. The simplest one, the bargaining cooperative, does not engage in handling milk or take ownership of it. Its main function is to bargain on a collective basis for the price of milk with the milk processors. In turn, it ensures that each producer receives an equitable share of the returns. It also encourages producers to adjust their production to the market needs and discourages seasonal fluctuations in production. Since bargaining cooperatives do not handle milk and have no milk-processing plants, the charge to its members is relatively low.

The operating or processing cooperative not only bargains for the producers but handles the milk. These cooperatives bargain with milk dealers for part of the milk and take responsibility for converting the excess milk into manufactured milk products in their own plants. Some operating and processing cooperatives also process and distribute fluid milk. This type of cooperative requires a great deal of capital because it must operate the milk plants.

Dairy cooperatives have four major objectives: (1) to guarantee a market for their member producers, (2) to bargain for the best price terms possible, (3) to market milk as efficiently as possible, and (4) to help achieve higher quality in producers' milk. Generally, they have been able to gain enough control of the supply of milk in an area to establish class I milk prices above Federal Milk Marketing Order minimum prices. Cooperatives have been able to add efficiency to the marketing of milk by balancing market supply and demand, develop more efficient hauling programs with lower hauling rates, improved traffic management in bulk milk assembly, and equalization of milk supplies across the market. Within recent years, the procurement of milk for processors has been shifted to the cooperatives, which in turn assures the processor of a steady supply of good-quality milk at a given price.

28.3 MILK PRICING SYSTEMS

There are four types or combinations of milk pricing systems now being used to market milk: (1) flat pricing, (2) classified pricing, (3) base-excess pricing, and (4) component pricing. Flat pricing is the simplest method; one price is paid for all milk regardless of use. The price is usually set by an individual processor for the milk used by the plant. The price is dependent on the supply of milk and the local demand for milk and milk products. Individual dairy farmers have little negotiating power with the processor. Only a small amount of the total milk supply is priced by this mechanism.

Classified pricing is the most common pricing system used in federal and state milk marketing orders; therefore, it is the most used pricing system. This system of pricing evolved because of the instability of fluid milk prices and because of the need for a mechanism to maintain orderly marketing procedures. This system provides adequate milk in a market and provides a market for producers when the supply of milk exceeds fluid needs. Other names for this pricing system are "pooled" or "blend" prices. Under this system, the producer is paid for milk on the basis of usage for fluid milk and manufactured products by the milk processor. Milk used for fluid milk commands a higher price than that used for manufactured products. Two or three classes of milk, depending on the milk marketing order, are defined. Class I milk is that used for fluid products. Class II milk is used for "soft" processed products, such as cottage cheese, sour cream, and yogurt. Class III milk is that used for "hard" products, such as butter, cheese, and nonfat dry milk. In some markets, only two classes are used, with class I designating milk used for fluid milk and cream and class II for the rest of the milk.

Class I milk is worth between $1.00 and $2.00 per hundredweight more than class III milk and class II milk is about 10 cents above the price of class III milk. The pricing system works as follows. Assume that the milk used in an area, such as the territory of a federal milk marketing order, has the following utilization by milk plants in the area: class I, 50%; class II, 20%; and class III, 30%. Assume that the prices per hundredweight of milk in that area for a given period are: class I, $13.00; class II, $11.60; and class III, $11.50. The price received by each dairy farmer in the area would be calculated as follows: ($13.00 × 0.50) + ($11.60 × 0.2) + ($11.50 × 0.3) = $12.27 per hundredweight. Each dairy farmer thus receives the same price for milk regardless of how the milk from an individual farm is used. In addition, corrections to the price would be made for variations in the milk fat content and possibly for hauling charges, depending on the distance from the center of the market.

Base-excess pricing is a modification of the classified milk pricing system. Class I milk is the amount of milk needed in the market for fluid use plus a small allowance to take care of the variation in the daily demands for fluid milk. The producers in the market are assigned a daily base for which they receive class I price. Milk in excess of the base receives a class II price. The base is usually established in the autumn when production is low and based on previous production by each dairy farmer. The base implies a responsibility of each dairy farmer to supply the class I amount so that the processor is assured of a daily minimum amount of milk.

The base-excess program has several advantages. The major objectives of this program are to prevent surpluses from lowering prices in the market and to ensure dairy farmers of a class I price for a fixed amount of milk determined as a fair share of the fluid sales in the market. This pricing system accomplishes both purposes. Also, it does not penalize an individual dairy farmer on the class I price or base if others in the market increase their milk production. The base-excess program has a tendency to smooth out the seasonal variation in milk production as well. Some cooperatives use a version of this in addition to the classified system in a market order in an attempt to reduce production in the spring and encourage it in the autumn. In doing this, a small amount of money is withheld from each hundredweight of milk marketed in the spring and the total amount apportioned to milk produced in the autumn.

Several problems exist when the base-excess is used as the primary pricing system. The program presents a hardship to dairy farmers wishing to expand their herd and for new dairy farmers. If the demand for milk is not increasing, the total base remains the same. In some markets the base is established each year based on previous years' production; however, in some markets the base is fixed for a number of years. Thus these farmers would be faced with producing milk for class II prices for one or more years before they received a class I base. In some markets the base can be transferred to another farmer.

Basically, all milk is marketed on its volume (weight) and its composition. Since the advent of the Babcock test for fat testing of milk, farmers have been receiving a price for milk based on its weight and fat composition. This gives recognition to the energy and nutritional value of milk since the protein, solids–not fat, and total solids composition of milk increase with the fat percentage. Component pricing of milk is used to describe the marketing of milk for more than one component. In some markets a differential is paid for the protein percentage of milk in addition to its fat percentage, whereas other markets might use solids–not fat or total solids in addition to fat percentage.

Component pricing is most suitable in those markets using a large percentage of milk for nonfluid uses, especially cheese. The yield of cheese from milk is primarily dependent on its protein, and more specifically its casein content; consequently, the pricing of milk for protein and fat is of economic benefit to the producer and processor. Fluid milk is not yet marketed for its protein or total solids content; therefore, there is less incentive to use component pricing of milk for primarily fluid markets. In the future, component pricing might help to change the public image of milk; that is, milk can be emphasized for its protein content and other nutritional value in the human diet. In some markets, the price received for milk is based on a cheese yield formula based on the amount of cheese produced from a hundredweight of milk.

28.4 GOVERNMENT REGULATION OF MILK MARKETING

During the depression in the 1930s, the stability of the cooperatives was undermined, which caused excess milk supplies, price wars, and producers being cut off from their market. The producers turned to the federal government for help. The federal government now has three basic programs that directly affect the milk industry: (1) the price support program, (2) milk marketing orders, and (3) import quotas for milk products along with the legislation governing farm cooperatives.

Price Support Program

The federal government's price support program directs the Secretary of Agriculture to carry out a program that supports the farm price of milk at a parity level set by Congress. The objective of the program is to support the price of milk to assure an adequate quantity of milk to meet current needs, to reflect changes in the cost of production, and to assure a farm income to maintain productive capacity to meet anticipated future needs. Parity refers to the level at which the price of milk used for manufactured products is comparable with other prices received by the farmer in reference to the base period 1910 to 1914. The prices for milk used for manufactured milk products were set between 75 and 90% of parity by congressional acts through 1979. Within recent years, Congress has set specific support prices for milk used for manufactured purposes instead of using the parity formula. The support price is attained by the Credit Commodity Corporation (CCC) buying the excess milk products of cheese, butter, and nonfat dry milk. Tax monies are used for the purchases and the commodities are put into storage. They are sold on the general market when insufficient products are available to meet commercial demands. A portion of the products are distributed to the needy, schools, and some overseas donations. The quantity of milk equivalent and the money spent by the CCC for the price support program are shown in Figure 28.1.

The price support program therefore supports a minimum national price for milk used for manufactured products. This price is quickly reflected throughout the dairy industry when the supplies of milk exceed commercial requirements for milk. The federal government's purchases of butter, cheese, and nonfat dry milk provide an outlet for the excess milk that does not move through commercial channels.

Milk Marketing Orders

In 1933 Congress passed the Agricultural Adjustment Act, giving the Secretary of Agriculture authority to regulate the marketing of milk and other farm products to stabilize markets and increase farm income. The Secretary was given authority to issue marketing agreements and licenses, using the classification pricing system

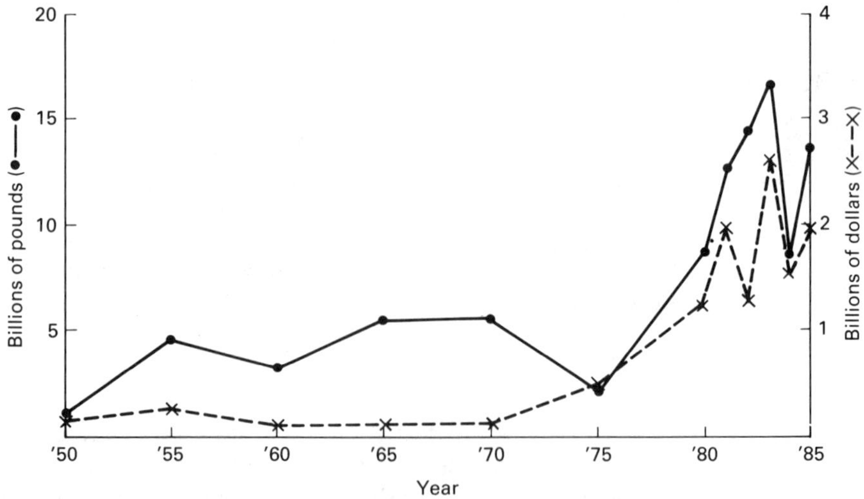

Figure 28.1 Amount of milk equivalent purchased (solid line) and dollars spent (dashed line) since 1950 by the Commodity Credit Corporation to support the federal milk price support program.

and pooling provisions that had been established earlier by negotiation between cooperatives and distributors. Adverse court decisions on the Act resulted in passage of an amendment in 1935, allowing the Secretary to issue marketing orders instead of licenses. The Secretary can issue an order applicable to a particular geographical area upon request of the producers in that area after a public hearing and after determination that it is in the public interest. The producers must also ratify the order in a referendum. If all of these steps are positive, the Secretary appoints a milk marketing administrator to supervise the local administration of the order. The order can be abolished by a vote of the producers in the area.

The terms of the federal milk marketing order specify the boundaries of the order and all the producers in the area are subject to the provisions of the order. The area usually includes more than one state; thus the order regulates the movement of milk across state lines in the order. Processors must pay producers the minimum uniform price for milk set each month as determined by the class prices and usage of milk in the marketing order. Each processor must weigh, test, and accurately account for the use of milk processed in the plant and report that to the milk market administrator. The administrator summarizes the records of each processing plant and determines the uniform price to be paid to the dairy farmers in the area. The salaries and expenses of the administrator and staff are financed by an assessment on the milk sold.

Federal milk marketing orders ensure the existence of a stable, dependable, and orderly market for producers. Although they specify a minimum amount to be paid to the producer for milk, the processor can pay a higher-than-minimum price or the cooperative can bargain for a price above the minimum. The orders assure consumers of an adequate and dependable supply of milk by establishing a minimum price to be paid to farmers to encourage future production. A market is not guaranteed for any dairy farmer in the order, but equity among producers is achieved because each producer receives a fair share of the higher-priced milk that is used for fluid milk and a share of the lower-priced milk used for manufactured products.

Most federal milk marketing orders use the classified price system, although the base-excess system can be used. In the classified pricing system, the class III price is usually determined by the Minnesota–Wisconsin (M-W) series index. This is the average price paid for manufacturing grade milk in Minnesota and Wisconsin and measures directly the current market value of this milk, since these two states produce approximately 50% of the manufactured milk products sold in the United States. The class II price of milk is usually the M-W class III price plus 10 cents. The class I price is set by adding a fixed amount to the M-W class III price. The actual amount, usually between $1.00 and $2.00 per hundredweight, reflects the cost differences in producing the grades of milk, the distance from Minnesota and Wisconsin, and the local supply–demand factor.

Some states also have milk marketing orders. These have provisions similar to those of the federal orders except that the area covered by the state order must be within the confines of a state and the legislation and processes are governed by each state. The area may be the entire state or it may be only the milk shed in a part of the state. In addition, some states also establish retail prices for milk and milk products. Some states also enact trade practice laws and control the movement of milk into the state. The states also license handlers, whereas the federal milk marketing orders do not have authority to do this.

Import Quotas

The world trade in dairy products is characterized by a large amount of government involvement and trade protectionism. Although agricultural exports now comprise about 20% of all U.S. exports, no fluid milk and only about 1% of U.S. milk production is exported as manufactured products. Imports account for about 1.5% of domestic milk production and are limited primarily to specialty cheeses and casein. Casein, a milk protein, is used for industrial, food, and feed uses. The United States at present has no import quota or tariff on casein, although the dairy industry is working to get an import quota established. Since the prices of milk products are usually lower in the major milk-producing countries in the world than in the United States and many countries have export subsidies, the U.S. government has had tight import quotas on most dairy products. Relaxation of the import quotas would reduce the price of milk received by dairy farmers and undermine the federal price support program.

REFERENCES AND FURTHER READINGS

BARTLETT, R. W. 1956. Governmental control of the dairy industry other than health measures and standards. *J. Dairy Sci.*, 39:892.

FALLERT, R. F. 1981. Milk pricing—Past, present, the 1980's. *J. Dairy Sci.*, 64:1105.

JACOBSON, R., H. HARRIS, F. WEBSTER, and P. CHRIST. 1978. Who will market your milk? *Tex. Agric. Ext. Serv. Bull.* D-1058.

PETERSON, W. E. 1950. *Dairy Science*, 2nd ed. J. B. Lippincott Company, Philadelphia.

SPENCER, L. 1956. Significant developments in the distribution and pricing of market milk in the United States, 1906–1956. *J. Dairy Sci.*, 39:884.

SWANTZ, A. 1962. How we came to have federal milk marketing orders: What they are and what they do. *J. Dairy Sci.*, 45:1397.

Index

Abnormal sperm, 387
Abomasum, displaced, 360
Abortion, 377–378, 381
  in blue tongue, 380
  in bovine virus diarrhea, 380
  in brucellosis, 378
  in infectious bovine rhinotracheitis, 380
  in leptospirosis, 379
  in listerosis, 379
  in vibriosis, 378
Abreast parlor, 103
Accuracy:
  of bull evaluation, 207
  of estimated transmitting ability, 199, 221
  of estimating real producing ability, 197
  of future progeny record, 222
  of genetic evaluation, 150
  of predicted difference, 207
Acetic acid:
  absorption from rumen, 273
  and appetite regulation, 279
  in corn silage, 318
  relation to milk-fat secretion, 50, 71, 362
  production in rumen, 71, 273, 362
  and rumen development, 410
Acetoacetic acid, 351
Acetone, 351
Acetonemia, 350
Acid detergent fiber, 71 (*see also* Fiber)
Acidified milk replacer, 414
Acidosis, 362
Acrosome, 387
Action lists, 405, 406
Acute calf septicemia, 418
Adenosine triphosphate (ATP), 47
Adjustments:
  for age of cow, 179
  for body weight, 189
  breed-age average, 183
  for calving interval, 189
  for days dry, 189
  example of, 191
  for management level, 190
  for preferential treatment, 214
  for selection of mates, 211
  for times milked daily, 191
Adrenal hormones:
  for edema, 131
  for initiation of milk secretion, 44
Adrenaline, 61
Adrenocorticotrophic hormone (ACTH), 44
Age correction, 180
  for breed-age-average, 183

Age, effect on:
brucellosis susceptibility, 377
brucellosis vaccination, 378
calcium absorption, 356
edema, 129
forage quality, 314
ketosis, 350
mastitis, 123
milk composition and yield, 67–68
milk fever, 356
nutrient requirements, 284
Agglutination test, 379
Agitation of milk, 92, 437
Agricultural Adjustment Act, 450
Agricultural Research Service, 403
Air conditioning for cows, 429
Albino, 137
Allantois, 366
Alleles, 133
Allergy to milk, 4
Almquist, J., 383
Alveoli:
development of, 44
structure of, 41–43, 46–49
Amantea, G., 383
American Society of Mechanical Engineers, 81
Amino acids:
absorption of, 286
in milk synthesis, 48, 52
synthesis in rumen, 288
Ammonia, toxicity, 308
Amnion, 366
Ampulla, 366, 371, 373
Amylase, 275
Animal fat and heart disease, 2–3, 17
Ankle strap, 399
Antibiotics:
for calves, 310, 414
feeding to cows, 310
for mastitis treatment, 116, 124
in milk, 126
for semen preservation, 384, 388
for vibriosis treatment, 378
Antibodies, 65
in calf's blood, 412
in colostrum, 411
Antifoaming agents for bloat, 356
Appetite, 278
Around-the-barn pipeline, 82, 96, 437
Artificial insemination, 366, 368–369, 373, 377–378, 382–390
in bull selection, 203
and lethal genes, 143
Artificial vagina, 383, 385–386
Ash (*see* Minerals)
Atherosclerosis, 2, 17
Automation in milking systems, 89, 107–112
Average rate of milk flow, 83
Ayrshire:
characteristics of, 26–27
curly coat, 136
gestation length of, 371–372
origin of, 19, 26–27
permanent identification of, 398

B vitamins:
absorption of, 275
for calves, 276
synthesis in rumen, 274, 296
Babcock, S. M., 12
Babcock test, 12, 449
Bacteria (*see also* Micoorganisms):
in mastitis, 116–120, 122, 124–129
Bacteriology, 11
Balance trials, 277
Bangs disease (*see* Brucellosis)
Bargaining cooperative, 447
Barley, 306
Barns:
free-stall, 439–441
loose housing, 438–439
remodeling, 442–443
stall, 429, 436–438
Barny flavor, 92–93
Basal metabolism, 278
Base, for genetic selection, 204
Base-excess pricing, 448–449, 451
Bateson, William, 133
Bedded pack, 438–439
Beef value of breeds, 425
Beet pulp, 309
Bentonite, 71, 310
B-hydroxybutyric acid:
in fat secretion, 50
and ketosis, 351
Bile, 274
Biological value of protein, 6
Black color, 136
Blend pricing (*see* Classified pricing)
Blind quarter, 36, 116
Bloat, 354
Blood:
composition, 52–53
flow to udder, 38–39
groups, 139
proteins, 53, 139
serum albumin, 53
sugar (*see* Glucose)
vessels to udder, 38–39
Blue tongue, 380
Bob veal, 424
Body condition:
effect on production, 60–67
score, 343
Body temperature, effect of thyroprotein on, 74
Body weight:
effect on milk production, 68
of heifers at breeding, 423
losses, 343
production adjusted for, 189
Borden, Gail, 11
Bovine virus diarrhea, 379–380

Breaker cup, 82
Breed:
characteristics of, 21–29
choosing of, 30
classification of, 19–20
crosses, 258
definition of, 18
distribution of, 20–21
improvement of, 29
origin of, 19
promotion of, 30
zoological classification of, 19
Breed-age-average adjustment, 183
Breed registry association, 29–30
Breeding:
area, 435
problems, 377–381
records, 400
value (*see* Genetic evaluation and transmitting ability)
weight of heifers, 423
wheel, 401
when to breed, 377
Brewers' grains, 309
Brisket tag, 399
Brown Swiss:
characteristics of, 25–26
distribution of, 20–21
gestation period of, 371–372
heat tolerance of, 70
maturity of, 26, 27
origin of, 19, 26
permanent identification of, 398
Brucella abortus, 378
Brucellosis, 91, 377–378, 420
Bucket milking machine, 80, 96
Bulbo-urethral gland, 373
Butter:
consumption of, 17
government purchases of, 450
Butyric acid:
absorption from rumen, 273
production in rumen, 71, 273
and rumen development, 410
By-product feeds, 308

Caecum, 275
Calcitonin, 292, 358
Calcium:
absorption, 358
in blood, 358
deficiency of, 292
in feeds, 292, 302–303
in milk, 52, 53, 72
and milk fever, 356
mobilization of, 358
requirements of, 2, 292
supplementation of, 292
Calcium carbonate, 71
Calf starter, 415
California Mastitis Test, 119, 121, 122
Calorimetry, 278
Calves:
dehorning, 417
digestive system, 409
diseases, 418
feeding of, 413
housing for, 416
liquid diets, 413
management at birth, 410
navel disinfection of, 411
raising of, 417
value for veal, 424
veal, 424
water for, 415
weaning of, 415
Calving, 410 (*see also* Parturition)
Calving interval, 74, 364, 374, 402
production adjusted for, 189
Calving record, 400
Calving stalls, 410, 436
Camel, 1
Cancer, 4, 115
Capacitation of sperm, 371, 377
Capper-Volstead Act, 446
Carboxypeptidase, 276
Carnucles, 366
Carotene, 296
Carriers, recessive genes, 142
Casein, 52, 53, 118–119, 452
Cash flow, 397
Cattle, importation of, 10
Cell wall, 286, 300
Cellulose, 300
Cervix, 368, 372, 377, 389–390
Cheese:
consumption of, 17
government purchases of, 450
yield pricing system, 449
Chinball marker, 376
Chlorhexidine, 126
Chlorine, 293
in milk, 53, 72, 119–120
Cholestrol and heart disease, 2–3, 17
Chorion, 366
Chromosomes, 133
Chute type parlor, 103
Chyle, 275
Chyme, 274
Chymotrypsin, 274
Chymotrypsinogen, 274
Circular milking parlor, 103
Citrate for semen extender, 387
Citrobacter bacteria, 118
Classification of cattle according to use and purity, 19–20
zoological, 19
Classified pricing, 447–448, 450–451
Clayton Antitrust Act, 446
Cloxacillin, 126–127 (*see also* Antibiotics)
Cobalt, 294, 381
Cod liver oil, 71

Codominance, 136
Colic, 422
Coliform mastitis organisms, 117, 122, 124, 128
Color:
  albino, 137
  black, 136
  red, 136
  roan, 136
  white, 136
Colostrum:
  for calves, 411
  composition, 64–65, 411
  soured, 412
Commodity Credit Corporation, 15, 450
Complete feeds, 344
Component pricing of milk, 449
Compositing manure, 430
Computers, 406–407
Concentrate:
  balancing rations, 333
  consumption rate, 339
  DHI feeding guides, 340
  economics of feeding, 347
  energy value of, 303
  feeding for calves, 415
  feeding guides, 337
  feeding tables, 339
  functions of, 331
  group feeding of, 344
  linear programming for, 340
  mixtures, 332
  requirements for cows, 333
  rules of thumb, 337
  substitution for forage, 332
Concentrate feeding system:
  complete ration, 344
  computer, 347
  conventional, 343
  door feeder, 347
  forage added, 344
  magnetic, 346
  parlor, 343
Conception rate, 371, 374, 376
  semen cost, 251
Conductivity of milk, 111
Confidence range, 222
  average of bulls, 226
  sire transmitting ability, 222
Contemporary comparison, modified, 195
Continuous grazing, 322
Contract raising of heifers, 426
Cooperatives, 446–447, 448–449, 451
Copper:
  feeding, 294
  in milk, 72
  and oxidized milk, 91
  and reproduction, 381
Corn:
  ear, 305
  high moisture, 306
  shelled, 305
Corn distillers' grains, 308
Corn gluten feed, 308
Corn gluten meal, 308
Corn silage:
  additives, 326
  for calves, 415
  consumption of, 325
  dry matter content of, 325
  feeding of, 304, 325
  harvesting of, 325
  intake of, 325
  maturity of, 324
  moisture content of, 325
  supplementation of, 325
  and urea feeding, 326
Corn stover, 305
Cornybacterium bovis, 118
Corpus luteum, 366, 369, 370–371, 393–394
Corrals, 442
Cottonseed, whole, 309
Cotyledons, 366, 378
Cow index, 195
Cowpers glands, 373
Cowy flavor (*see* Unclean flavor)
Crossbreeding, 258
  economics of, 264, 268
  experimental results, 263
  Illinois project, 264
  Jersey-Deshi, 269
  systems of, 266
  Zebu crosses, 265
Crowd gate, 107, 439
Cryptorchidism, 372
Curly coat, Ayrshire, 136
Cutin, 316

Daily hauling of manure, 431
Dairy cattle:
  numbers, 13–16
  production per cow, 14–16
Dairy Herd Improvement (DHI):
  feeding guides, 340
  records, 12, 29, 98, 121, 364, 376, 402–405
Dairy Herd Improvement Registry (DHIR), 404
Dairy records processing centers, 403
Dairying:
  development of, 11–13
  history of, 10–11
Date of forage cutting, 314
Days dry, production adjusted for, 189
Days open, 189 (*see also* Calving interval)
Dehorning of calves, 417
DeLaval, Gustav, 12
Dental cavities, 4
Deoxyribonucleic acid (DNA), 48, 75
Detergent analysis, 300
Diarrhea, 379, 380, 418
Diestrus, 370–371
Digestibility of feeds, 277
Digestion coefficient, 277
Digestion trial, 277

Digestive system:
of calves, 409
of cows, 271–275
Dipping teat cups, 126
Dipping teats, 90, 111, 117, 122, 126
Diseases:
of calves, 418
control of, 420
metabolic, 350
and milk quality, 91
reproductive, 377–381
Disinfection of calf's navel, 411
Displaced abomasum, 360
Diuretics for edema, 131
Dollar value, index for selection, 236
Domestication of cattle, 7
Dominance, 135
heterosis, 261
Downer cow, 357
Dropsy, 146
Dry lot feeding, 323
Dry matter intake, 334
Dry period:
effect on production, 64
feeding during, 342
infection rate during, 124–125
therapy for mastitis, 117, 122
Drying off cows, 57
Ductus deferens, 372
Dumping station (*see* Milk transfer station)
Dystocia, 381

Ear tags, 399
Economic value:
relative, 233
selection for, 229
traits for selection, 231
Economics:
of crossbreeding, 264, 268
net present value, 245
selection, 245
Edema, 115, 129–131
feeding and, 131
salt intake and, 129
Egg yolk extender, 387–388
Ejaculation, 372
Electroejaculation, 386
Embryo:
flushing, 390
sexing, 395
splitting, 391
transfer, 255, 382, 390–392
Embryonic membranes, 366
Embryonic mortality, 378
Endometritis, 381
Endometrium, 366, 368
Endoplasmic reticulum, 46–47
Energy:
digestible, 282
in feeds, 281, 302, 303
intake and ketosis, 351
measures of, 281–283
metabolizable, 282
net, 283, 304
requirements for, 285–286
total digestible nutrients, 281
use, 284
value of milk replacer, 413
Enterobacter mastitis bacteria, 118
Environmental mastitis organisms, 117–118, 127–128
Enzymes:
in abomasum, 273
in intestinal tract, 275
in pancreatic juice, 274
Epididymis, 372
Epistasis, 137
heterosis, 262
Eructation, 273
Esophageal groove, 409
Estimated breeding value, 150, 195 (*see also* Estimated transmitting ability)
Estimated real producing ability, 195, 198
Estimated transmitting ability, 195, 202
Estradiol, 369
Estrogen:
effect on lactation, 68
for mammary gland development, 43–44
for reproduction, 369, 370, 375
Estrous cycle, 43, 365, 369–371
Estrus:
detection of, 374–376, 377
synchronization of, 382, 390, 392–394
Ether extract, 290, 299
External pudenal blood vessels, 38

Fallopian tubes, 366
Fasting catabolism, 278
Fat:
body, 343
digestion of, 274
feed sources of, 290
mobilization, 343, 351
requirements for, 291
selection for, 244
Fat content of milk (*see also* Milk fat):
according to breeds, 21–24, 26–27
nutrient requirements for, 290
relation to feeds and feeding, 71–72, 291
Fatty acids:
absorption from intestine, 274
and appetite regulation, 279
in blood, 51–52, 273
essential, 290
metabolism of, 273
in milk fat, 50–51
and milk fat secretion, 50–52
production in intestines, 275
production in rumen, 273
and rumen development, 410
saturated, 17, 51

Fatty liver, 361
Fecal nitrogen, 290
Feed bowl covers, 111
Feed flavors in milk, 92–93
Feed intake, 278–280
Feed lot bloat, 354
Feeding area, 428, 429
Feeding frequency, 279, 347
Feeding records, 402
Feeding standard, 278
Feeding systems, 343, 438, 440
Feeding trial, 277
Fermentation losses, 316
Fertility, 239 (*see also* Calving interval)
Fertilization, in reproduction, 366
Fiber:
  acid detergent, 285, 300
  crude, 284
  neutral detergent, 286, 300
Fiber-bound protein, 300, 318
Fimbria, 366
Flat pricing of milk, 447
Flushing of embryos, 390
Follicles, 365–366
Follicle-stimulating hormone, 390
Food and Drug Administration, 76
Forage:
  analysis, 328
  bags, 320
  and concentrate ratio, 332
  consumption of, 334
  digestibility of, 314
  drying agents, 316
  feeding of, 304
  feeding value of, 328
  green chop, 323
  harvesting methods of, 317
  inventory of, 329
  leaf content of, 315
  legume content of, 304
  losses, 315
  moisture content of, 319
  stage of maturity, 314
  storage of, 316, 320
  systems, economics, 324, 327
  value of, 313
Free fatty acids, 275
Free martin, 142
Free stall barns, 439–441
Freeze branding, 398
Frothy bloat, 375
Frozen semen, 388
Furstenburg's rosette, 35

Galactose, 3, 51, 275
Gall bladder, 274
Gas accumulation in rumen, 354
Generation interval, 149, 154
Genes, 133
  map of, 138
Genetic evaluation:
  accuracy, 150
  bases, 204
  bulls, 203
  cows, 194
  natural service sires, 214
  young bull testing, 215
Genetic relationships, 169
Genetic superiority, 165
Genetics, definition of, 133
Genotype, 134
Gerber test, 12
Gestation period, 364, 371–372
Gillies, Alexander, 78
Gland cistern, 36–38, 54
Glucocorticoids for edema, 131
Glucose, 3
  absorption of, 275
  in blood, 52, 352
  use for energy, 52
  in lactose secretion, 51–52
  in milk fat secretion, 51
  and rumen development, 410
  treatment of ketosis, 353
Glycerol:
  for milk fat secretion, 51
  for semen extension, 383, 388
Glycogen, 352
Goats, 1
Goiter, 421
Golgi apparatus, 46, 47, 51
Gonadotropin releasing hormone, 369
Government regulation of milk marketing, 449–452
Graafian follicle, 366
Grade animal, 20
Grading up, crossbreeding, 267
Grain, addition to silage, 318
Green chop forage, 323
Group feeding, 344
Growth, nutrient requirements for, 284
Growth hormone:
  in mammary gland development, 44
  and milk section, 73–76
Guernsey:
  characteristics of, 24–25
  distribution of, 20–21
  gestation length of, 372
  origin of, 19, 24–25
  permanent identification of, 398

Hand stripping, 90
Harvest methods, 316 (*see also* Forage)
Hatch Act, 12
Hay (*see also* Forage):
  baled, 317
  barn dried, 317
  for calves, 415
  conditioners, 315
  harvesting of, 317

Hay-crop silage (*see also* Forage):
  additives, 318
  harvesting of, 317
  moisture content, 318
  preservatives, 318
  storage, 319
Haylage, 318
Health records, 400
Heart disease, 2–3
Heart rate, 74
Heat increment, 283
Heat mount detectors, 375
Heat, standing (*see* Estrus)
Heat tolerance, differences in breeds, 70
Heifers:
  contract raising of, 426
  feeding of, 420, 422
  growth of, 423
  health programs for, 421
  housing of, 424
  raising of, 420, 422
  vaccination program for, 420
Hemicellulose, 300
Hemoglobin, 295
Herd life and milk index, 241
Herd management, adjustment for, 198
Herd summary report, 405
Heritability, 150, 159, 163
  adjustment for predicted difference, 206
  economic traits, 232
  estimates, 160
Hernia, 422
Herringbone parlor, 103, 105–106, 113
Heterosis, 258
Heterozygous, 136
High-moisture corn, 306
Holding area for cows, 100
Holstein-Friesian:
  characteristics of, 21–23
  distribution of, 20–21
  gestation length of, 372
  heat tolerance of, 70
  origin of, 19, 22
  permanent identification of, 398
Homozygous, 135
Honeycomb (*see* Reticulum)
Hormones, effect on milk yield, 73–76
Horns, 135
Hospital area, 435
Hot iron branding, 399
Housing:
  for calves, 416
  for cows, 428–443
  for heifers, 424
  loose, 438–439
  open-lot, 441–442
Hulls, 310
Humidity:
  in calf barns, 416
  effect on hay drying, 317
  effect on milk yield, 70
Hutches for calves, 416
Hydrochloric acid, 273
Hydrostatic pressure, 129
Hypothalamus, 59, 278

Identification of cattle, 111, 398–400
Illawarra cattle, 28
Immune globulins, 53, 65, 411
Immunity, 411
Import quotas, 449, 452
Improvement, genetic, 165, 221
In vitro:
  fertilization, 382, 395
  methods, 301
Inbreeding, 170
  by breed, 175
  depression, 174
Incomplete milking, 58, 124
Index:
  cow, 195
  milk and herd life, 241
  product value and type, 239
Indexes:
  fertility and production, 239
  milk and constituents, 243
  product value, 236
  type-production, 235
Infectious and bovine rhinotracheitis, 380
Infectious pustular vulvovaginitis, 380
Inflating udder, 359
Infundibulum, 366
Inguinal canal, 38, 39
Initiation of milk secretion, 44–45, 51
Insemination, 389
Intensity of selection, 151
Interstitial tissue, 372
Intestines:
  absorption from, 274–275
  anatomy of, 274
  enzymes in, 274–275
  function of, 274–275
Iodinated casein (*see* Thyroprotein)
Iodine:
  in feeds, 294
  and goiter, 421
  in milk, 111, 294
  and reproduction, 381
  requirements for, 287
Iron:
  feeding and veal production, 424
  in milk, 72, 91
  and oxidized flavor, 91
  requirements for, 295
Isoacids, 310
Ivanoff, E. I., 383

Jamaican Hope cattle, 20
Jersey:
  characteristics of, 23–24
  distribution of, 20–21

Jersey (*Cont.*)
gestation length of, 372
heat tolerance of, 70
incidence of milk fever in, 356
origin, 19, 23–24
permanent identification of, 398
Johne's disease, 421

Keratin, 123
Ketone bodies, 351
Ketosis, 66, 350
Klebsiella mastitis bacteria, 118
Koch, Robert, 20

Labor:
efficiency, 100–102, 105, 107, 111, 113, 440–441
effort in milking, 96–97, 429, 436–437
Lactalbumin, 51, 53
Lactase, 3, 275
Lactic acid, 318
Lactoferrin, 123
Lactoglobulin, 53
Lactose:
content of milk, 51–52, 64
intolerance, 3, 18
secretion, 51–52
Lactose synthetase, 51
Land Grant Act, 12
Lardy, H. A., 383
Lasalocid, 311
Lead feeding, 342
Least cost ration, 340
Legume forages, 304, 313
Leptospirosis, 377
Lethals, 141
genetic, 142
Light, effect on production, 70
Lignin, 300, 314
Linkage, genes, 138
Linseed meal, 307
Lipase:
in pancreatic juice, 274
and rancid flavor, 92
Liquid manure, 432–434
Liquid protein supplement, 389
Listerosis, 379
Liver, 288, 351, 361
Loafing area, 424
Loci, 137
Longevity (*see* Herd life)
Loose housing, 438–439
Lush, Jay L., 154
Luteinizing hormone, 369, 370
Lymph, and udder edema, 130
Lymphatic system in udder, 39–40
Lysomones, 49

Machine idle time, 101
Machine milking, and mastitis, 109, 122–123, 129
Machine on time, 101
Machine stripping, 83, 84, 109–110
Magnesium, 292
Magnesium carbonate, 71
Magnesium oxide, 71, 310
Maintenance requirements, 284
Maltase, 275
Maltose, 275
Malty flavor in milk, 92–93
Mammary gland:
anatomy of, 32–40
blood to, 38–39
cytology of, 46–49
development of, 43–44
duct system in, 34–38
histology of, 41–43
involution of, 49
lymphatic system to, 39–40
nervous system to, 40
suspensory apparatus of, 32–35
Manganese, 295, 381
Manure:
collection and removal, 428, 429–435
compositing, 430
flushing of, 433–434
liquid, 432–434
ponds, 432–433
stacking of, 432
Map distance, 138
Mastitis:
abnormal milk from, 118–120
antibiotic therapy of, 116, 124, 126–127
effect of age on, 123
bacteria involved in, 116–120
effects of breeding and selection on, 129
control programs of, 116, 118, 124–129
diagnosis of, 119–122
infection and, 115–119, 123–124
inflammation and, 118–119
inheritance of, 123
kinds of, 118–120
losses due to, 115–116
effect of machine milking on, 122–123, 124, 129
effect on milk composition, 118–120
effect on milk flavor, 91
effect on milk yield, 115–116
effect of milking procedures on, 122–123, 129
effect of milking rate on, 123
overmilking and, 124
prevention of, 125–129
effect of sanitation procedures, 126–129
symptoms of, 118–120
effect of teat dipping on, 117, 122, 1126
relation to teat ends, 123
relation to teat meatus, 123
transfer of bacteria in, 122–123
vaccination for, 129
Mature equivalent, 179
Maturity:
differences among breeds, 22, 23, 24, 26, 67
effect on forage quality, 314

Maximum rate of milk flow, 83, 84
Mendel, Gregor, 133
Mertens, D. R., 300
Metabolic diseases, 350
Metabolic fecal nitrogen, 290
Metabolic rate, 73
Metestrus, 370–371
Methane, 282
Methionine feeding, 311
Metritis, 381
Microorganisms (*see also* Bacteria):
  in rumen, 273, 286
Milk:
  characteristics of, 1–2
  classes of, 448
  constituents and selection for, 243
  consumption of, 16–18
  cooling of, 92
  demand for, 16–18
  distribution of, 13–16
  exportation of, 452
  flavors, 92–93
  intolerance, 3
  marketing, 445–452
  nutrient requirements for, 1–2, 285
  nutritional value of, 1–2
  precursors of, 50–53
  pricing of, 447–449
  processing of, 447
  production of, 13–15
  proteins, 140
  quality, 91–93
  replacers, 413–414
  secretion, 49–53
  usage, 18
Milk composition and yield:
  effect of age on, 67
  effect of body condition on, 66–67
  effect of body weight on, 68
  changes during lactation, 64–65
  daily variation in, 65–66
  effect of disease on, 115–120
  effect of dry period on, 66–67
  effect of feeding on, 71–73
  effect of hormones on, 73–76
  effect of photoperiod on, 70
  effect of pregnancy on, 64, 68
  effect of season on, 70–71
  effect of temperature on, 69–70
Milk ejection, 58–61, 85, 86, 369, 375
Milk fat:
  composition of, 50
  depression of, 71, 362
  factors affecting, 63–76
  relation to feeds, 71–72, 363
  relation to yield, 63–65
  secretion of, 49–51
Milk feeding to calves, 413
Milk fever, 356
Milk let down (*see* Milk ejection)
Milk marketing orders, 447–448, 450–451
Milk-only record, 404
Milk ring test, 378
Milk secretion rate:
  effect of milking frequency on, 55–57
  effect of milking intervals on, 55–57
  effect of udder pressure on, 54–55
Milk transfer station, 96, 437
Milking frequency, 55–57
Milking intervals, 55–57
Milking machine, 78–85
Milking parlors;
  automation in, 107–111
  choosing a, 112–113
  concentrate feeding in, 98, 103, 343
  construction of, 97–100
  evaluation of, 100–102
  identification of cattle in, 111
  labor effort in, 96–97
  types of, 103–107
Milking procedures, 84–90
Milking rate:
  measurements of, 83–84
  effect of milk yield on, 91
  effect of milking machine factors on, 79, 84
  effect of teat meatus on, 90–91
Milking Shorthorn:
  characteristics of, 27–29
  origin of, 19, 28
Milo, 306
Minerals:
  macro, 291
  micro, 287
  requirements of, 287
  supplementation of, 339
Minnesota-Wisconsin price index, 451
Mitochondria, 46, 47
Modified contemporary comparison, 195, 204
Modified routine time, 102
Modified Whiteside Test, 119, 121
Molasses, 309
Monensin, 311
Morrison, F. B., 283
Multiple ovulation and embryo transfer
    (MOET), 255
Mycoplasma organisms, 118
Myoepithelial cells, 41, 42, 59, 369

National Association of Animal Breeders, 403
National Dairy Herd Improvement Association,
    403
National Research Council, 285, 286, 287
Natural service sire evaluation, 214
Navel, infected, 422
Near-infrared reflectance, 301
Neck chain, 399, 436
Nervous system:
  in mammary gland, 40
  in milk ejection, 58–59
Net energy (*see* Energy, net)
Net present value, 245
Neutral detergent fiber, 284, 300

New Zealand vacuum standard, 81
Niacin, 311
Nitrate poisoning, 326
Nitrogen balance, 277, 290
Nitrogen fertilization, 314
Nitrogen-free extract, 299
Non-fat solids:
  content of milk, 72
  government purchases of, 450
Non-protein nitrogen, 6, 288–289
Northeast sire comparison, 211
Nutrient:
  essential, 275
  deficiency of, 276
  definition of, 275
  digestible, 275
  requirements, 276–278

Oats, 306
Obesity, 3–4
Off color, 25
Off feed, 73, 362
Oils (*see* Fats)
Omasum, 273
Open-lot housing, 441–442
Operating cooperative, 447
Osmotic pressure:
  and edema, 129
  and milk secretion, 51
Osteoporosis, 2
Out-of-storage feeding, 323
Ova, 365
Ovary, 365
Overdominance, heterosis, 261
Overmilking, 66, 89, 90, 124
Oviduct, 365, 371
Ovulation, 366, 369, 370
Owner-sampler record, 404
Oxalacetate, 352
Oxidized flavor, 92, 297
Oxytocin:
  in blood, 54, 59–60, 87
  in milk ejection, 42, 59–61
  during parturition, 372

Pancreas, 274
Pancreatic juice, 274
Parasites, 421
Parathyroids, 358
Parentage test, 139
Parity, 449–450
Parturient paresis (*see* Milk fever)
Parturition, 371–372, 410
Pasteur, Louis, 11
Pasteurization, 11
Pasture:
  for calves and heifers, 415
  grazing systems, 322
  management of, 322
  supplemental feeding of cows on, 322
Pectin, 300
Pedometer, 375
Pelleted feeds:
  consumption rate of, 98, 344
  effect on milk fat secretion, 71, 363
Pendulous udder, 35
Penicillin, 127
Penis, 368, 373, 386
Pepsin, 274
Peptidases, 274
Peptides, 274, 275
Peptones, 274, 275
Perineal blood vessels, 38–39
Perry, E. J., 383
Petersen, W. E., 58
pH:
  in abomasum, 273
  in milk, 118–119
  in rumen, 71, 274
Phagocytosis, 123
Phase feeding, 341
Phenotype, 135
Phillips, P. H., 383
Phosphorus:
  in blood, 358
  deficiency of, 293
  in feeds, 302–303
  in milk, 52–53, 72
  and reproduction, 381
  requirements of, 273
  supplementation of, 273
Pinkeye, 421
Pituitary, 42, 369, 370
Placenta, 366
Pneumonia, 419
Polled, 135
Poloxalene, 356
Polygon milking parlor, 103, 105, 107–108, 113
Polymorphonuclear leucocytes, 121, 123
Pooled milk pricing (*see* Classified pricing)
Potassium:
  bicarbonate, 71
  in milk, 51, 53, 72
  and reproduction, 381
  requirements, 293
Predicted difference, 195, 205
Preferential treatment, genetic evaluation, 214, 217
Pregnancy:
  check of, 371, 377
  effect on milk yield, 64, 68
Present value:
  economics of selection, 245
  embryo transfer, 255
  MOET, 256
  semen, 250
  sexed semen, 254
Prestarter, 415
Price, milk equation, 237
Price support program, 449
Primary sprout, 43
Producing ability, 195

Product value:
  index for selection, 236
  milk fat, 238
  milk-fat-protein, 238
Production records, 401–402
Production testing, 401–404
Proestrus, 369, 370–371
Progeny testing, 215, 218
Progesterone:
  assay of, 377, 394
  in blood, 51
  effect on mammary gland development, 43, 44
  effect on milk yield, 68
  role in reproduction, 369, 370, 393–394
Progress, genetic, 165, 221
Prolactin:
  in blood, 45
  effect on mammary gland development, 44
  role in initiation of milk secretion, 44–45
Propionic acid:
  absorption from rumen, 273
  and appetite regulation, 279
  as a grain preservative, 311
  relation to milk fat secretion, 71, 362
  production in rumen, 71, 273
  and rumen development, 410
Propylene glycol, 311
Prostaglandins, 370–371, 393
Prostate, 373
Protein:
  biological value of, 6, 286
  blood, 140
  in calf starter, 416
  in concentrate mixtures, 322, 335, 337
  deficiency of, 4, 290
  digestible, 290
  in feeds, 302, 303
  functions of, 286
  measurement, 289
  microbial digestion, 286
  in milk, 64, 72, 140
  production, 6
  requirements of calves, 287
  requirements of cows, 286
  requirements of heifers, 287
  requirements of humans, 2
  selection for, 244
  supplements, 307
  synthesis, 47–48
  topdressing, 338
Prototheca organisms, 118
Proximate analysis, 299
Prussic acid poisoning, 323
Pulsation rate, 80, 84
Pulsation ratio, 81, 84
Pulsator, 79, 80, 81
Purebred, 12, 20
Purebred Dairy Cattle Association, 30, 403
Pytalin, 275

Rancid flavor in milk, 96
Real producing ability, 195
Recessive, 136
Records, 396–407
Rectovaginal insemination technique, 389
Red color, 136
Red Dane cattle, 20
Red nose, 380
Red Sindhi cattle, 20
Regression, estimate of heritability, 164
Reindeer, 1
Relationships:
  genetic, 169
  tabular computation, 171
Relaxin, 369, 372
Remodeling barns, 442–443
Renin, 274
Repeatability, predicted difference, 207
Reproduction:
  diseases of, 377–381
  organs of, 365–369, 372–373
Requirements:
  growth, 284
  maintenance, 284
  production, 284
  reproduction, 284
Residual milk, 54
Respiration:
  calorimetry, 283
  rate, 74
Retained placenta, 378, 381
Rete testis, 372
Reticulum, 273
Ribonucleic acid, 48
Ribosomes, 48
Ring worm, 421
Risk, sire selection, 225
Roan color, 29, 136
Rolling herd average, 405
Rotational grazing, 322
Routine time, 101
Rubber mats, 440
Rules of thumb for concentrate feeding, 337
Rumen:
  anatomy of, 272
  in calves, 409
  capacity of, 272–273
  digestion techniques, 277
  fatty acids in, 273
  function of, 271–274
  microorganisms in, 273
  pH of, 71, 273
  synthesis of amino acids in, 273
  synthesis of B vitamins in, 273
Rumination, 272

Salisbury, G. W., 383
Saliva, 273
Salt:
  feeding of, 339
  intake and edema, 129
  requirements of, 293

Salty flavor in milk, 92
Sanitation:
  for disease control, 129, 418
  in housing, 416
  for quality milk production, 91–92
Score card, 30
Scours, 419
Scrotum, 372–373
Season:
  effect on milk composition, 70–71
  effect on milk supply, 446
Secretin, 274
Seepage loss from silo, 319, 325
Selection:
  economic value, 229
  economics, 245
  gain per year, 165, 221
  index, 150
  intensity of, 151
  net present value of, 247
  for production and type, 233
Selenium, 295, 381
Semen:
  collection of, 384–386
  cost, present value, 250
  evaluation of, 387
  processing of, 387–388
  sexed, value of, 254
  sexing of, 395
Seminal vesicles, 373
Seminiferous tubules, 372
Separator, 12
Sertoli cells, 372
Services per conception, 251
Sex chromatin, 395
Shade, 442
Shadow prices, 341
Side opening parlor, 103–105
Sigmoid flexure, 39, 373
Silage, 317 (*see also* Corn and hay-crop silage):
  for calves, 415
  direct cut, 318
  field losses, 315
  grass, 317
  harvesting of, 317
  hay crop, 317
  and milk flavors, 92–93
  moisture content of, 319
  preservatives, 318
  storage losses of, 316
  wilted, 318
Silica, 300
Silo:
  air tight, 316
  bunker, 319
  tower, 320
Sire comparison, 208
Sire summary:
  Northeast sire report, 204
  USDA, 204
Slaughter trial, 278
Slotted floors, 433
Small grain forages, 305
Smith-Lever Act, 12
Sodium:
  in milk, 51, 53, 72
  requirements, 293
Sodium bicarbonate, 71, 311
Sodium propionate, 353
Soilage, 323
Solar radiation, 429, 442
Somatic cell count, 99, 116–117, 119, 120–121, 123, 402, 404, 405
Sorghum forage, 304, 322
Soybeans, 307
Spallanzani, L., 383
Sperm, 372–373
Spermatozoa, 372–373
Sphincter muscle, 35, 90
Sprinkler system, 429
Stall barn, 436–438
Stanchion barn (*see* Stall barn)
Standard deviation, 161
  genetic, 152
Staphylococcus aureus, 117–118, 120–122
Staphylococcus epidermidis, 117–118
Starch, 300
Straw, 305
Straws for frozen semen, 388
Stray voltage, 100
Streak canal (*see* Teat meatus)
Streptococcus agalactiae, 116, 117, 122
Streptococcus dysgalactiae, 117–118
Streptococcus uberis, 117–118
Streptomycin, 127
Stress, 438
Strip cup, 86–87, 119–120
Strip grazing, 322
Subcutaneous abdominal vein, 38–39
Subpelvic tendon, 34
Suburethral diverticulum, 368, 369
Sudan grass, 322
Sulfur 294
Superovulation, 390
Support price, 449–450
Supramammary lymph node, 39–40
Suspended type milking machine, 82
Syndactylism, 146

Tattoo, 398
Teat cistern, 36, 37, 38, 54
Teat cup liner, 80, 82
Teat end erosion, 123
Teat meatus, 34, 35, 36, 37
  relation to mastitis, 35, 123
  relation to milking rate, 35, 83
Temperature:
  in barns, 439–441
  for calves, 416, 419
  effect on forage quality, 314
  effect on milk composition and yield, 69–70
  effect on sperm production, 372
  effect on water intake, 297

Tendons, contracted, 421
Testis, 372
Testosterone, 372, 375
Tests, lethal carriers, 142
Thyroid gland, 73
Thyroprotein, 73–74
Thyroxine, 73
Total digestible nutrients:
calculation of, 282
in calf starter, 287, 416
in complete feeds, 287
in concentrates, 303
content of forages, 302
requirements of, 287
Transmitting ability, 195 (*see also* Estimated transmitting ability)
Trichomoniasis, 384
Trigon milking parlor, 103, 105, 107, 109, 113
Trypsin, 274
Trypsinogen, 274
Tuberculosis, 91
Twins, 141
Type:
classification, 29
and production index, 235, 239
selection for, 233

Udder (*see* Mammary gland)
Udder pressure:
and milking rate, 91
and milk secretion, 54–55
Udder washing, 85–86, 109
Unclean flavor, 92–93
Underfeeding, 72
Undulant fever, 378
Unit time, 101–102
United States Department of Agriculture, 15, 403, 405
United States Public Health Service, 445
Units per person, 102
Urea:
adaptation to, 308
addition to corn silage, 326
feeding of, 6, 307
toxicity, 308
Urethra, 373
Uterine milk, 368, 371
Uterus, 365, 366

Vaccination:
for acute calf septicemia, 419
for blue tongue, 380
for bovine virus diarrhea, 380, 420
for brucellosis, 378, 420
of calves, 378, 420
for infectious bovine rhinotracheitis, 380
for leptospirosis, 379
for mastitis, 129
for vibriosis, 379
Vacuoles, 51
Vacuum fluctuations, 123–124
Vacuum level:
effect on mastitis, 124
of milking machines, 79–80
effect on milking rate, 79, 83, 84
Vacuum pump capacity, 82
Vagina, 368, 371, 372, 389–390
Van Soest, P. J., 300
Variance, 161 (*see also* Standard deviation)
genetic, 155
Variation, genetic, 154
Vas deferens, 373
Vasa efferentia, 372
Vasectomized bull, 375
Veal calves, 424
Ventilation:
in barns, 438
for calves, 416
Vibrio fetus, 378
Vibriosis, 378–379, 384
Vitamin A:
administration, 296
in colostrum, 65
deficiency of, 296
in feeds, 296
in milk, 65
in milk replacer, 414
in reproduction, 381
requirements of, 296
Vitamin B (*see* B vitamins)
Vitamin C, 296
Vitamin D:
in calcium absorption, 358
deficiency of, 296
feeding and milk fever, 360
in milk, 2, 72
1,25 dihydroxyvitamin D (1,25-D), 360
Vitamin E:
in milk, 72
and oxized flavor in milk, 92, 297
requirements, 297
Vitamin K, 296
Vitamins:
absorption from intestine, 275
deficiency of, 296
in milk, 52, 65, 72
and reproduction, 378–379
requirements of, 287
synthesis of, 296
Vulva, 369

Warts, 422
Water:
for calves, 415
for cows, 297
function of, 297
intake of, 297
requirements for, 297
quality, 298
Water buffalo, 1
Weather damage to forage, 315, 316
Weather, protection from, 429